W0193409

EBURY PRESS

INDIA'S MOST FEARLESS 1

Shiv Aroor is senior executive editor and anchor at India Today TV, and has covered the Indian military for nearly two decades. He has reported from conflict zones that include Kashmir, India's North-east, Sri Lanka and Libya. For the latter, he won two awards for war reporting. Shiv also founded the popular, award-winning military news and analysis site Livefist.

Rahul Singh is senior associate editor at *Hindustan Times* and has covered defence and military affairs for over two decades. Apart from extensive and deep reporting from the world of Indian military, including several newsbreaks that have set the national news agenda over the years, Rahul has reported from conflict zones including Kashmir, India's North-east and the Democratic Republic of Congo. The first story on the ongoing India–China border conflict appeared under his byline in 2020.

INDIA'S MOST FEARLESS 1

True Stories *of* Modern Military Heroes

SHIV AROOR | RAHUL SINGH

EBURY
PRESS

An imprint of Penguin Random House

EBURY PRESS

USA | Canada | UK | Ireland | Australia
New Zealand | India | South Africa | China | Singapore

Ebury Press is part of the Penguin Random House group of companies
whose addresses can be found at global.penguinrandomhouse.com

Published by Penguin Random House India Pvt. Ltd
4th Floor, Capital Tower 1, MG Road,
Gurugram 122 002, Haryana, India

First published in Penguin Books by Penguin Random House India 2017
Published in Ebury Press by Penguin Random House India 2019

Copyright © Shiv Aroor and Rahul Singh 2017

All rights reserved

43 42 41

ISBN 9780143440444

Typeset in Bembo by Manipal Digital Systems, Manipal
Printed at Replika Press Pvt. Ltd, India

www.penguin.co.in

To every Indian hero who has lived and died

Contents

Introduction   ix
Foreword   xv

1. 'We Don't Really Know Fear'   1
    The September 2016 Surgical Strikes in PoK

2. 'They Didn't Know We Were There'   35
    The June 2015 Surgical Strikes in Myanmar

3. 'When He Awoke, Death Smiled'   65
    Lance Naik Mohan Nath Goswami

4. 'Even the Toughest Take Cover. But Not He'   87
    Havildar Hangpan Dada

5. 'Two Bullets Can't Kill a Commando'   105
    Captain Jaidev Dangi

6. 'Just Tell Me, Will He Live or Die?'   123
    Colonel Santosh Yashwant Mahadik

7. 'I Got Hit. I Can't Believe It'   137
    Major Mukund Varadarajan

8. 'Medical Science Cannot Explain This'   159
    Lance Naik Hanamanthappa Koppad

9. 'Everything Was against Us. *Everything*'   179
    Lieutenant Commander Niteen Anandrao Yadav

10. 'We Follow That Man. He Has Seen Death'   193
    Captain Varun Singh

11. 'This Is India's Honour. We Cannot Fail'   209
    Commander Milind Mohan Mokashi

12. 'You Think It'll Never Happen to You'   227
    Squadron Leader Rijul Sharma

13. 'Every Chopper Pilot's Worst Nightmare'   241
    Squadron Leader Vikas Puri

14. 'Could Taste the Blood on My Face'   257
    Wing Commander Gaurav Bikram Singh Chauhan

Glossary   269
Acknowledgements   271

Introduction

'Without heroes, we are all plain people
And don't know how far we can go.'
——Bernard Malamud

'Lead me, follow me,
Or get the hell out of my way.'
——General George S. Patton Jr

As we sat in an underground chamber with the young Indian Army officer, his beard hiding most of his face, it was with an overpowering sense of disbelief. Here was a man who had been trained for swift, unapologetic destruction of targets, a man who, only a few months before, armed with an assault rifle, night-vision goggles and a hand-picked group of India's most fearless warriors, had led his band of Special Forces (SF) men into Pakistan-occupied Kashmir (PoK) near the Line of Control (LoC). We needed to remind ourselves constantly that this was the first time Major Mike Tango was talking to journalists about the hair-raising mission he led into Pakistan-occupied

Kashmir (PoK) in September 2016. As a result, the book you hold in your hands contains the only first-hand account of that astonishing mission—by the very man who led it.

Maj. Tango's awe-inspiring tale is the first of fourteen stories we have the privilege of narrating in this book. The recounting of each story has been a journey into spaces that are usually both physically and emotionally out of bounds: where brothers in arms of fallen heroes still pick up the pieces of a glorious shared past; where widows, from the mountains of Uttarakhand and Arunachal Pradesh to the plains of Karnataka, resign themselves to a life that will forever be laced with a curious mix of pride and grief; where men who have demonstrated fearlessness beyond anything even conceivable to most of us explain it away as 'just another day on the job' . . .

These are stories like that of Lt. Col. Oscar Delta, who led a revenge mission on foreign soil just as his mother was being wheeled away for cancer surgery; or the young marine commando who would save fellow warriors only to have a grenade burst like a birthday balloon on his chest; or the Air Force pilot who decided, trapped in a screaming, shattered cockpit, that all those years of torture-testing needed to amount to something.

Not every one of the heroes you will read about in this book is alive. Telling their stories has meant that those who saw them fall, those who fought alongside them in their final moments, have permitted us access to what is for them a sacred place. It is a place where memories and trauma remain untouched and stowed away perforce so that the proverbial show may go on—like the hair-raising tales of Lance Naik Mohan Nath Goswami and Havildar Hangpan Dada, who are deified and worshipped by their units for acts of courage that even their fellow warriors say they will spend an entire lifetime coming to terms with. It is an irony we would encounter endlessly as we conducted interviews for this book—soldiers often do not have more than a few minutes to mourn their departed comrades.

The way we as citizens regard the lives and stories of soldiers today is another monumental irony. We live in times when we encounter almost daily a stream of grainy mugshots of deceased soldiers followed by photographs of their flag-draped caskets at military funerals. We feel exultant pride at their acts of bravery, fuelled by social and television media—only to be forgotten the next day. It occurred to us in the writing of this book that the levels of fearlessness displayed even routinely by the men you will read about—and countless more whose stories we hope to tell in future—reserve for them a place in the pantheon of immortals, legends. But never mind the details of their courage, how many soldiers can most of us even recall by name?

A third irony that we found ourselves frequently wrestling with was the godlike portrayal of military heroes in the media for the brief moments that they were remembered. As we journeyed through these tales, we were often struck by a violent collision—between the perception of these men as superhuman, and the frequent sledgehammer reminders that they are just like us, their lives back home just like ours, where PAN cards need to be obtained, home loan instalments to be paid, ageing parents to be taken care of, tiffs with girlfriends to resolve, decisions to be pondered over such as what cake to get for their daughter's birthday, and whether to order butter chicken or kofta curry.

Historically, an impulse has existed to revere the military and its heroes as a physically and mentally superior class of human beings (Plato called them 'guardians', for instance), closer to divinity than their human roots. It is possibly a way to offset our own feeling of inadequacy that such acts of courage and fearlessness are really possible by those among us. In the stories you read in this book, we have attempted to straddle both these worlds.

Becoming a part of the lives of the men we have written about, their units and their families, we found ourselves dealing with our own sense of trauma. Drawn into a world where life and death were literally just that and not a cliché, it was difficult for us to remain unscathed. We do not claim to bear wounds, but we also cannot claim to have been immune to the threads of heartbreak, fury, pride and disbelief that weave through all of these tales.

One often hears the phrase 'supreme sacrifice' being used to describe the death of a soldier in the line of duty. It is a paradoxical term, heavy with implication. Yet, it instantly conveys what it intends to: an act of selflessness so high that the most basic instinct—to survive—fades away and yields to the decision to fight to the death.

American writer and mythologist Joseph Campbell once said, 'A hero is someone who has given his or her life to something bigger than oneself.'

What explains that final act of giving? What is that inscrutable space where the will to survive gives way to an epiphany that death in those circumstances will serve a higher purpose? What is that purpose? The survival of fellow comrades, the extraction of a hostage, successful escape from a tightening cordon of a marauding enemy.

Where there is battle, there will be heroes. There is then as much the inevitability that most of them will not be remembered. How does one remember so many heroes, and so many acts of conspicuous courage?

On the rare occasions when they are recalled, Indian military heroes and legends in public consciousness are mostly from our wars: Captain Vikram Batra from Kargil; Major Kuldip Singh Chandpuri from the Battle of Longewala in 1971; the astonishing last stand of Major Shaitan Singh in 1962 . . . It would be the rare citizen who knows or recalls in any great detail the exploits of Second Lieutenant Arun Khetarpal in 1971, or Subedar Bana Singh in 1987, or even Naib Subedar Sanjay Kumar, who, like Vikram Batra, was awarded the Param Vir Chakra, but whose story remains obscure to most of us. These are stories that have fallen through the cracks, not a fraction of the well-known legends they ought to be. It is the odd story of military heroism that manages to penetrate and then pervade public consciousness with any degree of detail as to have instant recall. Capt. Vikram Batra's war-cry *Yeh dil maange more* during the Kargil War eased his passage into that rare public immortality.

The truth is that India remains constantly at war. Fighting terrorists in Jammu and Kashmir and the North-east means an endless state of combat for several units, including the Para Special Forces that you will read about in many of the stories that follow. We have chosen to tell fourteen stories that stood out to us as extraordinary tales of fearlessness in recent memory. By no means is this to suggest that this is an exhaustive or definitive list. On the contrary, we hope that this will be a small tribute to the large number of uncelebrated heroes in our military past, and the acts of invisible heroism that continue each day.

When we decided to call this book *India's Most Fearless*, there wasn't a moment of doubt that we had chosen correctly.

What, after all, is a more human emotion than fear? Yet, are we to believe that these men truly felt no fear at all?

As we present these stories, in the hope that they will mean as much to you as they do to us, we place them in your hands with a note on the astonishing generosity of the men we have written about. Asked what they owed their courage to, we encountered a perplexingly unanimous answer. We quote one of the men: 'I would say I owe it to the opportunity—being in that place at that time. I think any soldier in my place would do the same.'

It's true. Heroes walk among us.

Foreword

'Let me not pray to be sheltered from dangers,
but to be fearless in facing them.
Let me not beg for the stilling of my pain, but
for the heart to conquer it.'

—Rabindranath Tagore

The Indian Armed Forces personnel are the epitome of courage, valour and sacrifice. 'Peacetime' is an enduring misnomer for the Indian Armed Forces with a host of constant operational and training commitments.

Our soldiers, sailors and air warriors remain in a constant state of mission alert. The Indian Army's Special Forces are among the world's only elite units permanently deployed in hostile conditions and missions in Jammu and Kashmir and the North-east. For the preservation of peace, the war is constant. I am of the view that our citizens must hear about these stories of the unparalleled challenges and operational environment in which our soldiers operate fearlessly; their purpose to accomplish *any* mission, *anywhere, any time*.

Having seen combat through my years in one of the world's most professional and capable militaries, I can tell you that it is difficult to fully understand acts of heroism and fearlessness. What drives heroes to put everything they have worked so hard for—their families and their own lives—at stake during an antiterror operation? What compels a pilot to delay an ejection so he may save lives on the ground even though such a delay means certain death for him? What pushes a sailor to willingly venture into harm's way if it means the oceans are just a fraction safer? One of Mahatma Gandhi's lesser-known quotes perfectly captures how we in the armed forces regard acts of courage:

> Fearlessness is the first requisite of spirituality. Cowards can never be moral.

For all the decaying values we tend to be surrounded by, let nobody convince you that true heroes don't exist. It is therefore with great pride and anticipation that I recommend to you the true stories you are about to read. May India never forget her most fearless.

Jai Hind.

New Delhi   General Bipin Rawat, UYSM, AVSM,
                                         YSM, SM, VSM
                                    Chief of the Army Staff

1

'We Don't Really Know Fear'

The September 2016 Surgical Strikes in PoK

Uri, Jammu and Kashmir
18 September 2016

Final checks on the AK–47 rifles. Final checks on the stacks of ammunition magazines and grenades stuffed into olive-green knapsacks. The 4 men shoved fistfuls of almonds into their mouths, chewing quickly in the darkness and swallowing. Small, light and packed with a burst of energy, mountain almonds are as much a staple for terrorist infiltrators as their weapons are. The high-protein mouthfuls would have to sustain the 4 men for the next 8 hours.

At least 8 hours.

Dressed in deceptive Indian Army combat fatigues, and shaven clean to blend in, the 4 emerged from their concealed launch point below a ridgeline overlooking a stunning expanse of frontier territory. In total darkness, they trekked for 1 km down to the powerfully guarded premises of the Indian Army's Uri Brigade in Jammu and Kashmir's Uri sector, on the LoC.

The 4 men knew their mission was not particularly extraordinary. Indian military facilities had been attacked by Pakistani terrorists before. In fact, just 8 months earlier, in January 2016, an identical number of terrorists had infiltrated the Indian Air Force's base in Pathankot, where they had managed to kill 7 security personnel before being eliminated.

But there was something these men did not know. What they were about to do would change India like nothing else had in the past quarter century. It would compel India across a military point of no return that it had resisted until then.

Above all, it would awaken a monster that Pakistan had been arrogantly certain would remain in eternal slumber.

Infiltrating the Army camp at Uri before sunrise, the 4 men crept forward with an unusual sense of familiarity. Their Pakistani handlers had clearly 'war-gamed' the attack with maps and models of the camp. Wasting no time in familiarizing themselves with the camp's layout, they headed straight for a group of tents where the soldiers were sleeping.

By the time the sun was fully up and Special Forces (SF) commandos had been diverted to Uri as reinforcements, 17 Indian soldiers lost their lives. Two more would die later in hospital.

In a valley that has steadily numbed India with uninterrupted spillage of blood, the Uri terror ambush was special. Other than the horrifying scale of casualties the 4 terrorists managed to achieve, it was the hubris of the Uri attack that ignited unprecedented anger. It had come while families still mourned those who had died defending the Pathankot Air Force base only 8 months before.

Like the 4 terrorists, Pakistan was probably confident that India's ensuing wrath would be confined to public outrage and diplomatic condemnations, a standardized matrix of responses that it had learnt to handle with mastery. But Pakistan did make 1 devastating miscalculation. India was about to use precisely its reputation for inaction to exact a hitherto unthinkable revenge.

As blanket coverage of the Uri attack took over television news and the Internet on the morning of 18 September, a chill descended upon India's Raisina Hill in Delhi. Emergency meetings were held in the most secret 'war rooms' of the security establishment, one of them presided over by Prime Minister Narendra Modi along with National Security Adviser Ajit Doval.

It was at this meeting that the Indian leadership secretly took 2 major decisions: (1) the Indian military would take the fight to the enemy this time to deliver a brutal response to the Uri attack; (2) the country's ministers, including Modi himself, would play their parts to perpetuate and amplify India's reputation for inaction until such a time when the response had been delivered. An elaborate, carefully crafted political masquerade would thus begin the following morning.

Meanwhile, 800 km away and high up in the Himalayas, a young Indian Army SF officer sat grimly in front of a small television in his barracks. Uri was his area. His hunting ground. Away on a special 2-month mission to the Siachen Glacier with a small team of men from his unit, the calm of Maj. Mike Tango's demeanour belied the fury that consumed him within. He watched familiar pictures from the Uri Army camp flicker on the screen in front of him. And just as the Indian government was about to decide on an unprecedented course of action, a prescient warning rang in the Major's mind.

'We knew the balloon had gone up. This wasn't a small incident. There was no question of sitting silent. This was beyond breaking point,' he says.

As second-in-command, or 2IC, of an elite Parachute Regiment (Special Forces), or the Para-SF as it is called, Maj. Tango had spent a decade of his 13 service years in J&K. He had been part of over 20 successful antiterror operations. And yet, the morning of 18 September had sent a knife through the officer's heart. He could not wait to get back to the rest of his unit deployed in and around where the terrorists had struck.

Upon receiving the call from Udhampur that he had been expecting, from his unit's Commanding Officer, or CO, Maj. Tango gathered his men immediately for a quick return to the Valley. The team reached Dras that same night of 18 September—a date the men would never forget.

The next morning, as they began their journey to Srinagar, things were already in motion in Delhi. The first minister to make a statement was former Army Chief, Gen. V.K. Singh, who, after the traditional condemnations, made a remarkably generous appeal in the circumstances—he said that India could not act on emotion. It would be a critical spark to the success of the masquerade, followed shortly thereafter by Defence Minister Manohar Parrikar, who declared that the sacrifice of the Uri soldiers would not go in vain. Speaking to the Army in Srinagar, Parrikar sounded a familiar note, asking the Army to take 'firm action', but not specifying what such action needed to be. This was standard-issue *Bharat Sarkaar* (Indian Government) response after a terror attack.

However, to ensure that the government's messaging was not so measured as to rouse suspicion, junior ministers were tasked with adding some fire to the proceedings. That crucial bit was deftly served up by Manohar Parrikar's junior minister, Subhash Bhamre, who declared that the time had come 'to hit back'.

Two more top-level meetings took place on 19 September— one chaired by Home Minister Rajnath Singh, who had cancelled his visit to Russia, and the other by Prime Minister Modi at the PMO. Army Chief Gen. Dalbir Singh, who had dashed to the Kashmir valley just hours after the previous day's attack, had been conveyed the government's clear political directive. He arrived in Srinagar with the green signal that the SF had so far only ever dreamt about: permission to plan and execute a retaliatory strike with the government's full backing.

Over the next 24 hours, the Army would draw up a devastating revenge plan, with options for the government leadership to choose from.

The Army routinely simulates attacks on enemy territory during combat exercises and as preparation for possible hostilities. But as the COs of the 2 SF units (one of them being Maj. Tango's unit) began listing their options, they knew that history was being written then and there.

On 20 September, just as Maj. Tango and his team arrived in Srinagar, the Army's Northern Commander, or GOC-in-C of the Udhampur-headquartered Northern Command, Lt. Gen. Deependra Singh Hooda, had in his hands a final list of mission options and was preparing to present them to the government in Delhi through encrypted channels. The options were presented with remarkable detail.

'We just needed clearance. In the SF, we are war-ready at all times. When we are not in operations, we are preparing for them. There's a purpose behind everything we do,' Maj. Tango says.

At the Army Headquarters in Delhi, the mood was expectedly sombre, but focused. Aided by a team that had been galvanized by the attack, Vice Chief of the Army Staff (later Chief) Lt. Gen. Bipin Rawat was steeped in the planning phase, bringing decades of infantry training to what would be the most decisive operation he would help oversee. What happened on 18 September was personal for Lt. Gen. Rawat. As a young Captain, he had commanded a Gorkha Rifles company in Uri in the early 1980s and had gone on to command a brigade

in one of the most restive parts of the Kashmir valley. He
would return years later as a Major General to command the
Baramulla-based 19 Division. As he focused on the
unprecedented plans on his table, Lt. Gen. Rawat had no way
of knowing that a few months later, his experience in J&K and
his crucial role in planning India's response to Uri would be
high on the government's mind when it entrusted him with
leadership of one of the largest armies in the world.

The options were tabulated. The first column bore the
name of the location that would be attacked. The second
column provided its location represented by distance inside
PoK from the LoC. The third column provided information
about the location and the number of terrorists who
were likely to be encountered there. The fourth column
provided a detailed list of required resources in terms of
men, equipment, logistical and back-up support. There
was 1 final column. It provided a figure of the number
of casualties India could expect for that particular target.
Some of the targets listed predicted the possibility of zero
casualties if men and equipment were adequately ramped up.
Other options predicted definite casualties, in some cases in
double-digit numbers.

'The options provided were as specific as possible. The
government would have to take a decision based on these
inputs, which included probable casualty count. We spared no
details,' says Maj. Tango.

The target list was scrutinized along a top-secret chain
of command that numbered barely a handful of people, with
'need to know' rules applicable throughout. The options were
vetted by designated officers from the Intelligence Bureau and
the Research and Analysis Wing, before a final recommended
brief was presented to the government.

Meanwhile, arriving in Srinagar on the morning of 20 September, Maj. Tango went straight to a designated operations room to meet his CO, who had arrived after spending the previous day in Uri with his men.

'*Chhote, serious matter hai,*' the Colonel told Maj. Tango. The two men could cut to the chase like they always had. A decade ago, Maj Tango's CO had been Team Leader to a young, recently commissioned Mike as Troop Commander. Now, as 2IC, Maj. Tango would march into hell if his CO ordered him to. The Colonel had never hesitated before when speaking to Mike. But that morning, there was a clear trace of hesitation. And with good reason.

Orders had just arrived posting Maj. Tango out from the Kashmir valley and to a course in a different part of the country. While his departure was scheduled for a whole month later, Army units regard such situations jokingly as 'posted out, not interested' or PONI. Simply put, the sense is that once officers have received their next posting, they might not really have their heart in the current one any longer.

For Maj. Tango, that morning in Srinagar, there was no PONI, and no dilemma.

'I was given the option to either stay back and monitor the operation, or lead the operation and go in. It didn't take me a moment to decide,' Maj. Tango recalls.

Maj. Tango knew that the mission afoot was an operation all SF men dreamed of. The officer also knew there was no real option. His CO needed his best men in the lead. As the 2 men smoked, Maj. Tango reached out.

'*Sir, aap tension mein lag rahe ho.* (Sir, you look worried.) How can you be in a dilemma, sir? Just give me the order. No hesitation,' Maj. Tango said to his CO, his voice confident and unwavering.

'*Jis cheez ke liye SF join kiya tha, woh mauka ab aaya hai aur aap option de rahe ho* (The reason for which I joined the SF, that opportunity has now arrived, and you are giving me options),' Maj. Tango smiled.

Twenty minutes later, Maj. Tango and his 19 men bundled into squad vehicles to begin the 70-km dash to Baramulla. By midnight on 20 September, Maj. Tango's team arrived at a post on the LoC in the Uri sector.

The plans were so secret that even the teams tasked with executing the attack were not in the loop. The COs of the two Para-SF units had simply informed the teams to head to locations on the LoC in Uri, Kupwara and Rajouri sectors and await further orders. They were instructed to be on 12-hour notice to begin moving.

Maj. Tango and his warriors moved on foot. Being airlifted by chopper to the LoC post was out of the question. Apart from dangers posed by flying so close to Pakistan Army positions, the echo effect in the mountains would infinitely amplify the unsubtle whirring of helicopter blades.

Morale was high that night of 20 September as the soldiers crept up to the LoC. Their furtive arrival was an unmissable sign that offensive action was afoot. The SF are never deployed for defence. Their principal task is to attack and destroy. But remaining hidden would be an enormous challenge.

'When SF men get close to the LoC, alarm bells ring on the other side,' says Maj. Tango. 'No matter how much you try to mask your arrival, there's something about SF soldiers. They just know.'

Like Maj. Tango's team at Uri, 2 more Para-SF teams had been deployed—one in the Poonch area north of Jammu, and another at a post on the LoC in Kupwara in north Kashmir, each with a single launch pad to attack inside PoK. At these 3 locations, the warriors hunkered down and awaited their orders.

The weather at the LoC was mild and temperate in September, but would change only a few weeks later. The terrain, on the other hand, never changed, mountainous, undulating and hostile in every possible natural way—features that added immeasurably to the uninterrupted danger afforded by eyeball-to-eyeball perches of Indian and Pakistani border posts.

The weather, at any rate, was the only thing that was mild. The morale boost that came with the SF reinforcements had done nothing to blunt the fury that pervaded Army ranks after the Uri attack. In fact, the Army had decided to use soldiers from the units that had suffered losses in the Uri attack for the elaborate revenge mission. A Ghatak platoon was formed and soldiers from the 2 units that had lost men were roped in to man border posts and provide crucial terrain intelligence and support to the mission that lay ahead. Tactically, this was a smart move—few knew the lay of the frontier land better than they did. But there was another astute reason. Involving them in the mission would at least begin to lay the ghosts of Uri to rest.

But if the eagerness for payback was not already acute, Pakistan would gamely fan India's fury into an inferno the following day. And the man who would do it would be Pakistan's Prime Minister himself.

On 21 September 2016, Nawaz Sharif addressed the United Nations General Assembly (UNGA). The irony of his address would become apparent only in hindsight. A traditional platform for Pakistan to raise the Kashmir issue, Nawaz Sharif went a step further that evening in New York City, less than 4 days after the Uri attack:

> Peace and normalization between Pakistan and India cannot be achieved without a resolution of the Kashmir dispute. . . . Our predictions have now been confirmed by events. A new generation of Kashmiris has risen spontaneously against India's illegal occupation—demanding freedom from occupation. Burhan Wani, the young leader murdered by Indian forces, has emerged as the symbol of the latest Kashmiri Intifada, a popular and peaceful freedom movement, led by Kashmiris, young and old, men and women, armed only with an undying faith in the legitimacy of their cause, and a hunger for freedom in their hearts.[1]

The message was unmissable. Not only had Sharif dispensed with any bilateral decency of referring to the Uri attack, he had in fact found it fit to venerate a man India had designated a terrorist, and whose group had been responsible for hundreds of terror attacks on Kashmiri civilians. No one had expected Sharif to offer anything more than the usual diplomatic platitudes about how India and Pakistan are both victims of terror. But they had not expected the bare effrontery of choosing to invoke an enemy of the Indian state. In India, while public anger turned into a virtual call to war, Sharif's insolent speech was the confirmation the political leadership needed that their political masquerade was working. But the true master stroke in the elaborate theatre would be delivered

---

[1] *Hindustan Times*, 'Full Text of Nawaz Sharif's Speech at UN General Assembly', 21 September 2016, retrieved from http://bit.ly/2oBoRu9

3 days later. And the man to deliver it would be the Indian Prime Minister himself.

On 24 September, at 1755 hours, thousands gathered for a public rally in Kozhikode, Kerala. The Prime Minister, silent since the Uri attack except for tweets, had the media's gaze fixed on him. What would he say about Uri? Would he respond to Nawaz Sharif, a man he had cheerfully diverted his helicopter to visit in Lahore less than a year before? Would the Prime Minister satiate a public that was looking for Pakistan to be taught a lesson? It was time for Modi to play his part in a masterful facade that was now fully in motion:

> A leader [Nawaz Sharif] is reading the speech of a terrorist.
> I wish to speak to Pakistani citizens. Before 1947 your forefathers loved this entire land. India is ready to fight a war. A war against poverty. Let India and Pakistan fight a war to end social evils, illiteracy and unemployment. Let us see who wins.[2]

The media and the public were stunned. This was not anywhere close to the harsh, thundering rebuttal Modi was capable of. It provided not even the visceral satisfaction the crowds had come to expect from the Prime Minister when he spoke about or to Pakistan. The call to war on poverty was a feat of cunning that would be the penultimate step in the subterfuge. The messaging was calibrated precisely to accentuate India as a country high on political rhetoric and substantially low on political will.

---

[2] Transcribed from a video of the speech: Asianet TV News, 'Narendra Modi in Kozhikode Full Speech: BJP National Conference', 24 September 2016, retrieved from https://youtu.be/dopFyUeUnIg

At the LoC, Uri, Maj. Tango and his men were on their 4th day forward deployed. Unused to sitting in wait for long, the SF men were yearning for an order, whatever it was.

'We were very calm. Since we were so close to Pakistan Army posts, we had little or no movement. They may have suspected SF presence on the LoC, but being spotted was not an option,' says Maj. Tango.

As Team Leader, Maj. Tango had chosen every man himself, including the officers and men who would play a supporting role. He was also acutely aware of the fact that the lives of 19 men were, quite literally, in his hands.

The SF men waited, conducting brief reconnaissance patrols from the Uri post, but never straying too far. The wait was laced with tension, a numbing irony all commandos are familiar with—there is infinitely more disturbance in calm than in an actual firefight. Once 'contact' is made and bullets begin to fly, that's when calm truly returns.

Maj. Tango and his men had not been informed of their precise targets yet, but the team was by now certain of the nature of the mission they would be embarking on. The men were fully prepared. Each man knew the specific role he would play. Permutations of outcomes were discussed threadbare. The entire operation would be planned based on timing with the other 2 teams at the other 2 locations along the LoC. Given how terrain changes from south to north, inter-team coordination would prove to be a bit of a magical process.

From their post, Maj. Tango and his men had heard the Prime Minister's speech in Kozhikode. Maj. Tango was also aware that India's Minister for External Affairs Sushma Swaraj was scheduled to speak at the UNGA 48 hours later. That speech would be the final act of a national deception that had begun 8 days ago, a day after the Uri attack.

With signature indignation, Sushma Swaraj appeared to speak directly to Pakistan's Prime Minister. But the tone of her speech had been carefully sculpted to amplify not anger or indignation, but hurt and betrayal:

> Did we impose any pre-condition when Prime Minister Modi travelled from Kabul to Lahore? What pre-conditions? We took the initiative to resolve issues not on the basis of conditions, but on the basis of friendship! We have in fact attempted a paradigm of friendship in the last two years which is without precedent. We conveyed Eid greetings to the Prime Minister of Pakistan, wished success to his cricket team, extended good wishes for his health and well-being. Did all this come with pre-conditions attached? And what did we get in return? Pathankot, Bahadur Ali and Uri.[3]

Maj. Tango watched the minister's speech live on the evening of 26 September. There had been no warning in the address, no aggression. The chief emotion was disappointment. It would all make sense later, but the diversion was as unobvious to Maj. Tango as it was to Pakistan.

The officer did not know why, but he had an inkling that the team's orders were about to arrive. He was only half right. By midnight on 26 September, the warriors received word ordering them to be on standby, and identifying their targets: 2 terror launch pads in the area across from Uri. The 2 other teams were given a single launch pad target each.

---

[3] *Indian Express*, 'Sushma Swaraj's UNGA Speech', 27 September 2016, retrieved from http://bit.ly/2oVO1XT

A total of 4 terror launch pads operated by Pakistan's Inter-Services Intelligence (ISI) and protected by the Pakistan Army had just been selected for doom.

The Indian soldiers now needed to find out everything they possibly could about the targets in real time. An operational standby meant they would have nothing more than a few hours to roll out the mission once the final go-ahead was given. And they knew that the go-ahead could come any time.

A series of extremely furtive observation missions by Maj. Tango's team over the next few hours revealed that the 8 Pakistan Army posts overlooking the Uri post were not in an aggressive or overly defensive posture. Their guard was far from down, but it was clear that they had loosened up and were distinctly unaware of what was about to come their way.

The team knew that, for all the intelligence they were armed with, nothing could be quite as reliable as human intelligence on the ground in enemy territory. Through a series of masked communications over mobile, Maj. Tango's men contacted 4 'assets'—2 local villagers in PoK and 2 Pakistani nationals operating in the area—both moles in the dreaded Jaish-e-Mohammed terror group, men who had been turned by Indian agencies a few years before. All 4 assets separately confirmed the target information that was placed before them. In terms of intelligence, there was nothing further for the team to do on this side of the LoC.

Apart from the final preparations and reconnaissance missions from the Indian side, the men checked their weapons and equipment. Maj. Tango would be armed with his M4A1 5.56-mm carbine, the rest of the assault team with a mix of M4A1s and standard-issue Israeli Tavor TAR-21 assault rifles, Instalaza C90 disposable grenade launchers and Galil sniper rifles. Batteries on night-vision equipment were checked and other devices were charged too. As the hours went by and it was clear that a final go-ahead was about to come, Maj. Tango's chief worry bubbled to the surface.

'As Team Leader, my concern was to get all my men out safely. I had chosen the best men for the job. But the one thing bothering me was the de-induction—the return. That's where I knew I could lose guys,' Maj. Tango recalls.

This was no small worry. The 'induction' process, the trek down a steep ridgeline into PoK, would actually be the simplest part of the operation. Even the actual attack was not something that flustered the commandos. It was the return, an uphill trek to the LoC, that was the truly daunting part. Their backs would be facing a blaze of fire from Pakistan Army posts, belatedly roused from their slumber. And the dominant position held by the posts would make the escaping warriors easy targets to spot and kill.

Maj. Tango knew it was impossible to overthink the de-induction phase. And so, during those crucial hours, he rummaged through every bit of SF training he had had—to see if there was a tactic or stratagem he could employ to better ensure that every one of his men got out alive. Or, even better, not badly injured. Of all the uncertainties that faced them in the darkness that loomed over them from the mountain range, Maj. Tango knew the escape phase could be their true enemy.

The 2 terror launch pads identified as targets for Maj. Tango's team were well inside PoK and roughly 500 metres away from each other. Each launch pad is really a transit staging area for terrorist infiltrators before they are sent across the LoC. Both launch pads were close to Pakistan Army posts for logistical and administrative purposes. ISI handlers would often visit these launch pads before infiltration attempts.

Maj. Tango made a final call to his CO in Srinagar before ordering his team to turn off their radios. The men would now maintain total radio silence until they returned from the operation, with only the most fleeting data transmissions on their hand-held satellite equipment.

Shortly before noon on 27 September, the 3 assault teams received their final go-ahead to attack the 4 terror launch pads. Seven days after they had taken position on the LoC, they had finally been cleared to go into PoK. The mission brief was uncomplicated. The soldiers were expected to reach their targets, study the latest intelligence they could possibly access with their satellite devices and then proceed to wipe out every man they saw there.

Nobody in the military or the government knew about the mission apart from a carefully pruned chain of command that stretched from the Prime Minister at one end and 3 SF team commanders at the other. A bare handful of officers populated the top-secret chain, an imperative given Pakistan's remarkable human intelligence capabilities in the Kashmir valley. The assets Indian intelligence agencies had on the other side were far fewer in comparison.

Late on the evening of 27 September, with final confirmatory checks complete and darkness setting in over the frontier, Maj. Tango and his team from the Para-SF rolled out from their post at 2030 hours, putting the LoC behind them 25 minutes later and heading into PoK. They then began a 4-hour trek downhill into one of the most dangerous places in the subcontinent.

'This was meant to be total surprise action. And it was. But as we neared our targets, the Pakistan Army posts began firing illumination rounds to light up the area, as they normally do. This was a huge risk as we were inducting just then,' Maj. Tango recalls. 'If they even smelt us, we would have a fight on our hands. And their positions meant they could dominate us.'

Then, just as suddenly as the illumination bursts had begun, they died down. Maj. Tango took no chances, ordering his men to stay down, and waiting nearly 20 minutes before moving forward.

A kilometre from their targets, Maj. Tango split his team into 2—one for each target. Nine of the warriors followed the Major as they made their way in darkness towards their designated launch pad target.

Approximately 200 metres from the launch pad, Maj. Tango's team came to an abrupt stop. The men dropped down flat on their bellies—because the sound they heard from up ahead was the last thing they were hoping for: gunfire.

Out in front with his night-vision goggles, Maj. Tango quickly deduced that the firing emanating from the launch pad was speculative. It came in short, uncertain bursts rather than with the searing sureness of a targeted bullet hail. This was good news, because it meant that the Indian warriors had not been spotted. It was bad news too in a way, because it also implied that the terrorists suddenly had their guard up and may have been alerted that something was afoot. The intelligence network that Pakistan's military machinery enjoys along the LoC and in J&K is formidable. It was not beyond the realm of possibility that the news of the arrival of Indian soldiers had been shared with the Pakistan Army posts as a belated input, even as the final approach was in flow.

It was this possibility that made Maj. Tango pause. He knew that an enormously difficult decision needed to be made quickly—whether he would order the assault right then, or wait.

Maj. Tango craned his neck slowly, looking back towards the men behind him. He didn't need reminding that each of their lives was in his hands. Through the darkness, they looked straight back at him, waiting for the order. Crawling backwards on his elbows, Maj. Tango retreated a few metres to where his team lay in wait, informing them of his decision in the lowest, most hushed tone he had ever employed: he and his men would seek out a safe, hidden position in the area and hunker down through the night and the following day. Maj. Tango knew it was an enormous risk remaining in such hostile territory after sunrise. But the potential value of having a chance to visually study their targets before the final assault was difficult to ignore.

There was no question of sleep. Maj. Tango and his men retreated on their elbows about 200 metres in the darkness, finding a rocky crevasse wreathed by a thick clump of trees. They regrouped and waited for sunrise, a pair of look-outs keeping their night-vision-assisted gaze fixed in the direction of the terror launch pads.

The next 24 hours would be perilous in the extreme. With the cover of darkness gone, Maj. Tango and his men would need to do all they had to with only a bare fraction of the freedom that night afforded them. Switching on the sole satellite kit among them, Maj. Tango tapped on the touchscreen to first transmit his decision to his CO in Srinagar, and then to download two crucial bits of data: (1) a set of coded text messages from Srinagar with updated intelligence; (2) a pair of photographs of the area taken by an Indian Air Force Heron surveillance drone flying near the LoC.

Both sets of data confirmed that there was no major change in the situation, nor any reason to alter their plans. If their presence had been detected by the Pakistan Army posts or those manning the terror launch pads, the satellite data would have likely had indicators that their game was about to be up.

The data was useful, and a confirmation to proceed with the assault that night. But in those daylight hours, the team hadn't managed to visually study their targets first-hand.

It had been deemed simply too risky to the mission to send out reconnaissance teams in daylight. Maj. Tango made one last transmission to his CO before powering off his satellite kit: the operation would be carried out late that night.

In Delhi that evening of 28 September, the Indian Coast Guard commanders' conference was getting set to host its annual dinner. But its top invitees, Defence Minister Manohar Parrikar, National Security Adviser Ajit Doval and Army Chief Gen. Dalbir Singh, excused themselves from the event. The 3 would meet instead at the military operations room on the first floor of Army Headquarters in South Block—for one final look at the historic mission that was about to commence.

The media hadn't a whiff of what was afoot. The full focus was on India's diplomatic response to the Uri attack. As India publicly rallied support at the SAARC Summit, reports emerged that US Secretary of State John Kerry had spoken to his Indian counterpart Sushma Swaraj twice over the phone with a request that India not escalate tensions. In Delhi's upscale Gurgaon suburb, a scheduled concert by Pakistani singer Atif Aslam was cancelled following advice from the administration citing 'sentiments of armed forces/soldiers at the frontier'. The carefully crafted 'mess' of diplomatic, public and political indignation was the ultimate ruse.

At midnight, less than 1000 km away from India's capital, Maj. Tango and his men had emerged in the darkness from their temporary hideout, returning to their final positions from the previous night. There, Maj. Tango ordered his men to lie low and remain motionless until he gave the final word. Stilling the sound of his breathing, the Major peered through his night-vision goggles at the terror launch pad that lay literally a stone's throw away from where he and his men lay on their bellies, their weapons primed and ready. The speculative firing continued for many minutes, echoing through the valley. Then, finally, it stopped.

Maj. Tango once again told his men not to move. He needed to be absolutely sure that the terrorists were not waiting in ambush. If they were, this mission would end before it had even begun. The men in Maj. Tango's team were India's finest SF commandos. But even they could not face a terror ambush backed by dominant firing posts that could pour uninterrupted hell on them from positions of advantage. A few minutes after the firing had stopped, Maj. Tango ordered his men to spread out into 2 squads and follow him as he crept forward.

Fifty metres from the launch pad, Maj. Tango summoned his buddy to crawl forward towards him. He then pointed directly to the open space in front of a forested outcrop that made up the launch pad itself. Their silhouettes clearly visible in the green glow of their night-vision devices, they identified 2 terrorists who stood guard at the spot.

This was the moment the SF team had been waiting for— the moment when the fight would begin, when tension would dissipate and calm would return as bullets flew.

From a distance of 50 metres, Maj. Tango opened fire at the 2 terrorists, dropping them instantly. The officer then whispered urgently to the squads behind him to move towards the forest launch pad and attack the other terrorists who were certain to be there. Before they entered the forest, 2 commandos opened heavy fire from their hiding places directly at the launch pad, allowing the assault team to move in unchallenged. The men now sprinted in a crouch towards the hideout, immediately opening a blaze of fire as they reached. Nearly every bullet fired found a target.

The open area cleared by his initial attack, Maj. Tango now sped towards the forest hideout to join his men. As he reached the spot, he noticed 2 terrorists moving through the jungle in an attempt to attack the Indian warriors from behind. These were terrorists with commando-style training who employed movements and tactics that bore a disturbing resemblance to military strategy, a fact that was shatteringly brought home to India during the 26/11 attacks, and has been a staple with infiltrators ever since. The terrorists at the launch pads were trained like soldiers.

Realizing that the 2 terrorists would, in seconds, be in a position to shoot down his men with a few bursts of fire, Maj. Tango sprinted through the forest directly towards them. As he neared them, they saw him and leapt behind a tree taking offensive positions, ready to greet the incoming Indian team leader with a burst of fire that would drop him in his tracks. But before they could raise their AK-47s, Maj. Tango was upon them. The hectic dash had deprived him of the precious seconds he would need to raise his M4A1 carbine and fire. So in the few seconds that he had before they could fire, he had whipped out his Beretta 9-mm semi-automatic pistol. With a series of shots just 5 feet away from the terrorists, Maj. Tango felled both the men.

From the moment the firefight began until the last bullet was fired, it had been just over an hour. The frenetic pace of the assault meant the teams, now united after the split attack on 2 launch pads, would prepare to leave with only a very rough estimate of the number of terrorists they had managed to kill: 20. The figure would be corroborated days later by India's external intelligence.

A total of 38–40 terrorists and 2 Pakistan Army personnel were killed at the 4 targets. The 3 separate teams had simultaneously struck 4 launch pads across the LoC. Their entry into PoK had been coordinated and precisely timed.

The assault and exit were conducted in total radio silence; therefore, each team was entirely on its own. Their 2 designated launch pads cleared, Maj. Tango's band of warriors turned about and headed east, back towards the LoC.

The officer's biggest fear was about to come true in a far more frightening way than he had imagined.

The return, or de-induction, had to be carried out with a step beyond extreme care. The route was known, since the warriors had just used it to enter PoK, so it would take them less time to make the uphill trek back. But through the entire stretch, they would be dangerously vulnerable to Pakistan Army posts freshly aware of the intrusion and now doing everything in their power to stop the Indian soldiers from escaping back over the LoC.

A decision had to be made whether to risk taking the speedy escape route that would bring with it the certainty of angry firepower from Pakistani posts, or to trek a different path that would be longer and more circuitous, but offer safer passage to the warriors. Maj. Tango knew he didn't have more than a few seconds to take a decision. He was well aware that even a circuitous route wouldn't fully confuse the army posts that were now in an angry state of alert all along the LoC. Every alarm bell in Pakistan's military system in the field was now blaring. But aware of the immediate danger of interception by Pakistani quick reaction teams that were probably already in motion, the Major and his men decided on a longer, circuitous exit.

The return would need to be quick, but phased. One squad from Maj. Tango's team was designated as the fire support group. Men from this group would provide intermittent cover fire from the mountainside as their mates made onward progress. It would be a dangerous exercise, given that the Indian soldiers would be vastly outgunned and outnumbered by the dominant Pakistan Army positions. What they did have access to was continuous real-time guidance from the Indian drone that was back in the air over the LoC, helping the returning warriors chart as deceptive a route as they possibly could as they trudged back.

Maj. Tango was right about the risk of retaliation no matter which route they took. Enraged by the cross-border strike, the Pakistan Army posts opened fire with everything they had. From medium machine guns to rocket-propelled grenades, ammunition of every kind, short of heavy artillery, rained down on the earth around the escaping Indian warriors.

'At one point, the bullets were so close, they were whistling past our ears. There's a familiar *put-put* sound when rounds fly very close to your head,' Maj. Tango recalls. 'If I were a foot taller, I would have been hit many times over.'

During the circuitous escape, the men were frequently flat on the ground as trees in their path were shredded to bits by hails of ammunition. A particularly vulnerable 60-metre patch in the de-induction route gave the commandos their closest call. Still flat on their bellies, but with no natural feature hiding them, they needed to slither the full distance without being hit. Crossing in pairs as ammunition hit the ground inches from them, Maj. Tango's team made it to the LoC before the sun was up, finally crossing it at 0430 hours. But the men knew the LoC was not any sort of force field against Pakistani bullets—they still had some distance to go before they were fully out of range. But now, they were provided heavy covering fire by Indian Army border posts, and so the commandos could quickly cover the distance to the post they had departed from 36 hours before.

With 7.62-mm Pakistani sniper rounds still finding their way to targets within a few feet of the team, Maj. Tango and his men finally reached the Uri post. Maj. Tango made his first encrypted radio call to his CO. A few minutes later, Maj. Tango received a call from Lt. Gen. Satish Dua, commander of the Army's 15 Corps, headquartered in Srinagar and responsible for the entire Kashmir valley. Lt. Gen. Dua, himself a counterterrorism specialist, kept the call brief, merely informing Maj. Tango that a chopper would soon be on its way to get him.

Lt. Gen. Dua had been crushed by the Uri attack. As the man who oversaw all operations in the Valley, the ambush had taken place on his watch. If there was one senior officer who could not wait to get even, it was he.

In Delhi, Prime Minister Modi and the national security leadership was informed of the mission's success. A second act of making history would ensue shortly thereafter.

Meanwhile in Uri, as Maj. Tango waited, he took a full debrief on the other 2 strikes. Apart from a landmine blast that injured a commando in one of the other teams during de-induction, there were no casualties. Not only had the surgical strikes of 28–29 September been the most audacious and dangerous peacetime attacks mounted by Indian forces, they had also been the cleanest. In the table of options provided by the Army to the government leadership, the final list of chosen targets carried a probable casualty number of 1 or 2 for each target. By that measure, the mission would have been deemed successful even if 4–8 commandos had perished during the operation. With no man dead or left behind, the government would be thrust into a daze that such a surgically clean operation was possible. The reputation of the Para-SF, already inestimably high, would reach a historic zenith on the morning of 29 September 2016.

A few hours later, an Army Cheetah helicopter landed at a helipad near the Uri post. The helipad was on the leeward side of the mountain and facing away from Pakistani firing that still had not stopped. As Maj. Tango made his way towards the helipad, a series of sniper rounds smashed into the ground in front of him, killing a dog.

'The dog was walking a few feet in front of me. A bullet smashed right into the poor creature. And I was saved,' Maj. Tango says.

The officer rushed forward towards the helipad as a final blaze of sniper fire tried to cut him down. Taking off and following a terrain-hugging flight path, the Cheetah transported Maj. Tango to the headquarters of the 15 Corps, known in the Army as the Chinar Corps, after the rich, deciduous trees native to the Valley.

At the Corps Headquarters, Lt. Gen. Dua had skipped lunch in anticipation of Maj. Tango's arrival. At 1530 hours, the Cheetah landed at the 15 Corps' helipad. Maj. Tango was led straight to the operations room. Waiting for him at the door was his CO.

'*Chhote!*'

'Sir!'

The two men hugged, slapping each other on the back, pulling back and regarding each other wordlessly. Both knew what had just happened. There was no need for small talk.

Emerging from within the operations room was the Corps Commander, Lt. Gen. Dua. The no-nonsense General had a broad smile on his face as he approached the two officers. Maj. Tango straightened up immediately, saluting the senior officer.

As the Major and Lieutenant General shook hands, a waiter appeared bearing a tray with glasses half-filled with the rich amber of Black Label whisky.

'Bring the bottle,' the General ordered the waiter, 'these men eat glasses'—a fact Maj. Tango confirms as being true.

The waiter disappeared, quickly reappearing with a full bottle of Black Label. Lt. Gen. Dua grabbed the bottle, ordered Maj. Tango to open his mouth and began pouring. Then Maj. Tango, a full 5 ranks junior to the 3-star officer, returned the favour. It was only after the officers had had a chance to recover from the well-earned whisky celebration that an operational debrief took place.

Maj. Tango was now the secret centrepiece of the Indian military's modern history. An Army Dhruv helicopter arrived at the Srinagar Corps headquarters a few hours later, flying him straight to Udhampur, the headquarters of the Army's Northern Command. There he would meet Lt. Gen. Deependra Singh Hooda, the officer who vetted the final targeting options before they were presented to the Army Headquarters and government.

More whisky followed. Maj. Tango and his men hadn't eaten for a whole day. In his mind, he remembers thinking, '*Koi khaana de do. Saare daaru pila rahe hain* (Could we get some food too, please? Everyone's giving us only alcohol).'

In January 2017, 5 men from the 3 teams were decorated with the Shaurya Chakra, while 13 received Sena Medals for gallantry during the assaults. The COs of the two Para-SF units involved were awarded Yudh Seva Medals for their planning and leadership from Srinagar during the operation.

Maj. Tango went on to receive the highest decoration of the lot—a Kirti Chakra. His citation read:

By his decisive thinking, professional approach, warrior ethos, exemplary leadership and courage beyond the call of duty, Maj. Mike Tango ensured the execution of the task flawlessly with clockwork precision and eliminated 4 terrorists in close quarter combat.

Life changed drastically for Maj. Tango after the surgical strikes.

'Life has changed completely. It's more restricted now. But I cannot stop being an SF officer. That's who I am,' Maj. Tango says, referring to his inevitable status as a 'person of interest' for Pakistan and the terror groups his men smashed on the intervening night of 28–29 September.

Maj. Tango, 35 years old at the time this book was written in 2017, knew from the age of 6 that he wanted to be in the military. He remembers sitting wide-eyed on the edge of his parents' bed at their Mumbai home, watching the 1980s film, *Vijeta*, in stunned silence.

'I used to watch the movie once every day for months. I couldn't pull myself away from it. I knew I had to be in the military,' Maj. Tango says. 'My parents freaked out so much that they taped over the *Vijeta* tape.'

Over the next 12 years, Mike Tango's obsession with a future in the military would only intensify. In 2000, he joined the National Defence Academy (NDA) in Pune after failing to crack the test twice. While the Indian Air Force was a teenaged Mike's first choice, inspired by his memories of *Vijeta*, he would have to settle for the Army. He was not disappointed. He had just taken his first steps into the military, and that was all that mattered.

Over the next few weeks, Mike would be mesmerized by stories from J&K shared with him by a member of his directing staff, an officer from another elite Para-SF unit. Mike had already decided that he wanted to be in the infantry, clear in his mind that he would not fit into any other combat arm. And by the time he had finished at the NDA, it would be nothing but the SF. The young cadet's new ideal was cemented as he joined the Indian Military Academy (IMA) in Dehradun. His platoon commander at the IMA was from his future unit in the Para-SF. There would be no looking back for Mike Tango.

In 2004, Mike Tango was commissioned into the Army's Para-SF as a Lieutenant. The initial 6-month probation phase was a finely crafted period that would be the final boot camp before true SF operations. Over 3 months, Mike and other young officers were put through tests of mental toughness, integrity and honesty.

'In probation, *everyone* is assessing you. Are you a team leader? Are you a good support guy? Physically, everyone who joins the SF team is tough. They attempt to break you mentally,' Maj. Tango remembers.

None of the mental tests would of course preclude or replace physical trials. That would intensify dramatically during SF probation.

'The attempt is to try and break you, to find your breaking point, to see where you give up. The point is of course not to. But everyone has a breaking point,' says Maj. Tango.

The officer remembers occasionally considering giving it all up and quitting service during his probation. Sleep deprivation and stress tests had brought hell, in his words, to daily existence. Maj. Tango will never forget being thrown into a gutter or being ordered to dissect rotting carcasses of animals. It would dawn on the young officer that the seemingly sadistic rituals of probation were all part of the indispensable toughening-up that made the SF special.

'You can't freak out in a bad situation. No matter what happens, you have to deal with what's in front of you. That's what probation teaches you.'

A special memory remains of being dragged out of his bed at 0200 hours and being ordered to write a persuasive 1000-word essay on how the menstrual cycle of a former Pakistani leader affects the monsoon in West Bengal.

'The attempt is to throw anything at you and see how you deal with it. There are no options. You deal. Or you're out.'

Mike completed his 6-month probation in just under 4 months. He was dispatched quickly to the Kashmir valley to begin what would be an explosively active decade in the state. By October 2004, just a few months into service, the young officer had managed to prove beyond doubt that he would be a successful SF warrior. But the unit had decided that the young officer, high on his abilities, needed to suffer just a little bit longer. And so an elaborate plan was hatched by his seniors. It began with a summons to north Kashmir's Lolab Valley on Dussehra, 2004, and orders to embark on a mission fabricated to end without success. When Mike returned to the field headquarters that evening, he was roundly castigated.

'I was shouted at very harshly and told I wasn't fit for the SF,' Mike recalls. 'The next day in Srinagar, I got an even worse shelling by my Team Commander and CO. They said I lacked aptitude. I was shocked and angry. I had trained so hard for this.'

The prank was a meticulous one. Mike's seniors had even procured a movement order posting him out of the SF to a regular infantry unit.

'I was given a movement order to 18 Mahar Regiment and ordered to proceed immediately to a transit camp. I packed my bags and was on the verge of tears. I had never been so low.'

Just as Mike was leaving, a waiter from the officers' mess jogged up to him, informing him that the CO wanted to meet him one last time. Mike remembers being in no mood to meet his seniors, and simply wanting to leave as quickly as possible. Fighting back a tide of frustration, he decided to follow the waiter to the mess.

Mike's CO stood there, grim, staring, silent. A perplexed Mike was ordered to do 50 push-ups right then and there. Furious and in disbelief, Mike knew he could not disobey a direct order, so he fell to the ground to do as he was commanded. But as he rose to his feet, Mike saw his CO holding a brand-new maroon beret in his hand. The young officer had just earned the most iconic symbol of the Special Forces.

'I was beyond exhilarated. What followed was our traditional drink in the SF—every kind of alcohol mixed in a jug with our rank badges in there too. We drink it all in one go, and then the rank badges are pipped. I woke up two days later.'

Mike would see his first live firefight less than a year later in June 2005. Intelligence had just arrived about suspicious movements in Bandipora. Arriving on the scene with his squad, Lt. Tango and his men spotted the 3 'suspects', all in burkas. Their masculine voices while speaking on a mobile phone and the chance sighting of an AK-47 between them blew their cover. Mike and his men positioned themselves in a cordon around the suspects.

'It was the first and the last time my hands shivered before action. It happens only that first time. Never again,' Mike recalls.

He would go on to raise a covert/pseudo ops (operations) team for the Para-SF—a subunit dedicated to deep cover and intelligence gathering from the general population. It would allow Mike to begin understanding the level of intelligence in infiltration Pakistan had managed in the Kashmir valley, and how difficult it would be to conduct SF missions there. Not once during the 7 years he spent in covert operations did he imagine that he would one day be ordered to cross the LoC.

The Indian Army's September 2016 surgical strikes caused an immediate global sensation, facilitated and then fanned by a second stroke of history sanctioned by the Modi government—an official announcement. Just hours after

Maj. Tango and the other 2 teams crossed back over the LoC, the Army was given orders to hold a press conference to formally declare that the attacks had taken place, unheard of in special operations concerning Pakistan.

The honour of officially revealing the surgical strikes fell on the Army's Director General Military Operations, Lt. Gen. Ranbir Singh, at a joint press conference with the spokesperson of India's External Affairs Ministry. In front of a shocked crowd of media persons at Delhi's Jawaharlal Nehru Bhawan, the officer detailed the audacious mission:

> Based on receiving specific and credible inputs that some terrorist teams had positioned themselves at launch pads along Line of Control to carry out infiltration and conduct terrorist strikes inside Jammu and Kashmir and in various metros in other states, the Indian Army conducted surgical strikes at several of these launch pads to pre-empt infiltration by terrorists. The operations were focused on ensuring that these terrorists do not succeed in their design to cause destruction and endanger the lives of our citizens.
>
> During these counter terrorist operations significant casualties were caused to terrorists and those providing support to them. The operations aimed at neutralizing terrorists have since ceased. We do not have any plans for further continuation. However, the Indian Armed Forces are fully prepared for any contingency that may arise.[4]

---

[4] *Indian Express*, 'Surgical Strikes: Full Text of Indian Army DGMO Lt Gen Ranbir Singh's Press Conference', updated 29 September 2016, retrieved from http://indianexpress.com/article/india/india-news-india/pakistan-infiltration-attempts-indian-army-surgical-strikes-line-of-control-jammu-and-kashmir-uri-poonch-pok-3055874/

Four Pakistani terror launch pads had been annihilated. If the assault mission itself enraged Pakistan, at least Islamabad was not compelled to respond politically. But the audacious Indian press conference pushed the Pakistan Army, headed at the time by Gen. Raheel Sharif, into an embarrassing corner from which signature Pakistani obfuscation ensued. Pakistan's defence minister, Khawaja Asif, called India's claim a lie, while its military declared that only 2 Pakistan Army soldiers, Lance Havildar Jumma Khan and Naik Imtiaz, had been killed, and that too in a ceasefire violation.

On 20 March 2017, 6 months after the surgical strike, Maj. Tango received his Kirti Chakra from the President at Rashtrapati Bhawan. The Army had made efforts to play down the award ceremony's obvious links with the September mission across the LoC.

'By now they probably know who I am and where I am,' says Maj. Tango, then adds:

'But in the Special Forces, we don't really know fear.'

*

Note: In the interests of security, certain details have been masked in this account, and no sensitive operational details have been revealed. Some names in this chapter have been changed to protect the identity of Special Forces officers who operate in hostile territory.

2

'They Didn't Know We Were There'

The June 2015 Surgical Strikes in Myanmar

Imphal, Manipur
5 June 2015

A light breeze scattered the smell of burning flesh that morning as the bodies of 18 Army jawans lay charred and mangled in the remains of their convoy. The soldiers, all from the Dogra Regiment, were headed to their base after an operational deployment when their trucks were ambushed at 0830 hours, just over 100 km from Manipur's capital, Imphal. The surprise attack had been swift and unforgiving. As the 11 survivors were plugged with morphine drips and helicopered out of the site soon after, they knew 2 things for certain: this was one of the most expensive insurgent attacks in the restive North-east in decades, and that revenge would be a faraway castle.

On that second point, though, it would take less than a week for them to be proved more wrong than they had ever been before.

As the Army choppers with the grievously wounded survivors coursed through the air towards a base hospital in Manipur, in the country's national capital a team of India's toughest fighting men was making final preparations to depart for the strife-torn Democratic Republic of the Congo. The commandos from the Army's secretive Para-SF carried a fearsome reputation, and were looking forward to the UN peacekeeping duties they had been recently assigned in a country where daily brutality had made the Congo War Africa's deadliest in modern times.

In the Democratic Republic of the Congo, the Indian commandos would replace the coveted maroon berets of their Parachute Regiment with the blue headgear that sets UN peacekeepers apart from other soldiers. The UN Organization Stabilization Mission in the Democratic Republic of the Congo, known by its French acronym MONUSCO, accounts for the Indian Army's largest footprint on foreign soil. A fourth of the over 19,000 peacekeepers serving at MONUSCO, the world's most costly peacekeeping mission with an annual budget of $1.25 billion, come from India. In June 2015, a Para-SF team of about 100 men was to join a battalion-strength force of about 500 men that was on its way to North Kivu, a province blessed with a bounty of minerals in its earth, but which now had rivers of blood flowing through it as a result of war.

The Indian soldiers were ready for their African assignment. They would be part of an Indian brigade headquartered in North Kivu's capital, Goma, a fighting force controlling an area of 62,400 sq. km, encompassing a breathtaking landscape dotted with lakes, volcanoes, mountains, savannas and rivers. UN postings are sought after in the Army, and units go through rigorous screening before peace missions are assigned to them. There was no question of the men from the elite Para-SF unit not making the cut. In fact, they had been hand-picked to tackle the barbarism that had come to define daily life in the eastern part of the Democratic Republic of the Congo.

Just as they were about to leave from the Palam Air Force Station in Delhi, they got a call—from their home base in Jorhat, Assam. In the brevity that defines communication between commandos, the men were informed about the ambush in neighbouring Manipur. In 5 minutes, they had the story and fresh orders.

The soldiers did get on a plane that day. But it was the C-130J Super Hercules from the Indian Air Force's 77 'Veiled Vipers' squadron, which carried them at full throttle over 2350 km to Manipur. Everything had changed. Plans for the Democratic Republic of the Congo had been instantaneously put on hold.

The commandos knew what had happened, but would be briefed about their new mission only the next day. The Indian Army had been wounded—and deeply. Death on duty was nothing new, but the audacity of the ambush threw a blanket of unusual, simmering rage, not only over the Dogra battalion that had been targeted, but the entire security establishment as well as the government at the Centre.

The Army was breathing fire.

It soon emerged that insurgents from 3 outfits active in the state were behind the ambush: the Nationalist Socialist Council of Nagaland-Khaplang (NSCN-K), Kangleipak Communist Party and Kanglei Yawol Kanna Lup. The Army's Dogra unit had wrapped up its 3-year field tenure in Manipur and was in the process of relocating to a sprawling military station in the country's north when it had been attacked.

The insurgents had managed to slip back across the porous border with Myanmar to safe camps in the jungles along the border. Their hasty hit-and-run retreat wasn't surprising. In the past too, insurgents had been lulled into the comfortable routine of mounting attacks and fleeing across the border to safety. What they hadn't accounted for was that the blood spilt on that highway on 5 June would take the Army on a course of planning it had refrained from trudging on before.

A path down which there would be no turning back.

Lt. Col. Oscar Delta, 35 years old, was at Leimakhong, the Manipur headquarters of an Army Mountain Division, when news of the highway bloodbath came in. As with the death of any soldier or innocent in his area of responsibility, Lt. Col. Delta felt the blood run to his face. He closed his eyes for a moment, collecting his thoughts. Three words repeated themselves over and over in his mind.

18 soldiers. Killed.

Lt. Col. Delta couldn't remember a more savage action by insurgents in his military career, if not his entire memory. As the 2IC of his Para-SF unit, Lt. Col. Delta was well-known to the insurgent groups responsible for the ambush. They knew what he looked like. They were aware of what he was capable of. And they had just drawn blood, terrible amounts of it, in Lt. Col. Delta's own backyard.

'To say it was a big blow doesn't describe it,' Lt. Col. Delta, now a Colonel, remembers. 'When you lose 18 men like this, you have to figure out how you can hit back and what options you have. My first instinct was to launch an operation immediately with my team, hunt down those responsible and blow their brains out.'

A part of a commando's conditioning and training is how to keep his emotional responses in check. Emotions can be the enemy of every mission. Successful special operations require execution unclouded by sentiment—the knee-jerk quality of anger or sorrow. Suppressing the urges that welled up inside him after the ambush would be among the hardest things Lt. Col. Delta had ever done.

In the Army for 14 years at the time, he had spent a good part of it chasing and killing insurgents in the hilly jungles of the North-east. Among the many medals pinned to Lt. Col. Delta's chest is a Shaurya Chakra awarded to him for exceptional gallantry in action in 2004—over a decade before the Manipur attack. That year, a young soldier under his command was killed in an ambush laid by insurgents. Delta, then a young Captain, led a group of crack commandos that hunted down the insurgents and killed 8 of them.

'We killed 8 of the 11 guys who laid the ambush. We got them the same day. Their joy was short-lived. We were faster than them.' Lt. Col. Delta remembers the psychological significance of that swift retribution. It was a lesson that stayed with the insurgents for years.

Now, a decade later in 2015, Lt. Col. Delta was quick to grasp that the Manipur ambush would be a turning point in the conflict spread across the North-east. He also knew that his unit, specializing in jungle warfare, would be enlisted to deliver whatever retribution was deemed necessary. Over the next 36 hours, Lt. Col. Delta's squad, with the guidance and blessings of a grimly determined Army and government leadership, would chart out a spectacular mission of revenge.

Few know the North-east better than the men of this particular Para-SF unit. And nobody fights the way they do. The North-east has been a hunting ground for the unit for nearly 2 decades. From its Assam headquarters, commando squads are scattered across the region, primed and ready to jump straight into action anywhere they're ordered to.

Lt. Col. Delta knew he was a marked man. A well-built warrior standing 5 feet 10 inches tall, his reputation had already spread wide. For insurgents, he was the man to take down. Since 2006, the officer had been (and remains) on most-wanted lists of the Manipur-based People's Liberation

Army and United National Liberation Front. The fact affects neither his peace of mind nor his work.

'Militancy is all about fear. They have been seeing me for over 10 years. I think they should be afraid of me. *I* am the one authorized to carry an automatic weapon and walk freely on the streets in the area, not they,' he says.

But the paratrooper does not keep his family with him. For their own safety, they continue to live at an undisclosed Army base where they have been for several years. Their ancestral home in the North-east remains locked even today.

When Lt. Col. Delta got his battle orders, he didn't need to prepare his men. Even before the Manipur ambush, Lt. Col. Delta's unit had been preparing to strike NSCN-K camps across the border in Myanmar after suspected Naga rebels ambushed and killed 8 soldiers in the state's Mon district on 4 May 2015, exactly a month before the Manipur attack.

Had the 5 June ambush not happened, the Army would have gone ahead with the original plan. A team of Para-SF commandos would have infiltrated Myanmar on the night of 5 June and wiped out an NSCN-K camp that had been on their radar for a while.

Headed by Myanmar-based insurgent leader S.S. Khaplang, the notorious outfit fighting for the creation of Greater Nagaland ended a 14-year-long ceasefire with the Indian government on 27 March that year. It was clear that Khaplang did not want peace. After the abrogation of the ceasefire, the insurgent outfit launched a series of attacks against security forces. Retribution was already thick in the air. The punitive strike had been planned at the local level, but the central government wouldn't take ownership. It would have been executed swiftly and quietly, as such strikes mostly are.

But these sensitivities, along with the planned punitive strike from Nagaland, went straight out of the window when the Manipur massacre happened on 5 June 2015.

'We were told that the Army Chief was flying to Imphal on 5 June. Something bigger was unfolding,' Lt. Col. Delta recalls.

Something bigger and deeply more audacious was indeed being planned. The government at the Centre was absolutely certain that the Manipur massacre deserved immediate punitive action. But what precisely that action would be was yet to be determined.

There were already early signs that retaliation was being planned and it would be severe. Hectic developments followed. In Delhi, National Security Adviser Ajit Doval dropped out of Prime Minister Narendra Modi's tour of Bangladesh. Gen. Dalbir Singh, then Army Chief, postponed a visit to the UK. None of this was normal.

A day after the ambush, Gen. Dalbir Singh flew to Imphal where he was briefed on the developments by Lt. Gen. Bipin Rawat (who became Army Chief the following year), then commanding the Army's Nagaland-headquartered 3 Corps. Lt. Col. Delta and the involved Para-SF unit's CO were among the select group of men present at the briefing, where Lt. Gen. Rawat made it clear that the Manipur attackers were not beyond the Army's reach. Every man in that room knew that the attackers were now on foreign soil. The message couldn't have been clearer.

In the warm yellow glow of halogen lamps, the locations of insurgent camps were red-pinned on the maps showing the border areas spanning Manipur, Nagaland and Myanmar.

'We were told by Gen. Dalbir Singh that the operation had been approved at the highest levels of the government, and the defence minister would be controlling it. Full support. Full backing. This was the moment we had been waiting for,' Lt. Col. Delta says.

While most aspects of the plan remained fluid, one thing was clear by the end of that top-secret briefing: the Army would have its revenge within 72 hours. The decks had been cleared for a daring cross-border raid into Myanmar—surgical strikes that the government in Delhi would not deny.

7 June would be D–Day.

The senior Generals asked Lt. Col. Delta to stay back in the briefing room. The broad contours of the surgical strikes on insurgent camps emerged for the first time. The commandos from the lethal Para-SF unit would mount a 2-pronged attack. From Manipur and Nagaland, 2 teams would simultaneously infiltrate and destroy the jungle sanctuaries that were known to harbour and train insurgents.

The orders were grim but clear. Lt. Col. Delta and his squads were to hit camps with the largest numbers of insurgents, so the commandos could inflict maximum damage. The targets had to be chosen carefully before launching the mission. It would have served no purpose if the men found themselves at the doorstep of a thinly held camp. The stakes were enormously high and it wasn't just the government expecting results. A grieving, but steadfast Army was, too.

'We spent our first few hours selecting camps which the insurgents would have thought were beyond our reach— where they felt secure. It was not to be a token assault,' Lt. Col. Delta recalls.

Conventional military manoeuvres require a great deal of intricate planning. Special operations on foreign soil are something else entirely. They require a degree of preparation and detail that would make regular drills seem like child's play. Success of special missions depends on a range of factors, chiefly sound strategizing, accurate intelligence and scrupulous planning that takes every eventuality into account.

There was no shortage of targets to choose from across the border in Myanmar. Insurgents had found security in the border jungles. Hours after the Manipur attack, the Army had zeroed in on the targets that its highly skilled commandos would attack in Myanmar: 3 insurgent camps, which Lt. Col. Delta and his men would target. A similar squad would mount an assault from Nagaland.

But there was a problem. And it wasn't a minor one. Distance.

The camps across the Manipur border were located deep inside, and striking them on 7 June, just 2 days later, was a daunting prospect. D-Day would have to be rescheduled. And that, Lt. Col. Delta knew, was the only way to increase the chances of a successful strike. He spoke his mind to the Generals that evening. There was still much work to be done.

'The distance was too much. We needed more time. The attack could have been launched on 7 June from Nagaland as everything was in place, but the idea was to launch the teams simultaneously. And we couldn't have done it from Manipur in that time frame,' Lt. Col. Delta remembers. His assessment was as clinical as it could have been.

Distance wasn't the only hurdle. The Para-SF's strength in Manipur wasn't enough to mount the operation. Lt. Col. Delta had just 40 men on 5 June. And that was why the detachment of 100 Para-SF men on their way to the Democratic Republic of the Congo was stopped and diverted to their neck of the woods. They were in Imphal by dinner time.

Corps Commander Lt. Gen. Rawat had shifted base from his headquarters in Nagaland's Dimapur to Imphal to manage the new mission.

As a Brigadier, Lt. Gen. Rawat had helmed the Indian brigade in the Democratic Republic of the Congo in 2008–09 and the UN had credited him with defending a key Congolese province that could have been overrun by rebels. Lt. Gen. Babacar Gaye, then the Force Commander of the UN mission in the Democratic Republic of the Congo, wrote in a commendation later awarded to then Brig. (now Gen.) Rawat that it was due to his 'leadership, courage and experience' that North Kivu's capital Goma never fell, stability returned to the country's eastern region and the main rebel group was forced to come to the negotiating table.

The presence of a hands-on commander, known for his military acumen, helped the commandos prepare literally to defy death on the mission into Myanmar. They knew they were in able hands.

'The top bosses were in the loop, and arranging a special aircraft to fly the Congo-bound men to Imphal happened without a hitch. Lt. Gen. Rawat made sure all logistics were taken care of so that we commandos could focus solely on the mission,' Lt. Col. Delta recalls.

The timely arrival of the commandos was critical for the mission. The 3 insurgent camps were built next to each other. Intelligence inputs pegged the number of insurgents present at these camps at more than 120. Planning the mission with 40 commandos would have been possible, but a grave risk. It would have also limited Lt. Col. Delta's options to choose the best men for the mission.

For this mission demanded the best—of the best.

By midnight on 5 June, the raid schedule had been reworked and defined. Striking the targets on 7 June was officially ruled out. D–Day was now pushed 2 days further to 9 June, giving the commandos sufficient time to physically reach their targets. Lt. Col. Delta relaxed for the first time that day, exhaling as the men returned to their barracks for a few hours of sleep.

The plan had been discussed in as much detail as possible. Lt. Col. Delta had told his men that their objective would be to reach a wooded hilltop overlooking the targets by midnight on 8 June. The plan was to mount the final assault at the crack of dawn on 9 June. It sounded simple enough in theory, but no man listening to Lt. Col. Delta breathed any easier. They were looking forward to a tough new assignment on foreign soil. They had spent years training hard in the jungles of Mizoram and elsewhere for precisely this kind of mission. Lt. Col. Delta watched them as they retreated to rest. They were ready to prove to the country what India's Special Forces were capable of.

Alone in the briefing room now, Lt. Col. Delta stared hard at the map. A passion nurtured over the years, the officer adored maps of every kind, displaying an incredible ability to memorize locations and compute distances at terrific speed. His knack for understanding geography from a piece of paper had always been incredibly valuable to his squad. He switched off the light and walked back to his quarters, aware that this was possibly his final full night of sleep before they set off to exact revenge.

The Para-SF warriors were to reach a predetermined staging area on the border from where they would move into Myanmar. The movement of the commando team had to be kept as discreet as possible. There was no scope for the insurgents to get the slightest whiff that something was afoot. It would spell the mission's doom. Not only would revenge remain elusive, but more Army lives could be lost.

'We put our heads together and worked out a deception plan,' Lt. Col. Delta recalls. The commandos would be transported in Army trucks to make it look like regular infantry troops being moved, a routine affair in the North-east. Helicopters were out of the question—they would draw attention. Airlifting the commandos was therefore ruled out. Every detail was a closely guarded secret. Other than the commanders involved in the planning, not a soul knew about D-Day, the mode of transport or the composition of the 2 teams in Manipur and Nagaland. After a series of early morning briefings and presentations on 6 June, the commandos set out.

The men were very heavily armed. They were taking no chances. Sixty-four of them hand-picked for the job were carrying Carl Gustav 84-mm rocket launchers, Pulemyot Kalashnikov general-purpose machine guns, Israeli-built Tavor TAR-21 assault rifles, Colt M4 carbines, AK-47s and under-barrel grenade launchers.

Lt. Col. Delta was carrying his M4A1, the assault firearm he preferred. A smaller, fully automatic version of the venerable M16 assault rifle, the M4A1 carbine is finely tuned for use in special operations. Light and packing lethal power at both close and medium ranges, it remains the weapon of choice in elite squads around the world. Used across the US military as a primary infantry weapon, Lt. Col. Delta finds it perfect for quick reaction assault missions. The assault team's arsenal also included Israeli Uzi silenced submachine guns to take down sentries at the camp, and Galil 7.62-mm sniper rifles. SF battalions are usually the best-equipped units of the Army. And it showed that day. They were carrying enough weapons and ammunition to cause a great deal of damage. The rifles weren't slung across their shoulders. Commandos try not to use slings. A weapon sitting in your hands reduces reaction time. In a firefight, that could be everything.

Apart from the weapons, the men were carrying backpacks stuffed with extra ammunition, combat rations, medical kits and water. The soldiers were not travelling light. Each of them was carrying a personal load of 40 kg, excluding the weight of weapons. Carrying their heavy loads, the men strode towards the Army trucks that were waiting with their engines on to transport them closer to the Myanmar border.

Just as they were about to clamber into the vehicles, Lt. Col. Delta handed over his M4 to one of the men, who immediately realized what Lt. Col. Delta was about to do. The officer knelt down and kissed the earth, a ritual he followed before every mission: an invocation to the soil as a friend and guide.

The Army trucks dropped off the battle-ready team at the staging area at around 0300 hours on 7 June. The targets were across the border and still very far away. The men would have to cover a distance of more than 40 km on foot to reach a designated hilltop, from where they could scan the camps before beginning the actual attack.

Lt. Col. Delta had more than proven his worth as a soldier, but the next 48 hours would test his skills and leadership as a commander. On his shoulders rested the fate of the mission, the lives of the young men under his command and, above all, that intangible element that colours everything that soldiers account for in their line of work—national prestige.

At daybreak on 7 June, the commandos began their trek through hilly terrain towards the border, with Lt. Col. Delta setting the pace and ensuring there was no slow-down. He remembers thinking of nothing else.

Only very occasionally would he allow his mind to go to his mother who had been battling cervical cancer and was to undergo surgery 2 days later at a hospital near his hometown. Nobody in Lt. Col. Delta's chain of command knew this. It had not even crossed his mind to ask for leave to be with his family. The mission was literally the only thing that occupied his thoughts.

It was nearly dark when the Para-SF team reached the border. The temperature had dropped to about 25 °C and the men were greeted by a refreshing breeze coming from Myanmar. In 36 hours, this mission would be over. Everything was on schedule. After spending the night on the Indian side of the border, the men crossed into Myanmar at first light on 8 June.

There was no turning back now.

On foreign soil, Lt. Col. Delta and his men trekked through 6 km of hill and jungle. The men had done their homework. As part of the detailed planning that went into the operation, the team had hired 2 guides from an Indian border village in Manipur. The guides spoke Burmese fluently and knew their way through every bend in the thick woods that greeted the commandos. The guides' knowledge of the land, coupled with Lt. Col. Delta's photographic map-like memory, kept the squad precisely on course.

Prepared for anything, the commandos had not encountered a hurdle so far. But they soon would. Not long after the men had crossed a rivulet that demarcates the border between India and Myanmar, they stopped in their tracks. The route they were taking through the wilderness was supposed to be isolated, bereft of human settlements. Apparently not.

A group of men walked right into the squad, hauling a bounty of monitor lizards from among the 5 species found in the area. At being questioned by the guides, it turned out that the men were Burmese hunters who had killed the large lizards, considered a delicacy and aphrodisiac in some parts of the world. 'There were 5 of them. We weren't expecting them. But there they were, and something had to be done,' Lt. Col. Delta remembers.

The easiest option was to leave them alone and move on. But Lt. Col. Delta didn't want to leave anything to chance. Loose ends are frequently what cause the demise of precisely planned operations. There was always the possibility that the men were informants sent by the insurgents as eyes and ears. It would take one word of warning from them to destroy the entire mission.

The other option was to kill them. But Lt. Col. Delta knew he and his men couldn't ever bring themselves to do that—not without proof that they were helping insurgents. They could well have been hunters who were just in the wrong place at the wrong time.

'Or maybe it was us. We were on their turf,' Lt. Col. Delta smiles.

Lt. Col. Delta made his decision. He ordered his men to secure the hunters with the ropes they were carrying, and decided to take them along as captives right up to the point from where reconnaissance teams could survey the camps. They simply could not risk the mission.

Lt. Col. Delta glanced at his Mountain Hardwear watch, which had replaced his Fitbit activity tracker band that was his usual wristwear when not on duty. The men hadn't lost time. He told them they would make a brief halt soon for a quick meal. The team hadn't slowed down for a moment, moving briskly as they neared their targets. The tune playing in their heads could well have been a memorable military song from the film *Lakshya*. Lt. Col. Delta had played it on the sound system in their barracks the night before they left Imphal.

> *Kandhon se milte hain kandhe, kadmon se kadam milte hain*
> *Hum chalte hain jab aise toh dil dushman ke hilte hain*
> (When we walk shoulder to shoulder, and march in step as brothers
> When we move like this, our enemies tremble in fear.)

Like most Army men, the commandos loved the song and knew the lyrics by heart.

By noon on 8 June, the steep hill they were headed towards loomed into view. From its flattened crest, the commandos would get their first view of the camps they would obliterate the next day.

It was time for a much-needed hot meal. The commandos' backpacks held hexamine fuel tablets along with cooking stoves. A quick lunch was rustled up using the smokeless, odourless, long-burning tablets. The meals, ready-to-eat packs, contained *rajma*–pulao. The famished men ate in turns. A cordon of soldiers made sure they were well-guarded at all times. For dessert, they helped themselves to *shakkarpara* from a common polythene bag, a sweet that reminded them of celebrations back home.

Marching into thicker foliage at the base of the steep hill, their fatigues ensured the commandos blended in well. By the time the team began climbing up the hill, they realized they had almost run out of water. Each man had carried 7 litres in his backpack. It was the peak of summer and the men were parched. It wasn't something they were not trained for, but neither was it a favourable situation the day before a big assault. The sooner they finished their mission, the faster they could get water from the destroyed camps, Lt. Col. Delta joked to quiet laughter and cheers. He glanced at the men in his squad. Every one of them was under his charge. He was responsible. The mission had primacy in the men's minds, but Lt. Col. Delta knew that it was his responsibility to make sure no one got hit.

Or worse.

In a fast-paced situation, anything could happen. A casualty evacuation scenario would not only be a nightmare logistically, but also leave fewer men available for the mission. It was something the squad simply could not afford. Apart from the big picture, there was also something that only those in fighting units could fully comprehend.

'We knew if something went wrong, the battalion's honour would be sullied. People would say SF stands for *sabse faaltu* (most useless)! How could we let that happen? Never!' says Lt. Col. Delta.

As leader of the mission, Lt. Col. Delta was in touch with the Army leadership in Imphal. He was carrying a mobile phone and the signal strength was at 3 bars. It was known to the Army that Indian mobile phones continue to function up to a certain distance along the border. Missed calls based on a predetermined code were used to stay in touch during the operation. Through a combination of missed calls and communications over a secure satellite link, Lt. Col. Delta remained connected with the Army leadership.

Fully backing the Army's mission, the government in Delhi had kept its counterpart in Nay Pyi Taw, Myanmar's capital, in the loop.

For security reasons, Lt. Col. Delta cannot reveal the other modes of communication he maintained during the mission. The team would regularly receive intelligence updates.

Some inputs suggested the insurgents might have fled the camps.

'Delta, are you sure the guys are still there? Can you see them? There's a possibility that the situation may have changed,' said a voice at the other end of the line during one such transmission.

Lt. Col. Delta had been prowling the North-east for over a decade and had built an excellent network of informants. There was no doubt in his mind that the insurgents were at their camps, and in large numbers. He calmly told the caller not to worry. He and his men hadn't come all the way to raid empty camps.

As the commandos began climbing the steep hill, those carrying TAR-21 assault rifles took their slings out. The Israeli weapon is built butt-heavy and muzzle-light— the magazine and firing mechanism are located behind the weapon's trigger.

'SF guys don't use slings. We believe in carrying our weapon in our hands to react swiftly. But you can't climb a steep hill holding a Tavor due to its design,' Lt. Col. Delta explains.

A few hours later, as the sun set, the Para-SF squad arrived at the hilltop. Lt. Col. Delta and his men took carefully picked positions from where the targets were now in sight. The men felt their first big rush of combat adrenaline. The camps were barely 400 metres away. The reconnaissance elements from the team were at the highest positions to observe the camps, their layout and possible movement of insurgents within. The rest of the men had occupied positions below them. Every man on that hilltop had night-vision gear.

The camps they now gazed at belonged to the Manipur-based People's Liberation Army, but were used extensively by NSCN-K cadres.

As moonlight bathed the hilltop that June night, Lt. Col. Delta rehearsed and fine-tuned the final plan for the predawn assault. With his team leaders, he went over how the teams would be divided 4 ways. The first subunit would storm the camp in an initial direct assault. Two 'cut off' subunits would take down insurgents trying to escape. And a fourth team would form the crucial rearguard group.

The orders were clear. The assault would begin with devastating firepower before sunrise the next day, 9 June. At the same time, the corresponding Para-SF squad 100 km away in Nagaland would begin its own assault on an NSCN-K camp that was believed to be sheltering the insurgent group's notorious military adviser, Niki Sumi.

Just as the final assault plan was firmed up, the silence on the hilltop was shattered by a burst of automatic fire at around 2100 hours. The commandos sat up in their positions, their weapons ready. Had their cover been blown? Lt. Col. Delta hoped it hadn't. Had the insurgents discovered the commandos' arrival? Had they anticipated their route?

'I thought the whole plan had gone to hell. I thought it was finished,' Lt. Col. Delta remembers. He took a deep breath and in whispers ordered his men to remain statue-still in their positions. Not a sound could be heard but a mild breeze through the trees.

If the men had been detected, Lt. Col. Delta would have had to make an immediate decision. The mission would have been turned on its head from being primarily a surprise assault to a totally defensive, evasive escape mission. It would no longer be about prestige and retribution but about survival.

It was clear now that the insurgents had begun sending out patrols to secure the perimeter of the camps. They had ventured uphill, coming as close as 150 metres to the positions held by the commandos on the hilltop. The patrollers were singing at the top of their voices and flashing their torches. Their cheer, probably drunken, was a sign the patrolling was routine and that they likely did not have a clue about death crouching on a hilltop above them.

'They fired some shots in the air. It turned out to be speculative firing to provoke a reaction and confirm if someone was in the darkness in the woods around. They didn't know we were there,' Lt. Col. Delta says. A wave of quiet relief swept over the soldiers. Their plan was intact.

Some of the men had puris and chutney for dinner as they prepared to go without sleep for the third night straight. But even if the commandos had wanted to get a few hours of shut-eye, midnight brought a fresh surprise.

Just after midnight, a group of insurgents from the camp began firing in the air. The same worrying question about being robbed of the element of surprise exploded once again in Lt. Col. Delta's mind. A voice inside his head told him that Plan B was very likely going to be necessary. He didn't need to communicate this to the commandos. The men lay silent and motionless, rooted to their positions. Every last one of them expected a full firefight to break out at any moment.

The firing continued for 10 minutes. And then, just as abruptly, it stopped. The insurgents walked sleepily back into their camps, only to emerge again 3 hours later for another round of firing.

It was 0300 hours. This was proving to be confounding and frustrating. Had the insurgents really been tipped off about the stealthy advance of the Para-SF team? It became evident to Lt. Col. Delta by this time that the assault plan would have to be considerably altered. As things stood, the squad's solitary reconnaissance team was still perched on the hilltop, while the rest of the squad held positions on the slope below.

A commando officer on the team crawled from his position to where Lt. Col. Delta was. Then leaning in, he whispered that if the men could take their final assault positions on the hilltop without getting into a firefight with the insurgents, they should go ahead with the attack as planned, come what may. The mission commander had only 2 hours to decide.

'The conclusion we came to was that if we occupied our final assault positions without loss of surprise, then we would go ahead with a modified plan,' Lt. Col. Delta recalls. The modifications were significant. Instead of splitting the team into 4 subgroups, the mission commander now decided on 3.

Two teams would carry out the direct assault, while the third would cover the rear to prevent the commandos from

being encircled by insurgents. The third team's crucial task
was to ensure safe exit for the 2 assault teams. The plan to send
2 'cut-off' parties to the other side of the camp was dropped in
the changed circumstances.

By 0400 hours, the commandos were able to crouch and
crawl to their final assault positions on the hilltop without
giving themselves away. Lt. Col. Delta looked through his
night-vision glasses directly at the camps down the hill. The
insurgents had not returned to their enclosures after the last
round of firing an hour before. They were still around.

'They were not too far from us. We could have caught up
with them in about 10 minutes. They were walking towards
the camp,' says Lt. Col. Delta.

The commandos had reached the most critical phase of
their operation. They started moving cautiously in the direction
of their targets. Lt. Col. Delta took a deep breath and told his
assault teams that they would tail the insurgents right up to the
perimeter of the first camp. The strategy worked perfectly.

The insurgents had entered their camp oblivious to the
sudden creeping presence of 40 heavily armed Indian commandos
who had formed a semicircle around the site. The men were
positioned so that each one of them could directly fire at their
targets. The other 24 commandos stayed behind, keeping a close
watch through the techno-glow of their night-vision goggles,
ready to jump in if they were needed. Inside the camp, the
insurgents were apparently preparing their first meal of the day,
clueless that it would be their last. Their guard was at its lowest.

'These guys have 2 meals a day—at 0500 and 1500 hours.
The attack was timed with their first meal. We knew they
would all assemble in the dining area and would be the most
vulnerable at that time,' Lt. Col. Delta recalls.

The attack squad made a final assessment of the target. The first sentry post was empty. There were 2 posts behind it—one manned by 4 insurgents and the other by 2. With a sweep of his hand, Lt. Col. Delta quietly ordered the commandos equipped with Carl Gustavs to unleash their first rockets at the 2 sentry posts, blowing them to pieces in a wave of flame. There was nothing left of the 6 insurgents on guard duty.

The rocket explosions made the ground shudder. It was only then that the insurgents inside the camp realized they were being attacked. But they also knew they had nowhere to run or hide. The commandos then opened fire on the camp with their assault weapons. A single shot followed by a double tap, repeating the sequence till they emptied their magazines. The men would quickly slap in new magazines and continue raining rounds on the insurgents.

The commandos had not put the selector switch on their weapons to full auto mode for burst fire. 'The problem with bursts is that the bullets spread. Single shot and double tap is far more accurate and you also conserve ammunition,' says Lt. Col. Delta. Double tap is a technique where 2 shots are fired at the same target in quick succession.

The insurgents scattered, baffled and unable to understand what had hit them. The first camp was cleared without much effort. Dazed by the strength of the assault, the insurgents could not respond for the first 25 minutes. When the men were in the process of clearing the second camp, the insurgents started returning fire with automatic weapons. The layout of the camp had sprung a fresh surprise. The commandos had been under the impression that the camps didn't have bunkers. Not true. The insurgents had built Army-style deep-dug bunkers inside the camps from where they were now firing at the commandos.

The firefight had been on for about 20 minutes when the entrenched insurgents from the third camp began engaging the commandos in the rearguard, sending a deadly fusillade of bullets whizzing over their heads. The well-prepared commandos answered with overwhelming firepower: they opened up their automatic weapons and pumped grenades from their under-barrel grenade launchers (UBGLs), sending shrapnel scything through the air. The men also fired 2 rockets in airburst mode to cause maximum destruction over a wider area. As the earth burnt, it was time for another decision.

The mission commander realized that the assault teams had achieved their objective with 2 camps completely destroyed. It was time to pull back. 'When the rear party got caught in a firefight, I decided that it was time to move out ASAP. The assault teams also attacked the third camp with heavy weapons before leaving, but we did not enter it,' Lt. Col. Delta recalls.

On the ground, there was no time for arithmetic. It would emerge later that the insurgents had taken a huge beating. During that 45-minute assault, the commandos had expended almost 15,000 rounds, more than 150 grenades and a dozen rockets. The men had clear instructions not to stay back to conduct a headcount, but to return as quickly as possible after the mission was complete. They followed those orders.

Miraculously, not a single man on Lt. Col. Delta's team was hurt. Not even a scratch. The exfiltration route had already been planned. As the commandos pushed closer towards the Indian border, Lt. Col. Delta made the call that the Army had been waiting for. The magic words were spoken: 'Mission accomplished'.

In the operations room at Leimakhong, Lt. Gen. Rawat smiled broadly. Everyone in the room cheered. The message was immediately relayed up the chain of command. In the hours that followed, in a rare move that stunned the world, New Delhi would officially reveal India's military response to the Manipur ambush.

Lt. Col. Delta and his men returned to Manipur around noon on 9 June, trudging through forest for the next few hours before arriving at the first border village on the Indian side. It was 1500 hours. The sun was scorching and the men were exhausted, hungry and dehydrated, but smiling grimly in the glory of their mission's success. From a local store, Lt. Col. Delta and the men purchased bottles of ice-cold beer, treating themselves to hungry gulps of the beverage. Lt. Col. Delta noticed, as he drank, that the beer was from Myanmar. The men allowed themselves their first laugh in days.

Later that day, 2 Army Aviation Corps Dhruv helicopters clattered into a clearing near the village. They had taken off from Leimakhong an hour before to fly the victorious commandos back to base. On foot for days, the commandos were delighted that they didn't have to trek the 30 km more to the staging area where they had been dropped off for the mission. By the next morning, all 64 men on Lt. Col. Delta's team were back in Leimakhong.

The commando team assigned to target the NSCN-K camp across the border from Nagaland was not as lucky as Lt. Col. Delta's combat group. The insurgents there had fled hours before the raid. The men simply set the camp ablaze and returned.

Fifteen minutes after landing in Leimakhong, Lt. Col. Delta's phone rang. He was not expecting to hear from the man at the other end of the line. It was Defence Minister Manohar Parrikar.

'I hadn't even untied my shoelaces when the *raksha mantri* (defence minister) called to congratulate me and the team. It was exhilarating and absolutely incredible. It's not every day that you get a call from the Defence Minister,' Lt. Col. Delta, who went on to command a battalion in Manipur, recalls.

His work wasn't over, however. Lt. Col. Delta spent the day debriefing the Army leadership about the mission. He spared no detail. It missed nobody up the chain of command and government that history had just been made by the men of the Para-SF.

When he was finished, Lt. Col. Delta allowed himself the liberty of remembering his family. He wanted to drive to the hospital to meet his mother who had been operated upon that morning. Lt. Gen. Rawat and the other commanders were astonished to learn of Lt. Col. Delta's mother's cancer surgery just a few hours before. The commanders were not comfortable with Lt. Col. Delta's plan to drive down to the hospital alone in his Maruti 800. They knew hit squads would be after him. But Lt. Col. Delta politely refused a commando team for his personal protection. He spent 4 hours with his mother and returned to the base the next morning.

In Delhi that day, the Army would tell a packed news conference that the commandos had raided the insurgent camps on the basis of 'credible and specific intelligence' and inflicted significant casualties on the insurgents. The Army gave no specific numbers but estimates that emerged later pegged the figure in the range of 40–50. The Burmese Army is believed to have transported the bodies in 2 trucks to a nearby location for quick burial. Scores of badly wounded men were taken for treatment to a hospital south of where the commandos struck. The casualty figure would have been far higher had the cut-off teams been able to deploy on the other side of the camps as well.

Two months after the strikes, on the eve of Independence Day, the government announced a Kirti Chakra for Lt. Col. Oscar Delta, a Shaurya Chakra for a Havildar-rank commando on his squad and Sena Medals (Gallantry) for 5 other commandos who were involved in the cross-border raid into Myanmar. The strikes made Lt. Col. Delta a cult figure in a regiment already full of heroes. But you'd never know it if you met him.

'You are not a Rambo out there. My Kirti Chakra (India's second-highest peacetime gallantry award) belongs to every man who took part in that mission. We did our job and it's a fantastic feeling being recognized for it. But other than that, awards really do not matter,' says Lt. Col. Delta.

The officer does have one regret regarding the mission, though—not having brought back a flag or any other object from the camps as a 'war trophy'. Before heading out for the mission, Lt. Col. Delta had requested the Army leadership not to fix a deadline for exfiltration as the commandos were planning to return with a captured flag. 'Unfortunately, there was no opportunity for us to seize a flag. I guess we will have to wait for another mission,' he says, smiling.

But what happened to the Burmese hunters the commando squad had taken captive? Lt. Col. Delta instructed his men to set them free minutes before the commandos began their descent from the hill for the final assault. They were given Rs 5000 each to compensate them for the trouble. In appreciation, before they melted into the woods, the hunters offered the commandos some monitor lizard meat. Lt. Col. Delta remembers the flavour.

And the meat was just the way he liked it. Raw.

*

Note: Some names in this chapter have been changed to protect the identity of Special Forces officers who operate in hostile territory.

3

'When He Awoke, Death Smiled'

Lance Naik Mohan Nath Goswami

Udhampur, Jammu and Kashmir
1 September 2015

When he awoke each day, Death smiled.

It whispered in his ear, its arms outstretched. It stalked him with every crunch that his boots made over beds of dry pine needles—on the snow-blown peaks of mountains few have even heard of.

Death would tap him on the shoulder as he trudged forward eagerly through the thickest forests, never hesitating. Like the hundreds of times before, he brushed Death off like he would a speck of dirt on his camouflage combat fatigues.

L. Nk Goswami didn't have time for death. In the final 11 days of his life, he laughed at it straight in the face. And when it was finally time for their dance to end, it was he who let Death have him. Willingly, without wavering. And not on Death's terms, but his own.

Even among the countless tales of unspeakable courage in the Indian Army, the story of Lance Naik Mohan Nath Goswami of the Para-SF is legendary. It is a story that even the Army regards not just with pride, but also with a sense of awe and disbelief. It is the story of a man who volunteered for 3 operations over 10 days in 2015, killing 11 terrorists in all.

On 1 September 2015, L. Nk Goswami stretched his legs outside his quarters at his Para unit headquarters in

Udhampur, J&K. The unit, respected by other SF units and deeply feared by terrorists who infiltrated across the LoC, had a job that none envied.

Wearing his combat trousers, boots and a vest, the muscular warrior was fitting a few hours of rest into what had been a week bristling with action and peril. Just 4 days before, in an encounter on a mountainside not far from Rafiabad in Baramulla, L. Nk Goswami and his team had eliminated 3 Lashkar-e-Taiba terrorists over an extended, furious, 2-day firefight. The squad had even managed to capture a fourth terrorist, a Pakistani national, alive. L. Nk Goswami never breathed easy, but as he strolled back and forth that morning 4 days later outside his unit camp, he did the one thing he loved more than hunting terrorists.

L. Nk Goswami used his cell phone to call his wife, Bhawna, at their small home in Lalkuan village, about 30 km outside Uttarakhand's popular hill station of Nainital. The two would speak for a few minutes every day when possible. A gap, and this was often, always told Bhawna that L. Nk Goswami had set off on an operation. That morning, the phone didn't need to ring a second time for her to pick up.

'It was a very personal conversation on 1 September,' Bhawna recalls. 'Mohan spoke about the future, about our lives and how we needed to plan. He would constantly remind me of how uncertain a commando's life is, and how anything could happen at any time, and how we must prepare. But he said nothing like that in his last call. He was very matter-of-fact in his manner. He was building a new house for us across the path from where we live. All I wanted to know was how he was holding up. And all he would say was he was hoping to complete the construction of our new house the following March when he came home.'

The warrior had been home in the hills of Uttarakhand the previous month for his daughter Bhumika's seventh birthday. It had been 2 weeks of real rest before he took his train back to Udhampur and onward back to the unit he loved. Barely a week after his return to Kashmir, L. Nk Goswami would set off on the first of his final 3 operations.

Physically supremely fit, it hadn't taken L. Nk Goswami more than a few hours to work his muscles back into action mode following a fortnight of relative repose with his family. He had been raring to set off on another hunt the moment he stepped off the train at Udhampur on Independence Day, 2015.

On 21 August 2015, a Para-SF squad led by a young officer, Capt. Dipesh Mehra, was deployed to hunt terrorists near a remote village named Khurmur in north Kashmir's Handwara district.

The intelligence was solid: 3 terrorists were expected to arrive at a designated spot to receive 6 more who had freshly crossed the LoC. The warriors lay waiting. On the night of 22 August, while the squad descended the mountain towards Khurmur village, they spotted the shadows of 3 terrorists right in front of them in the dark. One was walking in a battle-ready position in front like a commando scout, followed by 2 behind. All 3 were carrying AK-47 rifles. Two had heavy rucksacks, presumably filled with ammunition and provisions.

Through the darkness, the commandos spotted them at a distance of no more than 10 metres. The warriors had night-vision glasses, but the foliage was very dense. The squad needed to make sure they were not civilians or woodcutters, or maybe even another Army unit that had accidentally stepped into the path of the Para squad on the hunt. They had to carry out a 'challenge protocol', an inherently dangerous task that involves calling out to the other.

'This was a face-on moving contact at very close range,' recalls a warrior who was part of the operation. 'As soon as they were challenged, the terrorists opened fire in our squad's direction. We returned fire. Capt. Mehra was hit by a ricocheting bullet. In the darkness it wasn't clear how serious the injury was.'

As a fierce, close-range firefight broke out that night in Handwara, back at the squad's headquarters in Udhampur, L. Nk Goswami and Maj. Anurag Kumar were getting set to depart in a squad vehicle on a separate mission with 2 more Para commandos. L. Nk Goswami was eager to begin. Looking at a piece of paper containing the input, while pacing excitedly in the small room where the men had met that night, L. Nk Goswami said over and over again: *'Pukka kuchh hoga, yeh solid input hai* (Something will definitely happen; this is credible information),' insisting they leave immediately.

When not on an operation, L. Nk Goswami was known to sit from morning to night near the unit's 'anchor'— the communication set that was used to convey inputs on terrorists, or summon the warriors on a hunt. In ways that even his comrades sometimes failed to fathom, living, for L. Nk Goswami, was the lull between operations.

The 4 men departed late that night in a squad vehicle. The plan was to gather reinforcements on the way and arrive at a contact site where intelligence had reported the presence of at least 10 terrorists.

They were 20 minutes away from the contact spot when they received a message on their portable communication.

'Mehra saab ko lag gayi (Mehra Sir has been shot),' said a voice at the other end. It was from Khurmur. L. Nk Goswami

immediately asserted to his colleagues, *'Mehra saab ghaayal ho
gaye hain. Wahin pe turant chalte hain* (Mehra Sir is injured. Let
us go there immediately).'

Maj. Kumar, the only officer among the four, paused,
wondering if this was wise. Wading into a live firefight could
be a terribly risky proposition for them. But L. Nk Goswami
implored the officer, saying there was nothing more important
than reaching their comrades in trouble.

'As team commander, I had to think for the whole squad.
Their lives were my responsibility. But L. Nk Goswami had
amazing persuading power. This was emotional work for him.
He would not let go,' remembers Maj. Kumar.

He asked L. Nk Goswami to calm down and think straight.
There were conflicting reports from the Khurmur site about
precisely what kind of trouble Capt. Dipesh Mehra's squad was
in. Did they really need to divert there instead of continuing
with their separate mission to hunt 10 terrorists? By now,
L. Nk Goswami was beseeching the officer.

'We have to go there. We will regret it forever if it turns
out that Mehra saab could have been saved if we had intervened
on time. We don't have a choice,' L. Nk Goswami had said.

By this time, word had also reached them that the terrorists
in Khurmur may have also had a Pulemyot Kalashnikov
machine gun, a brute force weapon that the Para commandos
would have found near impossible to dodge in such a close
engagement. Worse, communication from the squad at
Khurmur had fallen silent. Was it all over? It was the trigger
needed for a final decision.

Maj. Kumar ordered the vehicle to turn around and head
straight towards Khurmur at full throttle. As the vehicle tore
through the night at a speed of 120 kmph, L. Nk Goswami

sat up straight, every bit of his posture ready to leap out and into battle.

About 200 metres short of the contact site in Khurmur, L. Nk Goswami asked the vehicle's driver to switch off the headlamps, sensing they were exposed. He was right. The moment the lights were switched off, a hail of bullets clattered around the car, missing it by barely a few feet. The terrorists had seen the approaching vehicle and let loose a few shots directly at it. The four Para warriors scrambled out of their car and began the careful trek through the darkness in the general direction of the terrorists. L. Nk Goswami led the way.

'Mohan Nath always led from the front, in the dangerous but critical scout position,' recalls Maj. Kumar. 'I would often tell him he was senior enough now, and that he should let younger warriors play scout to get experience. His reply was always the same, and there was no arrogance in it, just plain, honest opinion: *"Sir, mujhse achha kaun kar sakta hai?* [Sir, who can do it better than I?]". He promised that if anyone proved more effective than him as a scout, he would personally invite that person to lead.'

Until a better warrior was found, L. Nk Goswami wouldn't risk the lives of the men he was with. He was clear about that.

Though the squad had just faced fire, they only had a general 6-figure grid reference, a rough geographical coordinate, on the location of the terrorists who had let loose the volley of bullets. The 4 men crept down the path towards the 100-sq. m area that would be their hunting ground in the dark. On the fringes of Khurmur village, as the land rose into a forested Handwara hillside, the 4 warriors got on to their bellies. In the darkness, 20 metres up that same hillside, were the terrorists with their weapons.

And 10 metres away from them lay an injured Capt. Dipesh Mehra along with the rest of his squad. Intermittent fire rang out at close quarters. L. Nk Goswami and the other 3 warriors had crept right up to the firefight, positioning themselves flat on the ground under a large tree.

The situation was ripe for disaster. With neither squad accurately aware of where the other was, the risk of shots being fired at mistaken targets was very real. L. Nk Goswami quickly slithered over to Maj. Kumar and told him to convey on the communication set to the other squad that he would be blinking his flashlight to indicate his position—a terribly dangerous move in the situation, but it was the only way to take the next step. Maj. Kumar once again cautioned him, saying the flashlight would paint a big fat target sign on them for the terrorists to fire at. L. Nk Goswami was nonchalant. 'Let them fire at us, sir. This tree will protect us.'

Maj. Kumar was right. As the flashlight was switched on, a burst of Kalashnikov fire exploded down the hillside towards the 4 men under the tree. And as L. Nk Goswami had predicted, the fat branches of the tree took much of the fire. Their position now revealed, the squad engaged directly with the militants, with L. Nk Goswami managing to quickly kill 1. Capt. Mehra's squad quickly relocated to join their 4 comrades at the base of the hill. As the terrifying close combat continued, L. Nk Goswami helped the injured Captain out of the area, his arm slung across his shoulder. All the while, he kept himself between the injured man and the firing terrorists, just in case a bullet managed to find them in the darkness. By morning, the other 2 terrorists had been killed too.

Operation Khurmur, as it was later formally code-named, was a victory, and a powerful message to the terror training camps in PoK. August was the heart of the waning summer season, and every soldier in the state knows what that means: that terror groups would be looking to crank up the infiltrations and replenish their hidden cells on India's side of the LoC before the unforgiving winter put a virtual stop to any such cross-border activity. The next 2 months would see a visible ramping up of scale and audacity. L. Nk Goswami and his team knew that squads like theirs were India's principal weapon against this annual rising tide.

The squad had barely squeezed in a 2-day lull when a fresh intelligence input trickled into their camp on the morning of 26 August. They hadn't expected much rest anyway. Para squads never do.

Five Pakistani terrorists had infiltrated across the LoC in the Uri sector. It was a difficult entry point, with some of Kashmir's most hostile terrain, but one that afforded several hiding places. The 5 had trekked over the Shamshabari range, scaled the formidable Kalapahar mountain and weathered the 4000-metre-high Kazi Nag Dhar as they made their way towards the more manageable lowlands of the Kashmir valley. One of the men had been shot dead in a brief encounter with men from the Army's 35 Rashtriya Rifles on a remote ridgeline on the Shamshabari. The other 4 had escaped at double speed, taking advantage of the evasive tactics that are a special part of terrorist training.

Minutes after killing 1 of the 5, the leader of the 35 Rashtriya Rifles team at the site had quickly called for a drone to track the 4 terrorists who had escaped. A Searcher Mark II drone was scrambled immediately from an airbase in the Kashmir valley, darting straight and high over the Shamshabari range, its cameras and thermal imagers switched on. Infiltrating terrorists make sure they limit their movement by day, using the darkness to cover ground and head to their destinations. But before sunrise on 26 August, the featureless crests of the Shamshabari gave the drone a perfect view of the terrorists in the dark, their bodies showing up as 4 'hot' blotches ambling over the featureless expanse of the range. The drone would remain in the air, tracking their every movement as they made their way towards Baramulla.

The 4 men could be seen climbing down a hillside littered with enormous boulders and, apparently, a network of caves and crevasses. At 0545 hours, with the sun now out, the drone witnessed the 4 men stopping and taking cover. This would be their last known location.

An hour earlier, over at the Para-SF unit's headquarters, L. Nk Goswami was among 12 men who were issued a 'warning order', placing them on operational standby to be dropped from a helicopter on to that boulder-strewn ridge to hunt and kill the 4 infiltrators. By 0700 hours, an Army Dhruv chopper lifted off from Udhampur with the 12 commandos, including L. Nk Goswami and Maj. Kumar, taking a deliberately circuitous route before finally identifying a landing zone on the ridge and dropping the warriors off at 0930 hours. The fully armed men grouped for a quick execution plan for a hunt that would later be code-named 'Operation Lidder Panzal'.

The landing zone was about 4 km away from the last known location of the terrorists as designated by the Searcher Mark II drone, which was still in the air above the range.

Twelve of India's most hardened warriors were on the hunt for the 4 terrorists. But even they knew that being transplanted to a 4000-metre altitude would substantially limit their true strength. The thinner air would bring on fatigue faster.

And this being an airborne operation, each man had a combat load of over 40 kg. As they gazed up at the ridge, none of the dozen knew just how long the hunt that lay before them would stretch on for.

As the men began their careful trek through a maze of boulders along the ridgeline, the drone kept the warriors in its sight, its pilot sitting far away at a secret location, updating the Para commandos in real time about how close they were to where the terrorists were believed to be resting. The men trudged on towards their quarry, crossing a glacial lake and uninterrupted mountain terrain. By noon, they were close.

As always, L. Nk Goswami led the squads as the scout out in front. Maj. Kumar remembers how L. Nk Goswami actually managed to keep the squads a step ahead of the technologically advanced drone that buzzed high above them.

'He used his terrain wisdom to detect a trail—something that the drone couldn't see. Mohan actually discerned what he was sure were footprints in the misty dew that settled on the patches of grass between rocks at that altitude,' Maj. Kumar remembers.

Some of the men were sceptical, wondering whether L. Nk Goswami was reading too much into what he saw on the ground. But the warrior was certain.

'*Issi taraf se gaye hai, sir. Contact humko mil jaayega* (They have definitely passed through here. We will find them),' L. Nk Goswami said, bending down to scrutinize a trail that was clear to him and apparently nobody else.

Not for the first time, team leader Maj. Kumar decided to go with L. Nk Goswami's gut. He dispatched 1 squad down the 'trail' L. Nk Goswami had found, and sent the second one higher up the ridge to take a commanding position in case the terrorists were waiting in ambush.

'I have never known a highlander like Mohan. If he hadn't spotted that trail in the mountain dew, we were almost certainly headed into an ambush situation,' remembers Maj. Kumar. 'We would have lost men.'

By this time, the Army drone pilot had communicated to the Para team that the 4 terrorists were hiding *inside* a cave at the target location. The 2 squads had split up through the boulders, agreeing to arrive at the cave at precisely the same time. The tactic was simple. Two teams arriving from 2 different directions would distract and divide the attention of the terrorists, halving the effectiveness of their possible retaliation.

An hour later, the squad taking the higher path along the ridge communicated to the trail squad below saying they had spotted something suspicious in the shadow of a huge 30-foot boulder. It was a grey pheran, the traditional woolly, robe-like garment worn in the state. The squad leader reported that he was about to fling a pair of grenades down at the boulder. He was asked to wait—because by this time, L. Nk Goswami had walked far ahead on the trail and reached the cave all alone.

A fierce firefight immediately erupted. L. Nk Goswami fired with his TAR-21 rifle, taking cover behind a low boulder in a small open area outside the cave's mouth. When the firing slowed, L. Nk Goswami jogged back up the trail to the rest of his squad, briefing them on what had just happened. Maj. Kumar remembers L. Nk Goswami's face.

'I had never seen him calmer. He had just come back to us after a few minutes of heavy firing at close range with 4 militants. But he wasn't even excited. All he said was

"*Sir, bande dikh gaye. Aur ek ko lagi hai.* (Sir, I have seen them. And one of them has been hit.)'"

The Major immediately cordoned off the area surrounding the cave, a 50x30-metre piece of land strewn with high boulders, creating an insidious maze. Twelve armed warriors stood waiting, their weapons trained directly at the mouth of the cave. The militants had been cornered.

'Storming the cave wasn't an option without incurring casualties,' remembers Maj. Kumar. 'We had rations and water. We had ammunition too. So we decided to wait the night and draw them out.'

L. Nk Goswami sat on the cave's right at a distance of 15 metres. If the terrorists emerged, they would be cut down by a C-shaped formation of soldiers. Night fell heavily on the tired dozen. There would be no rest, sitting as they were, flexed and primed for action through an uncomfortable, chilly night that brought waves of the same dew that had caught L. Nk Goswami's attention and led the men to their target earlier that day. In turns, they gratefully ate packed dinner of puri and pickle.

To send the holed-up terrorists a clear message that there was no way out, a soldier was ordered to fire 2 rocket-propelled grenades into the clearing outside the cave at night. In the concussive blast, 2 terrorists emerged from the cave in panic, trying to make a desperate run for it. The moment they emerged into the open area, 2 shots were heard. L. Nk Goswami, his TAR-21 in his favourite single-shot mode, fired 2 quick rounds, abruptly ending the escape bid.

'In such a situation, men usually fire their weapon in bursts. They want to be sure. Mohan fired just 2 shots,' recalls

Maj. Kumar, who had been sitting just a few feet away from L. Nk Goswami when he fired. 'Just 2 rounds. He was that confident of his skill.'

After all the firing, a deathly quiet descended on the ridge that night. The Para-SF squads were running low on support ammunition too, so they decided to wait till morning to see if the remaining 2 terrorists could be drawn out. It would be a long, sleepless wait.

Till noon the next day, 27 August, there hadn't been another sound. Four men, including L. Nk Goswami and Maj. Kumar, began a hectic search operation in the area around the cave. The mandatory task was to ensure that the terrorists hadn't, by some slim chance, emerged in the darkness and relocated to a different position.

With L. Nk Goswami once again as scout leader out in front, the 4 men crept between the boulders, their weapons cocked and ready. The 3 men heard a series of shots from L. Nk Goswami's rifle up ahead as a fresh firefight erupted. Creeping up virtually silently around a boulder, L. Nk Goswami had spotted the leg of one of the hiding terrorists behind another boulder. He had immediately aimed and fired at the leg, sparking a volley of return fire from a crevasse behind the boulder. L. Nk Goswami returned to the 3 others, and all 4 took position, their weapons now aimed directly up the path towards the boulder.

Moments later, screaming slogans in a hoarse shriek, the injured terrorist hobbled out into the clearing, firing his weapon, only to be cut down instantly by the 4 waiting warriors.

Only 1 terrorist now remained alive. And the squad knew he was fully cornered, deep inside a cave behind the boulder. Maj. Kumar remembers wanting to empty the last of his squad's rocket-propelled grenades into the cave, sealing the last terrorist's fate where he hid.

'We wanted to finish up and de-induct. We were out of rations too,' he recalls. But L. Nk Goswami broke the bristling silence. Stepping towards the cave he cupped his hands around his mouth and made a loud call.

'*Bhaijaan, abhi bhi mauka hai. Aap surrender kar do. Baahar nikal jao. Kis liye marne ke liye aaye ho?* (Brother, there is still a chance. Surrender and come out of the cave. Do you want to die here?)' L. Nk Goswami was speaking firmly, but on that windswept mountain ridge, there was a gentleness in his voice as he called into the gaping maw of that mountain cave.

The terrorists had been trekking for 5 days through the harshest terrain Kashmir could offer. L. Nk Goswami knew that they would be at the end of their reserves, their energy and perhaps their sanity.

'That was his presence of mind. He knew the terrorist would have no heart left for a fight. It was the perfect opportunity to capture him alive. That's what Mohan did,' remembers Maj. Kumar.

Many minutes later, the last terrorist finally hobbled out of the cave, his hands up. Emaciated, thirsty, with bloodshot eyes, he stood in front of the men, saying nothing. L. Nk Goswami turned towards Maj. Kumar. Both men knew they couldn't kill him.

The men secured the terrorist, giving him the last of their food and water. He was Sajjad Ahmad, alias Abu Ubaidullah. A young man from Muzaffargarh in Pakistan, he had been recruited into the Lashkar-e-Taiba and trained like a commando to infiltrate India and inflict casualties in population centres. His testimony would go on to add to the already enormous dossier of evidence India had in Pakistan-sponsored terror. With their very valuable captive, the men summoned a helicopter to fly them back to base.

'Mohan was far above the standards of even our battalion. Trust me, that's saying *a lot!*' says Maj. Kumar, who was later awarded the Shaurya Chakra for his leadership during the operation. 'He was as intelligent and emotional as he was fit.'

The men would get a 'bonus' of just 3 days before they got their next call.

L. Nk Goswami's next operation began on 2 September. Barely rested after the Lidder Panzal mission, L. Nk Goswami and his squad were put on scramble alert for the third time that week. It was a month that perfectly defined the relentless high-tempo operations that are the unique preserve of the Para-SF.

The intelligence this time was about 6 fresh infiltrators on their way across the remote Sutsalyar forests of Kupwara, one of Kashmir's densest jungles. Visibility through the foliage was never more than 3 metres.

The squad planned a long-haul operation of 96 hours with 6 squads of 36 men who would lie in wait for the infiltrators. The men were on location by first light on 1 September. Splitting into 2 groups, the teams deployed on either side of a wide mountain stream flowing through the forest. While the men knew the terrorists would arrive, the intelligence they had was not specific on precisely which direction they would come from. That explained the larger number of Para commandos in what the regiment calls a 'speculative ambush'. A bigger number had a greater tactical chance of knocking out the terrorists before they had an opportunity to act.

At 2030 hours, L. Nk Goswami's squad detected the terrorists as they arrived in the area. The intelligence had proved to be very good. There were 4 of them, as the input had said. The squads, positioned in buddy pairs through the forest trail, had set down Claymore mines. Unlike landmines, Claymore mines are detonated on remote command and designed to have their explosive effect pointed in a certain direction. But the terrorists arrived from an entirely unexpected direction, giving L. Nk Goswami and the other troops barely a few minutes to readjust and reconfigure.

But first, as always, the infiltrators needed to be challenged by the Para commandos. It remains one of the most risky standard operating procedures, but SF units have no way around it yet. Since such operations take place among Indian citizens and on Indian territory, protocol demands that every precaution be taken before hostile action. Challenging someone in such a situation instantly gives away your position, allowing the challenged party to take the first shot.

As the 4 figures were spotted in the darkness, it surprised nobody that L. Nk Goswami volunteered once again to step forward and challenge the intruders.

Taking cover behind foliage, L. Nk Goswami shouted into the shadows, demanding to know who the 4 men were. In the darkness of that forest in Kupwara, the response was instantaneous. The clatter of assault rifle fire immediately broke out.

In the first few minutes, 1 of the 4 terrorists was hit by a shot L. Nk Goswami fired, but not killed. The close fire exchange continued for several minutes. But then it began to drizzle, and the guns fell silent. The terrorists didn't dare move in case they gave away their locations. And the hunting commandos waited, watching through night-vision devices that were virtually useless in such foliage and weather.

As they waited, shortly after midnight, the squads were caught off guard by a loud explosion in a tree above them. It took seconds for the soldiers to realize that the terrorists had fired from an under-barrel grenade launcher attached to their AK-47 rifles. Splinters from the exploding grenade rained down on 2 commandos positioned under the tree, tearing open wounds on the sides of their faces.

Positioned a few metres away in the darkness, L. Nk Goswami was watching. And he knew what he needed to do. He always did.

If there was one impulse stronger than killing terrorists, it was L. Nk Goswami's overpowering need to make sure he evacuated a comrade injured in an encounter. The 2 wounded commandos were now pinned down, sitting ducks for whatever would come next. L. Nk Goswami and his buddy, Havildar Mahendra Singh, sprang from their location towards the 2 injured men, their assault rifles at the ready for what L. Nk Goswami knew would be a hail of bullets directed straight at them.

He was right. The terrorist who had fired the grenade moments earlier now opened a burst of fire with his assault rifle, cutting down Singh with a bullet that entered his abdomen and went straight into his spine, instantly paralysing the lower half of his body. L. Nk Goswami knew that in the next few seconds, his buddy would be torn to shreds in the volley of following bullets. His weapon now in burst mode, L. Nk Goswami sprang from his cover position, firing furiously at the terrorists. Two bullets tore through L. Nk Goswami's waist, 1 passing straight through him, the other lodged inside. As he crumpled sideways with a roar, he kept his weapon pointed straight, meeting the 2 advancing terrorists with a spurt of shots, killing them just as he crashed to the ground.

Bleeding from his gunshot wounds, L. Nk Goswami attempted to crawl towards Mahendra and the 2 injured jawans under the tree. But as the rain came down a little harder, washing his wounds, L. Nk Goswami seemed to decide to let Death have him.

With cover fire from Maj. Kumar, L. Nk Goswami crawled forward in the darkness towards Mahendra Singh. The injured soldier was quickly secured and pulled to safety for evacuation. Then Maj. Kumar, now on his elbows, pulled himself towards the fallen figure of his favourite soldier.

'I sat with Mohan, holding his hands through the night. There was no pulse. He would have done the same for me. He would not have left me.'

Maj. Kumar remembers those moments every day since it happened.

'Sitting there with him, his flesh cold, I don't remember anything feeling more unreal. A good part of me said he would wake up. This wasn't a man who could be felled by bullets. And yet here he was—cold.'

The disbelief stretched out through that rainy night.

'For many moments, I wondered if this was another one of his tactics. To play dead so he could spring up to kill the remaining terrorists,' Maj. Kumar says, smiling. 'I actually waited, hoping that was true. I have lost men before, but I couldn't come to terms with it. I waited till sunrise to see if he would wake up.'

The next day, the officers and jawans of the unit held a *bada khaana*, an alcohol-fuelled banquet at their headquarters. Through the disbelief and mourning, there was sweeping pride. A man who had walked among them, as jovial as he was fearless, had signed off in the only way he could have wanted—on his terms, doing what he loved. Over drinks to dull the disbelief and sorrow, the men raised their glasses to L. Nk Mohan Nath Goswami that night, hailing him loudly for his fierceness in battle and generosity in victory.

'He was the very core of an SF soldier. He loved his family and missed them every day. And he was married to the adrenaline of combat.'

As the national media stood mesmerized by the tale of L. Nk Goswami, his mortal remains would travel by road up the hills of Uttarakhand to his wife in their village. Viewers across the country would watch the stoic figure of Bhawna Goswami next to L. Nk Mohan Nath Goswami's flag-draped casket.

Through uncontrollable tears, his mother, Radha Devi, would appeal to the government: 'Build a school in his name—a playground in his name. This is what he cherished as a child. This is what he would have wanted for others.'

'He knowingly pushed himself into a hail of bullets to save his comrade,' a Lieutenant Colonel with the unit, who led L. Nk Goswami years ago as a troop commander, remembers. 'We are all working for the country. But in an SF unit, you're also working for each other. He was emotionally attached to us and his work. He was an ideal SF commando.'

That unwillingness to countenance suffering stretched beyond the battlefield.

'He had a very soft corner for those suffering,' recalls L. Nk Goswami's wife, Bhawna. 'If he saw a poor person, or someone mentally disturbed, he would give them his clothes. He would never give beggars a rupee or two; he would give them Rs 100! A yogi baba once came to him in a market and asked him to buy him some grapes. He bought him a kilo of grapes! Once, he took off his shirt and jacket and gave it to a naked beggar. He was like that.'

Four months after Operation Sutsalyar, Bhawna Goswami would make her way to Delhi to receive L. Nk Goswami's posthumous Ashok Chakra, the country's highest decoration for gallantry in peacetime. She steeled herself as it drizzled that morning on Delhi's Rajpath.

'Before we got married, he always said to me that I needed to be prepared for anything. I did feel scared, but it was written in our destinies that we would make our lives together,' Bhawna, who still lives in Lalkuan, says. 'In my heart, a voice would tell me that as a commando's wife, I needed to be as fearless as he was. If one has to die, it can be anywhere.'

Bhumika, L. Nk Goswami's daughter, is now in a boarding school not far from home.

'For us, he is always here. He will be with us for a lifetime. It's just that I can no longer see him. He walks with me. He even talks to me. But I cannot see him.'

A signboard, 'Shahid Mohan Nath Goswami', points to the couple's small home. After his death, Bhawna decided not to complete the construction of the bigger house they had been building across the path from where they lived.

'My nephews want to be like Mohan. All the children in our family want to take his legacy forward. They say, "He killed ten terrorists; we will kill twenty."'

In the many tales of inspiring courage in the Special Forces, an abnormal, nearly magical film coats the tale of L. Nk Mohan Nath Goswami.

'His call sign was *jaadugar* (magician),' remembers a fellow warrior who accompanied him on that final mission.

'I don't think I will ever see a braver person than him facing fire.'

*

4

'Even the Toughest Take Cover. But Not He'

Havildar Hangpan Dada

Shamshabari mountain range, Jammu and Kashmir
26 May 2016

His right hand, cold, blue and dead, was squeezed around the trigger of his Kalashnikov rifle. His face lay half in mountain snow, at a height of 12,000 feet. As life flowed out of him through a gunshot wound in the neck, the Havildar's dying hand had emptied a full magazine of bullets. A dead man had fired 30 7.62-mm rounds.

30.

There had always been something other-worldly about Havildar Hangpan Dada. Courage in combat is far from a rarity in the treacherous highlands of Kashmir. But nearly every officer and soldier who fought alongside Havildar Dada swears they have never seen a man so possessed by the fight. A man for whom the world revolved so absolutely around terrorist encounters, his courage was frequently alarming.

'He became a ghost long before he fell,' a soldier from his unit would later say. They would be referring to Havildar Dada's most unforgettable fight. His last.

On 26 May 2016, some 150 km north-west of Srinagar and across the LoC, a group of heavily armed terrorists did a final weapons and stores check as they walked out of their camp nestled in the lush Leepa valley in PoK. Their manner would get steadily less confident, but more furtive, creeping as they neared the front line.

They were departing from one of the most breathtakingly mellow parts of PoK. The Leepa valley draws huge numbers of Pakistani tourists in the summer months. Pictures of happy visitors in the splendid walnut and apple orchards, mountain brooks and meadows throng social media sites each year. But there is a side that few have ever visited, or even seen in pictures—a side infested with terror launch pads from where infiltrators are regularly dispatched across the LoC into J&K.

It was pitch dark and raining that May night as the 4 men began their hike towards the LoC. Flashes of lightning lit up a snowy ridgeline of the Shamshabari mountain range, the formidable natural barrier they would have to cross to reach the Kashmir valley. The barrier was where Indian soldiers held positions at heights of more than 14,000 feet, each post commanding a panoramic vista of unspeakable beauty—but also uninterrupted danger.

The posts, vacated in the winter of 2015, had been reoccupied by the troops barely a week before to choke off infiltration that begins with the onset of summer. For logistical reasons, it is not unusual for the Army to withdraw soldiers from desolate forward outposts in J&K to lower altitudes at winter's peak, before sending them back to control the positions once the weather improves. The unforgiving winter shuts down all infiltration routes, giving soldiers rare downtime in an unending cycle of combat.

As always, the terrorists had been well-trained in the route to follow and what to expect—the Indian Army's most hardened mountain warriors. They followed a carefully chosen path, sneaking across the LoC and heading east towards the ridgeline. A large part of their training would be expended in putting the Shamshabari range behind them. Once crossed, all the terrorists needed to do was duck into the many hiding places afforded by the gently flattening valley that would rise to greet them as they stumbled off the mountainside. If the terrorists could go undetected for a few more hours, they would be in a position to inflict a great deal of real damage.

Besides the deployment of its fighting units, India has spent hundreds of crores on building and maintaining a fence along the de facto border with Pakistan, but heavy snow causes severe damage to it every year. Apart from routinely taking advantage of such gaps, infiltrating terrorists, helped by Pakistan Army border units, also target vulnerable border stretches that have been left unfenced due to topographic factors.

This group of infiltrators had, on 26 May, found 1 such gap. The distance between the fence and the LoC can vary from 50 metres to over 2 km, depending on the area's geography. It was about 2 km in this sector.

Dawn was just breaking as the terrorists crept up the rugged ridgeline running from south-west to north-east when a sentry of the 35 Rashtriya Rifles battalion, a unit drawing soldiers largely from the Army's Assam regiment, saw something move from the corner of his eye. His eyes darted in that direction. There was nothing. He wondered briefly if the cold, as it frequently did in that stillness, was playing tricks on him.

It could have been an animal, he thought to himself. Leopards and bears were known to roam the snow-swept heights. But the Army post was on the lookout for something a good measure smarter, and incalculably more dangerous. Sepoy J.N. Baite, in his early twenties, quietly ventured out of the 'Meera post' atop the ridgeline to take a closer look.

Infiltration season was about to resume and nothing could be left to chance.

As the icy wind whipped across his face, the Assam Regiment soldier trudged through the snow to check if the movement he had noticed in his peripheral vision was something that needed more attention. Baite's finger rested on the trigger of his AK-47 as his company commander's words echoed in his ears.

'Never let your guard down. Not even for a moment. Split-second decisions can mean the difference between life and death.'

It was a tenet every soldier forward deployed in J&K operated by.

A soldier's instinct is seldom wrong. Baite had barely walked 30 metres from his post when he spotted human footprints in the pristine snow. He knew instantly that the boys of the 4th battalion of Assam Regiment, or 4 Assam, would have to skip breakfast that day.

Raised during the Second World War in 1941, the Assam Regiment draws its troops exclusively from the 7 North-eastern states, and the dauntless men are well-suited for deployments in the mountains. Maj. (later Col.) Sonam Wangchuk of 4 Assam was conferred the Maha Vir Chakra, India's second highest wartime gallantry award, for capturing an 18,500-feet-high cliff in Batalik's Chorbat La sub-sector during the 1999 Kargil conflict. It was the highest honour won by the regiment. Seventeen years later, the unit found itself in similarly hostile heights, waiting to face intruders.

Pretending that nothing was amiss, Baite, a Manipuri, retreated to his post and immediately raised the alarm about the presence of an unknown number of infiltrators in the vicinity.

Within moments, a radio operator relayed the message to the 'Echo Company' commander, Maj. K. Amirtha Raj, who was controlling all active operations on that patch of the Shamshabari.

The Major had just had his first sip of the hot, comforting chai. That was all he would have that day. The radio message made it obvious that a fight was on its way—the first one of the season in Kupwara's beautiful but deadly Naugam sector.

A familiar drill played out. The Army's posts near the LoC sit astride notorious infiltration 'highways' that terrorists use to enter the border state. The route that opened into Naugam was one of them. Soldiers manning these posts are tasked with carrying out patrols day and night, and to lay ambushes for terrorists, making sure no flank of the mountains is left uncovered even for a few minutes. Four posts on the ridgeline along with 3 posts sticking out from the snow-clad slopes below were alerted to the presence of infiltrators and ordered to tighten the net around them.

Havildar Dada and a few other soldiers were preparing to leave one of these posts, 'Sabu Post', 2000 metres below the ridgeline, to collect rations and stores from a logistics base down the mountain. That's how Army soldiers keep their forward posts stocked up on supplies. It would take the men 2 hours to reach the base on foot and another 2 to return to their post with provisions: food, fuel, ammunition, batteries and other rations.

Havildar Dada's patrol carefully negotiated snow-blown slopes as the higher reaches of the Shamshabari range had received more than 10 feet of snow over the previous few days, sending temperatures plunging to a bone-chilling −15 °C.

Just as the men were about to depart from Sabu Post at 0545 hours, in came the radio alert from Maj. Raj. The Major, a mechanized infantry officer on deputation with the 35 Rashtriya Rifles, had issued a simple order: Havildar Dada's unit, along with a handful of others below the ridgeline, had to immediately cordon off the area to prevent the terrorists from slipping away down the mountainside. No one knew how many terrorists there were—could have been 2, 4, 6 or even 8.

The numbers were of little consequence to Havildar Dada, at least at that juncture. Taking each of those terrorists down in the shortest possible time was the only thing that mattered to him as he met with 10 soldiers in the post for a quick operational briefing.

Dada was in charge of Sabu Post, a position he and his men had reoccupied barely a week before. Patrols are regularly sent trudging up to winter-vacated posts till the weather permits permanent positioning. Till that time, aerial reconnaissance is used as a stopgap measure. The desolation of Naugam was not entirely unfamiliar to Havildar Dada. The Shamshabari terrain resembled to a large extent some of the remote landscape of the state he had grown up in and the one he called home—Arunachal Pradesh. Havildar Dada had been sent up to the heights barely 3 weeks after he volunteered to serve in Kashmir with the Rashtriya Rifles battalion.

The night before, he had spoken to his wife, Chasen, on the phone. In that 5-minute conversation, Havildar Dada made a fleeting mention of the setting of his post and the challenging role he and his team were assigned. In another part of the country, holding the phone to her ear and closing her eyes, Chasen Lowang Dada tried to picture the view her husband had just described.

Eight hours later, Havildar Dada's universe would shrink to his area of responsibility, not uncommon on difficult operations in hostile conditions. Summoning all his experience as a soldier of 16 years, he swiftly ran his men through the possible routes the terrorists could take and ordered them to take positions at vantage points near the post. Before they spread out, Havildar Dada raised his voice in a familiar and commanding tone.

'*Woh hamaari post tak aa gaye hain. Bas, ab aur aage nahin. Yahin khatam kar denge* (They have come right up to our post. But we will not let them go any further. We will finish them off here).'

The soldiers who departed from Sabu Post with him that morning would have obeyed any order. Havildar Dada had proven his ability to motivate teams by example. There was nothing he demanded of them that he would not do himself, or that he did not do exceedingly well. He was an extremely skilled handler of weapons. When not on an operation, he would volunteer to instruct fellow soldiers on handling support weapons like rocket launchers, medium machine guns, light machine guns and multi-grenade launchers.

With Sabu Post some distance behind them, Havildar Dada and his men lay in wait. The terrorists were lying low near the ridgeline and the Army was yet to make contact with them.

From his mountainside position, the 32-year-old company commander radioed instructions to the posts above to open automatic fire with assault rifles and light machine guns in the direction where the terrorists were believed to be hiding. Bewildered by the sudden intensity of fire pouring around them, the terrorists made a panicked dash down the slope towards Sabu Post where Dada and his men had planned a grand reception for them.

Escaping the hail of ammunition from the ridgeline Army posts, the terrorists were now literally running for their lives, stumbling through snow down the mountainside, unsuspecting of what lay ahead. In the meantime, Maj. Raj had left the company operating base at Jatti, at a height of 12,500 feet, and was moving as fast as he could towards Sabu Post along with his squad to provide reinforcement and supervise the operation. It would take them at least 1 hour to arrive.

'I knew Dada was there and he was more than capable of dealing with any situation. He had the ability to work in teams and handle teams,' recalls Maj. Raj, who hails from Tamil Nadu's Dindigul district.

Havildar Dada, flat on his belly, half-hidden by a rock in the snow, lay very still until the terrorists had come very close. The terrorists had split themselves up into 2 groups to distract the soldiers and increase their odds of survival in the fight that was about to erupt. Two of them would soon be within range of Havildar Dada's AK-47. He didn't move a muscle. But something was amiss. The terrorists had stopped in their tracks. Havildar Dada and his men held their breaths.

Why had the terrorists stopped? The unthinkable, in what would have been a textbook ambush, had just happened. The terrorists appeared to have spotted the positions held by Havildar Dada's men. In bare seconds, it became clear that the terrorists were now readjusting themselves to draw first blood. Less than a second after that realization dawned on the men from Sabu Post, a burst of fire came clattering down the hillside from 2 terrorists towards Havildar Dada and his men, pinning them down in their positions.

What happened next is as difficult to explain for those who actually saw it, as it is for anyone else to imagine it. It would be the beginning of Havildar Dada's fearsome 'possession'. Watching his men duck for cover as the incessant rain of fire continued from the terrorists' rifles, Havildar Dada leapt from his position and charged up the slope towards the intruders with his fully loaded AK-47. Reaching them in a series of swift steps, the soldier then emptied his Kalashnikov magazine, 30 rounds, into both the men, cutting them down as they fired. It was a terrible risk, but one that Havildar Dada had clearly felt he had no choice in—it would have taken the terrorists only a few more moments for their bullets to find and shred Havildar Dada's men.

He swiftly thumped a fresh magazine into his AK-47 as he spotted 2 more terrorists sprinting in different directions to take cover behind giant boulders. They had seen what Havildar Dada had just done to 2 of their comrades, and had correctly concluded that they needed to handle this man with care.

By now, while his men remained in their cover positions, Havildar Dada had no cover at all. The soldiers behind him remember the silhouette that their leader cut against the snow. They could hear his breath as he finished loading his weapon and took the first tentative steps forward towards the 2 terrorists who had darted away. The men remember wondering if their leader would be doomed by his utter and total fearlessness that day.

Fear had given up on Hangpan as a teenager. In the early 1990s, he had dived into a fast-flowing river in Arunachal Pradesh's Tirap district to save a drowning classmate, earning the praise and admiration of an entire village. Later, shortly before he was recruited into the Army in 1997, young Hangpan risked his life to rescue passengers from a bus that had fallen into a gorge.

Havildar Dada's men knew they couldn't stop him from the course he had just chosen. But they tried.

'Dada sir, ruk jao, cover le lo (Dada sir, stop, take cover). Aage khatra hai (There's danger ahead),' one of the soldiers yelled through a cupped hand. It was futile. It became clear to his men that Havildar Dada had decided how he was going to finish this fight. And he didn't turn around once. By this time, his buddy, Lance Havildar Vareshang, had reached a spot 20 metres behind.

Havildar Dada, trudging through the snow, broke into a dash past rocks towards the third terrorist, with Vareshang trailing right behind. Vareshang would be 1 of the 2 men who saw what happened next.

The third terrorist sprang from behind the boulder, attempting to bring a marauding Havildar Dada down with a sudden burst of fire.

'It was a foolish miscalculation by the terrorist. And he would pay for it with his life,' recalls a soldier involved in the operation.

Havildar Dada twisted his body and flung himself sideways with lightning speed, incredibly managing to dodge a burst of bullets fired straight at him. The terrorist had taken cover behind the rock again, not realizing that those bullets had, far from deterring the advancing soldier, simply confirmed to him where the terrorist was hiding. The next few seconds would see Havildar Dada muster every bit of skill in hand-to-hand combat he had excelled in during his training, a skill he had demonstrated across deployments.

Creeping gently up and round the boulder, Havildar Dada lunged at the waiting terrorist, smashing him straight in the face with his rifle butt. He followed it with a series of swift, rapid blows. The bludgeoned terrorist dropped his weapon, struggling hard to wrench himself free from the Havildar's devastating grip. Havildar Dada made it quick. In a flash, the Havildar snapped the terrorist's neck, and let his body fall to the snow.

Havildar Dada had just dispatched 3 terrorists, all trained in military commando tactics and mountain warfare.

'His actions spoke volumes about his mental toughness, unflinching dedication and sense of purpose,' recalls Maj. Raj.

But Havildar Dada's fight was not over. There was still a fourth terrorist left, hiding a short distance up the hillside. As Havildar Dada began the final hunt, his AK-47 cocked and ready for a quick finish, a 7.62-mm round came whizzing down from behind a rock and ripped through the Havildar's neck. He had been sniped from close quarters as he stood exposed in his hunt. Havildar Dada collapsed in the snow, bleeding out. The sun was fully in the sky now, its brightness amplifying the blinding snow and crisp, clean air as Havildar Dada took his final breath.

The last terrorist had got him. It was 0945 hours—about the time when, in his village thousands of kilometres away, Roukhin and Senwang, Havildar Dada's 9-year-old daughter and 6-year-old son, would be getting ready for school. They knew that their father was on a mountain far away, and often wished they could visit him.

As Havildar Dada fell, a volley of shots spat out of his AK-47, the sound ringing through the deathly cold of the white landscape. Nobody knew it then, but Havildar Dada had managed to injure the last terrorist before he fell.

Havildar Dada's team and Maj. Raj's men, who had reached the spot, were in no position to retrieve the Havildar's body as the fourth terrorist was still alive, had a commanding view of the foreground and was possibly preparing to launch grenades. This would be a familiar situation for the Army. A beloved comrade had just been felled in combat, but the rush of sorrow and heartbreak had to be trampled, packed away, set aside at least for the duration of the operation. Fighting on with a broken heart is the most cruel, difficult thing a soldier ever has to do. Maj. Raj remembers feeling the hollow, dull grief of losing one of the most prized men in his charge.

But there was no time to rest. He quickly devised a plan to outflank the terrorist from the top before pinning him down and finishing him off. It would take the Major and his team 7 more hours to implement the plan. Finally, it was the Major himself who crept up on the last terrorist, and eliminated him with 3 quick bursts at 1645 hours.

The men could now finally reach Havildar Dada safely. It had started to get dark. The soldiers' faces were furrowed with tears when they saw him lying motionless in the snow, his hand still clutching the pistol grip of his Kalashnikov.

One of the soldiers knelt down to check for vital signs of life. He knew there would be none. But he had to. He pulled the magazine from Havildar Dada's rifle only to find that his team leader had fought till the proverbial last round. His lifeless hand, frozen on the trigger, had pumped out 30 bullets from the assault rifle. It was Havildar Dada's ghost that had injured the last terrorist, allowing the Major and his men to hunt him down later that day.

As they stood over his body, it was an overwhelming moment that the soldiers are unlikely to forget in their lifetime—one that is bound to find special mention in the Assam Regiment's historical records.

'Dada fought like a man possessed. And it is such men who win battles for an Army. He went after the terrorists *alone*. There are some things that cold logic can't explain,' recalls Col. Manish Agarwal, who commanded the 35 Rashtriya Rifles during the operation in Naugam. The mission was officially called off only the following day as the Army needed to comb the area and make sure it had taken down all the terrorists who had infiltrated. The numbers that hadn't mattered to Havildar Dada when he left Sabu Post mattered now.

The tale of Havildar Dada's final fight at the frigid heights of Kupwara would become a legend almost immediately. Hailed as a national hero, he was posthumously awarded India's highest peacetime military honour, the Ashok Chakra, by the President of India on 14 August 2016, for scripting an incredible tale of courage and sacrifice.

'I have come across lots of courageous soldiers during my 23-year military career. But when bullets are whistling over your head, even the toughest guy takes cover. Dada was something else. I feel honoured to have been his CO. The men cried, the officers wept when Dada died,' remembers Col. Agarwal.

Not a single soldier of Echo Company ate a morsel that night. And none slept. Havildar Dada's death had soured the taste of victory.

'We didn't celebrate the killing of 4 terrorists. We mourned the loss of a remarkably brave soldier, a brother. For 3 days, no one in the company spoke about the operation,' reminisces Maj. Raj, who was awarded a Sena Medal for gallantry.

Roukhin and Senwang miss the times they spent playing video games with their father when he visited them. The battalion misses Havildar Hangpan Dada also for the sermons he delivered as a pastor at the unit church, sermons that were attended by all the men.

Havildar Dada had visited home for the last time in early 2016. Chasen remembers he had bought her a new saree and a blazer, both items she considered expensive, but admired.

She remembers Hangpan telling her that the colour of the clothes—the blue saree and the white blazer—looked perfect on her.

'I think he liked the dress more than I did. Maybe it was his favourite dress on me,' smiles Chasen, 33, who never needed to ask Hangpan what he wanted to eat when he came home—it was Chasen's signature pork and *daal-bhaat*, each and every time.

Not surprisingly, she remembers every word of her last conversation with Hangpan, the night before the operation on 25 May. Nothing meant more to her than those rare calls from that mountain in north Kashmir.

'He asked me if the children and I had had dinner. He told me about his post. He said he would take us to Kolkata for Christmas. I wish we had spoken longer,' she says. '*Kahaani adhoori reh gayi* (The story was left unfinished).'

Havildar Dada and Chasen were married for 10 years. She remembers them as being friends as much as they were husband and wife. But he never spoke about the dangers soldiers faced in the field. Later she would learn that Havildar Dada always asked his brother, Laphang Dada, to offer prayers for the safety of his men.

Chasen also remembers reeling in horror when she learnt that Havildar Dada was an avid snake-keeper, a habit he had nurtured through tenures in the jungles of the North-east.

'Who keeps snakes as pets? I don't know of anyone! Had I known about it, I would have scolded him and asked him to mend his ways,' Chasen says.

More recognition would come for Havildar Hangpan Dada with the main office block at the Shillong-based Assam Regimental Centre being named after him. Chasen was invited to unveil the plaque in her husband's memory. The Arunachal Pradesh government decided to additionally immortalize the soldier by renaming the annual chief minister's football and volleyball championships after Havildar Dada. He is likely to also be commemorated closer to where he became a legend.

'There are plans to name an 8-km operational track from Jatti to Tiranga Post in the Shamshabari after Dada. We will never stop speaking about Dada's courage and how he fought that day. We will every day,' says Maj. Raj.

At the 2017 Republic Day parade on Delhi's Rajpath, Chasen received Havildar Dada's Ashok Chakra from President Pranab Mukherjee. Her head bowed in Namaste, she gracefully accepted the honour. Visibly steeling herself, it was one of Chasen's most difficult moments.

'Dil mein toh ro rahi thi, par sabko apne aansoo kaise dikhaati? Aakhir mai Dada ki wife hoon, aur Dada ki wajah se wahan hoon jahan hoon. (I was crying in my heart but how could I show my tears to everyone? After all, I am Dada's wife and I am where I am because of him),' she says.

She wore the blue saree and the white blazer that day.

*

5

'Two Bullets Can't Kill a Commando'

Captain Jaidev Dangi

Tral, Jammu and Kashmir
19 June 2014

The commando smiled.

It was Ritu.

Her name blinked on his battered iPhone screen.

The two had exchanged rings 4 months before, on Valentine's Day in 2014, as she had wanted. In 2 more months, he knew what would happen—he would be hoisted on to the shoulders of his comrades in his Para-SF unit and packed off to rural Haryana to be married.

Ritu sat in her hostel room in Rohtak, waiting for the commando to pick up the phone. She would not disconnect until the phone had rung through. If he didn't pick up, she knew he would call back.

From his team's operating base in the terrorist haven of Tral in south Kashmir, Capt. Jaidev Dangi, 25 years old, always called back.

He swiped to accept the call. Capt. Dangi and Ritu had not known each other before they had been brought together by their families earlier that year. It had been awkward in person, and since Capt. Dangi had no choice but to rush back to Kashmir after the engagement ceremony, they had only come to know each other over text messages and the daily phone call, which was sometimes as brief as a few seconds.

Capt. Dangi sounded relaxed that June evening, not something that Ritu got to hear often. She had found him calm enough not to ask about his work. Instead, she kept the conversation light, telling him what she had decided to wear for the wedding and the additions she was hoping to make to the guest list.

Capt. Dangi, who usually acknowledged every word she said with a sound, suddenly seemed distracted. Ritu found herself talking into a void. Something was amiss. The commando always gave her his full attention when he spoke to her. All too soon, he cut in abruptly.

'Ritu. Listen. I can't talk. I will speak to you later. Don't call back.'

He hung up. Ritu held the phone to her ear for a few seconds after it was disconnected. Was this what it was always going to be like? She waited for a few moments, staring out of her hostel window. Then she tapped out a text message and sent it: *Call me when you're free. Please take care.*

The silence was new, but Ritu had been faced with Capt. Dangi's hushed, distracted tone earlier. A month before, on 5 May, as the country braced itself for the big verdict of the national elections, her fiancé had fronted an operation that led to the killing of a Pakistani terrorist in a village near Tral. On that day too, he had cut their conversation short. Ritu knew that the man she was to marry was a soldier whose business was to hunt and kill terrorists. But that first time had made her blood run cold. It was something she would never quite get used to. Nor would Capt. Dangi's family.

The commando jogged to the operations briefing room that evening on 19 June 2014. Capt. Dangi and his team of Para-SF warriors had returned to their base earlier that day from a different operation. And just as Capt. Dangi had begun to tentatively unwind over the phone with Ritu, a fresh intelligence input had alerted the team to the presence of a terrorist inside a house in Buchoo village, less than 10 km away.

As the team spilt out of their field headquarters, the distinct scent of eucalyptus floated through the mild summer breeze. Young Kashmiri boys were enjoying a game of cricket at the playground a stone's throw away. The comforting sound of a ball against bats hewn from the willow trees abundant in the area resounded in the air.

The men did their final weapons check as an electric crackle passed through the team. Familiar to all soldiers, it is the frisson right before a hunt.

Capt. Dangi stroked his full beard. Normally clean-shaven and boyish, the lush growth on his face made the young commando look much older than his 25 years. Either way, it was crucial to his work. A beard helped conceal his identity as an outsider. There was now another incentive to keep it, though. Clean-shaven at his engagement, the commando had let his beard grow since, sending Ritu a stream of daily selfies that documented the steady shrouding of his sharp jawline with thick hair. Ritu reacted instantly. She forbade him to shave.

The hunt Capt. Dangi's team had set off on that June evening was a special one. The commandos had been waiting for that particular intelligence input for months. Their body language was brimming with quiet excitement.

The man whose whereabouts had been discovered was no ordinary terrorist carrying an AK-47 and a few magazines. He had been prowling south Kashmir for several years and was far more wily and dangerous than the men Capt. Dangi's team usually hunted down. It was Adil Ahmed Mir, a Hizbul Mujahideen area commander who the commandos had pursued unsuccessfully for several months. Adil had been mentor and trainer to Burhan Wani, then an upcoming social-media-savvy Hizbul commander, whose killing 2 years later in 2016 would plunge the Kashmir valley into a fresh cycle of bloodshed and turmoil.

Adil could not be allowed to slip through the net this time as he had several times in the past, to the intense frustration of the many teams that had been dispatched to capture him, dead or alive.

A Casspir mine-protected vehicle carrying the commandos rumbled off from Hardumir towards the location where Adil Mir was supposed to be hiding. The Hardumir company operating base offered a sweeping view of the area, including Buchoo village in the distance and the thick forest around it. Capt. Dangi quickly organized his thoughts. There was first a sense of disbelief that Mir had allowed himself to be spotted. Over 6 months, the terrorist had honed his skills of shaking his pursuers off into an art form. Capt. Dangi set everything else aside and focused his thoughts on the one thing he was certain of: Adil's killing or capture would deal a crushing blow to the Hizbul Mujahideen.

'We were 99 per cent sure it was Adil Mir and 100 per cent sure we would get him,' recalls Capt. Dangi. 'My team was looking forward to the action as this is what we were trained for. It was our chance to get the slippery fellow. You know, he never used a cell phone. He was so guarded.'

It was the height of summer and the sun would not disappear below the horizon for at least another 90 minutes, as the commandos sped towards the map location they'd been given.

As the Casspir rumbled swiftly past apple orchards, towering chinar trees and lush paddy fields, Capt. Dangi and his men conducted a quick tactical briefing. They made it to the spot in less than 10 minutes.

Capt. Dangi recalls the conversation he had with his men inside the vehicle.

'*Andhera hone se pehle operation khatam karna hai* (We have to finish the operation before darkness sets in).'

He did not need to tell them twice. The men knew that this was likely to be their one final chance to get Adil Ahmed Mir before he truly disappeared before the winter months.

It was 1700 hours when Capt. Dangi's 8-man squad reached Buchoo village with a team each of 3 Rashtriya Rifles and the J&K Police special operations group in tow. The intelligence input they had received was specific: it pointed to the presence of only 1 terrorist, Adil Ahmed Mir himself, in the village. This would soon prove to be a dangerous miscalculation.

Stepping quietly out of their Casspir vehicle in daylight, a short distance outside the village, the team was faced with a fresh quandary. The intelligence input had failed to factor in a crucial point. The man in whose home Adil was supposedly present owned 3 adjoining houses in the same compound. If Adil was actually there, he could be in any of the 3 houses.

In a matter of seconds, finding Adil Ahmed Mir had become thrice as difficult.

The men were rapidly assigned their roles before they headed towards the compound with the 3 houses. The Rashtriya Rifles soldiers would lay a cordon in front of the compound. Capt. Dangi and his commando squad would position themselves behind the wall at the rear end of the complex.

The young officer's instincts had told him that this was the escape route Adil was likeliest to take. Capt. Dangi and his 19-year-old buddy, Paratrooper Mukesh Kumar, took cover behind a eucalyptus tree with their Tavor TAR-21 assault rifles. They had chosen the spot for its unobstructed view of the compound's backyard. The 2 men did another weapons check as the sun sank a little lower on the horizon.

Almost 800 km away, Ritu was sitting in her hostel room, a silent prayer on her lips.

The eucalyptus scent wasn't just a familiar friend in the Kashmir valley. Capt. Dangi had grown up in Haryana's Madina village, where towering eucalyptuses lined the edges of fields owned by his father, who had died when Capt. Dangi was still in school.

For a boy who did not know much about the Army when he was a teenager, and who was coaxed by his physics teacher to sign up, Capt. Dangi's journey to Kashmir as a commando is an intriguing one. It was his instructor at the IMA, Maj. Kunal Rathi, who made him take the big step into the Paras. 'If you want to do what you are being trained for, then come to Special Forces,' the cadet had been told.

He had never regretted the decision.

Capt. Dangi and Mukesh strapped on their ballistic combat helmets. The remaining men had also formed pairs with their buddies and taken positions behind Capt. Dangi and Mukesh. This was to provide the lead pair cover, while sealing off alternative routes that could allow the terrorist to reach the stream and swim to the forest beyond. If he reached the forest, this mission was as good as dead.

The police team cautiously approached the entrance to the main house, ready to open fire if required. Seconds after

Major Mike Tango

Maj. Tango receiving his Kirti Chakra from then President
Pranab Mukherjee in March 2017.

Note: Some photographs have been intentionally blurred to protect the identity of officers operating
in hostile conditions.

Lieutenant Colonel Oscar Delta

Lt. Col. Delta receiving his Kirti Chakra from then President Pranab Mukherjee in August 2015.

Note: Some photographs have been intentionally blurred to protect the identity of officers operating in hostile conditions.

Lance Naik Mohan Nath Goswami

L. Nk Goswami and his wife, Bhawna, as newly-weds in Agra.

In full combat gear, L. Nk Goswami with his trusty TAR-21 assault rifle.

Havildar Hangpan Dada

Havildar Dada with his wife, Chasen Lowang.

Then Army Chief General Dalbir Singh meeting Havildar Dada's wife and children in Shillong in November 2016.

Chasen Lowang receiving her husband's posthumous Ashok Chakra from then President Pranab Mukherjee in January 2017.

Captain Jaidev Dangi receiving his
Kirti Chakra from then
President Pranab Mukherjee
in January 2015.

Capt. Dangi in combat fatigues
close to his unit base.

Capt. Dangi with his wife, Ritu, and their daughter, Inayat.

Colonel Santosh Yashwant Mahadik near his base in Jammu and Kashmir.

Service photo of Col. Mahadik after
he picked up the rank of Colonel.

Major Mukund Varadarajan in J&K.

Maj. Varadarajan with his wife, Indhu Rebecca,
and their daughter, Arshea.

Lance Naik Hanamanthappa
Koppad

L. Nk Koppad's wife,
Mahadevi, and daughter, Netra,
in New Delhi on Army Day,
January 2017.

One of the first photographs of the rescue teams at Sonam Post.

Prime Minister Narendra Modi at L. Nk Koppad's side, being briefed by then Army Chief Gen. Dalbir Singh on the soldier's condition.

L. Nk Koppad's daughter, Netra, visiting her father's samadhi in their village.

Photo courtesy: Anantha Krishnan M.

Lieutenant Commander Niteen Anandrao Yadav being decorated with a Shaurya Chakra in 2010 by then President Pratibha Patil.

Lt. Cdr Yadav and other members of INAS 315 'Winged Stallions' with an Il-38SD aircraft.

Lt. Cdr Yadav in the cockpit of an Il-38SD.

Captain Varun Singh at Wular
Lake, J&K, December 1999.

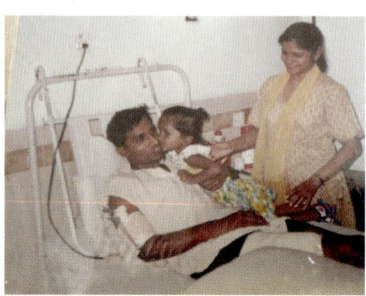

Capt. Singh meets his wife, Reena,
and daughter, Shivani, for the first
time since the encounter.

The commando with his daughter, Shivani.

Capt. Singh and his wife, Reena, in July 2016 when he took charge as
Commanding Officer of the INS Karna MARCOS base.

Commander Milind Mohan Mokashi

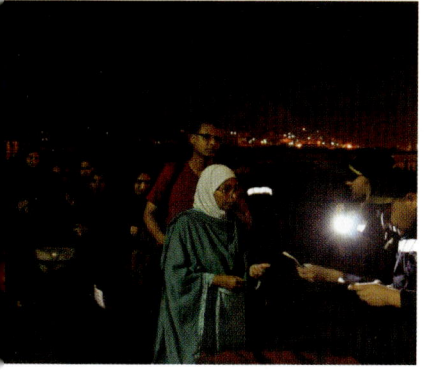

Evacuation operations at Yemen by Cdr Mokashi's ship *INS Sumitra.*

Squadron Leader Rijul Sharma with a
MiG-29 at Jamnagar, Gujarat.

Sqn Ldr Sharma with his wife, Deepika,
in a MiG-29 hangar at Jamnagar, Gujarat.

Squadron Leader Vikas Puri with an Mi-17 helicopter.

Sqn Ldr Puri and his helicopter crew.

Sqn Ldr Puri with his wife, Reshma, and daughter, Inayat.

Wing Commander Gaurav Bikram Singh Chauhan with a Su-30MKI at a forward IAF base.

Wing Cdr Chauhan with his wife, Avantika. His Su-30MKI is in the process of being armed with missiles in this picture.

they knocked on the door, Capt. Dangi spotted a well-built, bearded man in a pheran make a wild dash out the back door.

Not only was this a suspicious move, it was plainly hostile and confirmed the intelligence input. Yet, standard protocol had to be followed. The commandos had to be absolutely sure that the man who had rushed out of the house was not a civilian. In an icy calm voice trained not to alarm, Capt. Dangi called out to the man, asking him to reveal his identity, remove his pheran and drop his weapon to the ground if he was carrying one.

'You can surrender if you want to. There's still time,' Capt. Dangi warned.

This routine drill during the conduct of counter-terrorism operations exposes soldiers to enormous additional risks. But nothing is more important to the Army than eliminating the possibility of collateral damage. Never mind if it increases the chance of its own casualties, as it very often does.

The man did not respond to Capt. Dangi's call.

Instead, he jumped across the compound's back wall with the support of his left hand while the other grasped a now visible AK-47 assault rifle. As he landed on the ground 15 metres away from Capt. Dangi and Mukesh, the terrorist opened fire.

Dodging that first hail of bullets, the commando leaned towards Mukesh, '*Iska khel khatam* (His game is over).' The 2 men exchanged quick nods.

It is near impossible to describe the trust buddy pairs place in each other during operations. Placing their lives in each other's hands forms the basis of the relationship. When a firefight breaks out, buddy pairs are not just working towards eliminating an adversary; they also draw strength from and protect each other. It is a force multiplier system that creates the most basic human linkage at a tactical, instinctive and emotional level.

Right in front of them, the terrorist fired for 8 more seconds until he drained his first AK-47 magazine. Before he could reload his rifle, Capt. Dangi and Mukesh began their counter-fire, sending 6 single shots each from their TAR-21s whizzing through the air and straight into the terrorist's body, shredding him where he stood. The precision shots ensured swift death.

If the intelligence input was accurate, the mission had just been successfully completed—it lay in a bloody heap in front of the 2 lead commandos. And yet, somewhere in Capt. Dangi's reptilian brain, he knew it couldn't have been *this* easy. The men would know only moments later how the intelligence input had really only scratched the surface.

'We relaxed a bit for a few seconds thinking we had got our guy. But hell, we were wrong. Everything had only just begun,' remembers Capt. Dangi.

Less than a minute after the first man was shot dead, 2 more terrorists sprang from the house they were hiding in and immediately opened fire at Capt. Dangi's position. Firing their weapons on full automatic, the terrorists took the warriors by surprise. Not only had the intelligence input specified the presence of a single terrorist, the 2 men who had just emerged were not making any attempt to escape like the first one— they were in all-out attack mode.

It quickly dawned on the squad that the terrorists seemed to be following a well-thought-out plan. The first one stormed out of the house and made contact with the commandos. The ensuing firefight revealed the position Capt. Dangi and

Mukesh were holding. The next 2 terrorists emerged by surprise to finish the commandos off.

At this point, Capt. Dangi and his squad did not know that 1 of the 2 men firing at them was Adil Ahmed Mir.

The man they had just shot dead was only a foot soldier—either Abdul Ahad Shah or Tariq Ahmad Parray of the Hizbul Mujahideen. But the 2 men now firing aggressively at them were doing so with a worrying level of skill.

'The two were firing very accurately. They were extremely well-trained. The sheer intensity of the fire forced us to take cover with our heads down. We had to do something quickly,' recalls Capt. Dangi. He remembers feeling a spasm of anger at the abrupt turn of events that had put the commandos on the defensive. He felt a familiar dryness of the mouth as he wondered if the 2 terrorists would use their hail of fire from just 20 metres away to make a getaway. What he had not accounted for was actually taking a bullet while this happened.

The 2 terrorists had realized that they would need to get rid of Capt. Dangi and Mukesh, the lead pair, if they were to disappear into the tall grass and crawl down to the stream without being pursued. They were surrounded and that was the only exit route. By now, they were desperate to break out of the cordon laid by the commandos to prevent their escape.

The trunk of the tree behind which Capt. Dangi and Mukesh had taken cover was not wide enough to shield them both. Sensing an opportunity to pin the commandos down, if not hit them directly, the terrorists began moving towards them, their rifles blazing non-stop.

Capt. Dangi felt a sudden stab of pain in his left thigh as a Kalashnikov bullet ripped through it. Immediately after, a second bullet pierced his abdomen on the side. Under the hail of ammunition, Capt. Dangi inspected his injuries for a moment, but quickly looked to his buddy. To his horror, Mukesh had sustained nearly identical injuries to the right side of his body.

'It was only when I saw blood gushing from my wounds that I realized I had been shot. I was more worried about Mukesh as his injuries seemed to be worse. The tree trunk had covered our vital organs but some parts of our bodies were exposed,' recalls Capt. Dangi.

He quickly checked Mukesh's helmet. Fortunately, it was intact. No headshots.

'*Kuch nahi hua hai. Thodi si lagi hai* (Nothing has happened. It's a minor wound) You are alive. Stay that way,' he told Mukesh.

It had been barely a year since Mukesh had enlisted in the Army, and Capt. Dangi knew his first gunshot wounds would shake him up more than a little. His battle fatigues soaked in blood, the pallor on Mukesh's face showed that he thought the end was near. Over the deafening crackle of fire drawing towards them, Capt. Dangi whispered to his buddy. Pointing to his own wounds, Capt. Dangi told him 2 bullets were not enough to kill a commando. Mukesh smiled weakly, with a thumbs-up gesture. He was losing blood rapidly.

Capt. Dangi quickly dragged Mukesh to a position behind the tree that made him less vulnerable to the incoming fire. He then signalled to one of the other soldiers holding ground behind them to crawl to their position to watch over the injured Mukesh.

As the commandos provided him covering fire, Naib Subedar Tribhuwan Singh crawled on his hands and knees to reach the wounded man. Tribhuwan knew what had to be done. Reaching Mukesh, he quickly put pressure on the man's wounds with both hands to prevent further blood loss, telling him that the operation would be over soon and help was on its way. Mukesh's face was deathly pale by now, his breathing more rapid.

Blood oozing from his own gunshot wounds, Capt. Dangi stuck his TAR-21 out from behind the tree and opened fire at the advancing terrorists. In the tense crossfire, the officer's bullets hit one of the terrorists, sending him crashing to the ground.

It was Adil Ahmed Mir. But he wasn't dead yet.

Just as Capt. Dangi was about to open a final burst at the fallen terrorist to finish him off, a bullet came whizzing through the air and hit his assault rifle, jamming it and rendering it useless.

A more devastating situation could not have transpired. Capt. Dangi was now holding nothing but a piece of metal composite in his hands in the middle of a firefight at 10 metres. And he didn't have a moment to lose to think of alternatives.

He swiftly bent over to pick up Mukesh's weapon. Just then, the last standing terrorist decided to make a break for it, dashing full speed towards the tall grass about 30 metres away that led down to the stream behind Buchoo village. As he reached the grass, he dropped to his hands and knees and began crawling through it, trying to disappear in the undergrowth. Capt. Dangi checked the magazine of his new weapon only to find that Mukesh had emptied it in the firefight. Quickly slamming a new magazine in, Capt. Dangi stepped out of his position to finish off Adil Ahmed Mir, who he had hit a moment ago.

Mir, lying a few metres away, had enough strength to swing his weapon forward and open a fresh burst of fire directly at the now fully exposed Capt. Dangi. A bullet grazed Capt. Dangi's cheek. An inch to the right, and it would have been a direct headshot that would have instantly killed the young officer. Thankfully the bullet had only opened Capt. Dangi's skin, a mere trifle compared to the first 2 gunshot wounds.

That bullet was the last of the dying terrorist's ammunition. Capt. Dangi stepped up quickly, pumping 10 rounds into him. Standing over him to make sure it was over, Capt. Dangi bent closer to take a look at the terrorist's face.

'Before I took the headshot, I recognized him. It was Adil.'

Ten metres behind, Tribhuwan was still tending to the injured Mukesh and was unable to engage the third terrorist who had fled towards the stream. Capt. Dangi, bleeding heavily and losing strength, refused to be shifted out of the encounter site. Adil was dead, but this operation would be incomplete if the third man escaped.

'I didn't mind bleeding a little more. I knew I could deal with it,' Capt. Dangi recalls. 'If this guy made it to the stream, there was no way we were going to get him, at least not that day.'

Capt. Dangi quietly moved in the direction where the terrorist had disappeared into the undergrowth. The large rucksack on his back gave the man's location away.

'I could see something moving. His rucksack blew his cover. It was sticking out as he crawled on his belly towards the stream.'

His limbs stiffening from the pain and blood loss, Capt. Dangi stepped on the grass, his weapon ready. But this time there would be no firefight as the young officer crept up on the terrorist and killed him in a quick burst of close-range fire. It was time for Capt. Dangi to take another headshot.

'The third man didn't put up a fight. His commander was dead and all he wanted to do was escape. He had lost the will. It made my job easier,' says Capt. Dangi.

The officer did not stop to savour his victory. There were other priorities for now—Mukesh. Blood dripping to the ground, Capt. Dangi stumbled back to check on his buddy. Mukesh was still conscious.

'*Kaha tha na, do goliyan commando ko maarne ke liye kaafi nahi* (Told you, 2 bullets aren't enough to kill a commando),' Capt. Dangi grinned.

As the sun began its final descent, the two men smiled. From start to finish, the operation had lasted less than 20 minutes. From the time the first terrorist opened fire, till the time Capt. Dangi fired that final headshot, it had been just 8 minutes. The killing of Adil Ahmed Mir that evening in Buchoo village would be the definition of a lightning-quick operation. The 5 gunshot wounds suffered by 2 commandos were acceptable damage—every operation begins with the recognition that injuries are highly likely, if not fatal.

The Casspir vehicle carried the team back to its operating base in Hardumir where an Army helicopter was waiting to fly Capt. Dangi and Mukesh to the 92 Base Hospital in Srinagar.

In Rohtak, Ritu was close to panicking. Every passing minute had seemed like an eternity to her as she fought to keep away frightening thoughts. Like a good professional, Capt. Dangi had shared broad aspects of his work with his family, but never the specific details—as much for their own safety as his.

What Ritu did know was that her fiancé operated in south Kashmir. Unable to bear the worry any longer, she contacted a paramilitary officer also deployed in Tral. The officer was a friend of her cousin's. It was through him that Ritu learnt of Capt. Dangi's mission to hunt a deadly terrorist commander. That piece of information was enough for Ritu to throw Capt. Dangi's promise right out of the window—that he would call her back and not to worry. She began dialling his number repeatedly, but there was no response. The young commando's phone happened to be with his team commander, a Major, who finally picked it up at 2100 hours.

Ritu felt her skin crawl. In a calm tone, the Major informed her that Capt. Dangi had suffered minor injuries in the operation and had been admitted to hospital. Ritu phoned nobody else that night, neither her family nor Capt. Dangi's mother. And she didn't sleep. Through the night, she tried to contact Capt. Dangi, hoping his phone had somehow found its way back to him.

But it wasn't until the next morning that she finally heard his voice.

Capt. Dangi's CO and team commander arrived at the hospital to see him early on 20 June. The Major dialled Ritu's number and handed the phone to Capt. Dangi.

'I was too shy to speak to my fiancée in the presence of my CO and the team commander. I just told her I would speak to her after 2 days. She was silent. She wanted to talk. But she understood the situation I was in,' he says.

Their wedding had to be postponed from August to December as Capt. Dangi's injuries needed time to heal. Both commandos were discharged from the base hospital after 3 weeks.

Six months later, Capt. Jaidev Dangi was awarded India's second highest peacetime gallantry award, the Kirti Chakra, on 25 January 2015 for showing 'dauntless courage and extraordinary valour under heavy fire from close quarters in face of certain death.' The award citation would note that 'despite his near fatal injuries, the officer refused to be evacuated till the termination of operations'.

He would receive the award from the President of India at the Rashtrapati Bhawan on 21 March 2015. With him was Ritu, his bride of 3 months, his mother and his best friend, Capt. Pradeep Balhara of a sister SF unit.

Capt. Dangi's example has inspired many in and around Haryana's Madina village to join the military.

'Jaidev was the first boy from our school to join the NDA,' says Ravinder Dangi, director at the Ramakrishna Paramhansa Senior Secondary School where Capt. Dangi had studied. 'But now a steady stream of students joins the academy every year following in his footsteps. He will remain an inspiration for current and future generations of students.'

School friends talk fondly of Jaidev, who still meets them on the rare occasion that he has time off from his duties. They remember how he once limped to school with his foot in a plaster cast as he did not want to miss classes.

'I told him to take leave and rest. I loved his answer. He said he would rather sit in class than sit at home,' the school director says.

After the Buchoo operation, Capt. Dangi entered a Para refresher course that commandos have to undergo every year to hone their skills. It was during the toil of this course that Capt. Dangi would receive news about the gallantry decoration the recently elected Narendra Modi government had decided to pin on him.

'No words can explain the joy and pride I felt when I was told I had been awarded the Kirti Chakra. I had never imagined in my dreams that I would come so far in life,' says Capt. Dangi.

The operation that killed Adil Ahmed Mir would be the first major SF win in Kashmir under the new Modi government, barely a month old at the time. It would become a touchstone referred to by the country's security top brass as India's Parachute Regiment units prepared for devastatingly more ambitious operations in the months that followed.

SF officers rarely have time for leisure or other interests. In Capt. Dangi's case, the binding nature of his work allows him hobbies not far removed from his professional duties. In the little spare time that Capt. Dangi gets, he pursues interests deeply related to the profession he has chosen for himself. He likes to fire different types of weapons, research tactics employed by global special operations units and read about military leaders. Mongol leader Genghis Khan, who rose from humble beginnings to carve out the largest empire in history, tops his list of favourites.

Capt. Dangi's beard was shaved off at the hospital to treat the wound on his cheek. As they cleaned him up, he remembers the mixed feelings he had at the time.

'Ritu really loved that beard. I would have to grow it back really fast,' he remembers.

Following the operation in south Kashmir, Capt. Dangi was posted as an instructor at the NDA outside Pune, where cadets, like he once was, prepare for a military career. At the time this book was written, the young officer was waiting to complete the posting and return to the Kashmir valley to do what he likes the most.

Hunt.

*

6

'Just Tell Me, Will He Live or Die?'

Colonel Santosh Yashwant Mahadik

Subaya, Jammu and Kashmir
16 November 2015

'*Papa jaldi aao. Hum wait karenge.* (Papa, return soon. We will be waiting for you.)'

Kartikee Mahadik, 11 years old, threw her arms around her father's neck. It was near midnight that night in May 2015. Kartikee's little brother, Swaraj, waved from behind her. Their father was headed, with a group of his men, to Muhri, not far away, where a group of terrorists who had infiltrated across the LoC had been spotted.

Kartikee was old enough to wonder why her father, the CO of the 41 Rashtriya Rifles, needed to personally lead every operation into the beautiful, unforgiving forests of north Kashmir that surrounded his field headquarters. She was old enough to know that battalion bosses were not required to physically front every mission, but play the nerve centre with command and control.

Yet, she never once wondered.

Which is why, 8 months later, when Col. Santosh Yashwant Mahadik phoned his wife, Swati, then staying 300 km away at the Army's Northern Command Headquarters in Udhampur, she didn't blink.

It was freezing that night of 16 November 2015. Col. Mahadik and a team of his men had dashed by road from

Kupwara to a thick sector of forests in the frontier hamlet of Subaya. It was the red alert the Colonel had been waiting for.

Over the first half of November, along with the first snow, the trail of a small group of terrorists wearing white snow jackets and bearing suspiciously heavy backpacks had gone cold. They had been spotted by civilian porters on 3 November in Trumnar, a village within the 41 Rashtriya Rifles' area of responsibility. Rushing to their destination at Kamkari, the porters had immediately notified officers at the 57 Rashtriya Rifles unit headquartered there about the suspicious men in snow gear. The men they saw were definitely not locals, the porters insisted, before they went on their way.

An officer at the 57 Rashtriya Rifles immediately relayed the tip-off to Col. Mahadik's unit, advising him to make preparations to intercept the suspicious group of men as they came down the Shamshabari mountains. The CO of the 41 Rashtriya Rifles immediately deployed teams to place ambushes along points near the range. For 3 days, the men searched, patrolled and waited. A fresh batch of snow, several feet deep, fell on 7 November—4 days after the hunt had begun.

The soldiers were then redeployed on surveillance missions in the area. Two days later, on the afternoon of 9 November, fresh intelligence reached Col. Mahadik's men that a body had been recovered from Trumnar, the village where the suspects had first been spotted. As always, the Colonel rushed to the site himself.

The body had clearly been dumped there. The dead man's weapon, an AK-47, lay by his side on the banks of a thin stream. A closer inspection revealed that the man's legs had rotted from gangrene. Speaking to his contacts among Trumnar's locals, Col. Mahadik made a quick deduction. The dead man was likely from the same group that had been spotted by the porters 3 days earlier. Snowfall and frostbite had clearly got to them. The 4 suspects had probably passed through the village, and this 1 man had been left behind because he could walk no further. He had then probably died in the care of someone ordered to look after him. Locals, fearing they would be held responsible for harbouring a terrorist, then probably dumped the body outside the settlement in a nala so sniffer dogs would not be able to detect it.

The sighting of the corpse was confirmation that at least 3 others in the group had survived and were moving towards the Kashmir valley. On 10 November, Col. Mahadik launched operations at 2 sites—Kupwara's Manigah and Baramulla's Jugtial. Meticulous patrolling through snowbound forests and ridgelines revealed nothing. The hunt continued for 48 hours.

At 0800 hours on 13 November, Col. Mahadik received word that one of the suspects had descended from a jungle and entered a hamlet near Trumnar to ask for food. Teams were rushed to the spot, but they were too late. The suspect had collected food and was already headed back up towards the jungles. Locals showed the Army team the direction in which he had gone.

Following the trail up the hill, soldiers noticed someone attempting to conceal himself with a shawl. It was a hill track frequented by people from the village. When challenged, the man in the shawl turned around, revealing an assault rifle and opened a burst of fire. The spray of bullets hit 2 jawans, injuring them. As the Army team took cover, the weather and visibility worsened abruptly, bringing fresh rain and snow, allowing the man to escape.

Over the next 2 days, the team searched grimly for the man who had got away. By 15 November, the team was exhausted after an uninterrupted phase of combat alert, and their morale had dipped considerably. They had effectively been on the hunt for 7 days and achieved nothing. Troubled but determined, the men were asked to return to their headquarters.

As they moved back from the location, another group of soldiers took their place to continue the hunt, this time in the adjoining hamlet of Subaya. The team, from the 160 Territorial Army battalion, began scouring forests near the village. The 160 TA, comprising infantry soldiers, looked after intelligence operations and was co-located with Col. Mahadik's 41 Rashtriya Rifles. The unit had been deployed just to make sure nothing was amiss. But as they entered the area on the night of 16 November, they walked right into a firetrap.

A soldier in the party, Rifleman Mohammad, received a gunshot straight in the leg, and the team was now pinned down by sustained fire from at least 2 quarters. The sun had set, and the clatter of Kalashnikov fire echoed through the forests that night.

Back in Kupwara, Col. Mahadik was getting ready to meet local officials as part of Operation Sadbhavna, the Army's ambitious mission to build bridges and, as the government officially states, 'win the hearts and minds' of the local population. At 2000 hours, he received word about the 'contact' that had been made in the general area of Subaya village.

Col. Mahadik summoned his men, which included a few J&K Police special operations group jawans, climbed into his vehicle and sped towards Subaya. He and his teams reached the area and formed a wide cordon to hem the suspects in. Col. Mahadik climbed to a vantage point and established a quick surveillance post that often proves crucial in such operations. He needed to ensure that the entire area could be seen. His weapon ready and the cordon in force, they carefully extracted injured Rifleman Mohammad from the forest and transported him to safety. Up in his perch, the Colonel spent an uncomfortable, cold night, his gaze scanning the swathe of dark, nebulous forest before him.

Early the next morning, on 17 November, Army reinforcements had been called in to join the hunt. Now an infantryman, Col. Mahadik had a special strength urging him on—he was a commando himself, originally from an elite Para-SF unit. In 2003, he had earned a Sena Medal for bravery in action against the United Liberation Front of Assam terrorists in the North-east. Twelve years later, he had volunteered to lead a counter-insurgency unit in the restive Kashmir valley. He knew what the Paras brought to any fight. They were most welcome.

As the search continued with renewed aggression, Col. Mahadik received a call on his cell phone. It was from a local source, one of many he had cultivated as friends during his time in the Valley. The voice at the other end of the line kept it brief: 1 suspect had come down a hill in the Kashmiri Manigah area nearby and requested food from villagers there. The Colonel had no reason to doubt the information he had just been supplied. And once again, with his men deployed in the active operation, he decided to act on the information himself.

He gathered half of his Quick Reaction Team (QRT) and the J&K Police jawans who had accompanied him, jumped into 2 squad vehicles and made straight for Manigah. When they arrived, it became clear that the suspect had headed back up the mountainside and into the jungles. Col. Mahadik reconnected with his source from the ground. There was more information. The suspect likely had at least 2 more men for company up in the jungles. The officer smelt the end of an operation that had stretched for a full 2 weeks now.

The 14 men were quickly split into 2 teams of 7 and sent in 2 different directions to begin their search. The terrain was familiar and taxing. A nala cascaded down the mountainside, with ridgelines ascending on both sides. An hour-long search revealed nothing. The teams patrolled along the nala on their way back to the village. Just outside the village, they stopped for a quick break to plan their next move. It was there that one of them spotted it.

A bottle of mineral water, its top half cut off to make a tumbler, sat precariously on a rock. Next to it was a battered cooking utensil with some freshly cooked food—rice and meat. Commanded by Col. Mahadik, the men immediately took cover positions. Two things were immediately and disturbingly clear. The suspects were definitely nearby. That was the good news. The bad: they had likely abandoned their precious food because they had spotted the Army team, and were therefore almost certainly in a position of advantage.

'It was an extremely tense situation. We had no time to readjust. We needed to quickly make our next move, or we stood the risk of being ambushed and massacred,' remembers a jawan who was in the team that morning.

Col. Mahadik wasted no time. He quickly sent one half of the team up the ridgeline to gain a vantage position, crucial to a situation where they didn't know where a burst of fire would come from. Col. Mahadik and his team of 7 moved laterally across the ridgeline. As they did so, one of the police jawans noticed a black pheran behind a tall bush right ahead. He held the Colonel back, pointing straight ahead and alerting him to 2 men hiding behind the foliage. The men were ordered to cock their weapons and pull the chain levers on their rifles to 'rapid' mode, which allowed for a burst of fire. Single shots weren't going to be of use in this fight.

As the team inched forward, one of the terrorists abruptly changed position. The other remained where he was. Stopping in his tracks, and motioning to the other men to stand back, Col. Mahadik raised his weapon and fired a few rounds directly at the bush. Four of his men were in cover fire positions, while 3 stood with him as he fired. But it was the other terrorist, who had shifted position moments earlier, who fired back. And this was from a position the Colonel and the 3 men next to him didn't have their eyes on. A hail of rifle rounds tore through him as he fell to the ground.

Nobody could see where the firing had come from. The 3 men with Mahadik were in the open and totally vulnerable. They dropped on to their bellies, waiting for a certain follow-up volley of bullets. Their CO knew that if he didn't act fast, his men would be butchered in the open. Despite his gunshot wounds, Col. Mahadik crawled forward towards the ridgeline.

Word about the CO having been shot hadn't yet reached the other teams because Col. Mahadik was the one with the radio set. Teams attempting to make contact with him thought he was silent so as not to alert the terrorists. With blood gushing from his many wounds, Col. Mahadik heaved himself across the ridgeline to the other side. Hauling himself up over a rock, he opened fire on the positions where he assumed the terrorists were hiding. The bullets he fired didn't find the terrorists, but effectively pinned them to their positions.

By firing continuously while still exposed and wounded, he basically ensured that the terrorists did not fire at the men behind him. He was running out of ammunition and began pacing his fire to keep the terrorists on the defensive for as long as possible so that his men could get to safety.

By this time, other Army teams had reached the location, but there was nobody to brief them on the situation because the CO was down. Desperate to make contact with the team up on the ridgeline, a Havildar in charge of the QRT team called one of the Police special operations group jawans on his mobile phone.

'Saab ko goli lagi hai (Sir has been shot),' he reported back. It was the first information from the fight.

The Havildar rushed to the ridgeline where Col. Mahadik lay. As the other men provided covering fire, he picked up the CO and carried him down the hill to a location 1.5 km from the roadhead. Evacuation by helicopter wasn't possible because the Manigah area was located in a narrow valley. Pale from blood loss, his body already turning cold, Col. Mahadik was driven to the nearest site of hope—the 168 Military Hospital in Drugmulla, just off the Sopore–Kupwara highway. Doctors at the hospital quickly declared that they couldn't revive the Colonel and suggested he be flown immediately to the Army's Base Hospital in Srinagar 85 km away.

'While Col. Mahadik was being airlifted to Srinagar, I received a call from his wife. She had already heard,' remembers Maj. Pravin, then Adjutant at 41 Rashtriya Rifles headquarters. 'Her question still haunts me. All she asked was 'Zinda rahenge ya nahi rahenge? Bas itna bata do (Will he live or not? Just tell me that).'

Maj. Pravin did not know what to say. He knew that she knew the truth. But he still hoped that the doctors at 92 Base Hospital, often magicians in their abilities, could bring Col. Mahadik back.

'She called again a short while later. This time, she asked me a question that haunts me even more,' remembers Maj. Pravin. 'She asked how many rounds had hit her husband. I mustered my strength to inform her that he had taken 7 bullets and that he was unlikely to survive. She said nothing further and hung up the phone.'

Col. Mahadik was declared brought dead at the 92 Base Hospital in Srinagar.

At the time this was written, Col. Mahadik's wife was at the Officers Training Academy in Chennai, gearing up to become an Army officer. Less than a year after her husband's passing, she decided she wanted to wear the olive-green uniform. It was a difficult decision that forced her to send her 2 young children away to boarding schools in Maharashtra and Uttarakhand. Like they never did when their father picked up his weapon and ventured out at night, Kartikee and Swaraj never once wondered why their mother, well past the age to be a cadet, has decided to be an Army officer.

'We didn't think this was the right way forward. But she is an incredibly courageous lady. She wishes to take her husband's unfinished work forward,' says Maj. Pravin, who continues to be deployed in the Kashmir valley.

The Colonel's unfinished work has nothing to do with the terrorists who managed to escape that day. They were perhaps only incidental to a larger mission he had assumed for himself in a beautiful and dangerous land. His business was war, but accounts of what Col. Mahadik was engaged with in Kupwara suggest he was looking to sow every last bit of goodness he possibly could in the terrain and the people around him.

From sessions on inspirational leadership for children in Kupwara to yoga camps and lessons in adventure tourism, Col. Mahadik took the task of winning the hearts and minds of people as seriously as he took his fighting. Described by his men and peers as a visionary, the Army has spent the months since his demise studying the work he did in Kupwara.

Warned by local leaders that yoga would not be accepted by the predominantly Muslim population, in 2014, the Colonel decided to send a group of 15 citizens to Pune to attend a camp conducted by the Siddha Samadhi Yoga programme. The group returned with a request that the organization set up a special camp in Kupwara. A group from Pune arrived shortly thereafter to conduct special sessions for children and local traders in Kupwara, with a promise to institutionalize yoga in the town.

'He had an outstanding rapport with Kupwara's citizens. He met locals very often and sometimes kept an open house. He did for Kupwara tourism what local government officials haven't even tried to,' says an officer who was deployed with him at the 41 Rashtriya Rifles.

Maj. Pravin concurs. 'Col. Mahadik never really believed that the military was a lasting solution in this area, and that we are only here temporarily—we must leave this beautiful land even better than we found it,' he says.

During his time in Kupwara, there was a discernible change in the public perception of the Army, say men who served under the Colonel.

'Separatists from the Hurriyat would often attack Col. Mahadik through the local media or through statements issued

in public,' recalls a jawan of the 41 Rashtriya Rifles. 'They
didn't like how he was reaching out to the locals and having
an impact on their lives. They warned him to simply do his
work and get out. All that never bothered him.'

Col. Mahadik's leadership by example would become
legend well beyond his unit, and has already become a
touchstone of what inspiring COs do.

'Even as a CO, he would go and sit in ambushes for 2–3
days, which is unusual for someone of his seniority and rank.
He enjoyed being in the field with his men,' says Maj. Pravin.
'He always said that every officer is a soldier, and every soldier
is a leader. He truly believed that. He never differentiated
between the two.'

Soldiers in his unit remember the unusual level of personal
interest the Colonel took in the well-being of men under his
charge. They speak of a particular jawan from Col. Mahadik's
state, Maharashtra, who was distressed following bitter marital
discord. As divorce proceedings began, Col. Mahadik invited
the jawan's wife and children to Kupwara, acquiring special
permission from the Army so they could stay on-site for a
month. Over that month, Col. Mahadik counselled the
couple. The two decided not to separate and are still together.
A few weeks later, the domestic problems of another jawan
were similarly resolved.

'It is very rare for officers to get so deeply involved in the
personal lives of their soldiers. It earns them loyalty of a kind
that cannot be put into words,' Maj. Pravin says.

Five days after Col. Mahadik was lost, 1 of the terrorists
was killed in an encounter with soldiers from the 160 TA at
Haji Nakah. What became of the remaining 2 terrorists is as
yet unknown.

An image remains imprinted in the mind of an Army officer who accompanied the family as Col. Mahadik's remains were transported by air from Udhampur to Pune that November.

'Swaraj sat on the coffin, playing. He was too young to know what had happened,' the officer remembers. 'He was oblivious. I just sat and watched him play.'

*

7

'I Got Hit. I Can't Believe It'

Major Mukund Varadarajan

Yachu Guchan, Jammu and Kashmir
June 2013

Single 7.62-mm shots rang out through the air. The Jaish-e-Mohammed terrorist commander was cornered. Two terrifying odds loomed before him as he crouched with his AK-47. One, he had a single ammunition magazine left. And two, the 12-man Indian Army team that had him cornered was led by Maj. Mukund Varadarajan.

It was a warm June evening in 2013 and Altaf Baba knew the end was near. One of the most fiercely hunted terrorists in Kashmir, he knew he did not have a choice but to fight until everything faded to black. The apple orchard in south Kashmir's Yachu Guchan village he had chosen as his final hiding place was in bloom, but the fruit would not be ripe for picking until 2 months later.

Altaf Baba had not been following a cardinal rule he had learnt from his Pakistani handlers. According to that rule, terrorists should keep their weapons in full automatic mode during a firefight with Indian forces to maximize the possibility of inflicting damage. Now down to his last magazine, Altaf Baba had no choice but to fire 1 bullet at a time to draw the encounter out for as long as possible. Hemmed in by a dozen of the Indian Army's most determined hunters, the terrorist seemed to know that this was his final fight, and that escape was impossible.

But he was not the only one counting his bullets. Several feet away, taking cover with his men from the 44 Rashtriya Rifles battalion, Maj. Varadarajan was counting them too. Each and every bullet fired from within the orchard was duly noted.

'He is running out of ammunition,' Maj. Varadarajan whispered to his buddy, Sepoy Vikram Singh, correctly guessing the reason why the terrorist was not firing a spray of bullets. 'I will take him down after he has fired 30 rounds.'

The arithmetic was crucial. A regular Kalashnikov magazine holds 30 7.62-mm rounds. And Maj. Varadarajan knew how many rounds from his last magazine the terrorist had already fired.

17 . . . 16 . . . 15 . . . 14 . . .

Half a magazine was left. Maj. Varadarajan knew it was more than enough ammunition in the hands of a cornered, determined, military-trained terrorist to kill at least 5 men before being stopped. He waited, his finger on his weapon's trigger, as the shots continued to ring out.

9 . . . 8 . . . 7 . . . 6 . . . 5 . . . 4 . . .

With deliberate pauses, Altaf Baba expended the last of his bullets.

As Maj. Varadarajan had predicted, the firing stopped. The young Major had spent the previous 25 minutes taking cover, and was fully ready for his next move. With Vikram providing cover fire, the 6-feet-3-inches-tall Maj. Varadarajan emerged from his position and charged directly at the terrorist's position inside the orchard. Sprinting the short distance in a few long strides, the officer arrived with his weapon blazing. Altaf Baba was thrown off the ground in a hail of point-blank fire, landing with a crunch in the leafy undergrowth, dead.

Standing over the remains of the Jaish commander, Maj. Varadarajan took off his bulletproof headgear and fished out a Motorola handset from a pouch in his combat fatigues. He had to report the operation's success to his CO. Altaf Baba, a native of Pakharpora, oversaw all terror operations for the Pakistan-supported Jaish-e-Mohammed in south Kashmir. This was a big kill.

'Sir, there's good news. We got Altaf,' Maj. Varadarajan said as he wiped the sweat off his forehead before placing his headgear back on. The terrorist had been killed, but officers and soldiers can rarely afford to let their guard down.

'How can you be so sure it's him, Maddy?'

The voice at the other end of the satellite line was Col. Amit Singh Dabas, the battle-hardened CO of 44 Rashtriya Rifles who was at his headquarters in Zawora Manlo, near the apple town of Shopian.

'I am standing over his corpse, sir,' Maj. Varadarajan replied. 'I'm looking at his face.'

Col. Dabas, a decorated SF officer, quickly realized that Maj. Varadarajan was no ordinary soldier. The Major had been posted under his charge in the Kashmir valley barely 3 months prior to the incident. It was proving to be difficult for the Colonel not to like him. From the day they first met, Maj. Varadarajan was christened 'Maddy' by the CO.

'I thought he looked like the spitting image of the film actor, Madhavan. Maj. Varadarajan was also from Chennai. So I started calling him Maddy and the name stuck,' says Col. Dabas.

Maj. Varadarajan's tactic of counting Altaf Baba's final round of bullets soon became the talk of not just the

44 Rashtriya Rifles, but other battalions operating in south Kashmir as well, a hotbed for Pakistan-backed terrorists. He became known as the '44 RR Major' who was so remarkably composed during a firefight that he could actually keep count of the bullets that were fired in his direction.

The legend would also be a source of amusement at the unit. Mathematics had never been Maj. Varadarajan's forte and he would often share with fellow officers how he had a hard time not failing at mathematics in school. On one occasion, when the Altaf Baba encounter came up for discussion during a round of drinks at the unit mess, a fellow company commander joked, 'Thank God you didn't goof up with your counting, Maddy. Look at your size! You think that bugger would have missed you?'

The jokes at Maj. Varadarajan's expense were fine in the atmosphere of brotherhood and bonhomie at 44 Rashtriya Rifles. But not one of them had any delusions about just how crucial the killing of Altaf Baba was for the security forces. The terror commander had been steering an effort to establish linkages between the Jaish-e-Mohammed and the Hizbul Mujahideen to synchronize and amplify the terror machine's effectiveness in south Kashmir. His killing would be a crucial step forward in the fight against established Jaish and Hizb networks in the area. Apart from the dead terrorist, mobile phones and coded matrix sheets found on his person left a trail for investigators. This eventually provided vital leads to help identify routes used by terror cadres and the civilian overground operatives supporting, protecting and facilitating them.

In the months that followed Altaf Baba's killing, Maj. Varadarajan's focus was on deciphering the codes he had found on the terrorist. And for that, he made repeated visits to the Army's electronic warfare detachment in Srinagar.

'He was in Srinagar every second day for several weeks to find out what progress the electronic warfare detachment had made,' remembers Col. Dabas. 'He knew that the codes masked solid details.'

Cracking the codes became an obsession. The young officer had become unusually convinced that the information that lay encrypted in the codes would lead to bigger terrorist targets and plans. The belief consumed Maj. Varadarajan for weeks—officers at his unit recall how there was little else he would speak about.

Three months after the encounter, Army specialists finally cracked the codes, allowing Maj. Varadarajan to piece together several vital details crucial to counterterror operations in and around Shopian. Maj. Varadarajan's suspicion had been proven true—the codes provided extremely specific information. Chief among the secrets they held was a specific reference to a group of houses on a hilltop in Qazipathri village overlooking Shopian town. These houses sheltered terrorists on their transit from the Yarwan forests to Shopian and beyond. As soon as Maj. Varadarajan received the classified report, he went to his CO with a satellite map identifying the houses.

'Sir, these houses have to be kept under surveillance at all times. We will get something big there one day,' Maj. Varadarajan told Col. Dabas. 'But we have to be seen to be totally inactive in that area.'

Seven months later, in the summer of 2014, as India's longest-ever general election reached its peak, Maj. Varadarajan's prophecy would come true.

At 1430 hours on 25 April 2014, Maj. Varadarajan had just sat down to lunch at the 44 Rashtriya Rifles' Charlie company headquarters at Shajimarg in Baramulla when he received a tip-off from one of his ground sources. It was the kind of input that meant lunch would have to wait. Jaish-e-Mohammed commander Altaf Wani was in Qazipathri village. And he was in one of the houses Maj. Varadarajan had identified from the codes. There was not a moment to lose. Ordering his QRT to arm up for the operation, Maj. Varadarajan dialled his CO.

'Sir, I have A-1 intelligence. Wani is there. I am rushing to the village with my QRT.'

Altaf Wani had replaced Altaf Baba as the Jaish's Divisional Commander and was trying desperately to take his predecessor's work to the next level—in uniting the suicide squads of the Jaish and the commando-style units of the Lashkar-e-Taiba. Wani had popped up on the Army's radar not long after Altaf Baba was killed.

Qazipathri, Jammu and Kashmir
25 April 2014

In their squad vehicles, Maj. Varadarajan and his men arrived at the village in less than 30 minutes. By 1500 hours, the QRT had split into 6 buddy pairs and set up a cordon around the two-storey brick house about which the men had received detailed targeting information. Maj. Varadarajan then proceeded to do what he always did before operations in civilian areas—he ordered his men to move residents out of the area for their own safety.

The house Maj. Varadarajan and his men had their eyes on that afternoon was the biggest in the village. Its spacious compound included a sprawling orchard and 2 outhouses. Maj. Varadarajan scanned the setting quickly. Heavy-calibre weapons could not be used as the men were not clear how many civilians were still inside the house. Army snipers arrived a few minutes later and took positions on rooftops of neighbouring houses. But the sharpshooters had no clear view of their target. They would remain in position, but had no idea what they were aiming at.

As Maj. Varadarajan began a final briefing with his men, he received a shattering additional piece of information from a resident of Qazipathri. Terrorists hiding in villages routinely endanger the lives of citizens, most of whom are too afraid to speak. Others are desperate not to become pawns in the violence and at times break their silence. Maj. Varadarajan immediately radioed his CO.

'Sir, it's not just Altaf. There are 2 Lashkar terrorists with him. We are about to make contact.'

Barely had Maj. Varadarajan signed off when Col. Dabas heard gunshots. The battalion headquarters was not far from Qazipathri village. The sound of gunfire is usually the last thing that alarms a soldier. And in this case, Col. Dabas knew that one of his finest officers was on the job. Even so, he immediately ordered more soldiers from the unit to rush to the location and strengthen Maj. Varadarajan's cordon.

The gunshots that Col. Dabas heard were the starting point of what would become a fierce firefight. From well-entrenched positions within the house, the terrorists fired at Maj. Varadarajan and his men, who returned tentative fire as they squinted at the house, trying to figure out where the terrorists were hidden. The gun battle raged on for an hour but the terrorists' positions could not be pinpointed.

Maj. Varadarajan checked his watch. It was well past 1700 hours. He knew that something had to be done before daylight faded. Darkness would give the terrorists tremendous advantage. They were probably already planning to draw out the encounter until the sun set so they could slip away from the cordon under the cover of night.

Maj. Varadarajan winced. He knew that was not an option. Allowing Altaf Wani to slip away after having him surrounded would mean the security forces could effectively forget about getting anywhere close to such an opportunity soon. It would be a psychological blow to the men, and a huge morale boost for the terror cadres. If Altaf Wani managed to escape this encounter, his image would be inestimably inflated across propaganda material as a 'miracle man' who had outfoxed the Army's most hardened soldiers.

There were other reasons why that evening's hunt was so important. The terrorists hiding in that house in Qazipathri had killed a polling officer and injured 5 others a day earlier in voting at the Anantnag Lok Sabha constituency. The polling staff was on its way to Shopian when the terrorists opened fire on their bus. Eliminating the men responsible for the murderous attack on the already vulnerable democratic process in the Valley went far beyond just kills by a Rashtriya Rifles squad.

The day rapidly ended, and with no real forward movement in the fight, Maj. Varadarajan came up with a plan, whispering it quickly to Vikram, who nodded back that he was ready to go.

As the other men provided covering fire, Maj. Varadarajan and Vikram dropped to their bellies and crawled through the orchard. As bullets flew in 2 directions over their heads, the 2 reached the front entrance of the house. Maj. Varadarajan quickly planted an improvised explosive device, armed it and then motioned to Vikram to retreat with him to a safe distance.

At 1730 hours, in a deafening blast, the front portion of the house came crashing down in a cloud of debris. Not waiting, literally, for the dust to settle, Maj. Varadarajan and Vikram switched their weapons to burst mode and stormed the house, straining through the murk for their targets. Instantly, they were greeted by a hail of bullets fired by a terrorist who had been half buried in the debris. One of the bullets grazed Maj. Varadarajan's forearm. A microsecond more and the 2 soldiers would have been shredded, but they both reacted fast, turning their own AK-47 rifles on the terrorist, pumping an unwavering stream of bullets at him.

Maj. Varadarajan took a step closer to the body in front of them. Bullet smoke rose from the dead terrorist. He bent down to get a clear look. Maj. Varadarajan knew what Altaf Wani looked like. And this definitely was not him. Two more terrorists were still in the house. They had mounds of rubble now providing them with cover. And 1 of them had to be Altaf Wani.

Before Maj. Varadarajan and Vikram could decide on their next course of action, grenades came flying through the air from a dark corner of the shattered ground floor of the house. Well-trained in room-clearing techniques, Maj. Varadarajan and Vikram dived to the ground, their hands protecting their ears. The grenades exploded feet from the 2 soldiers, shrapnel smashing off the debris, missing them by inches.

The terrorists followed the grenades they had thrown with fire from their assault rifles. Using this as cover, one of the terrorists bolted out of the house and towards an outhouse in a far corner of the compound. As he exited, the terrorist had briefly looked Maj. Varadarajan's way. And Maj. Varadarajan saw his face. It was Altaf Wani.

Maj. Varadarajan had gained a formidable reputation for dominating firefights, aggressively ending them with sheer power.

'His aggression was cold and calculated. There was nothing brash about it. It was Maddy's belief that the man who takes initiative and packs aggression in a firefight is the one who triumphs,' recalls Col. Dabas.

It hardly surprised Col. Dabas, therefore, when one of the officers at the encounter site radioed him with the update that Maj. Varadarajan and Vikram were now approaching the outhouse to take a cornered Altaf Wani down. The QRT continued to fire bullets at the house where they believed the third terrorist was still hiding. But there was no return fire.

Maj. Varadarajan lobbed a grenade into the outhouse. The blast should have debilitated, if not killed Altaf Wani. But as the 2 men stormed the outhouse, a volley of fire came smashing into Vikram. It became clear what had just happened—the grenade had killed 1 terrorist, but there was another with him. Altaf Wani wasn't alone when he fled from the debris of the residence and into the cement outhouse—the second terrorist had fled with him. It was this second terrorist who had been killed by the grenade. Wani had survived.

The sepoy returned several rounds of fire at the terrorist, but Wani was shielded by a row of logs stacked in the outhouse and was able to fire his weapon from the confined space he was in.

Maj. Varadarajan saw his buddy collapse to the ground. Vikram had taken 2 bullets: 1 had sliced his neck open, while the other had penetrated his jaw. A gunshot through the neck usually spells certain death. Maj. Varadarajan knew he was about to lose one of the most courageous and dependable soldiers in his team. He knew that not only was his buddy through with this fight, his life too was about to end. As in all encounters, there was not a moment for emotion or mourning. Without pausing for a moment, Maj. Varadarajan lunged forward with his AK-47 and sprayed bullets at Altaf Wani, killing him instantly.

In those final seconds, some of Altaf Wani's shots hit Maj. Varadarajan.

'He walked out of the outhouse. He looked okay. We thought he was fine,' recalls an officer in the cordon outside the house. 'But then he suddenly collapsed.'

The soldiers were not sure if Altaf Wani was dead. And they didn't know what had happened to Vikram. Maj. Varadarajan was breathing heavily when he was pulled from the site, but nothing about his demeanour betrayed that he had 3 gunshot wounds and was losing copious amounts of blood from all 3.

'Yaar, we got him. But sheer bad luck, we lost Vikram,' Maj. Varadarajan said to an officer who was removing him from the site, barely a grimace on his face. 'And I got hit too. I can't believe it.' Then Maj. Varadarajan lost consciousness.

An ambulance had arrived to dash Maj. Varadarajan to the Army's 92 Base Hospital in Srinagar—the only place equipped to handle the injuries he had suffered in the firefight. Col. Dabas had alerted the Pulwama civil hospital en route about the officer's critical injuries and asked them to make arrangements to stabilize and possibly revive him before sending him onward to Srinagar. But Maj. Varadarajan would not make it beyond a few kilometres from Qazipathri village. He died in the arms of his unit's 2IC and the regimental medical officer.

Maj. Varadarajan had celebrated his 31st birthday on 12 April 2014, a fortnight before the Qazipathri operation. It is a

day his CO will never forget. The Colonel was on his way to
Srinagar airport to drop off an officer, Maj. Aashish Dhankar,
who had completed his tenure at the unit. He decided to pick
Maj. Varadarajan up from his company operating base and treat
him to an extra special lunch at a luxury hotel.

As the officers drove towards Srinagar, one thing needed
to be resolved: should they head to the Taj or the Lalit?
Col. Dabas remembers telling Maj. Varadarajan that while the
Taj offered a stunningly beautiful view, the other hotel was
reputed to serve better food.

'Sir, pehle Taj chalte hain, phir Lalit (Sir, let's check out the
Taj first, then the Lalit),' was Maj. Varadarajan's instant reply.

The Colonel couldn't say no. The officers went to the Taj
first where they had chocolate brownies and steaming cups of
cappuccino, before driving to the Lalit where they cracked
open a few bottles of beer and finished the afternoon with a
sumptuous meal.

That afternoon was a rarity in the cloistered life that
soldiers who participate in operations lead. Maj. Varadarajan
wanted the unit's adjutant, Maj. Ankur Datta, to feel
miserable about missing out on the jaunt. So he suggested
to his CO that they take some pictures and share them with
others in 44 RR.

'When he saw the snaps he decided against it. He thought
he looked too bulky. Of course I shared them!' Col. Dabas
later wrote in a letter to Maj. Varadarajan's young wife, Indhu
Rebecca, 3 days after his death. In the same letter, he wrote:

Now, he [Maddy] is the pillar on which the unit history
will rest. The paltan and the Army are indebted to him.
As his commanding officer, I am grateful to him for the

moments of trust and laughter we shared. Whenever he is mentioned, I will walk tall and say, 'I knew him. He was my finest officer.'

Men from the unit remember how no matter how long or hard Maj. Varadarajan's day had been, he was always up for making others feel special on their birthdays or anniversaries. It was the CO's birthday on 24 March when Maj. Varadarajan arrived at his house at midnight along with his QRT and the neighbouring company commander. 'He would do that for everyone. That was Maddy. Personal relationships meant a lot to him,' remembers Col. Dabas.

Officers of the 44 Rashtriya Rifles say it was Maj. Varadarajan's strength and sheer bulk that kept him alive for over an hour even after taking so many bullets.

'The way he was bleeding, I knew he would not come out of this. But it didn't stop us from praying for a miracle,' says an officer who was part of the Qazipathri operation.

Another buddy pair pulled out Altaf Wani's body from the outhouse after Maj. Varadarajan was dispatched from the site. He had gunshot wounds in his head, neck, abdomen and limbs. Maj. Varadarajan had made sure that there was nothing left of him.

Maj. Varadarajan's last words would haunt his CO. Col. Dabas would spend days wondering why the young officer had said, 'I can't believe this has happened to me.'

'The only thing I can conclude is that Maddy was so sure of his battle craft and fighting skills that he couldn't believe they had got him,' says Col. Dabas who went on to become an instructor at the Defence Services Staff College in Wellington, Tamil Nadu.

The day after the operation, Col. Dabas called the 44 Rashtriya Rifles' Subedar Major, the most senior enlisted soldier in the unit, and told him he would like to meet Maj. Varadarajan's Charlie company. Their morale would have been crushed from the loss of 2 of their bravest. At dinner, Col. Dabas broke bread with Maj. Varadarajan's QRT. Few words were spoken. Grief overwhelmed pride that night.

'In Maddy, I saw a special operative. I could relate to him,' says Col. Dabas. 'He had the patience to cultivate sources and the aggression to influence the outcome of a firefight. His Charlie Company was very much like an SF unit,' says Col. Dabas. And Col. Dabas would know. He was commissioned in the Para-SF and later moved to another Para-SF unit. Col. Dabas would beam with pride as Maj. Varadarajan's QRT would always bag top position during division-level competitions involving 6 teams each from the 9 battalions that were part of Victor Force, which oversees operations in the Anantnag and Pulwama areas of J&K.

Maj. Mukund Varadarajan was posthumously awarded the country's highest peacetime gallantry award, the Ashok Chakra, on the eve of Independence Day 2014. Indhu arrived in Delhi to receive the award from the President of India on Republic Day the following year.

'India should see the man Mukund was, not my sorrow,' Indhu said that evening in a television interview on NDTV.

Indhu has since moved with her daughter to Australia to pursue her 'new-found passion' for teaching.

'I often try my best to explain to my 6-year-old how wonderful he was, and we recently came to an agreement after watching *Baahubali* that Appa is our Baahubali because he was the strongest and he did the best he could to do his duty. It is true, he was perfect for me, but perfect or not, he did his best, always,' says Indhu.

Maj. Varadarajan's citation said he personally led the demolition team and used the resources available to him in a critically short time period to bring down the target house. The citation made a special mention of his 'aggressive action' and the avenging of the attack on election officials 'within 24 hours' that had 'restored the faith of public in democracy'.

'*Maj. Mukund Varadarajan exhibited most conspicuous bravery and exemplary leadership and made the supreme sacrifice while fighting with the terrorists,*' the citation added. His buddy, Sepoy Vikram Singh, was awarded a posthumous Shaurya Chakra, the country's third highest peacetime gallantry award.

Mukund would often ask Indhu to promise him that she would never cry if something happened to him in the line of duty. Indhu took that promise very seriously. As Maj. Varadarajan's mortal remains were brought to Chennai, television viewers across the country saw Indhu's calm, stoic figure next to her husband's casket, their 3-year-old daughter, Arshea, by her side.

'There is this moment that I will never forget. Mukund told me that his parents and I were his gods. I could understand him extolling his parents, but I was scared when he put me up on that pedestal. I tried to explain to him that I am just human,' Indhu recalls.

'He saw it differently. He called our daughter his god too on the day she was born. He really did love and respect us as much as God. How can you not be blown away by that kind of love?'

'She (Indhu) would often say she would pull through because she had made a promise to him. We draw strength every day from Indhu's composure and demeanour,' says Col. Dabas.

The Ashok Chakra that Maj. Varadarajan was awarded posthumously was not the first decoration for gallantry that came his way. His CO had originally planned to recommend Maj. Varadarajan for an award in June 2013 for the Altaf Baba operation. But Maj. Varadarajan would not hear of it. Such recommendations are always confidential, but the young officer had learnt about the honour in store for him. He marched to the CO's office and insisted that if someone had to be recommended for an award, it needed to be 2 young soldiers from his company who had played the role of scouts during that operation.

'Sir, bahut badhiya kaam kiya un donon ne. Woh award ke haqdaar hain (Both of them did a tremendous job and deserve the award),' Maj. Varadarajan pleaded. He got what he wanted. The 2 men were awarded the Sena Medal for gallantry shortly thereafter.

The 5-page letter Col. Dabas wrote to Indhu was drenched in the emotions of a CO who respected and admired a soldier 8 years his junior.

'Each one of us who knew him will find our own way to overcome the loss. Grief will fade away and when it's gone, only one thing will remain—pride,' he wrote.

'To be honest every day is a struggle for his parents and for me,' says Indhu. 'It should not come as a wonder that some days are tougher than others. For his parents, he was the son who loved them to bits; he had pet names for all of us and we yearn to hear him calling us by those names. We yearn for a sighting of his dimpled smile that could warm our now frozen hearts. It has not been easy and it is not going to get any easier.

The trick is to conceal the struggle and will yourself through a day as happily as you can because that is what he would want. The toll it has taken is no joke and can be seen in my daughter too. She often hides Mukund's pictures because she is worried that either I or his parents will feel sad. It proves that she has definitely taken after him in terms of concern and love and that is the silver lining in an otherwise cloudy situation.'

Mukund's favourite quip over drinks would always be that he had already won the toughest battle of his life—convincing his traditional parents to accept his love for and intention to marry a woman from outside his community. In August 2009, aged 26 and barely 3 years into the Army, Mukund and Indhu were married.

'It wasn't easy getting married,' recalls Indhu. 'There are words that were thrown at both sets of parents by some relatives and the community that were difficult to stomach, but we were selfish in our love and had parents who stood by us irrespective of their misgivings. My parents took a little longer because they were giving away their precious daughter, but what held me through it all is the faith that we had in each other and his acceptance and understanding that I couldn't do without my parents. I knew he would wait and I knew his ideals were the same as mine. It took us 5 years to convince my parents and then tie the knot. To have lost all that in a few moments is never going to be easy to come to terms with. There are days when it gets really difficult to accept that the man who said that he will stand beside me and show the world how our relationship will survive whatever life throws at us is no longer with me.'

Three years after they got married, in December 2012, Maj. Varadarajan would arrive in the Kashmir valley to become part of the 44 Rashtriya Rifles.

Folklore surrounding Maj. Varadarajan's memory remains.

If unit officers had something tricky to discuss with their CO, they would get Maj. Varadarajan to articulate it to the boss. 'He would give it an operational twist and leave me with little choice but to agree to whatever was put forth to me,' recalls Col. Dabas.

Apart from being an outstanding operations man, Maj. Varadarajan is remembered for his ruthless wit. Everyone in the unit knew that if Maj. Varadarajan was looking particularly happy, somebody was certainly having a miserable day.

The young officer missed his parents, Varadarajan and Geetha, tremendously. He knew he could not speak to them every day, but when he did, he would devote an hour if he could to his mother.

Maj. Varadarajan had dreamed of shifting his parents to a brand-new apartment in Chennai. The purchase had been finalized a month before his death. After he was gone, and as the family grieved, it would be the first thing that Indhu ensured.

Indhu will never forget her last conversation with Mukund that April morning when he departed on his final operation.

'We had just had a silly fight a few days ago and the fight was, as always, about how I needed more time with him. We got over it the same day as both of us would rarely leave a fight unresolved. On the morning of 25 April, we had a short and ordinary conversation. He asked for Arshea and spoke to her over the speaker and we told him that we missed him. He told me to take Arshea out and enjoy the day, do some shopping and cheer up, and that he loved us too. He hadn't been getting much rest, often coming back early in the morning after patrolling and calling me to wish me a good day before getting a few hours of sleep before he had to be up and about again. We said our goodbyes a little earlier than usual that day because I knew he was busy and needed some rest. Ironic, isn't it?'

Col. Dabas's WhatsApp profile picture is not his own. And his status message reflects the image: 'Mukund and Vikram. Salute.' He hasn't found the heart to change those settings for more than 3 years. And they remain a constant reminder of 2 of the finest soldiers the SF Colonel had known—2 soldiers he thinks of every day, 2 young men he will remember all his life.

'I think of them every single day. They were lost on my watch,' says Col. Dabas. On the Colonel's study wall are framed pictures of Maj. Mukund Varadarajan and Sepoy Vikram Singh.

Indhu remains in touch with her husband's battalion.

'The army never leaves anyone behind and we are a family more than just a group of people tied together by a common profession. They have been with me with words of comfort and all forms of support. I am confident that they will stand by me as I will stand by them in the future as well,' she says.

'Death on the battlefield is the ultimate privilege for a soldier. We don't go there to die. But if it happens, we have to turn it around into a celebration,' says Col. Dabas.

'Only soldiers who have bled together in combat will understand that.'

*

8

'Medical Science Cannot Explain This'

Lance Naik Hanamanthappa Koppad

9 February 2016, 1226 hours

Of the 10 soldiers presumed dead in #Siachen, one has survived. Lance Naik Hanamanthappa is critical. Pray for him. What a miracle!

The late-night flash on social media by newspaper correspondent Rahul Singh galvanized a country that had spent a week in mourning. TV news channels broke out of their regular recorded late-night programming. Across the Internet, a ripple of disbelief churned into a tide.

Was this even humanly possible?

Prime Minister Narendra Modi was nowhere close to turning in for the night. Two hours earlier, at his Race Course Road residence, he had received a call from the then Army Chief, Gen. Dalbir Singh. The conversation lasted barely a few minutes. And when it ended, it was not necessary for the Prime Minister to say it. He knew that the Army, already grieving over the loss of its men, would do everything in its power to save the superhuman they had pulled out alive after 6 days of being buried under more than 25 feet of snow— 6 days at temperatures of −40 °C under a terrifying block of blue, unforgiving ice.

From the moment it happened, the Prime Minister had demanded a daily briefing of the rescue operation 20,500 feet high on the northern glacier near Siachen.

Siachen Glacier, Jammu and Kashmir
2 February 2016

The 10 Army men, including 8 from the 19th battalion of the
Madras Regiment, had hiked up to Sonam Post just 2 months
before. Sonam, one of the highest permanently manned
military posts in the world, sits way up on the Saltoro Ridge
that overlooks the Siachen Glacier to the east and Pakistan-
occupied territories to the west. Named for Havildar Sonam,
an intrepid Ladakhi soldier who braved unspeakable weather
and Pakistani fire to occupy the point in 1984, the Sonam Post
offers soldiers a magnificent position of advantage, but is also
fully exposed to what is literally the worst weather on earth.
Not to speak of insidious crevasses and devastating avalanches
similar to the one that came crashing down on the 10 Army
men early on 3 February.

While some of the men were on observation and guard
duty, the others were in their tents. And none of them even saw
it coming. An enormous block of ice shattered the ridgeline
above them and came rumbling down the mountainside,
completely burying the post.

The men at Sonam Post had arrived in the Siachen
area in October 2015. Including nursing assistant Sepoy
Sunil Suryawanshi of the Army Medical Corps, the team
comprised team leader and head of the post, Subedar
Nagesha T.T., Havildar Elumalai, Lance Havildar S. Kumar,
L. Nk Sudheesh B., Sepoy Mahesha P.N., Sepoy Ganesan
G., Sepoy Rama Moorthy N., Sepoy Mustaq Ahmed S.
and L. Nk Hanamanthappa Koppad. The men had been
hand-picked in December to take position at Sonam. The
choice was not random. Col. Um Bahadur Gurung, CO of

19 Madras, had chosen them for what everyone on the glacier knew was the most demanding deployment possible. Soldiers deployed for high-altitude warfare are frequently the most resilient men. The 10 chosen to pitch their tents at 20,500 feet for a few months represented the cream of the crop.

Politics and tragedy have occasionally thrust Siachen into the national discourse. But away from the diplomatic aggression over the northern glacier areas, it remains the Army's enduring regret that few truly understand what it means to even operate in such terrain, far less engage in combat. It is not without reason that Siachen has earned the epithet 'frozen hell'.

Soldiers deployed to high-altitude posts do not only have to be rigorously trained in the art of warfare in the most devious, unforgiving terrain imaginable, but they also need to be highly skilled in survival and sustenance. In addition, every man has to be psychologically conditioned so he does not run the risk of losing his mind at those altitudes. The training regimen and deployment schedules for India's glacier units have been tailored over time to account for the worrying effects that spending long periods at those heights can have on men.

'They were all incredibly brave men,' says Col. Dinesh Singh Tanwer, who, as 2IC of 19 Madras when the avalanche struck, operated from the operations room established at Siachen base camp established under the base commander Col. Hari Haran, about 90 minutes away from Sonam Post by helicopter.

'These guys were the best of the lot. Motivated and fit. When you go to the glacier, the fear of the unknown overpowers all other fears. We know where the enemy is, but we don't know where avalanches will come from, or where the crevasses are. We can't take anything for granted,' he says.

The monstrous visitor at Sonam Post on the morning of 3 February was a slab avalanche, the most sinister kind. Formed by an enormous 800x1000-metre block of snow fracturing away from the mountain, it had buried the post up to 25–30 feet deep in mere seconds. Where the tents once stood, now there was nothing but scattered debris of blue ice boulders harder than rock. The wind dropped suddenly, as it ironically always does after an avalanche, filling the thin air above Sonam Post with a cold silence.

That was all Maj. Vipin Kumar, the Company Commander posted a short distance below Sonam Post, heard on his radio set. The Major, who would get a radio report from the men at Sonam Post each morning at 0400 hours, had heard nothing that morning. He was not immediately worried. The sub-zero temperatures frequently paralysed equipment, and radio sets sometimes needed to be warmed up artificially before they would work again. Over an hour later, there was still no word.

Then, at 0515 hours, a feeble voice cracked through the radio.

'*Saab, hum dab gaye hain* (Sir, we have been buried).' It was Havildar Elumalai.

Originally from Adukumparai village in faraway Tamil Nadu, Elumalai was now buried under a wall of snow 20 feet deep. He had miraculously been able to reach for his radio set and transmit the news.

Maj. Vipin knew instantly what had happened.

Exactly a month before, on 3 January, 4 Army men had perished in a similar avalanche on the Siachen Glacier. Havildar Dorjey Gason, Havildar Tsewang Norboo, Rifleman Jigmat Chosdup and Rifleman Mohammed Yusuf had died instantly. The voice from Sonam Post, however, suggested there was hope.

The Major did not waste another moment. He immediately formed a rescue party before alerting his CO, Col. Gurung, who was at the Kumar Post at an altitude of over 15,000 feet. He in turn relayed the message up the chain of command via the Siachen Glacier's Independent Infantry Brigade, on to the 14 Corps in Leh, the Army's Northern Command Headquarters at Udhampur and finally to the Army Headquarters in Delhi.

An early riser, Army Chief Gen. Dalbir Singh was about to go for a morning run when he received word about the avalanche. Cancelling his exercise routine, he immediately began to get ready to dash to his headquarters at South Block. Ten Army men at this lofty military post had been deluged in a terrifying flood of ice. The government leadership needed to be notified immediately. It was the first thing he needed to do.

As Maj. Vipin and his men gathered troops from other posts and headed straight for Sonam, lower down at Siachen Base Camp, a makeshift coordination centre was set up under the unit's 2IC, Col. Dinesh. Soldiers at the base camp had been preparing for their own induction into high-altitude posts. Everything would now have to focus on the tragedy at Sonam Post.

Two hours later, Maj. Vipin and his team arrived at the site. It would take them whole moments to digest the scene.

The enormous debris field from the avalanche had gorged on every visible aspect of the post. There was nothing left to see. It was a scene of devastation the men would never forget.

Helicopters soon brought in more personnel, this time doctors and ace Army mountaineers. They arrived bearing metal detectors, excavation equipment and specially trained avalanche rescue dogs. By noon on 3 February, a team of nearly 50 personnel were at the site of the tragedy.

This in itself was an enormous logistical challenge. Sonam Post, which was suitable for no more than 10 men, was now not only completely destroyed and buried, but had 50 men who could not leave any time soon. In coordination with Col. Dinesh's team at Siachen Base Camp, the rescue teams needed to organize tents, water, medicine and rations for themselves. There could be no oversights at this juncture.

'We had to plan very carefully. These men were about to engage in hard labour at 20,500 feet. If we made errors, we could easily suffer further casualties,' says Col. Dinesh.

Accompanied by dogs and machinery and with prayers on their lips, the men began digging through the vast icy sheet of debris at Sonam Post in temperatures that forced them to take frequent breaks just to be able to flex their fingers and limbs. If frostbite or altitude sickness set in, it would not just damage the rescue effort, but would also place an exponentially bigger burden on the base camp to rescue the rescuers themselves.

By 1455 hours on 3 February, the media got wind of the disaster that had struck Sonam Post. Having just reported the deaths of 4 Army men a month before, the newsflashes sounded grim.

At 20,500 feet, the rescue effort continued till sundown. The night brought fresh snowfall and winds of unspeakable ferocity, forcing the large rescue team to hunker down into their tents. There was no question of searching through the night. The men spent an uncomfortable night, their tents buffeted by a howling draft, fully aware that every hour they spent unable to search was an hour closer to the end for the

10 soldiers buried under several feet of ice. If the end hadn't already arrived for them, that is.

At first light on 4 February, the men set to work again. Shortly after 1100 hours, the Army Headquarters in Delhi put out the first of several updates on a rescue effort that would, in a matter of days, mesmerize the entire country.

The men saw and heard nothing as they continued to dig and scour through the icy debris on 4 February. Trained not to yield to despair even in the most hopeless conditions, the rescue teams could not help but be sceptical about the men surviving a full day under all that ice. Was this really a race against time any more, or simply an exercise to find 10 corpses? It did not matter. The Indian Army never leaves its men behind.

Then, when they were least expecting it, a radio set with one of the rescuers crackled, a broken voice emerging from it.

It was Sepoy Rama Moorthy N.

Like Havildar Elumalai the previous day, the sepoy from Gudisatana Palli village in Tamil Nadu had managed to find his radio and make a call. The rescue team had not been able to raise a sound out of Elumalai since they arrived on the scene, making them fear the worst, but not slowing them down. The call from Rama Moorthy exploded through the rescue team at Sonam Post, dusting every bit of despair from their shoulders.

The men doubled up, focusing every resource on trying to pin down Rama Moorthy's location under the ice.

'He stayed in contact with us for 2 hours, but he was unable to tell us exactly where he was,' recalls Col. Dinesh, a steady sadness in his voice. 'He tried very hard, but he couldn't direct us. He could have been upside down for all we knew.

Both he and the teams knew his radio set would be out of batteries soon.'

Shortly before 2100 hours that night, Sepoy Rama Moorthy went silent.

A team from the Siachen Battle School had arrived that day too, carrying special sensor systems that were capable of seeing through walls of ice. The new team had also brought tree cutters to slice through the ice with greater efficiency as they raced against time. By the time Rama Moorthy's radio transmissions died out from under the snow, the rescue team had swelled to 110 strong. A team of 50 civilian porters had also been employed to help transport equipment to the site.

The voice from below had acted as a booster dose of hope to the rescue team. But the silence that followed threatened to drag them down again. Search operations continued till late that night and for the following 2 days. For over 48 hours, the men saw nothing and heard nothing. Every pair of ears over the debris field strained to pick up any sounds. Some men put their ears directly to the ice, hoping to hear something, *anything* from below.

By the evening of 7 February, 5 days after the avalanche, the team had dug 3 40-feet-deep holes straight through the avalanche remnant.

'We were fighting to pinpoint any location that could give us a clue about the men below,' recalls Col. Dinesh. 'Then at 1800 hours on 7 February, we saw something.'

A team suspended on one of the shafts had spotted a cable sticking out from the side of the hole. It was a communication cable. In 5 days of digging, it was the first physical object the team had found below the devastation.

'It was obvious we had to follow the cable,' says a mountaineer member of the rescue team at Sonam Post. 'Digging further vertically was very difficult. The shaft was unstable too. We lowered barrel-halves above us so the shaft wouldn't cave in with us inside.'

With the blue ice threatening to crumble and bury them inside the shafts any moment, the men began carefully excavating a horizontal tunnel to follow the cable they had found. Night arrived quickly, bringing with it the most vicious winds the men had encountered that week at the site. Rappelling up the shaft, they rushed to their tents, with no choice but to suspend the search for another night. The team knew it had found something that could lead to the buried soldiers. Few slept that night.

At first light on 8 February, shaft work resumed. They were chilled to the bone, but members from the team remember an unusually calm morning bringing with it a palpable sense of hope in the shadow of the fractured ridgeline. They trudged to the shafts, lowering themselves once again to continue tunnelling, carefully following the communication cable through the icy darkness. Special halogen lamps on their helmets cast a grim glow on the featureless, solid ice they were cutting through. It was not until noon on 8 February that the team had its first big breakthrough.

Through the dark portal, created by ceaselessly drilling horizontally through the ice slab, they spotted a piece of tent. They did not hear a sound, but they knew they had to be very close to the men they were looking for.

The excavation continued for another 5 hours until they had gained access right into the snow-filled tent. As the shaft groaned once again with the shifting of ice, they spotted the first body. After 6 days in the snow, this was the first of the 10 men they had seen. Carefully, the body was removed from the tent. It was cold, stiff, unmoving. Another 3 hours passed before the rescuers reached the next body deep within the tent. But as they took hold of the body to move it through the shaft, the rescuers realized something that stopped them in their tracks.

The soldier was breathing.

Calling for an emergency evacuation, the body was quickly prepared and eased out through the horizontal tunnel, and then carefully up the long shaft to the surface, strapped to one of the mountaineers.

The man was L. Nk Hanamanthappa Koppad. Far from his home town in Karnataka's Dharwad district, the soldier had been pulled out alive—and conscious—after 144 hours under nearly 25 feet of murderous ice. The disbelief among his rescuers had to wait—they needed to get him to safety as quickly and delicately as possible.

'We immediately rushed him to the medical tent that had been set up with 3 doctors. He was provided humidified oxygen, warm intravenous fluids and passive warming from the outside. We could not risk anything else at that point, as he was clearly in shock,' says one of the rescuers.

L. Nk Koppad was as medically critical as it was possible to be, but his body had sustained itself for 6 days under the avalanche debris. Not a soul slept that night at Sonam Post. Inside the medical tent, the Army doctors took turns to watch over L. Nk Koppad to help him survive the night. Now in an open space, they feared his body would sink into further shock. If he did, it would be impossible to revive him.

By midnight on 8 February in Delhi, the news of L. Nk Koppad's survival had broken on social media, rapidly becoming an overnight sensation. It would be the biggest story the following day. Across newsrooms, journalists were designated to track the rescue operation and feed the news machine as rapidly as possible. On the Internet, L. Nk Koppad was instantly hailed as a miracle man and a hero. A soldier from the hot, tropical South, selflessly deployed at death-defying altitudes and temperatures in India's extreme north, had survived while standing guard in the service of his country's sovereignty.

In equal measure, L. Nk Koppad had stunned, shamed and captured the nation's imagination. Shamed because his survival had come as a shattering, unexpected reminder of how little public or political attention was ever paid to soldiers who stood sentinel at the country's most dangerous forward areas.

Late that night in Betadur village, the Lance Naik's home town, L. Nk Koppad's family would learn about his miraculous survival from television news. His wife, Mahadevi, had spent the week in mourning, hopeful but quite certain her husband had not survived. A neighbour who visited the Koppad household that night later told a Kannada TV news station that Mahadevi was momentarily paralysed by the news, too terrified to believe it could really be true, wondering if someone was playing a cruel prank on her. She picked up her baby daughter, Netra, and wept as late-night celebrations began outside her house.

Nearly 2000 km away, her husband spent the night in the green glow of the medical tent, an oxygen mask over his face.

Early the next morning, barely conscious as he was strapped to a stretcher with a cylinder of oxygen by his side and a doctor for company, L. Nk Koppad was flown in a Cheetah helicopter to the Siachen Base Camp. Col. Tanwer recalls the unforgettable moment he met L. Nk Koppad in the speciality medical tent he was shifted to.

'He was conscious,' Col. Tanwer says. 'I held his hand and he pressed my hand. His eyes were open. He had energy in him even after everything he had been through. I strongly believed he would survive.'

The Colonel did not know what to say, but found himself telling L. Nk Koppad, '*Tambi, tujhe kuch nahi hoga, tu theek ho jayega. Koi dikkat nahi hai, Tambi, tu bilkul theek ho jayega.* (Nothing will happen to you, young brother. There is no problem. You will be absolutely fine.)'

After a rapid medical check to make sure the flight down from Sonam Post had not further compromised his vitals, L. Nk Koppad was quickly loaded into another Army Cheetah and flown at full speed to Thoise, a military airfield that functions as the lifeline and gateway to the northern glaciers. Nestled in Ladakh's Shyok Valley, 'Thoise' is an acronym for 'Transit Halt of Indian Soldiers En Route (to Siachen)', and is barred to civilians.

When the Cheetah landed at Thoise, an Indian Air Force C-130J Super Hercules from the 77 'Veiled Vipers' squadron was already waiting for it. It had made the 90-minute flight from the Hindon Air Force base on Delhi's outskirts early that morning and was carrying specialized life-support equipment and a critical care specialist from the Indian Air Force. Its 4 engines whipping up dust over the high-altitude airfield, the aircraft roared off the tarmac minutes after receiving perhaps its most special passenger till date.

Shortly post noon, the C-130J landed at Delhi's Palam Air Force base, the military and VIP terminal co-located with the Indian capital's Indira Gandhi International Airport.

A super-speciality ambulance from the Army Hospital Research & Referral was waiting to receive L. Nk Koppad. In a convoy with priority access, the ambulance tore down the short 5 km to the sprawling hospital complex, carrying L. Nk Koppad strapped to his stretcher in a thick sleeping bag, passing through a crowd of television cameras and journalists that had collected in large numbers outside the hospital to capture the media's first glimpse of Siachen's heroic survivor.

Lt. Gen. S.D. Duhan, then commandant of the Army R&R Hospital, had seen nothing like this in his 38-year medical career in the military. And he had seen some amazing things. He stared at L. Nk Koppad in disbelief, wondering how the 32-year-old had defied all odds to stay alive.

'Medical science simply cannot explain how he survived at that altitude. It cannot. Remember, 9 others had died,' Lt. Gen. Duhan says. 'L. Nk Koppad was a man of incredible physical and mental courage. People have survived in those conditions for a day or 2. But 6 days? I have been through volumes of literature but have found no such instance,' says Lt. Gen. Duhan, who supervised the specialists attending to the soldier.

The soldier was immediately wheeled into the hospital's ICU. Thirty minutes after treatment began, Prime Minister Narendra Modi and Army Chief Gen. Dalbir Singh arrived at the hospital. The soldier had slipped into a coma. At his side in the ICU, the Prime Minister listened in silence as the doctors explained what L. Nk Koppad had endured.

The Lance Naik had been recovered alive and conscious, but he was severely dehydrated, hypothermic, hypoxic, hypoglycaemic and in shock. Comatose and in a precarious situation, the soldier also had severe pneumonia. Making matters worse, he had developed signs of liver and kidney dysfunction. One saving grace, the Prime Minister was told, was that L. Nk Koppad had miraculously escaped frostbite and bone injuries. The bodies of some of the other men in that same tent had been recovered crushed by the block of snow that buried them. In an enormously unlikely stroke of luck, L. Nk Koppad had managed to be caught in a tiny space between the ground and the ice slab, giving him a small bubble of oxygen to live on. Surviving in that position for a few hours, maybe even a day, was conceivable. How he did it for 6 days is something medical specialists, mountaineering legends and his comrades at 19 Madras can only guess about.

'It wasn't just about physical endurance. I like to think Hanamanthappa's belief in his team, his seniors and the Army pulled him through,' says Col. Tanwer, who, at the time of writing this book, was deployed in a different sector of Ladakh, this time along the border with China. 'He believed in his training and didn't let despair or exhaustion force him to give up.'

On prime-time television that night on 9 February, anchors and reporters drafted medical experts into their studios to make sense of the miracle man of Sonam Post. A sense of hope and disbelief would pervade coverage late into the night.

'We can talk all we want to about what happened,' says Lt. Gen. Duhan. 'But, for a moment, picture being buried in total darkness under a mountain of ice with nothing for company but the sound of your own voice.' He paused, nodding. 'Now imagine the same thing for 6 days.'

Late that evening, L. Nk Koppad's wife, daughter and extended family landed in Delhi. Received by an Army vehicle and once again thronged by journalists, they were driven straight to the Army R&R Hospital where L. Nk Koppad lay.

A smiling but anxious Mahadevi would speak more confidently to television journalists this time. 'It is a day of joy for me and my family, but let us never lose sight of the tragedy for 9 other families.'

The spotlight had shifted to Delhi, but search operations for the other men continued at Sonam Post and would not stop until all 9 bodies had been recovered.

By midnight, the team of doctors attending to L. Nk Koppad decided on a more intensive course of medical action to revive the soldier. It was a single notion that powered them on: if he had survived so much, surely he could make it. Surely, he was special.

Enlisting himself with the Army at a recruitment camp in 2002, L. Nk Koppad had immediately proven himself a tough infantryman. The year after he joined the Army, he was sent on counterterror duties in the Kashmir valley where he spent 3 years. He would return on a 4-year tour in 2008. In 2010, he was packed off to the North-east for counter-insurgency operations. It had become clear to his seniors that L. Nk Koppad was a tough soldier ready for greater operational challenges. Which was why, when a 10-man team was to be selected in December 2015 to man Sonam Post, L. Nk Koppad's name was a shoo-in.

Mahadevi and her family visited a temple and prayed late into the night on 9 February. They had been reassured by the Army doctors, but knew that L. Nk Koppad's condition was extremely delicate. The hope that L. Nk Koppad's rescue had brought to Mahadevi had both energized and terrified her. On the one hand, it was a magical moment of joy and relief. On the other, the reality of his condition made it likely that her joy would be short-lived.

The next morning, the Army provided a brief update on the soldier's condition:

'L/Nk Hanamanthappa Koppad continues to battle the odds and his condition remains very critical. The medical team treating him is monitoring his situation continuously and is treating him with expertise available in the world.'

By this time, a panel of medical experts from Delhi's renowned All India Institute of Medical Sciences had arrived to inspect the soldier's condition, joining critical care specialists, a nephrologist and a senior neurologist.

In a span of a few hours, the news from inside the hospital took a sharp turn for the worse. Doctors had discovered evidence of oxygen deprivation in L. Nk Koppad's brain from a CT scan. The tone of the new medical bulletin that followed strongly suggested that the soldier's condition was no longer within medical control:

'There is evidence of pneumonia in both lungs. His multi-organ dysfunction state continues unabated. His condition has deteriorated despite aggressive therapy and supportive care.'

Prime Minister Modi received a briefing on the soldier's health that evening. The Army informed him in no uncertain terms that saving L. Nk Koppad was going to prove next to impossible. When the Prime Minister inquired if there were any other medical avenues that still existed for the soldier, the Army assured him that L. Nk Koppad was receiving the best possible treatment that existed anywhere in the world.

Early next morning, the Army released its final medical bulletin on L. Nk Koppad. It would spare no details and soften no words:

'Extremely critical with worsening multiple organ dysfunction. His circulatory shock is now refractory to all drugs in maximum permissible doses and his kidneys remain non-functional. His pneumonia has worsened and his blood clotting disorder shows no sign of reversal despite blood component support. He is on maximal life support with aggressive ventilation and dialysis. He has slipped into a deeper state of coma.'

Mahadevi and the family were escorted to the hospital before noon.

At 1145 hours, L. Nk Koppad stopped fighting for his life.

An unusual eruption of nationwide mourning broke out that Thursday in February 2016. Amplified by a media that refused to move away from the story, L. Nk Koppad's death had, for a moment, sobered millions.

Later that day, L. Nk Koppad's remains were placed in a casket and taken to Brar Square, the ceremonial Army compound in Delhi reserved for the last rites of military personnel. Politicians do not normally attend wreath-laying ceremonies for soldiers. For L. Nk Koppad, they all did. The long line of government and state leaders would include the Prime Minister himself. Before he arrived, he tweeted a brief message:

He leaves us sad and devastated. RIP Lance Naik Hanamanthappa. The soldier in you remains immortal. Proud that martyrs like you served India.

'L. Nk Koppad's wife was amazingly strong. For her, the soldier had died for the second time in a matter of a few days,' recalls Lt. Gen. Duhan.

When news of his death was broken to her at the Army hospital, Duhan recalls her saying, 'Don't tell his mother that he's gone. She will not be able to bear it.' His mother, Basamma, had not stopped praying.

'Courage has many shades. The unbelievably strong-willed L. Nk Koppad added another shade to it,' says Lt. Gen. Duhan.

The soldier's body was flown to his home town in Karnataka, where his final resting place, a *samadhi* near his home in Betadur, became a site protected by a metal gate and fence.

Eleven months later, on Army Day 2017, Mahadevi arrived in Delhi to attend the parade and receive her husband's posthumous Sena Medal. With little Netra in tow, she was briefly the centre of attention at the Army Chief's residence that evening, at a reception attended by the President of India and members of the government. Mahadevi had been sitting quietly in a corner with her daughter.

'She can do what she likes, but I hope she will grow up to be an Army officer and be posted in Siachen,' Mahadevi smiles. 'I know they don't send women up there. But Netra is still young. That will change when she's old enough.'

*

9

'Everything Was against Us. *Everything.*'

Lieutenant Commander Niteen Anandrao Yadav

Dabolim, Goa
22 May 2009

It had been 6 months since 10 Pakistani terrorists had entered Mumbai from the sea in November 2008, holding the city hostage for 4 whole days and massacring 166 people across 3 locations. The 26/11 attacks would occupy India's national security system for years to come. But just 6 months later, it was by far the biggest thing on the country's mind.

By May 2009, as a wounded country attempted to make sense of the invasion and horror, the burden of keeping a watchful eye over the Arabian Sea had increased dramatically. India, it was clear, had let inevitable gaps in surveillance and intelligence be terrifyingly manipulated by foreign terror machinery focused on spilling blood on Indian soil. In the months after 26/11, it became clear that for all its ambitions as a regional power, India had let slip from its mind one of Jawaharlal Nehru's most memorable quotes on strategy: 'To be secure on land, we must be supreme at sea.'

The terror attack was a devastating jolt. But it served to amplify the inevitable—that India's problems from the sea could only multiply. The steady audacity of pirates on the high seas, pushing ever closer to Indian waters, had been keeping the Indian Navy feverishly busy throughout 2008. And when Ajmal Amir Kasab and his fellow terrorists stepped ashore from their rubber dinghy, India was about to be violently roused to the dangers the sea could present.

As Mumbai burnt, hundreds of kilometres away, at Goa's Dabolim airbase, groups of naval officers sat rapt in front of television screens, taking in the enormity of what was happening on their watch. Even before it became clear that the terrorists murdering people in Mumbai had come from Pakistan, military personnel across the country automatically knew they would be expected to be in a heightened state of readiness.

For anything.

From Air Force bases to Army infantry units and naval strike formations, explicit orders did not need to be given for men and women to brace themselves for what they were trained for. Among them was 38-year-old Lt. Cdr Niteen Anandrao Yadav. Lt. Cdr Yadav knew instantly that his unit would have an exponentially bigger role to play in what lay ahead for India.

He was right. Just days after 26/11, India officially bestowed its Navy with greater defence responsibilities. Already stretched thin by virtue of being the smallest of the 3 armed forces, but with the largest domain of responsibility— including a 7517-km coastline—the Navy collectively flexed its muscle as it assumed the role of guardian against every conceivable threat from the sea.

By May 2009, the Navy had possibly never been busier. Along with the gruelling after-effects of 26/11 on maritime security, the Indian Navy was preparing for another event with far-reaching ramifications—the expected delivery of a Chinese-built warship to Pakistan, the first of a new generation of lethal ships Islamabad had purchased for nearly $1 billion. The F-22P Zulfiquar-class missile frigate was expected to sail from Shanghai to Karachi in 2009. At Dabolim, Lt. Cdr Yadav and other naval aviators with the INAS 315 'Winged Stallions' had their mission cut out for them.

Early on 22 May, the base received word that 2 Pakistan Navy warships were expected to transit southward through the Arabian Sea, possibly in preparation to escort the brand new Zulfiquar back to Karachi a few weeks later.

Snooping on each other's ships in the Arabian Sea has been standard operating procedure for India and Pakistan for decades. So the 'search and surveillance shadow' mission that Lt. Cdr Yadav and his crew were tasked with that May morning was not much more than routine. But volatility in relations between the two countries post-26/11 meant that even routine surveillance missions carried substantially greater risk.

The aircraft Lt. Cdr Yadav and his crew would be flying that day was IN305, a Soviet-era Ilyushin Il-38, a large ocean-surveillance aeroplane with 4 turbo-propeller engines. Fitted with advanced sensors to detect ships and submarines, this eye-in-the-sky aircraft can also fire torpedoes to destroy submarines or missiles to sink ships.

The Il-38 is little known or recognized beyond the military and the world of aviation enthusiasts. It is not a particularly arresting sight. And being tucked away at the Goa naval air station has helped to maintain a low profile that suits its typically classified surveillance duties. In 2002, however, the Il-38 entered public consciousness in a devastating manner when 2 aircraft collided mid-air over Goa during a flying display to celebrate the squadron's silver jubilee. Twelve Navy personnel perished in the freak tragedy. Lt. Cdr Yadav could well have been on board one of the doomed planes. It was something he would never forget each time he strapped himself in before a flight, just as he did 7 years later, that morning in May.

After a brief chat with his navigator and systems operators, Lt. Cdr Yadav and the rest of his 8-man crew clambered into their Il-38 through a hatch in the aircraft's belly—the aircraft has no other entry points. The 2 pilots strapped themselves into their seats in the cockpit. A flight signaller and a flight engineer took their seats in a small space behind the pilots. And in a cabin in the rear, 4 navigators cum sensor and systems operators sat at their electronic consoles. The crew's mission that morning was to fly straight to a patch of ocean where they expected to encounter a pair of Pakistan Navy warships—a tanker and an armed frigate—identify both and gather as much information as possible about them, including their visuals and electromagnetic signatures.

In a routine surveillance mission, Lt. Cdr Yadav and his crew would have had authorization to 'buzz' the Pakistani warships—swoop down low and make it obvious that they had been spotted, a maritime equivalent of 'Gotcha!', part of the endless cat-and-mouse chase at sea. But during this mission, the crew only needed to locate and shadow the Pakistani warships. Under no circumstances would the crew put the aircraft in any danger. Pakistan Navy warships were known to be fitted with Chinese-built surface-to-air missile systems capable of easily hitting an aircraft as big and slow as an Il-38.

With permission to depart on that blazing hot morning, Il-38 IN305 roared off the Dabolim tarmac and climbed out over the Arabian Sea, heading due east towards its target. On board the aircraft, all systems reported normal. While the 2 pilots gently eased the aircraft higher, the navigator and operators made preparations for the mission objectives, corroborating their map coordinates and tuning up their sensors, which they would switch on only when they were close to the 2 Pakistani warships.

On board the plane, the crew wore intercom headsets to communicate with each other and with ground control. Without the headsets, the Il-38's 4 Ivchencko AI-20M propeller engines would drown out every other sound in the cockpit and cabin. Even a pressurized, reinforced cabin could only keep so much of the sound out.

The crew checked in with ground control, indicating that they were now over the high seas. An hour had passed. The Il-38 was flying at 21,000 feet over deep, blue ocean, about 550 km off the coast of Goa.

The flight navigator signalled to the crew that they were now very close to their target, alerting the mission systems operators to get ready for the task. But as the crew prepared to descend slightly and begin its shadowing mission, they felt a small shudder pass through the 70-tonne aircraft.

It took Lt. Cdr Yadav a few seconds to realize what had happened. And it couldn't possibly have come at a worse time.

The Il-38 had 8 electric generators providing power to virtually every system in the aircraft. Like a pack of dominoes, each generator abruptly sputtered and died. In a matter of seconds, the entire aircraft was stripped of electrical power, and every single one of its systems was suddenly suspended right before they were needed the most. But it was not just the mission that had been jeopardized by stalled equipment. The very lives of those 8 men now hung in the balance.

Lt. Cdr Yadav instantly realized that he and his crew were sitting on a ticking time bomb. Oil fed into the engines was controlled electrically. So was the temperature of the oil. When the generators failed, the oil-cooler shutters got stuck and couldn't be moved since they were electro-mechanically operated. Lt. Cdr Yadav stared in horror at his dying cockpit instruments. He knew the ideal temperature of engine oil was 70 °C, with an emergency limit that stretched up to 100 °C. The gauges informed him that the oil temperature in the engines had crossed 150 °C. It was very simple. The 4 screaming engines could explode at any moment.

The aircraft now completely stripped of electrical power, and the cockpit and cabin lit by the red glow of a solitary emergency lamp, the pilots lost engine indicators, crucial data necessary to keep the aircraft safe and not turn into a ball of fire hurtling into the Arabian Sea.

With all instruments either dying or dead, Lt. Cdr Yadav took his partner's altimeter—his own was electrically powered and rendered useless.

'I had no navigation, no communication and limited control over my engines. I was doing my best not to touch the throttles. I knew that a single abrupt move could cause the engines to explode,' Lt. Cdr Yadav remembers.

The engines weren't the only things screaming on board IN305. With their intercom headsets now rendered useless, the men on board had to literally scream instructions at each other to be heard.

The crew of IN305 had few friends in the air that afternoon, and many adversaries. Starting with the generators, virtually everything turned against them in a chain reaction of terrifying circumstance. Every minute brought with it a fresh piece of reality that pushed the crew ever closer to giving up.

The Il-38, built for extended missions over sea, was full of fuel that morning—26 tonnes of aviation kerosene for the 4 hungry engines. The mission commanders at Goa had wanted to give the aircraft crew enough endurance and range to shadow the Pakistani warships adequately before returning to base. That mission, of course, had now gone straight out of the window.

The one thing Lt. Cdr Yadav knew he needed to do was descend so he could allow the navigator to get his bearings. But there was a dilemma. Diving to a lower altitude meant feeding the engines thicker air, which could put an additional strain on them and accelerate a disaster. Lt. Cdr Yadav waited, making a series of rapid calculations in the air—something he had learnt to do as a young Lieutenant.

Using the artificial horizon, a thankfully analogue instrument that tells pilots the orientation of their aircraft relative to the earth's horizon, Lt. Cdr Yadav took charge and gently steered IN305 towards India's west coast. In his lap was an emergency compass, the size of a rupee coin. Every other navigational aid was dead. With the gentlest possible touch he could muster for a large aircraft that had nothing gentle about it, Lt. Cdr Yadav pushed it into slow descent to about 7000 feet. As the aircraft lumbered uncertainly through the thicker air, Lt. Cdr Yadav looked out of his side window at the whining propellers. Then he looked over at the younger pilot next to him, and the other men standing behind him. He knew he had to do it.

'I briefed the crew in no uncertain terms. I told them that if the engines started exploding, we had no choice but to bail out. No thinking twice, on my command, we would jump out through the aircraft's belly hatch with our parachutes over the sea,' Lt. Cdr Yadav recalls.

Jumping out of an Il–38 was a daunting proposition. Unlike other aircraft that had rear ramps or exit doors that were clear of the aircraft's propellers, exiting from the belly hatch of IN305 would send the crew careening through space between the 2 inner propellers on both wings. A tiny bit of abrupt turbulence could send them head first into the equivalent of a giant blender. There would be nothing left of them.

His crew stared back at him. He was the most experienced man on the plane. His word would be final. If he ordered them to jump, that was precisely what they would do. There was no doubt in their minds. What they would do once they landed with their parachutes in the middle of the Arabian Sea was a problem they would tackle once they got to it.

The crew waited. Every one of the 8 men hoped they wouldn't need to exit the aircraft over the ocean.

A set of 4 emergency batteries on board the aircraft were in a precarious state after the generators perished, but still had a few wisps of voltage left. Lt. Cdr Yadav picked up his high-frequency radio transmitter and beamed out a message he hoped would be caught by airliners in the surrounding airspace. The emergency battery indicators suddenly plummeted to near zero, ruling out that last vestige of communication with the outside world. A minute later, the batteries were dead too. IN305 was now 100 per cent bereft of any direct or stored electrical power.

'We were flying by the seat of our pants and hoping to make it back. The worry for me was, if we make it back, how do we land?' says Lt. Cdr Yadav.

An Il-38 that had finished its mission and expended its heavy load of fuel was automatically rendered light enough to land safely. But without the ability to dump fuel off the aircraft, the crew of IN305 was locked into what was a veritable missile. Lt. Cdr Yadav wiped his brow with the back of his hands. About 450 km off the coast, he lowered the aircraft's wheels, a hydraulic process.

A fresh dilemma presented itself. And Lt. Cdr Yadav knew he had bare minutes to make a decision.

'With all our fuel on board that we couldn't get rid of, we were 18–20 tonnes heavier than permissible landing weight.

The wings had flaps to lower the approach speed, but wouldn't budge an inch without electrical power,' he says.

A safe landing speed on the Il-38 is no more than 200 kmph. Without his flaps to slow the aircraft down, IN305 would hurtle towards the runway at 300 kmph, an unacceptably high speed. As if a dangerously heavy, unacceptably fast and almost uncontrollable aircraft coming in for a landing was not enough for the crew to deal with, they also had to contend with the fact that it was pre-monsoon season. The runway at Goa's Dabolim airfield was soaked with rainwater.

'Everything was against us,' Lt. Cdr Yadav recalls. 'Everything.'

For a brief moment, he brought up the possibility of ditching in water—the act of landing the plane in a controlled manner on the sea surface, and exiting rapidly in an inflatable boat. The navigator was asked to see if he could spot some ships for IN305 to ditch close to, so chances of a rescue would be quicker. Lt. Cdr Yadav hated the idea from the moment it left his mouth. He swallowed and waited for a few more minutes, holding the aircraft steady. Then he saw it.

The west coast loomed into view through the weather haze. Glancing out of the cockpit, Lt. Cdr Yadav realized they were about to fly over Karwar in Karnataka, a town about 100 km south of Goa and home to one of the Indian Navy's largest warship bases.

IN305 was now over familiar territory. But that brought no comfort at all to the crew on board. If those engines exploded now, they were too low to bail out. And if they were forced to, it was land below them. Well past the point of no return, Lt. Cdr Yadav used both hands to mechanically steady the aircraft. Bereft of electrical power, the control column strained against every bit of pressure from the pilots. Every input was manual. The sweat dripped off the pilots' faces as they fought to control the aircraft in its descent. Then, Lt. Cdr Yadav heard one of the men behind him yell something through the noise. He turned around to see one of the crew brandishing a mobile phone.

'I grabbed the phone. By some stroke of luck, there was a mobile signal,' Lt. Cdr Yadav recalls. The pilot used the mobile to call the Dabolim air traffic control, screaming into the phone a description of IN305's situation and calling for full preparation on the ground for a possible disaster.

Using the mobile phone on board was against procedure, like with civil flights. But Lt. Cdr Yadav knew it was his one lifeline in the event his aircraft hurtled down the Dabolim tarmac with a collapsed landing gear and in a ball of angry flame.

Grimly aware of just how slim their chances were, Lt. Cdr Yadav and his crew steadied IN305 in its final seconds as the runway emerged through the haze. The engine roar increased as the temperature abruptly spiked, threatening to explode just a few feet off the ground.

'I knew the engines could burst at any time. I shouted to the crew that we had only one chance to make it to the runway. This was do or die,' he says.

There was no turning back now. The crew of IN305 held its breath as the aircraft slammed down on the runway. Just seconds before it did, a fresh realization dawned.

'I remembered we didn't have hydraulics to stop the aircraft. So how do we control the speed? I couldn't shut the engines down either, because they are electrically operated too,' Lt. Cdr Yadav says. But there was a little-known last resort that he recalled out of nowhere.

An emergency option had been designed into the Il-38 that allowed pilots to pneumatically 'feather' the engines, a method to decrease drag and stop the propellers. The option, though, was a terrible risk. The Il-38 has 4 engines, 2 on each wing. If even a single engine was feathered before the other, the aircraft could be thrown violently off its path and into a wreck of mangled metal and flame. Lt. Cdr Yadav looked to his flight engineer, directing him to pull the feathering levers at precisely the same moment.

'We could have handled a small bit of yaw (oscillating movement on the vertical axis). But anything extreme would have meant the end of all of us,' Lt. Cdr Yadav recalls.

The feathering worked. As soon as IN305 touched the tarmac, Lt. Cdr Yadav rushed to switch off first the outer 2 engines and then the inner ones. And with one final vestige of hydraulic power that had built up in the dead aircraft, he controlled it on the tarmac's centre line, wrestling to keep it from veering off the runway. Keeping IN305 on the runway was crucial not only to their own lives but the safety of other personnel and aircraft at Dabolim.

After an unforgiving hour in the air, the crew of the Il-38 met their first friend: Dabolim's unusually long runway, one of the longest in India, that gave the hurtling aircraft the extra distance it needed to slow down and stop without brakes.

'Around 9000 feet after touchdown, the aircraft came to a stop. I could see the valley beyond the runway come into view. Luckily we were able to stop the aircraft,' he recalls.

The roar of the engines had died down, giving way to the sound of emergency sirens as a medical crew pulled Lt. Cdr Yadav and his men from the Il-38. It was only then, Lt. Cdr Yadav says, that he fully grasped what had just happened.

'When you're in the air, all you're doing is solving the problem in front of you, first one, then the next. You've got no choice. You can't see the big picture, only the problem in front of you,' he says.

That evening, after a mandatory medical check, Lt. Cdr Yadav made his way to his residence at the base. His wife and 2 young children were at home. The family had made plans to dine at the Naval Officer's Institute that evening. But Lt. Cdr Yadav was exhausted. When he told his wife he'd prefer to relax at home, she wasn't surprised. Her husband often came home tired after long airborne missions.

It was only the next day, when she was asked by other families about the incident, that she learnt about what happened on board IN305.

The crew of IN305 had flown raw, with all possible odds stacked against it. Three years later, when Lt. Cdr Yadav travelled to Russia on military work, an old designer from the Ilyushin design house cornered him at an event to ask about the now-legendary flight. In rapt attention, the Russian listened, taking notes, amazed by how the Indian crew had saved themselves and their aircraft.

A crucial mission to shadow Pakistani warships remained unfinished on that day. But the lives of 8 naval aviators had been saved by what the Indian Navy described as Lt. Cdr Niteen Yadav's 'high degree of maturity, composure and a sense of resolve in the face of impending peril'. A year later, his leadership on board IN305 won him the Shaurya Chakra.

While IN305 was grounded for investigation into the failure that nearly doomed its crew, 4 days later, Lt. Cdr Yadav strapped into another Il-38 on an identical mission. This time, a pair of Pakistani warships was transiting the other way.

The mission is reported to have been a success.

*

10

'We Follow That Man. He Has Seen Death'

Captain Varun Singh

All that he remembers seeing, lying paralysed in that stretcher and slipping in and out of consciousness, was the blur of the Cheetah's rotor blades as the tiny helicopter dashed through the airspace over the Kashmir valley. His chest had been ripped open by a hundred metal splinters. A handful of them had punctured his right lung, smashed his sternum and broken several ribs. Another had pierced his heart. His right arm hung from its skin. His thick beard was matted with drying blood, with more forming scarlet rivulets down his front, staining his assault vest and combat fatigues.

Capt. Varun Singh was as physically close to death as a man could possibly get. The first coded transmission from the encounter site in J&K's Bandipora area had confirmed the Marine Commando officer's death. But his eyes had fluttered momentarily while a combat medic bandaged his blown-open chest. As he was carried into the waiting helicopter, an officer updated the status: the commando was alive, but he was as good as dead.

On 12 July 2016, 16 years after that encounter, Capt. Singh still felt the pain—not an ambiguous discomfort that came in waves, but a real, steady, unrelenting pain. It was agony that was overwhelmed, however, by the honour that came his way on that searingly hot day in Visakhapatnam. In the shadow of a 20-feet-high human hand hewn from stone clutching a jagged hunting knife, Capt. Singh was appointed CO of the country's first dedicated Marine Commandos (MARCOS) unit, INS Karna. Every man and woman at the ceremony that morning knew that Capt. Singh's salute didn't just carry with it a sense of achievement. It exuded the energy of a miracle, and a sense of disbelief that Capt. Singh would feel every moment of his life after that incident 16 years earlier.

It was snowing heavily across the Kashmir valley in December 1999, 5 months after the Kargil conflict, when Capt. Singh, then a Lieutenant, landed in Srinagar. The young MARCO had just been deployed to the Wular Lake, a vast wetland reservoir on the Jhelum, about an hour by road, west of the state capital.

As beautiful and placid as the lake was, flanked by gentle hills, it was a veritable highway for terrorists that had infiltrated from PoK. Using the lake, terrorists easily cut 40 km from their journey to Srinagar. Devoid of the nuisance of Army and police checkpoints along the roads, the lake proved a perfect conduit for them. It was for this reason that in 1996, the Indian government decided to deploy a squad of the Indian Navy's MARCOS to secure the lake and patrol its environs. The Valley got a fresh set of highly qualified guardians. The MARCOS, in turn, got access to what was then their first and only live operational battleground in the country.

Capt. Singh remembers thinking it was a perfect opportunity for operational exposure.

'The MARCOS spent hard hours training, but we simply didn't have a place to demonstrate our skills in the real world. This was it,' he says.

Like the squad they were replacing, Capt. Singh and the other MARCOS would be based in a pump house overlooking the massive lake that provided water to surrounding villages. The position provided a commanding vista from Bandipora to Sopore, including the famed Baba Shakur-ud-Din mosque, and prayers wafting through the crisp, clear air would become an accompaniment to the daily commando patrols on and around the lake.

On that day in December when he arrived, Capt. Singh stood staring out over the Wular Lake, its frozen surface glistening in the dying light. A memory that would visit him often inevitably came flooding back. His father, a retired naval sailor, had named him after the Hindu god of the sea because the infant's *janamkundli* (astrological birth chart) said he would die in water. A star at the Navy's diving school, he was trained to operate at depths of up to 180 feet. As he stared out over the lake, he told himself, and not for the first time, that if the lake claimed his life, it would all make perfect sense.

Capt. Singh hadn't always been drawn to the water. His teenage dream was much more specific. He simply wanted to kill terrorists. Growing up certain that he would join the military, he entered the hallowed campus of the NDA in 1989, fully prepared to become an Army officer. It was his father's coaxing that persuaded the young cadet to switch to the Navy.

'Like a good son, and to honour my father, I joined the Navy. It was an enormous thing for him. As a sailor in the Navy, to see his own son become an officer couldn't be described in words,' Capt. Singh remembers.

Honouring his father was one thing. But Capt. Singh knew that his dream of combating terrorists had receded considerably. The Navy would take him out to sea and far from the dangers he truly wanted to fight. As he had expected, soon after he was commissioned into naval service, Capt. Singh was posted as a navigating officer on a warship.

The years rolled on. Capt. Singh got married, and in 1997 his first child, Shivani, was born. As he held her in the hospital, he had the thought all fathers do—that he had never seen anything as beautiful. Watching his wife and newborn girl sleep that night, the sense of well-being and joy that pervaded his mind was also a reminder—that his long-cherished dream of going into combat was no longer a decision he could take easily. It was his wife, Reena, aware of what was bothering her husband, who pushed him to try.

Soon after, Capt. Singh enlisted as a trainee combat diver—his first voluntary brush with the element that he had been told would one day kill him.

'Everyone told me, "Why now? You've got a family, Varun. Life at MARCOS is disturbing and dangerous. You know you will get in harm's way." They were right,' Capt. Singh says. 'In which other profession do you knowingly get into harm's way?'

And there was no shortage of harm hidden in the expansive view Capt. Singh got that evening in December 1999 as he gazed out over the lake. In only 5 months, the young commando would experience first-hand the most dangerous kind of harm the restive valley could offer.

The MARCOS at Wular Lake operated under the Army's 15 Rashtriya Rifles, its headquarters in Watlab on the lake's banks. The unit's CO, Col. S.K. Jha from the Gorkha Rifles, loved the MARCOS under his charge, and used his discretion to constantly send the young naval commandos on patrolling duties beyond their designated areas of responsibility just to give them exposure and experience.

'He had a great deal of faith in us. And we promised him our undying support no matter what the requirement.'

In the winter months at the dawn of the new millennium, there was little the men around Wular needed to face beyond the rigours of daily patrols in hostile weather. By April, the heavy snow would melt and make way once again for infiltration across the LoC to the west. By the end of April, the harsh realities of Kashmir would have flooded back full force. Which is why, when Capt. Singh and 7 of his MARCOS mates were summoned at 0200 hours on 2 May to investigate an intelligence tip-off about possible terrorist bunkers on a mountainside, they sprung up ever-ready.

The 8 MARCOS joined about 100 Army men to search for hideouts across 3 mountains. In single-line formation, the men trekked up to each peak, then rolled downhill, checking each crevasse for possible weapon caches or hidden stores, or perhaps even terrorists.

The men found nothing that night. By lunchtime, they had assembled at a secure location in the foothills when they were alerted by radio static. A crackling voice announced that an encounter was on in Bandipora, not far from Wular Lake, in a village called Puttushahi. Three Al-Badr terrorists were reported to be hiding there inside a two-storey house.

Capt. Singh stopped eating and walked over to the Army team leader, asking him if he and the MARCOS could dash to the encounter site. This was the first terror encounter since the snow had melted. Capt. Singh remembers feeling every fibre of his most enduring dream explode at once. Here was a chance to fight actual terrorists. There was another reason he

wanted to get to the site without any further delay. The voice
on the radio belonged to an Army officer who happened to
be Capt. Singh's course mate at the Academy. And he was
leading the antiterror hunt.

In minutes, Capt. Singh and his buddy, Lead Mechanical
Engineer Vijay Singh Rawat, were cleared to proceed to the
location by road. The 2 men arrived at the encounter site in
45 minutes. Right outside the village, they noticed that the
15 Rashtriya Rifles CO and their boss, Col. Jha was standing
near his vehicle. The actual encounter was taking place a short
distance inside the village. The Colonel had decided to stand
back so there was no pressure on the team leading the hunt. As
soon as he saw the 2 MARCOS emerge from their vehicle, he
ordered them to proceed to the spot inside the village.

'He knew we had knowledge of explosives and demolition
stores. He had faith in us. We ran into the village without a
second thought,' Capt. Singh remembers. The 2 commandos
received a quick briefing on the situation from the officer
heading the operation as they took positions.

Surrounded by a 6-feet boundary wall, the house in which
the terrorists were hiding was about 18 feet high, with an attic.
The compound, which also had a small cowshed in the rear, was
surrounded on all sides by 2-storey houses just feet apart, each one
cleared of all residents for their safety. The closest house was to
the right. Capt. Singh and Vijay sprinted into the house and up
the stairs to a second-floor balcony that afforded a direct vantage
point overlooking the house the terrorists were hiding in.

In radio contact with the Army team leader outside the
compound, Capt. Singh relayed that he was providing fire
cover as his buddy tossed a pair of grenades into the house on
2 separate floors. The ensuing chaos, he thought, would allow

him and Vijay to spot the terrorists and take them out with precision shots. The grenades were tossed, exploding loudly in the main house. But nothing moved. The men heard nothing but the sizzle of air after explosion.

Vijay had just used both his grenades. Just as Capt. Singh was handing over 1 of his own 2 grenades (MARCOS carry 2 grenades in their vest pouch), the Army team leader radioed in asking for help with an RDX explosive he wanted to plant in the house as the next course of action. Capt. Singh placed the grenade back in his pouch and went down to the staging area to inspect the bomb that had just been built. A resident of the area was enlisted to walk into the compound and plant the bomb quietly on the ground floor of the house.

'The assault vest I was wearing was given to me by my brother-in-law, an Army officer posted in Srinagar. He had been presented the assault vest by a foreign Army when he was deployed there,' Capt. Singh notes. The vest was made of cotton, not the polyester assault vests that were standard issue to the MARCOS at the time. That difference in material would play a shattering role in the minutes that followed.

The RDX explosive was placed and detonated, causing a clean implosion of the house the terrorists were hiding in. Not a single brick blew outward. The entire structure collapsed on itself in a cloud of dust. Shock waves from the blast shattered every windowpane in a 20-metre radius. Capt. Singh and Vijay emerged from behind a mattress they had propped up for protection. Stepping back out into the balcony, they tried to peer through the dust. When it had settled, they noticed the body of 1 terrorist hanging from the first floor. The other 2 weren't visible. Capt. Singh looked at his watch. It was 1700 hours. The sun would set soon and that would give the other terrorists a chance to escape, if not attempt an attack. The operation needed to be completed before sundown at all costs.

The MARCOS could not spray bullets into the dust even if they wanted to.

'We're trained for *ek goli, ek dushman* (one bullet, one enemy). We conserve ammunition and use as little of it as necessary to do the job. We couldn't fire at something we couldn't see,' Capt. Singh remembers.

Leaving Vijay on the second floor, Capt. Singh went downstairs and jumped over the wall and into the main compound of the detonated house. Three Army soldiers came in from the front. Capt. Singh stepped over the smouldering debris, reaching up to check the pulse of the terrorist who hung down from the first floor. He was dead. Capt. Singh removed the AK-47 the terrorist had with him and confirmed over the radio that the first terrorist was confirmed dead.

But in the next few seconds, there would be a momentary, but fatal loss of judgement, a split-second collapse of command and control, and of the unwavering focus required during such operations.

'Since this was our first kill after the snows had melted, there was a sudden outburst of excitement by the jawans. It was a matter of great pride. There was some chaos. About 14 soldiers suddenly scaled the wall into the compound to see the dead terrorist. This was a big mistake,' Capt. Singh remembers. 'We didn't know where the other 2 terrorists were. In those few seconds of excitement, this was overlooked.'

Amidst the impromptu outburst of cheer, 1 soldier from the group ventured towards the cowshed in the backyard. Standard operating procedure would have dictated that the men destroy the cowshed with a rocket or bomb, since there was no way to safely inspect its interior. In a fatal lapse of training, Army jawan Reddy opened the cowshed door and peered within.

The clatter of Kalashnikov fire rang out as the 2 terrorists hiding inside the cowshed cut down the jawan in a hail of bullets. Hearing the fire, most of the jawans in the compound ran for the boundary wall, hoping to scale it and take up safe positions to return fire.

The 2 terrorists crept slowly out of the cowshed, their weapons at the ready. It was the first time that Capt. Singh had seen a live terrorist in an operation. In that moment, ironically, he now recalls, his dream of killing terrorists could not have been further from his mind. This was no dream. This was an operation, and he had little over a second to act. As he was raising his weapon to take aim, he noticed that 2 Army jawans had frozen in their tracks, unable to move, stunned by what had just happened. Wasting no time, Capt. Singh leapt towards them, forcing them to the ground, landing on his knees, his AK-47 pointed directly at the 2 terrorists.

'The 2 terrorists were slowly stepping past me about 5 metres in front. But they hadn't noticed me,' Capt. Singh recalls. His weapon in single-fire mode to conserve ammunition, the commando opened fire. The next few seconds are a sequence Capt. Singh remembers in devastating slow motion.

'As they were stepping past me, their right profile was facing me. I fired 7 rounds at them. That's all I remember. When one of the terrorists fell, his finger was on his rifle's trigger and his right hand was stretched out towards me,' Capt. Singh recalls. 'As he fell, his finger squeezed the trigger, spraying bullets all over. One of those bullets came straight at me and smashed into the grenade I had placed in my pouch a short while ago.'

If the grenade had exploded, there would have been nothing left of Capt. Singh and the soldiers around him. But that 7.62–mm bullet penetrated the upper part of the grenade, splintering it but not detonating it.

'The bullet had hit the shoulder of the grenade, the top part of the device that houses the closing cap and striker. The pin was out. Had I been wearing my usual polyester assault vest, which was more spacious, the lever would have been activated and the grenade would have detonated, tearing me to pieces. The cotton vest I was wearing was so tightly packed that despite the grenade lever breaking free, it was seized by the vest.'

If a grenade detonation would have indisputably killed the young commando, the splintering grievously injured him. Just as it registered that he had fulfilled his boyhood dream of killing terrorists in live combat, Capt. Singh felt himself slip into darkness. Consciousness came in short, blood-tinged flourishes.

As he lay there, his body broken, the Army team leader ordered Vijay, who was still in the balcony of the neighbouring house, to destroy the cowshed with a rocket. Respectfully, the marine commando protested.

'*Mere saab andar hain, mujhe unko leke aana hai* (Sir is still in there. I need to get him out),' he said over the radio. The Army team leader was furious by this time, ordering the commando to blast the cowshed and get out of there as quickly as possible. But Vijay stood his ground. Exasperated, the Army officer gave the commando 5 minutes to retrieve what both believed was Capt. Singh's body.

Vijay scaled the wall, stepping slowly towards the fallen figure of his MARCOS partner. First, he fired shots at the terrorists' bodies to make sure they were dead. Then he reached down towards the officer's bloodied torso and took hold of the splintered grenade. Holding it with both hands and pressing the lever, he removed the detonator and disarmed the device. Then he checked Capt. Singh's wrist, feeling the faintest pulse. He shouted to the men positioned outside to begin the emergency evacuation.

Bandaged and strapped into an Army vehicle, Capt. Singh was dashed to the Army's 15 Rashtriya Rifles headquarters 12 km away. Those with him, including Vijay, knew it would only be minutes before the officer bled out and died. They had seen men die from far slighter injuries in combat. This commando's chest had been burst open, the cavity within visible, every organ impaled by unforgiving little shards of metal. There was literally no hope as the jeep carrying Capt. Singh accelerated down the highway towards Watlab.

An Army Cheetah helicopter at Sharifabad was alerted for a possible emergency airlift from Watlab to Srinagar, and the pilots had been briefed about who they would be carrying and the condition he was in. But they were finally told to stand down because the chopper could not take off in the dying light. With a 5-minute twilight window, the 2 pilots, who had met Capt. Singh at a unit party just 2 days before, decided to make a break for it. Cleared for the mission, they sprinted to their helicopter and were airborne in minutes, dashing straight to Watlab.

'I only recall what happened in flashes. I remember being shoved into the helicopter. Then, it was landing at 92 Base Hospital in Srinagar where they had a helipad. The third was being pushed into the X-Ray machine there,' Capt. Singh recalls.

The Army doctor in charge at the Srinagar base hospital, Lt. Col. Ravi Kale, took less than a minute to decide that it wasn't worth trying to save Capt. Singh—he was too far gone to be retrieved. To this day, Kale says he has no idea what stopped him from walking out of that ward. Something compelled him to stay and try to save the dying commando in the gurney in front of him.

'There was a Christmas tree of shrapnel inside my chest. Splinters were scattered everywhere. I was pushed into the operating theatre at 2030 hours. I was finally stitched up the next day. I didn't regain consciousness until the evening of 4 May,' Capt. Singh recalls.

Still extremely critical, the commando couldn't speak. He remained on forced ventilation since an entire lobe of his left lung needed to be amputated. By 5 May, he was fully conscious and able to speak very slowly.

'When the doctor came, he just pressed my hand and went away. He later told me that no matter how much anaesthesia I was under, whenever he asked me how I was, my thumbs were up. He couldn't explain it,' Capt. Singh says. 'And neither could I.'

In July 2016, 16 years after he decided to operate on a commando against his best judgement, Lt. Col. Ravi Kale (who retired as Major General) would write a letter to Capt. Singh. The man he had helped save was about to take over as CO of India's first MARCOS unit. He wrote:

What kept you alive was your cool, not crying or taking big gasps when your chest was shattered. This kept the rest of the lung okay. If you had shouted with pain, the chest cavity outside of the lung would have got filled with air and compressed your lung and heart.

You also had a large grenade splinter going from your right
chest to embed in the left second intercostal space. Again
if you had jumped with pain this would have torn your
heart or the aorta with no chance of recovery. So friend,
it was your cool which saved you. I was just a small cog
in that wheel.

Take care.

(Private communication, 12 July 2016)

After the encounter, Capt. Singh spent a week in intensive care
in Srinagar before being declared stable enough to be airlifted
to Delhi. In his own words, he spent the next 2 years begging
nurses for morphine injections at both Delhi's Base Hospital
and the Army R&R Hospital. His arm shattered, doctors
had installed external fixators with metal rods bypassing the
smashed humerus.

Teams of doctors fought valiantly to match the courage
and inexplicable cheer the young commando displayed in his
hospital bed. But even they knew that Capt. Singh would
never fight again. His injuries had taken an enormous toll.
Eventually, he was medically downgraded by the Navy—a
process that keeps officers and soldiers in service, but for
duties other than combat. If Capt. Singh was not already
dealing with the agony of his injuries, he now had to stomach
never being able to go to war again. It was a devastating blow
for any soldier, especially a young commando.

When his family visited him for the first time at the R&R
Hospital in Delhi days after the encounter, Capt. Singh's
3-year-old daughter, Shivani, clambered up on the bed
her father lay in. Holding his forearm, she said, 'My daddy
strongest!' It was one of only a handful of occasions in which
Capt. Singh allowed himself to weep.

'My wife is an amazing human being. She preserved the sanity of our family when everyone thought I was dead. She rushed from Rewa in Madhya Pradesh to neighbouring Satna where my parents lived. They had nearly collapsed. She kept the family together. I can never thank her enough,' Capt. Singh says.

It soon became clear to Capt. Singh's superiors that the wounded commando would not be confined to a hospital bed. His right arm nearly useless, he trained himself to play squash and fire a weapon with his left hand. In and out of hospital for cardio-thoracic therapy that would never really stop, Capt. Singh proved instrumental in setting up a training wing for the MARCOS in April 2002 in Mumbai. Three months later, his son, Sarthak, was born.

'Sarthak wants to join the MARCOS. He's still young, but he seems determined,' Capt. Singh says. His daughter, Shivani, who would become more aware of her father's mission as she grew up, would be filled with a ferocious anger that her parents fought to calm.

'She started hating Pakistan. She was at a very impressionable age. She could never understand why it was I who got hurt for a national cause. She had lots of personal questions. Some of those we could answer. Many of them we couldn't,' he says.

For Capt. Singh, being given command of INS Karna was an unspeakable privilege, one he still cannot fully personally fathom.

'The men I train are my brothers. Unless we train together, we cannot fight together,' Capt. Singh says.

'When they look at me, I know what they're thinking: This man has seen death. We will follow him to our deaths if necessary.'

*

11

'This Is India's Honour. We Cannot Fail'

Commander Milind Mohan Mokashi

Port of Aden, Yemen
12 October 2000

It was an attack that shook the world—one that may have grown deeper roots in wider public memory had it not been overshadowed less than a year later by the 9/11 attacks. In more ways than one, what happened that day was a shattering trailer to 9/11. As the United States Navy's formidably armed Arleigh Burke-class destroyer *USS Cole* docked for fuel at the harbour, a small fibreglass dinghy approached the 500-feet, 6000-tonne warship. On board, its sailors were lining up for an early lunch.

The boat, steered by 2 Al-Qaeda suicide bombers, carried a 300-kg explosive device that the 2 men detonated once they reached the *Cole*. The explosion killed 17 American sailors and tore a frightening 40-feet hole in the ship's hull—a ship that no force in its right mind would mess with. Images of the fearsome vessel being helplessly towed away for repairs scarred a nation that depended on battle groups featuring ships like the *USS Cole* to project power far from American shores, and support the most difficult foreign military operations.

Two decades later, the bombing of *USS Cole* is a relatively vague memory, if it is remembered at all. In navies around the world, however, it remains a terrifying touchstone—a reminder of just how vulnerable menacing ships of war really

are in the face of a threat that doesn't play by any rules. If a
ship armed to the teeth with enough weapons to level a small
town could be paralysed by 2 men in a tiny motor boat, then
the game had changed forever.

Arabian Sea
31 March 2015

It was the inevitable first thought that crashed through
Cdr Mokashi's mind that morning, as he captained the Indian
Navy's patrolling warship *INS Sumitra* on an escort mission for
vulnerable merchant vessels transiting through pirate-infested
waters and out into the Arabian Sea. Cdr Mokashi and the
150 men under his command had arrived in the Gulf of Aden
just 20 days before and had already ensured the safety of a
continuous line-up of freight ships making their way through
the internationally recognized trade corridor.

That day in March, Cdr Mokashi had just received orders
to immediately pull out from the escort mission, turn his ship
around and steam at full speed towards Yemen's Port of Aden.
His instructions were clear: he was to reach Aden and hold
position 20 miles out at sea.

Like all good military men, Cdr Mokashi was fully prepared.
When his ship had arrived in the Gulf earlier that month, the
situation in Yemen had already turned volatile, with a violent
civil war drawing air strikes from a Saudi-led coalition. On
26 March, as Cdr Mokashi and his men were sailing towards
the tiny African nation of Djibouti across the Gulf of Aden to
refuel and stock up on supplies (termed 'operational turnaround'
in military parlance), the crew was told that Saudi Air Force
Typhoon jets had begun bombing runs in Yemen. Cdr Mokashi
was focused on his anti-piracy mission. But he also knew that

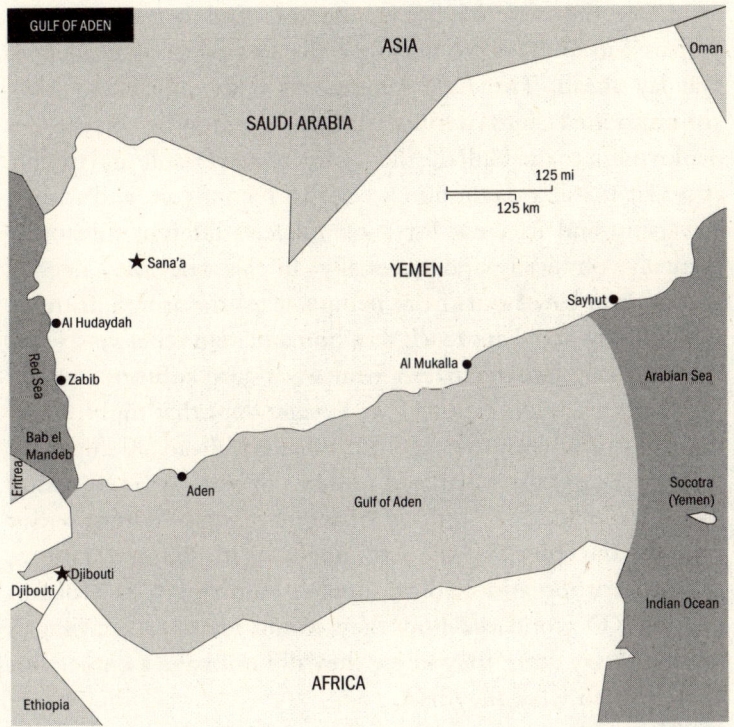

his ship and crew were in the best possible position to dash to Yemen to rescue the huge numbers of Indian citizens who worked there, if things got worse.

On the night of 29 March, as a fully fuelled *INS Sumitra* sailed out of Djibouti on its escort mission, Cdr Mokashi summoned the ship's executive officer (XO). While the 2 men stood on the ship's deck watching the sun sink beyond the horizon, the skipper spoke his mind.

'We need to do our homework. We need to be ready in all respects for Yemen,' he said.

Over the next 36 hours, the ship and its crew quietly prepared itself. This would prove the first of many challenges that lay ahead. The *INS Sumitra* was practically brand new, commissioned into service barely 6 months before its deployment to the Gulf of Aden. Built to be versatile and nimble at sea, it is still a daunting exercise to reconfigure and realign a warship and its crew for a completely different mission— mentally, physically and materially. In this case, they needed to switch in bare hours from being a fearsome armed platform that nobody dared approach to a humanitarian relief vessel that would rescue Indian citizens from a war-torn country.

Very few men on the *INS Sumitra* slept that night. Over the hours, the crew worked out what lay ahead. A ship built to carry a carefully calculated number of battle-trained young men would soon see a flood of women and children, senior citizens and possibly the wounded. With limited rations, accommodation and medical supplies on board, Cdr Mokashi and his XO wondered how they would manage. But like all good military men, they knew they did not have a choice but to fight with what they had.

When the orders finally came on 31 March, there was a quiet lack of surprise on board the *INS Sumitra*. It took no more than 10 minutes for the ship to set course for Aden. With no charts and maps, essential to safely and easily sail into the port, Cdr Mokashi decided to wait until he had brought his warship to waters off Aden. Then he would deal with the very serious predicament of making his way into an unfriendly port without the essential charts for such sailing.

Now mentally prepared for the mission that lay ahead, Cdr Mokashi decided to use the few hours he had before arrival at Aden to sharpen all preparations. Speaking to his crew over the ship's intercom system, he briefed them. The crew were split up into units with specific tasks, including baggage screening, personnel screening and documentation.

But that would come later. Inevitably, the chief concern was security. Every single man on the *INS Sumitra* knew about the *USS Cole*. And now they knew they were headed to precisely the same place where that American ship had been ambushed 15 years before. Worse, they were headed there at a time when the country was mired in open hostilities. Frayed nerves became apparent on board the Indian vessel as it sliced through the waters of the gulf towards Aden.

The team of 8 MARCOS on the *INS Sumitra* was prepared with weapons and boats. Armed and ready for antiterror operations, this crack team would 'sanitize' the area surrounding the ship and provide a formidable layer of security. Other protective measures included a group of armed sailors who prepared themselves for possible ground combat. Six months old and on its first humanitarian mission, the *INS Sumitra* had just stepped into a war zone.

On the afternoon of 31 March, Aden loomed into view, the forbidding jagged rim of the ancient dormant volcano of Kraytar visible on the horizon. Thirty miles out and approaching, Cdr Mokashi and the rest of *INS Sumitra*'s crew heard the first sounds of war. The low thud of shells and the reverberating shatter of airdropped weapons wafted across the water. The old port in the distance was swathed in wisps of smoke.

About 22 km out, the *INS Sumitra* stopped. The port was closed for normal operations. And without clearance from Saudi Arabia, which now controlled airspace over the port and whose jets were carrying out uninterrupted strikes, moving the ship any closer could be catastrophic. Looking through binoculars at what had become one of the most dangerous ports in the world, Cdr Mokashi waited. The Saudis were taking their own time to respond to his request. With no eyes on the ground and very little information, the CO signalled his headquarters.

India's diplomats in Yemen, key facilitators of information and data crucial to a humanitarian mission, were all in the country's capital, Sana'a, over 400 km away. The only representative they had in Aden was the principal of an Indian school there, who doubled up as honorary consul. Cdr Mokashi dialled him, praying the call would connect.

It did, opening the first and very welcome conversation with an Indian on the ground and amidst the hostilities in Aden. Things would hopefully move faster now, Cdr Mokashi thought, as he ordered his crew to prepare a final approach into the port. But the Indian at the other end of the line was tense. 'Anything beyond 1730 hours is dangerous. We cannot be on the jetty beyond that time. It is extremely risky,' he told the Indian warship's Captain.

It was already 1600 hours. And there were still no orders to move closer. Cdr Mokashi waited, wondering if he should move in anyway. But a quick calculation in his mind told him it was futile.

The ship's crew watched as the sky began to darken at dusk. A planned daylight rescue operation had just been sunk. With the spreading darkness, the crew of INS Sumitra knew that their mission had just become infinitely more dangerous— for the ship and for them.

In that darkness, as flashes of light erupted from Aden 14 miles away, Cdr Mokashi ordered his MARCOS to lower their vanguard boat and set out for the port. Led by an officer and armed with assault rifles and sidearms, the 8 commandos sailed through the darkness of the ancient harbour to 'sanitize' the ship's intended path. Peering through the darkness and careful not to give themselves away, the 8 men scanned the area for suspicious boats that may have entered after sunset. Terror groups or rebels would have anticipated the arrival of foreign warships and may have wanted to spring another *USS Cole*-like incident, or worse.

The waters were quiet, almost lake-like placid, their surface disturbed only gently by the distant rumbling. Reaching the barely lit port, the commandos docked their boat and stepped quietly on to land, their weapons cocked in every direction. The team leader signalled back to the ship that the area was tentatively secure, but they needed to be prepared for any eventuality.

As dusk gave way to night, the *INS Sumitra* was finally ordered to proceed towards the port. The deserted city was shrouded in a darkness that was lit up intermittently by gunfire and flares. Guided by a rudimentary map, Cdr Mokashi's crew downloaded a steady stream of inputs from the Indian Navy Headquarters in Delhi and the Western Naval Command in Mumbai.

Four miles off the coast of Aden, Cdr Mokashi got a call from his MARCOS team leader at the port. The commandos had found groups of Indian citizens sheltered inside shipping containers. This was good news. But the commando wasn't finished. The numbers were more than double of what the crew of *INS Sumitra* had prepared for or expected. There were 350 people waiting to be rescued. And not one of them could be left behind.

At 1945 hours, *INS Sumitra* docked at the deserted port of Aden. A sole Yemeni individual at the harbour, possibly a port official, approached the ship as it docked. Aware of why the Indian ship had come, he issued an ominous instruction.

'You have 45 minutes to dock fully, get your people and leave,' he said.

Cdr Mokashi and his XO glanced at each other. *45 minutes.* Both knew that demand was beyond the realms of the possible. Even if the crew screened and embarked 1 Indian per minute, it would take 6 hours to board the 350. Of course, it would take far longer than a minute per person. Apart from regular screening for contraband, the ship's crew needed to deal with another threat—the possibility of a suicide bomber sneaking on board with the crowd. If the pressure at hand was not enough, the Navy Headquarters in Delhi called Cdr Mokashi to board the Indians and get out of Aden as soon as possible.

Wasting not a moment more, the crew of *INS Sumitra* began their work. Unused to the logistics of humanitarian screening, the crew started tentatively, gradually increasing the tempo and getting into the rhythm of their mission. Minutes before they began, their Captain had a word of advice for them.

'This is a mission of honour. These are our people. They've just been through a lot of trauma. They've had to leave their homes and belongings. Many of them will be women, children and the elderly. We have to deliver them out of this place in a very short time. But let's be as gentle with them as we can. No high-handedness. No harshness. Be firm, but not impolite. Let us imagine what they have been through,' Cdr Mokashi said to his men, as he formally called for the embarkation process to begin.

Around the berthed *INS Sumitra*, there were 2 layers of security. The 8 armed MARCOS who had arrived earlier formed a wide outer cordon ready to engage with any threat that emerged from the darkness. An inner layer of armed naval sailors formed a QRT near the embarkation point. On board the ship itself, 2 MARCOS sat perched on one of the ship's masts. While 1 kept a watch with his binoculars, the other sat with his Israeli Galil 7.62-mm sniper rifle, his head bent to the telescopic sight, scanning the port for anything he potentially needed to take down.

The boarding finally began, with the Indian citizens being scanned, photographed and boarded. Their luggage was scanned separately and loaded into a different part of the ship that could be accessed once boarding was complete.

The Indian Navy does not yet deploy women on board warships. So to screen the many women among the 350 people, the crew of INS *Sumitra* enlisted senior women from the group who appeared relatively calm. Several of the Indians boarding needed immediate medical assistance for heart ailments and diabetes. Others needed to be calmed down after the trauma of escaping a bombed city. Many hadn't eaten in over 24 hours.

The crew had already vacated their cabins and quarters for the rescued Indians. Women, young children and senior citizens were given priority accommodation in the officers' and sailors' mess decks, carefully chosen so they did not need to climb the steep ladders between decks that make up the inside of any big warship.

A galley equipped to cook for 150 men was now fired up beyond its capacity to churn out meals for 500 persons. No sailor ate a morsel until every one of their 350 guests had been served and made comfortable.

While the journey from Aden to Djibouti was only 7 hours, Cdr Mokashi knew the Indians under his care needed rest and access to facilities on board his ship. Strictly functional and based on community existence between brother sailors and officers, privacy on board the warship needed to be taken care of with sensitivity. The ship's crew, all 150 of them, made it a point to make the women on board as comfortable as possible, staying out of their way and only appearing when someone needed help.

After the last Indian had been boarded, the MARCOS and naval sailors finally relaxed from their ready-to-fire positions and got back aboard their ship. As the ship pulled away from its berth, several people remained at Aden awaiting rescue. But the *INS Sumitra* had permission to board only Indians—a restriction that would change a few days later.

Through that night, *INS Sumitra* sailed at full speed across the Gulf of Aden. After a few hours of adjustment, exhausted from worry and exertion, most of the rescued Indians fell asleep. Safe in the confines of the 2200-tonne warship from home, they were likely too tired to dream that night.

The *INS Sumitra* arrived in Djibouti at 0700 hours on 1 April. The arrival and disembarkation of 350 rescued Indians became a broadly televised event in the Indian media, with the Minister of State for External Affairs, Gen. V.K. Singh waiting at the port and famously addressing the crowd. Pictures of the ship, the tearfully relieved citizens and the smiling sailors would appear on many front pages. Cdr Mokashi finally afforded himself the luxury to exhale. He knew that his work had just begun.

The crew would have only a few hours to rest as their ship was fuelled and restocked—this time with more suitable equipment, including disposable plates and spoons. That afternoon, they received fresh orders to make a dash for a different part of Yemen—one that was literally in flames.

On 1 April, a dairy factory owned by Yemen's massive Thabet Brothers business conglomerate in the Red Sea port city of Al Hudaydah had been bombed, killing over 30 workers. Given the large number of Indian nationals living in Al Hudaydah and known to have been employed at that factory, it seemed likely that Indian blood had been spilt in that attack. It was towards this port town that Cdr Mokashi and his men accelerated their ship.

The approach to Al Hudaydah would prove even more challenging than Aden. While the latter could be navigated with some care, there was a standing advisory to all ships approaching Al Hudaydah to arrive and depart in daylight. With very limited room and once again without navigational charts, the port would prove a dangerous gamble. But Cdr Mokashi and his men had their orders. As they sailed slowly down the narrow approach channel to the familiar roar of fighter jets overhead, the crew of *INS Sumitra* did not know that Al Hudaydah had already fallen into the hands of Houthi rebels. This meant dealing directly with an armed rebel group in order to secure safe passage for Indian nationals.

At noon on 2 April, the Indian ship docked at Al Hudaydah. On board, it was action stations. Every armed member, including the MARCOS, was in a state of maximum alert with every weapon primed and ready. It was daylight, but the crew of *INS Sumitra* had just sailed into what was easily one of the most dangerous zones in the world that day.

Houthi rebel officials who had taken over the port communicated directly with the ship as it docked, ordering crew to embark Indian citizens and leave within 4 hours. As the sound of bombing and air strikes got closer, the Indians waiting at the port were rapidly boarded. With the practice at Aden 2 nights ago, the ship's crew breezed through the procedures. But an unexpected challenge arose—an Indian woman stepped up for the screening with her 7 children. And while she had the relevant documents, none of her children did.

Cdr Mokashi knew he did not have the time to conduct a background check. Nor could he bring the woman on board without her children. With strict orders not to board individuals who couldn't be screened completely, it came down to the young officer's discretion to allow the family to board.

When the ship was finally ready to leave with 317 Indians, rebel authorities would not give them permission to cast off. From India came the constant, and alarming, input urging the ship to leave as quickly as possible. Every conceivable threat to the ship had been imagined. Cdr Mokashi needed to make a decision. And once again, the day had given way to night.

Once again in pitch darkness, *INS Sumitra* inched its way out of Al Hudaydah's treacherous harbour. Cdr Mokashi looked out from the bridge as his ship felt its way out from another bombed and broken town. The moon's phase ensured that there was no natural light to give the ship's crew a hand. They sailed out blind.

Clipping at 25 knots (about 46 kmph) through the darkness, the ship's crew needed to ensure that the many children on board remained safe as they leaned over the ship's railings. As the adults slept, the crew remembers the children who fanned out, fascinated by the vessel and the new experience that had been thrust upon them. This passage was a long one. After hours of play, the children were escorted back to their parents' cabins, where they proceeded to fall asleep. Twenty hours later, on the night of 3 April, the ship once again arrived in Djibouti.

After Aden and Al Hudaydah, it was the third Yemeni port that would prove the trickiest of the lot. The crew had just been ordered to steam at full speed to Al Mukalla, 480 km from Aden on the Arabian Sea. The capital city of Yemen's Hadhramaut region, Al Mukalla had just witnessed a terrifying series of attacks the day before.

On 2 April, terrorists from Al Qaeda in the Arab Peninsula (AQAP) had stormed several buildings in Al Mukalla, including a large prison, freeing hundreds of inmates (including 2 AQAP commanders) and looting millions from the central bank. A street war between the AQAP, the Hadhramaut Tribal Alliance and the Yemen Army was raging, with at least 15 people having been killed over 2 days. The AQAP had seized full control of the port town. As the *INS Sumitra* zipped across the gulf in response to a rescue call for 250 terrified Indian nationals there, the infamous Battle of Mukalla had just begun.

Sailing at full speed for nearly 40 hours, Cdr Mokashi and the crew of *INS Sumitra* arrived off the coast of Al Mukalla on the morning of 5 April. But there was no talk of docking, not even the suggestion of it. The port was closed and completely under the control of Al Qaeda terrorist affiliates, groups not disconnected from those who had bombed the *USS Cole*. Cdr Mokashi rounded up the ship's commandos and armed teams for a quick meeting. The men agreed. Getting any closer to Al Mukalla would heighten the possibility of a direct attack on the *INS Sumitra*.

With almost no intelligence about the capabilities of the terrorists who had seized control of the city and port, Cdr Mokashi would take no chances. The ship then received an input about another port town, Ash Shihr, about 63 km further north, with a PetroMasila oil terminal not far from the border with Oman. The company employed hundreds of Indians who also awaited rescue.

Unlike Al Hudaydah and Aden, where Cdr Mokashi and his crew managed with rudimentary maps to help them inch into unfriendly ports, they had nothing on Ash Shihr. Not a port for regular freight operations, it did not have a jetty, so berthing the ship was ruled out. *INS Sumitra* would therefore have to hold position a short distance off the coast. And in an extremely risky exercise, stranded Indians would have to be brought to the ship 1 boatload at a time.

The MARCOS team leader on board had his hands full that afternoon. His squad noticed a number of small boats in the sea not far from their ship as it arrived off Ash Shihr. Having done their homework about the area, they were aware that in 2002 in this precise area, an oil tanker, *MV Limburg*, chartered by Petronas, was attacked by Al Qaeda suicide bombers in a dinghy, killing 1 man from the crew and spilling a huge amount of crude oil into the gulf.

The Indian warship's crew knew that stopping or slowing down off Yemen's Hadhramaut region was possibly the most risky thing they could do in what was the world's most dangerous zone at the time. By now, the Indian Navy's rescue operations were well known around the world. Clear intelligence inputs warned the ship's crew that terror groups were almost definitely expecting them. There was no clarity about which of the many boats were friendly and which could possibly be carrying enough explosives to sink *INS Sumitra*, or at least paralyse her a very long way from home and turn her into a sitting duck for a bigger attack.

Over the next few hours, in one of the tensest humanitarian operations ever, 2 boats from *INS Sumitra* and 2 from the oil terminal delivered 203 Indian nationals to the waiting warship. A third boat carried the MARCOS and a fourth stood by for search and rescue with naval divers. Not for a second during those hours was a finger off the trigger of any of the weapons on board.

Boats neared *INS Sumitra* that day, but veered away. None came too close, saving the Navy men from having to use their weapons. Throughout the operation, Cdr Mokashi remained in contact with his headquarters in an effort to squeeze as much real-time intelligence as he possibly could for the operation at hand.

He knew he could leave nothing to chance. After many tense hours, the 203 Indians, including an infant, were safely on board. The ship couldn't have steamed out faster that night.

By the time *INS Sumitra* docked at Djibouti to deliver its third load of rescued Indians on the evening of 7 April, 2 more Indian Navy warships—frigate *INS Tarkash* and destroyer *INS Mumbai*—had arrived in the Gulf of Aden from Mumbai. While the burden of operations could now be shared by 3 warships, none of the ships could dock at Aden again. The situation there had deteriorated considerably. It had also been decided by this time that Al Hudaydah was logistically the most accessible port, both for the warships as well as stranded Indians spread across Yemen.

That evening in Djibouti, the crew of *INS Sumitra* got their first chance to speak with their families in India. Cdr Mokashi spoke to his wife and children, answering their anxious questions with the usual reassurances that are second nature to military personnel. His work wasn't over yet, and he wouldn't see them for the next 2 months.

INS Sumitra would conduct 2 more rescue operations from Al Hudaydah on 9 and 15 April, rescuing 349 and 403 persons on those final missions. In total, the ship pulled out 1621 stranded persons, which included over 600 foreigners from 26 countries. The Indian government had given *INS Sumitra* clearance to receive foreigners on board after being flooded with international requests following the first Aden rescue.

An elderly couple from Pakistan were among the foreign nationals Cdr Mokashi's crew rescued from Al Hudaydah. Cdr Mokashi remembers not knowing they were Pakistani until they introduced themselves personally to him in order to thank him. He was only doing his job, he told them.

Of the 9 rescue missions, *INS Sumitra* conducted 5, with 2 each by the other 2 Indian Navy ships.

On 16 April, after delivering their final load of Indian nationals to Djibouti, Cdr Mokashi and his crew, already heroes back home, got fresh orders. Seventeen days after they were diverted from their original mission in the Gulf of Aden, they were ordered to return to their anti-piracy patrol. After 2 more months of securing ships and making its way up or down the gulf, *INS Sumitra* was summoned home to Chennai.

On Independence Day 2015, Cdr Mokashi was awarded with the Shaurya Chakra for 'unparalleled valour, conspicuous gallantry, bold and daring decisive actions beyond the call of duty'. But Cdr Mokashi is uncomfortable with the individual decoration.

'This operation wasn't handled by me alone, or by one man. The success of the mission was by us as a team,' says Cdr Mokashi, who was promoted shortly after his gallantry decoration to the rank of Captain. The government appeared to recognize the young skipper's sentiment and *INS Sumitra* became the Presidential Yacht during the International Fleet Review and received a unit citation in 2016.

Cdr Mokashi's father, who was seriously ill through 2015, tracked the brave operations under his son's command through reports on television. In August 2015 when the government announced that his son would be decorated with the Shaurya Chakra, it became an overpowering wish for him to attend the ceremony and see his son receive the medal. But his health had dipped drastically. With great effort, the senior Mokashi was transported to Delhi for the ceremony on 22 March 2016.

He would pass away 20 days later.

*

12

'You Think It'll Never Happen to You'

Squadron Leader Rijul Sharma

Sqn Ldr Rijul Sharma, 30 years old, woke up early like he always did. It was a warm Wednesday at the Jamnagar Air Force base in Gujarat's Gulf of Kutch. A light breeze from the north-west whipped around the dusty skin of the station. If there was one thing Sqn Ldr Sharma knew as he squinted out of the window of his quarters, dressed in his shorts and T-shirt, it was that there couldn't have been a better day for a flight.

The young pilot stepped into a patch of sun in front of his residence, stretching and twisting the sleep out of his bones. Like hundreds of other military aviators across the country that morning, he would be going through the paces of an exercise that keeps the Indian Air Force nimble, alert and ready for active operations.

But this was peacetime. The last time an Indian combat jet had been used in hostility was during the Kargil conflict 17 years earlier. Sqn Ldr Sharma had been 13 then, a boy who had just discovered his athletic gifts. He remembered the stories his father, a former Air Force pilot himself, told him about Indian fighter jets cruising into hostile territory to blast away Pakistani intruders in Kargil. News channels, then in their infancy, beamed out grainy photographs and videos to a mesmerized Indian public. His father, who had flown

DC-3 Dakota transport aircraft and Canberra light bombers in the restive North-east sectors decades before, was a living inspiration to the teenaged Rijul. The Kargil War merely sealed his fascination for flying.

Sqn Ldr Sharma had got married only a few months before. Pulling on a fresh T-shirt and jeans, he smiled goodbye to his wife, Deepika, and made his way to the operations briefing room at his unit, a MiG-29 squadron, about 3 km away. The squadron was constituted a year after the 1962 Indo-China war and code-named 'First Supersonics' because it was the first to operate supersonic MiG-21 jets. It has, since the 1990s, operated another type of aircraft from the same legendary Russian Mikoyan-Gurevich stable: the MiG-29 Fulcrum. It was the jet that Sqn Ldr Sharma flew.

The young pilot received his flying orders that June morning from his Squadron Commander. He was to conduct an 'airframe and engine sortie', a kind of torture test to ensure the aircraft is in top shape and capable of stretching itself to the physical limits it is designed for—a structured workout to keep the nuts and bolts humming, and pilots in sync with their machines.

Strapping a G-suit over his dark blue flight overalls, Sqn Ldr Sharma made his way out to the flight line to climb into the MiG-29 he would be flying that day. A final check of all systems and weather confirmed it was an excellent day to stretch the MiG-29's limbs out over the shimmering Gulf of Kutch. He climbed into the familiar cockpit and strapped in, putting his helmet on. He would lower the visor later to keep out the harsh sun as he soared.

Power on, the cockpit came to life in a low wheezing hum as the MiG-29's twin Klimov RD-33 jet engines were gradually brought to a ground idle position. Releasing the brakes, the pilot eased the jet out of its parking bay and taxied it on to the apron towards the end of the long tarmac. Sqn Ldr Sharma sat in his cockpit, waiting for permission to take off. He glanced around, as he always did, at the base.

The Jamnagar Air Force station is an old, venerable combat base. It came up shortly after India gained independence, initially as a weapons training wing for pilots to hone their bombing skills. It was after Jamnagar proved enormously useful as a base to launch strikes into Pakistan during the 1965 and 1971 wars that it was upgraded to the status of a premier fighter aircraft base in 1979. In the early 1990s, the base received its first Soviet-built MiG-29 Fulcrums.

The Indian Air Force calls its MiG-29s 'Baaz'—Urdu for 'falcon'. Sqn Ldr Sharma had been flying MiG-29s for years, and like his squadron mates, loved the eager agility the aircraft demonstrated in the air. Built as a highly manoeuvrable dogfighter during the Cold War, the jet had matured well across Air Forces, displaying a capacity for missions well beyond frenetic close combat. In Indian hands, the MiG-29s at Jamnagar proved worthy of a variety of missions, from defending airspace against testy airborne intruders from Pakistan to projecting power over the Arabian Sea armed with anti-ship missiles. Sqn Ldr Sharma and his jet had been scrambled several times before on such missions.

But on that June morning in 2016, there was no apparent threat. No intruders in the air or suspicious ships at sea to ward off. Just a clear, blue sky, and the comforting crackle of static in the earpiece embedded in his helmet. Sqn Ldr Sharma waited.

A few minutes later, right before 1000 hours, Jamnagar ground control gave him permission to take off. Sqn Ldr Sharma gently pushed forward the throttle, throwing the MiG-29 into a growl, then a steady roar, as the engine's afterburners created 2 licks of orange flame from the jet's twin nozzles, propelling the MiG-29 off the ground and into the air. He quickly put the jet into a steep climb to an altitude of 1000 metres, his peripheral vision under the fighter canopy noticing the base peel away from beneath and behind him.

Flattening out after his steep climb, Sqn Ldr Sharma did a quick systems check on the aircraft's airframe and twin engines. Then he fed the engines some more fuel and steered the jet upward to an altitude of approximately 11,000 metres through a tenuous cloud deck that had blown in from the Gulf of Kutch. The Perspex glass of the fighter canopy glinted in the harsh sunlight at that altitude. Sqn Ldr Sharma levelled out his jet, bringing it into steady flight.

The test points he needed to achieve on that flight included stretching his MiG-29 to its limits in the 'supersonic corridor', where the aircraft would be flying at just over the speed of sound at that altitude, while executing a series of manoeuvres, all the while checking airframe response and engine performance. Sqn Ldr Sharma lowered his visor as the sun came up at him from the left. Several kilometres behind him, the Jamnagar airbase silently dropped away over the horizon.

Now cruising, the pilot increased throttle up to Mach 1.1 (1358 kmph), crossing the sound barrier, and proceeded to kick-start his routine of systems checks. The cockpit instruments beamed out their comforting figures, telling the pilot that all was well, and predictable. This is the one thing combat pilots love above all else: Predictability. Everything checked out.

About 110 km from base, just as Sqn Ldr Sharma was getting ready to begin his next set of manoeuvres, he noticed a whistling sound in the cockpit. it was a sound he had never heard before.

In a pressurized, air-conditioned fighter cockpit, a pilot only really hears 3 things: the steady hum of his engine, the radio voice from ground control and the sound of his own breathing, amplified by snug headgear designed to tune out any other sound. The sharp whistling sound immediately stood out as odd.

Sqn Ldr Sharma raised his tinted helmet visor and took a look around. As he did so, the whistling abruptly stopped. And then it happened again.

'I looked up. The entire canopy had shattered and a part of it had blown off, with some parts crashing into the cockpit. I felt something smash into my shoulder and a sharp pain. It was a moment of shock. It took whole seconds for me to fully understand what had happened,' says Sqn Ldr Sharma.

It was a situation that is as difficult to describe as it probably is to imagine. Sqn Ldr Sharma, still strapped into his cockpit, was flying at a screaming velocity in a jet that had no canopy. He was now totally exposed to a headwind that smashed him straight in the face, pinning him back in his seat with brutal force. And the terrifying roar of the wind at that speed brought with it a new evil—since he was still flying faster than sound, much of the sound was still 'behind' him. By now, only one thing had become totally clear to Sqn Ldr Sharma: he could barely move his right shoulder from the pain, and the rest of his upper body was quickly sinking into numbness from the sub-zero temperatures at that altitude.

Shaking away the shock, Sqn Ldr Sharma gathered himself and made a quick series of calculations, drawing on every bit of emergency training he had received as a flying cadet and rookie pilot, while his body steadily sank into a near-unresponsive state from the trauma and the temperature. He first did the one thing he knew he needed to before anything else: drop speed. The MiG-29 was still flying steady but shuddering now from the aerodynamic turbulence caused by the open canopy. It slowed down shakily as Sqn Ldr Sharma pulled back on the throttle.

'Once I had gathered some of my thoughts, there was only one thing on my mind. I needed to recover the aircraft,' Sqn Ldr Sharma says. 'I remember thinking, "This is what we prepare and train for years. You never think it'll ever happen to you. Then you realize why you learnt what you learnt."'

The pilot continued to pull back on his throttle, hoping he could regain some of the physical faculties that had been rendered numb by pain and cold by this time. Slowing down to a subsonic speed, a loud, shuddering bang jolted Sqn Ldr Sharma, but also allowed him to push himself into a higher state of alert. Sound had now caught up with the jet. And it was ever more deafening. Cold and pressure crushed the pilot, hitting him in the ears, making him feel that painful pinch that only thin, high-altitude air can. His upper body was now completely numb, having been subjected to whole minutes of wind blast. His head was being thrown around in every direction with every twitch of the jet, every whim of the air that roared into the cockpit.

(The Indian Air Force's formal description of the incident describes what Sqn Ldr Sharma went through at this time simply as 'discomfort'.)

Tumbling inside the cockpit and desperately trying to regain control, Sqn Ldr Sharma was now flying at about 500 kmph and had managed to descend to about 10,000 feet. He was still flying way too fast for comfort and there was literally nothing he could do about the cold—steady, insistent, like an icy sledgehammer against his face, neck and ribs.

Then, for the first time since the canopy blew off his jet, Sqn Ldr Sharma tried to make contact with ground control. There was no way he could have tried earlier. There was simply too much of a wind roar to hear anything else. Even at this slower speed, he could hear nothing, as he repeatedly radioed his controllers, relaying what had happened in a high-pitched scream, hoping to somehow convey his situation to anxious colleagues in the control tower. Over and over, Sqn Ldr Sharma bellowed into his radio talkie, screaming that he was returning to base for an emergency landing. The pain in his shoulder was now so severe that his right hand had become virtually useless. It hung limp, and there was little he could do with his fingers. Not one muscle would flex.

With his left hand, he continued to throttle down to 400 kmph and an altitude of a little less than 10,000 feet. Suddenly he realized he had another problem on his hands. The aircraft had proven capable of flying steady after the canopy blew off, but that didn't mean it was safe for landing.

Landing an aircraft puts a special toll on its airframe and metal skeleton. Sqn Ldr Sharma needed to be absolutely certain that the destroyed canopy hadn't damaged any other part of the jet, including its tail, wings and crucial movable control surfaces. There was no way he could tell for sure. It was impossible for him to twist around in the cockpit to take a look. He would have to take a chance, he told himself. What he didn't have to tell himself was that if something went wrong during the final approach or touchdown, he would have no time to punch out. Not a single moment.

At this point, the Squadron Leader could have taken a decision to eject from the damaged MiG-29. Ambiguity over whether the jet was safe for landing was solid justification to abandon the aircraft and punch out. Nothing is more important than human life, and pilots know that. Sqn Ldr Sharma tried once again to see if he could be a little more certain that his aircraft wasn't damaged. But he just couldn't do it. He waited 10 seconds, quickly rehearsing his next move. Then he made his decision—to stay with the jet.

With controllability checks barely complete, Sqn Ldr Sharma shaved the aircraft throttle back a little more. Ironically, slower and lower, the amount of discomfort and disturbance in the cockpit had only increased. The winds at this stage were more violent, the turbulence peaking near ground level as a result of denser sea-level air.

'I was slapped left and right in the cockpit by the turbulence. I couldn't hear much outside or on my radio talkie. I simply told ground control what I wanted to do,' he remembers.

The MiG-29, pretty much like a convertible now with its hood blown completely off, descended into final approach mode, the tarmac finally in sight. Sqn Ldr Sharma had begun to feel groggy from the pain in those final moments as he steered the aircraft into position for a landing. Miraculously, his yells from the wind-blasted cockpit had been heard, and airspace had been fully cleared around the normally busy fighter base that doubles up as a civilian airport.

At about 6500 feet, Sqn Ldr Sharma realized with no small measure of delight that he was able to get a burst of warmth when he flew through clouds. He did as much of this as he could before taking the aircraft down for its final approach.

As he came in to land, the ground controllers chimed in, informing Sqn Ldr Sharma that he would face 20 kmph head-on winds. 'Feels like 200 kmph straight on my face,' he screamed back at them, before lowering his wheels and executing a perfect landing on the Jamnagar tarmac. A rescue team and crash tender received him at the end of the strip, immediately pulling him from the jet and away from the area.

In his stretcher, Sqn Ldr Sharma glanced up smiling at the men who carried him away. Beyond their silhouettes was the still-glistening Jamnagar sky. The warm air wore the numbness off Sqn Ldr Sharma's injuries, bringing back a hot pain.

Straight to the base hospital, the Sqn Ldr was given a full medical check-up for concussion and his damaged shoulder. It was a blunt-impact injury with internal consequences, but no flesh wound. From his hospital bed, Sqn Ldr Sharma phoned his wife, Deepika, at their home on the base. Up until that point, she had had no idea what had happened—and he was thankful for that.

'She panicked. Anyone would. She rushed to see me. And it was only then that she knew everything was going to be okay,' Sqn Ldr Sharma says. The injured pilot called his parents in Delhi. His father, Wing Cdr Sandeep Sharma, and mother, Neeta, were shocked and anxious about their son's injuries, but they also hoped he would be able to fly again soon.

'I heard only a full hour after Rijul was back on the ground,' says the pilot's father. 'There was not a moment of pain or anxiety in Rijul's voice. He was totally cool and calm. He flies high but keeps his feet on the earth.'

His mother, Neeta, who had lived a life of anxiety waiting for her husband to return from his combat flights, braced herself before she spoke to her son.

'I know my son. When he's up there, he knows what he's doing,' she says.

'Rijul's mother is very brave. She reacted the same way I did. She was worried, but only I could see it. Nobody else could have said she was worried,' says Wing Cdr Sharma.

Back at the Jamnagar base hospital, Sqn Ldr Sharma's condition suggested that it might be a long stay in a recuperation ward. Amazingly, the young pilot proved fit enough to be discharged after only a few days. And a week later, he was back in a cockpit.

'The first sortie after I recovered was special. There was that overpowering feeling that I had just been through something. But then you keep talking to yourself. There's anxiety. But then you tell yourself it won't be the same. And if it is, I'm trained for it. I've defeated it before,' says Sqn Ldr Sharma.

How did Sqn Ldr Sharma decide against ejecting from his stricken aircraft when the possibility of an unsafe landing was very real? A landing that could have killed him or rendered him incapable of flying for the rest of his days? Sqn Ldr Sharma is thoughtful. 'Ejecting could have been an option. I mean, it definitely was an option. But I wanted to first figure out if the aircraft was controllable. If I had lost control, I would probably have had to eject. I thought if I could save myself and the aircraft—that needed to be my priority.'

India's MiG-29 fleet is currently in a phased upgrade programme that makes the aircraft an ever more formidable multirole warplane. The aircraft that Sqn Ldr Sharma flew that day hadn't been upgraded yet. While safety is something the Indian Air Force takes very seriously, and spends enormous resources in zeroing in on the reasons for accidents, investigations are never simple affairs.

Sqn Ldr Sharma has a message for new pilots and those who aspire to join the Air Force.

'I have to convey a message to my younger brothers and sisters. I'd say that fighter flying is a wonderful profession, perhaps the best anyone can ever choose. There is an inherent risk that makes the profession challenging. Don't ever let a situation overwhelm you. You have been trained to tackle any eventuality. These are the times to put all those years of hard work and training to use. Even when I'm born again, I would like to be a fighter pilot.'

Sqn Ldr Sharma has been flying regularly since the incident. He was awarded a Vayu Sena Medal (Gallantry) on Republic Day for his courage, skilfulness and fortitude in the cockpit, saving not only his own life, but an expensive piece of national property—the MiG-29.

'I had no hesitation in jumping back into a cockpit after what happened,' says Sqn Ldr Sharma.

'As my father told me when I was joining the NDA, "Lead your life. Don't let life lead you."'

*

Note: This chapter appeared first as a post on LivefistDefence.com

13

'Every Chopper Pilot's Worst Nightmare'

Squadron Leader Vikas Puri

'Her voice is soothing to the ears. But nothing she says is pleasant—especially if you're airborne.'

At a height of 4500 feet above the Brahmaputra, Natasha's voice pierced through the helicopter cockpit. Her tone was tailored to just the right pitch of matter-of-factness to not jolt the pilots, but with a polite insistence that could not be ignored. As the voice of the cockpit warning system built into the helicopter by its Russian makers, Natasha's was the one female voice no pilot ever wanted to hear.

'Service tank pump failed,' the voice, inflected with a Russian accent, called out.

Sqn Ldr Vikas Puri, lead pilot of the Mi–17 helicopter, heard Natasha clearly. On reflex, he glanced down at his cockpit instrument panel for details about her warning. A pump feeding fuel to the helicopter's 2 main engines had malfunctioned.

In a second, Sqn Ldr Puri had jogged through the flying manual in his mind, like all pilots do. Natasha was programmed to call out any problem the helicopter's systems encountered, big or small. And this was not a big one. The pilot knew that a service tank pump failure was not a major emergency that required drastic action like an emergency landing. Built with

243

several layers of rugged safeguards, the Mi-17's engines had additional pumps to draw fuel in. Everything was okay. There was no need to worry.

That would change in the next 30 seconds. The warning he had just dismissed would escalate into a terrifying anomaly that would see the 13-tonne helicopter abruptly fall out of the sky like a stone, headed straight for the marshy Brahmaputra valley below.

It had been a perfect morning to lift off that Saturday, 12 March 2016. Sunny but mild, just after dawn, an uninterrupted, clear sky loomed over the Shillong airfield. The beautiful state capital of Meghalaya, also headquarters to the Indian Air Force's Eastern Air Command, is nestled in the Khasi Hills at an altitude of nearly 5000 feet. The weather that day could not possibly have been better for the flight bound to the North-east, to forward areas in the frontier state of Arunachal Pradesh and its long border with China-controlled Tibet. It was just another mission for Sqn Ldr Puri and his crew that morning, but their cargo, and intended destination, was anything but ordinary.

The man who was on board the Mi-17 that day was none other than the head of the Indian Air Force's Eastern Air Command. Air Marshal Chandrashekharan Hari Kumar was to be flown to 2 newly revived and rebuilt advanced landing grounds at Ziro and Along in Arunachal Pradesh to inaugurate them and declare them open for flying operations.

Having the Air Marshal on board that particular flight gave the helicopter its special radio call sign that day: EASTERN 1.

The 2 airfields on the China border in Arunachal Pradesh were among 8 that the Indian Air Force had identified in 2009 to bring back to life after they had fallen into disuse half a century before. Apart from giving military planners welcome additional air access to the North-east, the revival plan was also intended to be a direct counter to China's aggressive border infrastructure build-up. A valuable benefit was that airfields provided new avenues for tourism into one of the most arrestingly beautiful parts of India. The Rs 1000-crore project involved giving the 8 defunct airfields new runways, air traffic control towers, buildings to handle cargo and passengers and other infrastructure for flight operations.

Air Marshal Hari Kumar, a highly qualified fighter pilot who had once commanded a front-line combat aircraft squadron, was fully aware of the significance and value of the inauguration he was to oversee that day. With him were his wife, Devika, and several officers from his staff, including Air Vice Marshal Manvendra Singh, a qualified chopper pilot.

Sqn Ldr Puri and his crew had risen early that day in Shillong, and headed straight for the airfield to prepare the Mi-17 before their VIP passengers arrived. A 'ground run' was conducted where the engines were switched on and the helicopter repositioned on the apron, ready for its guests.

'The weather was great that morning. But we didn't have a moment to lose,' Sqn Ldr Puri remembers. 'A violent western disturbance was likely to hit the Arunachal valley the following day, so we literally had only that day to carry out the commitment. If we missed the window, it was certain to be delayed by about a week.'

The workhorse Mi-17 successfully passed its ground tests, with all systems reporting normal and ready for flight. As the crew waited, Sqn Ldr Puri did a final check with the meteorological office at the Shillong base for a weather update along the path they would be flying to the border areas of Arunachal Pradesh.

'Though the reported weather at our intended destination was still turbulent but improving, we decided to play it safe,' says Sqn Ldr Puri, who briefed his crew and reported a final flight plan to the Shillong air traffic control. The Mi-17 would head towards Arunachal, but would make a stop at Tezpur in Assam along the way and wait there until the weather at the forward landing grounds had improved. This was as normal a 'Plan-B' as there ever could be for an air force fully used to contingency plans driven by fickle weather and terrain in their area of operations.

By 0720 hours, Air Marshal Hari Kumar, his wife and the 10-member staff boarded the Mi-17. Fifteen minutes later, the helicopter lifted off, climbing quickly to 6500 feet. It safely exited the Khasi Hills and arrived in the airspace over the plains of the Brahmaputra. Since this was an early flight and the airspace was largely clear, ground control in Assam cleared Sqn Ldr Puri and his crew to fly directly to Tezpur without any cautionary diversions.

The Mi-17 in military configuration is not particularly suitable for VIP transport. Rugged and functional, its large cabin does not shield its occupants from the steady roar of the main rotor or the intense vibrations that course through the machine during flight. But the Indian Air Force loves the Mi-17's toughness, a factor that has seen the government make this Russian-built machine the backbone of India's military chopper strength. With its familiar grey silhouette, the Mi-17 is deployed across the country and on every conceivable mission— from flood rescue in Srinagar to being used as a gunship for assault drills during military exercises and casualty evacuation operations in Maoist hotbeds. It has proven tough enough to even be deployed abroad for India's UN peacekeeping missions in Africa. The Mi–17 cruising that Saturday morning over the Brahmaputra was as reliable a chopper as they come.

'Everything was copybook perfect. I handed over controls to my co-pilot, Flight Lt. Adarsh Gupta, and picked up the map I had organized for the flight,' Sqn Ldr Puri remembers. It was a highly detailed map comprising a thick stack of laminated sheets of paper that opened out to the size of a bedsheet. It was far from handy.

In the cabin behind the cockpit, a member of Sqn Ldr Puri's crew welcomed the passengers on board and passed around bottles of water and Tetra Paks of juice. Noise from the rotor blades ensured that conversation of any kind was impossible in the cabin. The 12 passengers sipped water and juice, and periodically looked out of the Mi–17's blister windows at the glistening river 6500 feet below.

The helicopter had been airborne for 25 minutes and was flying above the river plains when Sqn Ldr Puri put it into a gentle descent to 4500 feet. Almost immediately, the chopper flew into a pall of haze and visibility dropped drastically.

On the controls, co-pilot Gupta quickly checked with Sqn Ldr Puri if he could escape the haze by descending further. The request was immediately declined. There was a reason why Sqn Ldr Puri wanted to 'hold height' till the helicopter had crossed the Brahmaputra plains. Several reasons, actually.

Firstly, the higher they flew, the better their chances of being in range for radio contact. Sqn Ldr Puri wanted to remain in touch with Tezpur as much as he possibly could, since he was transporting the area commander and it was likely that ground control would want frequent updates. Secondly, higher flying altitude also had the benefit of greater airspeed and fuel efficiency. Finally, descending further would make the cabin uncomfortably warm. In the absence of air conditioning, the higher the helicopter flew, the cooler the cabin would be. Descent was therefore ruled out.

The Mi–17 was halfway to Tezpur and cruising comfortably at 4500 feet when Natasha broke her silence once more.

'We were about 55 km from Tezpur, flying over marshy land and paddy fields south of the Brahmaputra, when the distress sequence started,' Sqn Ldr Puri remembers. He had correctly deduced that the warning was not a grave emergency, but what happened next made it clear to him that the situation on his hands was what was classified in helicopter flying as 'rarest of the rare'.

Sqn Ldr Puri instructed his flight engineer, Sergeant Surjeet Singh, to investigate the failed service tank pump to confirm Natasha's warning. Just as he did so, Sqn Ldr Puri noticed that the helicopter was not flying steady and straight, but was slightly inclined to one side.

'The artificial horizon ball (a cockpit instrument that reveals the aircraft's orientation relative to the earth's horizon) had jumped to the left at a steeper angle than normal. There was a feeling of being rotated to one side,' Sqn Ldr Puri remembers. This was definitely not good. Quickly, he cautioned his co-pilot to steady the machine and fly straight so their passengers would not experience any discomfort.

The noise of the Mi–17's rotor might be too loud for any conversation that is not conducted through a headset, but it is also the most reassuring noise when you are suspended in mid-air.

'The noise made by the constant churning of 2 engines, the main gearbox grinding and the rotors beating the air into submission is deafening,' Sqn Ldr Puri remembers. 'But imagine being in a helicopter when all that reassuring noise suddenly falls silent.'

The silence arrived in a heart-stopping moment. Sqn Ldr Puri threw the map to the floor and stared down at his cockpit instruments in alarm. The main rotor's revolutions per minute had dropped below 88 per cent and the power generated by both the engines was rapidly winding down. In seconds, the helicopter would be stripped of all engine power. Sqn Ldr Puri had flown hundreds of hours in Mi-17s and other types of helicopters in service. He instantly knew what was coming.

'One of the worst nightmares for twin-engine pilots. That's what it was.'

Taking back control from his co-pilot, Sqn Ldr Puri pushed down the collective lever to his left. In a helicopter flight, lowering 'collective' would lessen the burden of staying airborne that is imposed on the rotors. With both the engines dying, the control input Sqn Ldr Puri had just provided pushed the Mi-17 into autorotation, a situation where the main rotor turns by virtue of the air moving up through it, rather than powered by an engine.

'For a fraction of a second, it was hard to believe that both the engines of our aircraft had failed and we were auto-rotating. It was something which happens next to never in a twin-engine aircraft. Both the engines going off together is perhaps the rarest of rare emergencies,' Sqn Ldr Puri says.

Jostled in the cabin with his wife and staff, Air Marshal Hari Kumar stepped up to the cockpit. A professional pilot himself, he was fully aware of what an in-flight emergency could be like. Cool but serious, he asked the pilots if everything was normal. Sqn Ldr Puri nodded back, but he knew that things were anything but.

The other senior officer being flown that day, Air Vice Marshal Manvendra Singh, was a veteran Mi-17 pilot. Singh had been Sqn Ldr Puri's station commander at Udhampur in J&K and his contingency commander during a deployment to the Democratic Republic of the Congo for UN peacekeeping duties.

The crew knew, therefore, that their passengers would be quite aware of how dire the emergency was. All-out panic appeared imminent.

Meanwhile, Flight Lt. Gupta, in disbelief, busied himself attempting to figure out if it was the autopilot that had malfunctioned to push the helicopter off its steady bearing. But if the 2 dying engines were not enough of a nightmare, the situation was about to deteriorate exponentially.

Two generators of the Mi-17, which drew power from the spinning main rotor, were now failing too. This meant that every instrument and system on board the helicopter would soon suffer the equivalent of being simply unplugged. Natasha chimed in with the inevitable announcement.

'The cockpit voice warning came on announcing the failure of the first and second generators, with the chattering sound of warning lights flashing in the background. Autopilot had failed and our compass system had also switched off. Every system on board was dead,' Sqn Ldr Puri remembers.

The chopper had begun to plummet through the sky at a rate of 10 metres per second. The steady death of rotor power made it an unwieldy, un-aerodynamic object that was jerked about by wind currents as it fell. Sqn Ldr Puri ordered Flight Gunner Ajeesh to ensure that the passengers were seated and secured as he decided on his next course of emergency action—a possible crash-landing in a paddy field. As he fought to steady the lopsided, falling Mi-17, he scanned the ground, which was now 3500 feet below, for a safe clearing. He knew that touchdown at this rate of descent would be violent, if not a complete disaster.

'I hunted for a landing ground. We were flying over marshy land and flooded paddy fields. It had rained too. I could not identify any spot for an emergency landing,' Sqn Ldr Puri remembers. The in-flight emergency had become a matter of life and death. Flying the 'dead' chopper now, he fought to keep it steady and in a gliding path for a few more kilometres in the hope that a clearing would present itself. He reduced airspeed to give him and his crew a few more seconds in the air as they fell towards the ground.

All preparations were being made on board for an emergency crash-landing. Flight Engineer Singh began turning off all systems to lower the chances of a fire breaking out on impact with the ground.

But Sqn Ldr Puri stopped him.

'Hold on,' the pilot called through his headset to his 3 crew mates. They stared at him. The Mi-17's rate of descent had crossed 12 metres per second. Sqn Ldr Puri paused for a moment, running the scheme in his head once more before issuing the order. It was a huge risk, but it needed to be attempted.

'Let's try to relight and restart the engines.'

Drawing on years of training in the most critical in-flight emergencies, Sqn Ldr Puri was convinced that the Mi-17's 2 engines had died from fuel starvation. He ordered the Flight Engineer to switch on the fuel bypass valve. But for the actual reignition of the engines, the crew would need a small on-board auxiliary power unit to be primed and ready—a process that would consume 30 full seconds, a frightful expense in the circumstances. The Mi-17, now at below 2500 feet, was falling so fast that it would smash into the ground in less than 90 seconds.

Those 30 seconds felt like the longest the crew had ever experienced as they waited for the auxiliary power unit to be charged and readied. Not a moment too late, Sqn Ldr Puri called out for the first engine to be relit and restarted.

The helicopter was now descending at an even more alarming rate. They were barely 600 feet above ground and less than 20 seconds from impact when they realized it.

Their plan had worked. The first engine had been successfully revived. In a flash, Sqn Ldr Puri ordered his co-pilot to pull up his levers to crank up power to the main rotor. Seconds later, the rotor began feeding on fresh power and spinning faster, rising to 98 per cent full speed. Raising the collective lever, the helicopter was now flying on a single engine brought back from the dead. Fifteen seconds from impact with the ground and near-certain death or crippling injury to all 16 on board, the Mi-17's rapid descent was slowed down to a shuddering hover.

From the moment the engines died till the time the Mi-17 stopped falling, just 2 minutes had passed. In that time, the helicopter had dropped 3500 feet. Sqn Ldr Puri remembers being able to clearly see power lines and cattle grazing in fields from that height. But the emergency was not over yet—the helicopter's terrifying fall had been halted, but it was still only flying on 1 engine. Spotting a rice field that was flooded with water but clear of obstructions and power lines, Sqn Ldr Puri carefully manoeuvred the Mi-17 towards this possible emergency landing zone.

As the helicopter recovered delicately from its auto-rotating flight and began to hover, the second engine too came alive.

'On hearing the Flight Engineer confirm that the second engine had started up, I asked my co-pilot to neutralize the levers. We had just witnessed a miracle,' says Sqn Ldr Puri.

With both engines back to their steady hum and the reassuring roar of the rotors coursing back into the cabin, the Mi-17 slowly climbed back to 2000 feet.

'An emergency landing was still an option, just to be sure that nothing else happened to the helicopter. But I decided against it,' Sqn Ldr Puri remembers. 'My co-pilot had informed me that the nearest location was in fact our destination, Tezpur.'

Breaking their 120-second radio silence forced by the emergency, Sqn Ldr Puri and his crew now contacted Tezpur air traffic control, informing a dumbstruck officer at the other end that their Mi-17 had just recovered from a twin-engine failure. After a stunned pause, the officer asked if the Mi-17 needed assistance.

'I laughed to myself. What assistance could we get up there?' Sqn Ldr Puri remembers thinking.

The revived engines were spinning the Mi-17's main rotor at a reassuring rate. But the fuel bypass valve that Sqn Ldr Puri had ordered to be activated before the emergency recovery procedure had a disturbing effect. Aviation kerosene fumes were now billowing directly into the passenger cabin. This was a problem. With not long to go before they landed, Sqn Ldr Puri was dead against changing any of the emergency settings that had saved the helicopter from crashing. And yet, he could not ignore the comfort and safety of his VIP passengers. Deciding against fiddling with the engine parameters, Sqn Ldr Puri ordered Flight Gunner Ajeesh to open some of the cabin's blister windows for ventilation. The roar of the rotors became even louder, but in came some welcome fresh air to dissipate the fumes.

Carefully controlling the Mi-17 at 2000 feet, Sqn Ldr Puri and his crew flew the helicopter straight to Tezpur, landing 20 minutes later safely amidst a ring of crash tenders and firefighting vehicles that had been deployed by the panicked ground staff.

Smiling and relieved, Air Marshal Hari Kumar, his wife and staff disembarked from EASTERN 1. While Sqn Ldr Puri and his crew sealed their helicopter and handed it over for a mandatory investigation by an Air Force inquiry team, the passengers proceeded to their destination a few hours later on board 2 Indian Air Force Dhruv helicopters.

Sqn Ldr Puri and his men would spend the rest of the day with bewildered officers at Tezpur, wondering how the crew of EASTERN 1 had saved an Mi-17 from twin-engine failure. The pilot's command and control over those 120 seconds would go on to become a legend, not only at his squadron, a helicopter unit at Mohanbari, Assam, but among the larger community of chopper pilots in service as well. It reaffirmed the Indian Air Force's flying training as being among the best in the world. And while a flight safety team from the Air Force would go deeply into the reasons why the dangerous failure took place at all, it would also ironically burnish the Mi-17's credentials as a machine rugged enough to return from the jaws of disaster.

Helicopters are not just tricky to fly; when they go out of control, it is virtually impossible to recover them. The fact that the crew was able to bring their chopper back just 15 seconds from a certain crash-landing was testament not just to the skill of those flying, but to the machine itself. The incident was therefore considered a vindication of India's decision to bet big on the lumbering Soviet-era design for present and future operations.

Those 120 seconds on EASTERN 1 would also be textbook material for the Indian Air Force flying cadets, a disturbing but powerful indication of just how important their training could prove to be. For Sqn Ldr Puri, this was a poignant reminder. His own training had been anything but smooth.

As a young cadet at the NDA in 2000, shortly after the Kargil War, Vikas Puri had failed maths in his first semester. Certain that the cadet was headed for humiliation, Vikas's Divisional Officer, Maj. Sajan Moideen, had sent a letter to his father informing him that his son, with no real mathematical acumen, needed to switch from science to the social studies stream if he wanted to pass out with his batch.

Vikas's father, Vinod Kumar, was furious, but decided to respond to the letter. In a letter to Maj. Moideen, he began by saying:

> *Girte hain shahsawaar hi maidaan-e-jung mein*
> *Woh tifl kya girenge jo ghutnon ke bal chale?*
> (It is only those who ride a horse in the battlefield who fall
> How will they fall who crawl on their knees?)

It was a letter filled with hurt, but with a no-nonsense instruction. Quoted by Maj. Moideen years later in a blog post, Vikas's father had written,

> *My son has dreamt to become a pilot and by shifting his stream to social studies you are taking him away from his dream. Vikas may fail, but he will learn. I have handed my son over to you. Do what is required. You can kick him, kill him, but I want to see him as an air force pilot.*

His new licence to 'kick' Vikas into shape worked. Maj. Moideen applied an enormous amount of pressure on the cadet. It worked, but only partially. Vikas managed to barely pass his physical training tests, but his academic performance that semester was an unqualified disaster. He had failed completely. His marks in the mathematics paper were so low that the NDA decided to put him on a relegation list— equivalent to detaining a student for the semester.

Maj. Moideen received the news with horror. What would he say to Vikas's father? What explanation was possible, given that the man had put his son's future in someone else's hands? Maj. Moideen decided he could not see his cadet relegated. It would destroy whatever will he had to work hard. Every bit of work he had put into the young man would be a waste. Mustering every emotion at his command, Maj. Moideen made a case for Vikas to be allowed to reappear for the exam instead of being held back for the whole term. The Academy agreed.

Vikas took the test again a few days later—and passed. He went on to graduate from the Academy 3 years later and proceeded to flying school, choosing to be a helicopter pilot. Vikas Puri would fly for 13 years without an incident before the death-defying mid-air encounter 4500 feet above the Brahmaputra.

Ten months after his leadership in the cockpit saved EASTERN 1, Sqn Ldr Vikas Puri was decorated with a Vayu Sena Medal for gallantry. The citation read:

In this fearless and courageous effort, he not only saved 16 invaluable lives but also a precious war waging asset. He showed exemplary valour, bravery, maturity, exceptional professionalism and situational awareness in tackling one of the gravest emergencies in a Mi-17 helicopter.

*

14

'Could Taste the Blood on My Face'

Wing Commander Gaurav Bikram Singh Chauhan

Blood flowed down one half of his face from a deep gash between his closed eyes. He lay flat on a stretcher, his head tilted back by a neck brace that had been strapped on a few minutes earlier. Wading through a clutch of officers at the Jodhpur tarmac, Avantika Agarwal, 6 months pregnant, was calm in the only way she knew to be. As the Air Force personnel watched cautiously, ready to step in if she needed them, she made the final few steps to the stretcher, not pausing once. She would hold herself together in those minutes. And in that late winter dusk, through the bandages and blood, Avantika noticed something else. He was smiling.

Wing Cdr Gaurav Bikram Singh Chauhan had barely seen his wife that week. The Air Force station in the heart of the Thar Desert was in a high state of alert and activity. Combat aircraft from across the country had arrived there to prepare for a first-of-its-kind show of strength, code-named 'Iron Fist' a few days later.

Pokhran, the playground for the event, had been made famous by India's underground nuclear tests several years before. Iron Fist had nothing to do with nuclear weapons, but would involve an extended, relentless demonstration of brute air power that would include the use of real live weaponry.

An audience comprising the country's leadership, diplomats from several countries and the global press had been invited to witness Iron Fist from a safe distance. The purpose was to serve a reminder to India's neighbours and the world that a peace-loving country could still wreak considerable damage on those who mistook that quality for weakness. If the exercise was provocative politically, as military drills often are, it was an enormous task for the Air Force. Over 100 aircraft were set to fly at Iron Fist in razor-sharp corridors that afforded crews zero margins for error.

Aircraft showing off their strength at the desert firepower show would include the MiG-21, Mirage 2000s from neighbouring Madhya Pradesh and Jaguar fighter bombers from Ambala. Raising the pitch at the exercise with a list of live precision bombing runs would be the Indian Air Force's most formidable front-line jet, the Russian Sukhoi Su-30MKI. It was one of these that Wing Cdr Chauhan and his flying mate, Sqn Ldr A.R. Tamta, strapped into earlier that Tuesday, 19 February 2013.

With Sqn Ldr Tamta flying and Wing Cdr Chauhan the designated weapons systems officer in the rear cockpit, the big fighter, India's largest, was fitted out with 18 100-kg bombs—6 on each wing and 6 slung on to hard points on the aircraft's belly. The 2 men had been cleared for a night training flight that involved a bombing run from an altitude of 7000 feet.

It was still cold in the Thar and the sun was almost out of sight when their jet roared down the main runway of Jodhpur's Air Force Station. At their home in the desert base, Gaurav's wife, Avantika, knew he would be flying that evening. He flew every day. Like most family members of pilots, she heard the roar of the Sukhoi's twin monster NPO Saturn AL-31FP engines and made a mental note that her husband was now airborne. A veneer of anxiety would creep in, and stay until she could hear the sound of the jet returning to base. This was routine.

In the air over the Thar, Wing Cdr Chauhan and his mate quickly put the Su-30 in a climb to about 6900 feet, their designated cruising altitude. As Sqn Ldr Tamta manoeuvred the jet from the front seat, in the rear cockpit, Wing Cdr Chauhan quickly programmed parameters for the bombing run. The drill would see the bombs released over the Chandan desert range.

The Su-30 wasn't alone over the Thar that evening. Other aircraft were also airborne, rehearsing for the final event 4 days later. Wing Cdr Chauhan knew that he was sharing airspace with 2 confirmed aircraft. One was a Jaguar, piloted by Wing Cdr Chauhan's course mate from the Academy, also on a bombing run. The other was the Israel-built Heron surveillance drone that had been deployed to film the bombing runs with its thermal night-vision camera and synthetic aperture radar.

Wing Cdr Chauhan used the aircraft's mission computer to select waypoints, giving the jet map markers to navigate through for the weapons release. The markers would be carefully chosen to account for the long arc the bombs would trace before hitting the ground. The navigation data punched in, the jet was switched to autopilot for the run. For the duration of the weapons release, the pilot, Sqn Ldr Tamta would be required to press the fire trigger on his flight stick. When Sqn Ldr Tamta pressed the button and held it, the first bombs should have begun to drop.

They didn't.

The Su-30 is a big truck of a jet. It doesn't shudder easily. But when Sqn Ldr Tamta pushed down on that trigger, Wing Cdr Chauhan experienced 2 things: (1) an extremely bright flash of light—bright enough that he could only see white when he closed his eyes for a moment; and (2) the heavy jet was jerked violently off its level course. It became instantly clear to both men that something catastrophic had just happened.

The bombs on the fighter's right wing had detonated without detaching, instantly destroying much of the wing and sending high-speed debris smashing at the fighter canopy and into the cockpit.

These few seconds would later be described by the Indian Air Force in a report thus:

> [This was followed by] the aircraft being engulfed in a large ball of fire and breaking up into several parts in mid-air. The explosion on the right wing caused it to be ripped off at the root and the aircraft viciously spiralled downwards in an uncontrolled trajectory with a very high rate of descent.

Wing Cdr Chauhan felt shards of the shattered canopy crash into his face. His helmet visor had shattered too, with a piece of it cutting him right between the eyes, but he wouldn't know it at the time. The thing that changed the most in the cockpit was the noise. Through the vortex of the fractured canopy, a deafening whoosh of high-speed wind made all communication between the pilots impossible. Sitting a few feet apart, the pilots could not talk.

As Wing Cdr Chauhan and Sqn Ldr Tamta fought to regain control of the Sukhoi, which had by this time begun to break up mid-air, something was heading straight for them through the darkness over the Thar. If there was one thing worse than sitting strapped into an aircraft that had just lost a wing and nearly all its structural strength, here it was: an aircraft, heading straight in their direction.

The two would later learn it was the Heron drone that was about to crash right into them. But before the drone overshot them, the Su-30 had lurched into a steep nose-down posture, turning in a loose rightward spiral, heading towards the Thar below. The wind through the canopy fracture brought with it the whiff of explosive—the first real confirmation to Wing Cdr Chauhan that the weapons had detonated on the aircraft.

By this time, Wing Cdr Chauhan had attempted at least thrice to punch out of the doomed jet. But the heavy turbulence and wind blast had put him fully out of reach of his ejection handle. The fighter had by now attained a dangerously high rate of descent. In a final effort, Wing Cdr Chauhan pushed with everything he had against the railing of the cockpit, burning hot at the time, and managed to pull the ejection handle. Seconds later, both pilots blasted out of the Su-30 in their NPP Zvezda K-36DM ejection seats, sideways and outward, their parachutes deploying instantly at that depleting altitude.

Wing Cdr Chauhan held on tight to his parachute cord, too shaken to even try shifting his position to locate Sqn Ldr Tamta who, as it turned out, was not far behind him and descending from a little higher. As they sank into the darkness over the desert, a fresh wave of fear engulfed them.

Wing Cdr Chauhan remembered the Jaguar his course mate was flying in the area at the time, and knew that it was probably just about primed for its own bombing run. He said a silent prayer, hoping that ground control had managed to figure out what had happened and instructed the Jaguar to keep its bombs and peel away from the area. Fortunately, officers on the ground had immediately latched on to the Su-30's catastrophic air incident and cleared airspace over the Chandan range. In his Jaguar, Gaurav's course mate, worried by what ground control had just informed him, returned to base without releasing his bombs.

The unmanned Heron had been a terrifying near miss, but was kept in the air, every single one of its cameras and sensors pointed straight at the crashing Su-30. From the darkness above the Thar, the drone had silently managed to film the terrifying incident: the blazing right wing, the Sukhoi in its howling downward spiral, the awkward ejection and, most disturbingly, the flaming debris that rained down around Wing Cdr Chauhan and Sqn Ldr Tamta as they parachuted downward, some of it dangerously close. One touch was all it would have taken to destroy the synthetic material of the parachutes.

Descending in the darkness with little or no depth perception to tell how far away from the ground they were, the 2 pilots separately and coincidentally recalled what they had seen the previous day during a 'paradrop' from a C-130J Super Hercules transport aircraft over Pokhran. The Army paratroopers had landed and quickly rolled over to the front to avoid injuries from the faster-than-it-looks descent. Both pilots decided that this is precisely what they would do. Except, they couldn't see the ground as it rose towards them from below.

As Wing Cdr Chauhan finally grasped depth bearings, he noticed a well bang in the middle of his descent path. The emergency parachute was a life-saving device and wasn't highly manoeuvrable. *Great!* he remembers thinking, *I've punched out of a flaming Su-30, and now I'm headed straight for a well in the middle of the desert.*

He made a strenuous effort to coax the chute in a different direction. And he thought he was imagining things when he saw the well actually move with him. He was 30 feet from the ground when he realized what that well really was: the shadow of his parachute. Tamta would later confirm he had the precise same sequence of hallucinations. Both pilots rolled forward in the sand when they landed.

By this time, Wing Cdr Chauhan could taste the blood on his face. He did a quick check to make sure he was okay. No injuries to his limbs. His back was okay. No apparent compression fractures to the spine, a common effect of ejection from a fighter. He was, as they meaningfully say in the military, in 1 piece.

Wing Cdr Chauhan then pulled out a cell phone from the thigh pocket of his flight suit to quickly take a video of his face. Blood flowed from the deep gash between his eyes. His left hand was badly burnt, probably while holding on to an air scoop that was spewing burning hot air during the final attempt to eject.

Shaken and bleeding, but assured that he was safe and had survived, Wing Cdr Chauhan wanted to let his wife know. Hoping she would hear it from him first, he texted her: 'Ejected. Am OK.' In Jodhpur, Avantika hadn't heard. She called her husband back that second. Over and over she called his number, but Gaurav wasn't picking up. Out there in the desert, his phone's battery was low and he needed the light from his cell phone to signal to a rescue chopper that, with guidance from the Heron still buzzing above, had zeroed in on the pilots who had landed about 1 km apart.

Overshooting the 2 pilots a few times, the helicopter couldn't seem to zero in on their precise location. Wing Cdr

Chauhan remembers standing in the desert rejecting a barrage of incessant calls—most of them from Avantika—so he could signal to the chopper. Exasperated by the non-stop ringing, he finally picked up. It was Avantika. She tried not to let the worry show in her voice, but remembers fighting hard not to break down. Her husband's message had pre-empted panic, but he was still out there. And wasn't safe just yet. At the other end of the line, Wing Cdr Chauhan gently told his wife that he was okay, and to *stop* calling!

The 2 pilots were picked up and transferred back to base on stretchers, where the crash had created an enormous buzz, coming as it had right before a nationally televised event of political and diplomatic importance. With only superficial cuts and burns to treat, the 2 pilots did not have to face the unthinkable of never flying again.

On Independence Day 2013, Wing Cdr Gaurav Bikram Singh Chauhan was decorated with a Vayu Sena Medal for gallantry. The award carried with it a citation that spared no description:

In face of such [an] unprecedented situation wherein the aircraft bursts into flames with no warning of impending failure Wing Commander Gaurav Bikram Singh Chauhan displayed exceptional courage, situational awareness, uncommon reflexes, in extricating himself and crew member from a distressed aircraft.

If anything could define the cliché of a life-and-death situation, an in-flight emergency in a combat aircraft would do it better than most things. Pilots like Wing Cdr Chauhan are thrust into situations where they have to make a choice in less than a few seconds. In a flaming jet screaming towards the ground, a decision postponed by milliseconds could make all the difference. Sqn Ldr Rijul Sharma fought a terrifying emergency in his MiG-29 to arrive at a choice that could have killed him. But he felt he was equipped with the instinct and skill to justify that decision. For Wing Cdr Chauhan, the choice to stay with the aircraft never came up. Facing a similar catastrophe, but with an aircraft that could in no way be saved, his decision was whether to submit to the situation and go down with the jet, or fight through a physically unimaginable position to depart from an aircraft headed for certain doom.

Gaurav and Avantika's son was born 3 months after the accident. It would be many months before Wing Cdr Chauhan was cleared to fly again. It was a worrying time for the young fighter pilot.

'I remember being surprised at how keen he was to get back into a cockpit again,' Avantika remembers. 'It was more than a year after the accident that he was allowed back into an aircraft again. And he was so thrilled! I guess, once a fighter pilot, always a fighter pilot.'

The strain with which a pregnant Avantika held herself together to deal with the accident would have a lasting impact on her.

'I didn't think I was affected much by the accident. I only realized how much I was when the phone rang unexpectedly the first time Gaurav went night-flying after the accident. Palpitations and a cold sweat broke out. It was just Gaurav calling to chat because he hadn't gone flying. I remember telling him *not* to call me at a time I was expecting him to be airborne because I expected it to be bad news and I really couldn't handle the stress any more.'

The Indian Air Force court of inquiry would take years to figure out what went so catastrophically wrong over the Thar that night. While it was finally put down to a technical glitch, only the Indian Air Force would know just how close both men got to going down with a fighter on fire that moonlit night.

Avantika has advice for the families of pilots and warriors: 'Having seen how our fighter pilots function, the stress they work under, all I can say is that families need to be supportive and think positive. The usual rules don't apply to our fighting forces.'

They never do.

*

Glossary

2IC	Second-in-command
AQAP	Al Qaeda in the Arab Peninsula
Col.	Colonel
Gen.	General
IMA	Indian Military Academy
J&K	Jammu and Kashmir
LoC	Line of Control
Lt.	Lieutenant
Maj.	Major
MARCOS	Marine Commandos
MONUSCO	United Nations Organization Stabilization Mission in the Democratic Republic of the Congo
NDA	National Defence Academy
NSCN-K	Nationalist Socialist Council of Nagaland-Khaplang
Para-SF	Parachute Regiment (Special Forces)

PoK	Pakistan-occupied Kashmir
PONI	Posted out, not interested
QRT	Quick Reaction Team
R&R	Army Hospital Research and Referral
Sqn Ldr	Squadron Leader
Thoise	Transit Halt of Indian Soldiers En Route (to Siachen)
UN	United Nations
UNGA	United Nations General Assembly
Wing Cdr	Wing Commander
XO	Executive officer

Acknowledgements

The force behind much of this book was the Chief of the Army Staff, General Bipin Rawat. He opened his doors to us with generosity and heart. When we met him to tell him about our book, he smiled and said he couldn't wait to read it. We hope we have met his towering standards!

Gen. Rawat's team of officers, Maj. Gen. A.K. Narula, Brig. Manoj Kumar, Col. K.S. Grewal, Col. Saket Jha and several others across departments who wished not to be named embraced this book and helped take it from a mere idea to the book you now hold in your hands.

Chief of the Naval Staff, Admiral Sunil Lanba, and Chief of the Air Staff, Air Chief Marshal B.S. Dhanoa, allowed these two journalist-turned-authors into the restricted world of the men under their charge with the sort of faith that humbled us. We hope we haven't let them down.

Captain Dalip Sharma, a shining star of the Indian Navy, was one of the earliest believers in this book and us. Wing Commander Anupam Banerjee, helped us choose the air force heroes whose stories we have told.

Our editor, Swati Chopra, a true hero herself, employed an infectious brand of enthusiasm and passive aggression to ensure we scaled this hill.

And to Mriga Maithel for her patience as we fought to the finish.

We are indebted to our families, whom we cannot thank enough. They often read drafts of these stories with tears in their eyes and an enormous sense of disbelief that such courage was humanly possible.

But above all, we salute the heroes themselves. And their comrades and their families. It is unlikely that we will ever again encounter the same generosity, humility and grace as we did during the course of writing this book. Thank you for being India's most fearless.

INDIA'S MOST FEARLESS 2

More Military Stories *of*
Unimaginable Courage
and Sacrifice

SHIV AROOR | RAHUL SINGH

EBURY
PRESS

An imprint of Penguin Random House

EBURY PRESS

USA | Canada | UK | Ireland | Australia
New Zealand | India | South Africa | China | Singapore

Ebury Press is part of the Penguin Random House group of companies
whose addresses can be found at global.penguinrandomhouse.com

Published by Penguin Random House India Pvt. Ltd
4th Floor, Capital Tower 1, MG Road,
Gurugram 122 002, Haryana, India

Penguin
Random House
India

First published in Ebury Press by Penguin Random House India 2019

ISBN 9780143443155

Typeset in Bembo Std by Manipal Digital Systems, Manipal
Printed at Replika Press Pvt. Ltd, India

www.penguin.co.in

MIX
Paper | Supporting
responsible forestry
FSC® C016779

Contents

Forewords   vii

Prologue   xiii

Introduction   xxiii

1. 'Killed, Maybe, but Never Caught'   1
   Major Mohit Sharma

2. 'He Avenged Them, Didn't He?'   33
   Corporal Jyoti Prakash Nirala

3. 'Fire when You Can See Their Faces'   57
   Lieutenant Navdeep Singh

4. 'I've Been Ready since the Day I Was Born'   77
   Major David Manlun

5. 'Get to the Upper Decks, Don't Come Back'   99
   *Lieutenant Commander Kapish Muwal
   and Lieutenant Manoranjan Kumar*

6. 'There Are More Terrorists Inside, Sir!'   119
   Captain Pawan Kumar

7. 'I Rust when I Rest'   141
   Major Satish Dahiya

8. 'Climb over Me, Get to the Submarine!'   165
   Lieutenant Commander Firdaus Mogal

9. 'Just Tell Me when to Begin, Sir'   183
   Captain Pradeep Shoury Arya

10. 'What's Higher than Saving Someone's Life?'   213
    Captain P. Rajkumar

11. 'Half of My Face Was in My Hands'   237
    Major Rishi Rajalekshmy

12. 'I Repeat! Fire in My Cockpit!'   259
    Squadron Leader Ajit Bhaskar Vasane

13. 'Not a Sound until They Enter the Kill Zone'   279
    Major Preetam Singh Kunwar

14. 'You Cannot Sustain Fear of Death'   307
    Flight Lieutenant Gunadnya Ramesh Kharche

Acknowledgements   329

Forewords

'Either I will come back after hoisting the tricolour, or I will come back wrapped in it, but I will be back for sure.'

—Captain Vikram Batra, Param Vir Chakra

These immortal words of one of India's bravest warriors encapsulate the spirit of the Indian Armed Forces. Our fine men and women epitomize the highest standards of honour, courage and commitment. They meet the myriad challenges of our volatile security environment with an unflinching sense of purpose, fully prepared to make any sacrifice required to protect the nation.

Over the last four decades, there have been innumerable acts of valour by our men and women from the three services. Many a times, in challenging situations at sea, one has seen extraordinary feats being performed to overcome seemingly insurmountable odds. Similarly, stories about the heroic acts by our soldiers and air warriors, against the enemy and in protection of the citizenry, abound. These chronicles are an affirmation of the strength of our ethos, conviction and resolve.

This sequel to *India's Most Fearless* offers the reader a poignant insight into a few such instances. The willingness and confidence of the individuals to surmount all fear and push the limits of the possible, described in each of these stories in the book, will leave an indelible mark. The stories about our heroic submariners give a unique insight into the obscure, hostile and unforgiving realm in which these silent warriors operate.

Readers will undoubtedly gain a deeper understanding of the culture of our forces, as well as an appreciation of the sacrifices made in the defence of the nation. In this day of incessant 'Breaking News' and short-lived societal memory, accounts such as these serve to rekindle our memory of those who made enormous sacrifices for India's security and our tomorrow . . . 'lest we forget'.

Our heroes, through their sacrifices and bravery, have continued to strengthen the very foundations of our great nation. By remembering them, we ensure that their example continues to guide us in all walks of life.

Jai Hind.

Admiral Sunil Lanba
PVSM, AVSM, ADC
Chairman, Chiefs of Staff Committee, and
Chief of the Naval Staff

'Always do everything you ask of those you command.'

—General George S. Patton, Jr

3 June 1999 seems like yesterday to me. I remember taking off from Srinagar with my Flight Commander in a pair of MiG-21 jets, soaring over the Drass-Kargil heights and dropping 250 kg bombs on Tiger Hill. Our squadron had moved from Punjab to the Kashmir Valley a fortnight earlier to help hunt down and destroy the enemy intruder positions.

Being deployed for war was a dream. As difficult and dangerous it was, this was what my squadron and I had trained tirelessly for. What could be more fulfilling than to be called upon to do what you had joined the Indian Air Force for?

Those few weeks threw up some of India's best known and most beloved heroes. Men whose actions have deservedly won them a valuable price in public memory. But as I have always held, India has never needed a full-scale war or conflict for its heroes to step up. In a country that faces threats from across the spectrum, the demand for courage and gallantry remains high. History is rife with acts of courage and valour by the three services.

Air power and employment of the Air Force is frequently seen as an indisputable act of hostility in a conflict. The nature of military aviation and the many other roles of the Indian Air Force mean that our men and women are uninterruptedly in difficult and demanding lines of duty. From a fighter cockpit that's on fire, to a transport plane headed to a dangerously small airfield high up on a mountain, from daring helicopter rescues during floods to the extensive ground operations that occupy thousands of our ranks, it does not stop.

In *India's Most Fearless 2*, you will gain insight into the story of an air warrior, amongst others, in a role that you would not normally associate with the Indian Air Force as you know it: Anti-Terror Operations. The other accounts, equally, must occupy us with questions not only about the will to survive and the skill of our three services, but also about what it takes to make peace with one's own likely demise, if in the bargain many more lives are saved.

My compliments to the authors for the sequel, for very aptly highlighting the valour of our heroes once again.

Jai Hind.

Air Chief Marshal B.S. Dhanoa
PVSM, AVSM, YSM, VM, ADC
Chief of the Air Staff

The wide readership of the first book, *India's Most Fearless*, has been a source of great satisfaction to the Army fraternity. The book undoubtedly brings to light the bravery against all odds of a few of the 'heroes' of Indian Army, amongst many such stories of unparalleled guts, glory and courage.

The book and its popularity is an affirmation that despite many distractions in our modern and hyper-connected society, millions are still interested in the stories of our men and women in uniform. The soldiers keep themselves professionally abreast and ensure that they are available for the 'call of duty' at all times to uphold the sovereignty and integrity of our great nation.

Our men and women in uniform have never let the nation down in the highest traditions of the Armed Forces. After having seen the entire spectrum of challenges in which Indian Army soldiers operate, I can say with conviction that we will continue to put the nation first, always and every time.

We are grateful to the citizens of our nation who continue to acknowledge the bravery and sacrifice of our men and women in uniform in any manner they can, through messages, cards, letters, social media or even standing ovations at public places. We are sanguine of the continued support of our countrymen in the times to come.

Compliments to the authors for having continued with the sequel to highlight the actions of our 'heroes'.

Jai Hind.

General Bipin Rawat
PVSM, UYSM, AVSM, YSM, SM, VSM, ADC
Chief of the Army Staff

Prologue

Just after 3.30 a.m. on 26 February 2019, climbing abruptly to 27,000 feet in dark airspace over Pakistan-occupied Kashmir (PoK), an Indian Air Force (IAF) pilot flying in a single-seat Mirage 2000 fighter jet pushed a button on his flight-stick. A few feet below him, from the rumbling belly of his aircraft, an Israeli-made bomb silently detached itself and dropped away to begin a journey—first gliding and then careening—towards a target over 70 km away. The bomb, fed with satellite coordinates and an on-board guidance chip, had all the information it needed to hurtle to its destination.

The Mirage 2000 was far from home. It had taken off from the Gwalior air force base over 1000 km away earlier that night along with at least six more Mirage jets from the three squadrons based there. Over the hour the jets flew over central India and into the northern sector. Following in their wake, five more Mirage 2000 jets took off in the darkness from an air base in Punjab.

The dozen Mirages, flying in three separate and unequal formations, weren't alone in the air. Two airborne early warning jets, an Embraer Netra from the Bathinda air base and a higher performance Phalcon jet from Agra were already in the air, their powerful radars and sensors on full alert to the mission ahead. Communications between aircraft were kept to a minimum. This was a mission with almost no room for deviation unless absolutely necessary. And it needed to last for as little time as possible.

As the three Mirage formations flew in a circuit at low altitude, very much in the manner of night flying training sorties conducted by squadrons, ten jets more roared off the tarmac from two more air bases, including Sukhoi Su-30 MKI fighters from the forward air base at Halwara. It was this pack of Su-30s that would play a crucial role in what came next.

With a total of twenty-two IAF fighters in the air, the jets slowly mixed their formations to create three separate packs— two mixed packs of Mirage 2000 and Su-30 fighters. And a third pack comprised only of Su-30s. While it's tempting to think of these three packs as neat little jet formations in the sky, it was nothing quite like that. The jets in each pack flew tens of kilometres from each other, and were only bound by a loose common flightpath and mission profile.

Shortly after 3 a.m., the mission began with a pre-planned deception.

The third fighter pack, consisting of big, heavy Su-30 jets, turned south, heading out of Punjab and into the Rajasthan sector, all the while ensuring it remained prominent and visible to Pakistani radars on the other side of the international border. Turning around over Jodhpur, the fighters began provocatively flying in the direction of the international border north of the Chandan firing ranges, their noses pointed towards a Pakistani city that couldn't possibly have been on a higher alert at the time—Bahawalpur, 250 km to the north, the city that was home to the Jaish-e-Mohammad's (JEM's) headquarters and largest terror training facilities. The IAF planners had counted on Pakistan's 'hair-trigger' state of alert to provoke a reaction. It happened within minutes.

The Pakistan Air Force scrambled a group of F-16 jets from the Mushaf air base in Sargodha about 320 km to the north of Bahawalpur. Just as the jets were getting airborne and moving south to fend off any possible attack by the Indian Su-30s, the second IAF pack, comprising Mirage 2000s and Su-30s, broke away from its circuit and turned south over Jammu along a radial pointed towards Sialkot and Lahore in Pakistan, both large and commercially important cities. This second pack split further, with one part flying along a radial that would pass through Pakistan's Okara and lead once again to Bahawalpur.

The twin air manoeuvres from two directions doubled the air threat to the 'capital city' of the JeM. More F-16s departed Sargodha to engage with this second Indian threat. Pakistan's instantaneous scrambling of fighters wasn't surprising to Indian radar controllers and sensor operators on the two airborne early warning jets. The country's air defences would have been on their highest state of readiness since the 26/11 Mumbai terror attacks, an act of carnage terrible enough that it got India to seriously consider retaliatory air strikes for the first time.

And now, for twelve days without pause, Pakistan's military had cranked its alertness levels to maximum.

Eleven days earlier, at 9.30 a.m. on 15 February 2019, the chiefs of the Indian armed forces and intelligence agencies, top ministers and the National Security Advisor arrived at Delhi's leafy 7, Lok Kalyan Marg compound where the Prime Minister of India lives and sometimes operates from. It was far from a routine weekly meeting for the Prime Minister to take stock of national security.

Eighteen hours earlier, 800 kilometres north, in the Lethapora area of Jammu and Kashmir's Pulwama district, a vehicle packed with explosives and driven by a young man named Adil Ahmad Dar, had managed to snake between vehicles of a large convoy of Srinagar-bound trucks carrying 2500 troops from the Central Reserve Police Force (CRPF), and rammed it. The explosion killed forty troops, spattering the highway with their blood and body parts. Minutes after the blast, a stream of pictures of the mangled vehicles and sickening carnage taken from mobile phones of locals and first responders flooded social media.

With the Pakistan-administered JeM terror group claiming responsibility for the attack, the Prime Minister had convened this meeting of the Cabinet Committee on Security (CCS) solely to assess how India could respond. Forty minutes later, the meeting was finished. Asked if air strikes on a terror target were a viable option, IAF Chief Air Chief Marshal Birender Singh Dhanoa responded in the affirmative, also briefing the Cabinet Committee that the country's jets would be ready to strike with confirmed targets in a matter of days. He was given two weeks.

From 16–20 February, the IAF worked with intelligence agencies at the operations room in Delhi's Vayu Bhawan. With National Security Advisor Ajit Doval receiving a daily update on proceedings, the deliberations were honed by satellite imagery, human intelligence from the ground in Pakistan and PoK, and photographs from a pair of Heron drones flying daily missions along the Line of Control (LoC).

On 21 February, the IAF presented a classified set of 'target tables' to the government via the National Security Advisor.

The first in the list of seven separate target options was a JeM terror training compound that sat on a hill called Jabba Top outside the city of Balakot in Pakistan's Khyber-Pakhtunkhwa province. The IAF recommended Balakot, just 100 km from Pakistan's capital Islamabad, since it was a secluded target with the lowest probability of non-terrorist casualties. The two other 'viable' targets presented to the government were in PoK—Muzzafarabad, 23 km south-east of Balakot, and Chakothi about 70 km away. But these two, along with Bahawalpur, carried not just the risk of collateral damage, but a slightly higher chance of being hindered by Pakistani air defences. Among the remaining options was Muridke, north of Lahore, the city that held the headquarters of that other dreaded India-focused terror group, the Lashkar-e-Taiba (LeT). This too was deemed a highly risky target to consider.

By midnight on 22 February, a highly controlled chain of command decided that the Indian jets would strike the first target in the list—the one outside Pakistan's Balakot. Every man and woman in the secret chain was aware that if such a mission went through, it would be India's first air strike on Pakistani soil since the 1971 war. What amplified the mission ahead was that the two countries weren't at war in 2019. Could such a mission change that?

There was another important reason why Balakot was chosen. Unlike Muzzafarabad and Chakothi, Balakot was in Pakistan and not PoK. As an international message, an air strike on sovereign Pakistani soil—as opposed to PoK, which India considers its own territory—would make all the difference in the world.

The target dossier submitted to the government also contained pages of data detailing the latest intelligence assessments of the kind of damage that could be caused to terrorist infrastructure in each case. In the case of Balakot, apart from satellite imagery and some medium-grade electronic intelligence, the Indian intelligence agencies had also been able to procure invaluable human inputs from Balakot town. The intelligence, obtained from Indian 'assets' on the ground, provided invaluable shape to the target, and was the original source of a number that would later be the subject of much controversy and debate. India's assets in Balakot had reported that there would be at least 300 terrorists and terror trainees on site at Jabba Top at any given time. In other words, a facility that was known to house a significant enough number of handlers, terrorist recruits and ideologues, to justify a high-risk air strike from airspace peppered with and primed for anti-air defence.

As a fully intelligence-based operation, it was imperative that India chose targets that involved not just terror infrastructure, but the presence of a significant number of terrorists at any given time. Apart from the National Technical Research Organization's (NTRO) signal intelligence inputs, it was this human intelligence that helped guide and lock India's choice of target.

It wasn't the first time India was using such human assets for an offensive operation in hostile territory. In September 2016, during the Indian Army Special Forces 'surgical strikes' in PoK, Indian assets[1] in the JeM had confirmed the terror launch pads as viable targets, revealed first in the first book of the India's Most Fearless series.

A data analyst with one of India's intelligence agencies told the authors, 'An operation of this kind is very difficult without human intelligence on the ground. It would have been a huge risk to do so without a conclusive word to corroborate your other inputs, whether satellite or electronic.'

An Army officer who served on the composite intelligence team that formulated the target packages during the 2016 trans-LoC strikes says, 'The question is not about whether ground assets were used or not. They 100 per cent were. The

---

[1] See *India's Most Fearless 1* (New Delhi: Penguin Random House India, 2017).

only question, might I add that nobody needs to ever know about, is whether these were the same assets that helped in 2016 or similar assets—or assets of a totally different kind. That will hopefully remain guesswork. Let films and books (!) do the guessing.'

On 24 February, pilots of the Mirage 2000 squadrons in Gwalior were briefed about the mission. That same day, aircraft would be airborne over central India for a short mock air drill alongside a Phalcon AWACS jet and Ilyushin-78M mid-air refuelling tanker from Agra. The jets taking part in the drill didn't return to Gwalior, instead landing at a base in Punjab. They would remain at the base all of the next day.

The IAF was about to take a violent break from history, but in Delhi, every effort was made to ensure that it was business as usual. On the night of 25 February, hours before the Mirages took to the air on their mission, the IAF hosted a customary farewell banquet for the outgoing chief of the Western Air Command, Air Marshal C. Hari Kumar—he was retiring three days later. The sit-down dinner was organized at the Akash Air Force Officer's mess near Delhi's India Gate, where just a few hours earlier Prime Minister Narendra Modi had inaugurated the country's National War Memorial.

In his speech, IAF Chief Dhanoa regaled the audience with stories of how he and Air Marshal Hari Kumar had gone to the same school—Rashtriya Indian Military College (RIMC), Dehradun—and were from the same house. It was a typical military evening of mirth and nostalgia. The banquet had over eighty senior air force officers in attendance. But only a handful of them, IAF chief Dhanoa and Air Marshal Hari Kumar included, were in the 'need-to-know' loop on what was about to happen. Those who *weren't* in that loop confirm to the authors that there was absolutely no indication that evening that some of their service personnel were about to soar out across the border to drop bombs inside Pakistan.

After farewell speeches and dessert, the banquet wound up at 11 p.m. IAF Chief Dhanoa was driven back to his official residence on Delhi's Akbar Road. He tells the authors he received a final update on preparations before turning in for a quick couple of hours of rest—everything was in control by a team he knew he could trust his life with. Thirty minutes before the Mirages took off from Gwalior, Dhanoa woke up to plug back into the secret proceedings.

Four kilometres away at 7, Lok Kalyan Marg, Prime Minister Modi was awake too. He received his final pre-mission brief 20 minutes before the jets departed Gwalior. There would be communication silence for the next half hour—covering the most crucial part of the mission. The intrusion.

As the second and third fighter packs flew menacing flight paths on the Jodhpur–Bahawalpur and Jammu–Sialkot radials, the first pack, comprising six Mirage 2000 jets, crossed the LoC at low altitude in the Keran sector in Kupwara. Flying over the Athmuqam town in PoK, the six jets spread out further and climbed, crossing between 12–15 km into hostile airspace.

With the Jabba Top hill now in effective range, and given the all clear from radar controllers in the Phalcon jet, the aircraft dropped their bombs one by one. Five munitions from five aircraft dropped away in the cold dark, whooshing west out of PoK and into sovereign Pakistani airspace.

Tracking the weapons as they closed in on Jabba Top, pilots in the air and controllers on the ground knew history had already been made. The IAF weaponry was about to hit targets on Pakistani soil for the first time in over forty-seven years—and, crucially, for the first time when the two countries weren't involved in a full-blown war. The very act of pushing the button and letting those bombs loose was a message, the IAF leadership would tell pilots in a debrief later.

Seconds after the weapons release, a warning call went out to the Mirages as they tracked the bombs screaming towards their targets using infrared sensors. Three Pakistani jets had been scrambled from the Minhas air base in Pakistan's Kamra town, just over 60 km north-west of Islamabad. Tracked by the Indian Phalcon jet, the Pakistani fighters, believed to be Chinese-origin JF-17 Thunder jets, flew at full throttle towards PoK. It was near impossible for Pakistani defences to know that Indian bombs were headed towards Balakot.

With Pakistani jets inbound, Indian controllers on the Phalcon jet instructed the Mirage pilots to turn around immediately, drop altitude quickly and return across the LoC. With the Pakistani jets well over 50 km away, the six jets would cross back between Chowkibal and the Leepa Valley, flying close to Chakothi, one of the targets that had been considered but dropped in favour of Balakot.

The intrusion into hostile airspace had lasted only a few minutes. The six Mirages, their backs watched by the Phalcon jet, landed safely in Srinagar. The other Mirages and Su-30s from the second and third pack would also be summoned back to bases in Punjab. A debrief of the full mission would later affirm that the second and third packs had very ably lured Pakistani 'first responder' F-16s away from the area of attack to the north, and kept them engaged and 'on edge' until the strike mission was complete. The Netra early warning jet from Bathinda would record the deception, providing compelling battlespace imagery for post-mission discussions.

Thirty hours after the air strikes on Balakot, the historic mission would be overshadowed briefly by Pakistan's retaliation attempt over the LoC in the Sunderbani sector near Jammu, using a pack of fighters that included F-16s, Mirage IIIs and JF-17s from the Sargodha and Rafiqui air bases. While the Balakot air strikes had passed without the names of any of the Indian pilots or personnel involved reaching the media, one name would per force become public the following morning. Wing Commander Abhinandan Varthaman, in his MiG-21 Bison jet, would be shot down while chasing a Pakistani intruder back across the LoC. With its pilot repatriated to India barely 48 hours later, the IAF would publicly credit Varthaman with having shot down a Pakistani F-16 jet during the air joust, triggering an uninterrupted storm of claims and counterclaims, with questions likely to linger indefinitely.

The true history of that late winter week, however, would be in the work done in darkness by a group of IAF pilots—many of them young—who, under instructions, had flown into the most hostile airspace imaginable, to conduct a mission never done before. Each of those pilots will have known the risks and the substantial chances that they could be shot down by ground fire or surface-to-air missiles in an area that was on high alert. Just how the Indian Mirages managed to make their way so deep into PoK unchecked will likely be a low-profile introspection within Pakistan's military. Just as the events of the following morning will be one for the IAF.

While the authors have interviewed several officers involved or familiar with the 26 February mission, the pilots must remain nameless. At the time that this book is published, the historic air strike on Balakot remains a classified operation.

Introduction

At 2.45 p.m. on 11 January 2019, just as we had finished writing much of this book, an improvised explosive device (IED) was remotely detonated at the LoC in Nowshera, north of Jammu, an area infamous for Pakistani ceasefire violations and infiltration attempts. Two Indian Army personnel on patrol were instantly killed in the explosion, believed to be the work of a hybrid infiltration unit comprising Pakistan Army commandos and terrorists, better known as the Border Action Team (BAT).

The soldier killed in the blast was Rifleman Jiwan Gurung, twenty-four years old and at the start of his life. The officer, Major Sashidharan Vijay Nair, wasn't much older—thirty-three. For a few days after the incident, their deaths would merely add to the familiar statistics of mortality from that part of the country. But within four days, by 15 January, journalists would discover a numbing back story that would push the deceased Major into the news headlines.

The back story would radiate from a young figure in a wheelchair first seen at the Pune war memorial and later at the city's Vaikunth cremation ground. Maj. Sashidharan had met and fallen in love with Trupti six years ago. Only months into their engagement, Trupti was diagnosed with an autoimmune disease of the central nervous system that manifests itself with progressively intensifying symptoms that can be managed but never cured. On the threshold of marriage, friends and family are said to have advised Maj. Sashidharan—then twenty-seven—to reconsider his future. Friends say that even Trupti told her fiancé she would understand if he were to break off the engagement and move on with his life.

Maj. Sashidharan would hear none of it. Trupti and he were married a few weeks later. Months after the wedding, Trupti suffered another serious health setback that rendered her paralysed from the waist down, permanently consigning her to a wheelchair. The young officer would devote himself to Trupti, ensuring that they never missed out on the army life he had signed up for. Days before his death, while on leave in Pune, he had calmed a worried Trupti about being deployed at the LoC.

In life as in death, said a news report about him.

But would Maj. Sashidharan be more than a flag-draped casket on the inside pages of a newspaper had it not been for these details about his personal life? What if they had never been discovered? What about the soldier, young Rifleman Gurung, who died with him? Does he too have a crushing back story that burnishes his heroism? Is such a back story even necessary to amplify the heroism of those who put their lives on the line as a matter of daily routine? These aren't loaded black-and-white questions, but ones we have continuously grappled with through the writing of this book, the second in the India's Most Fearless series.

A month after the explosion that killed Maj. Sashidharan and Rifleman Gurung, on 14 February, an election-bound India was shatteringly interrupted by a suicide vehicle attack on a convoy of the CRPF in Jammu and Kashmir's

Pulwama. For a country that has become inured to periodic Pakistan-sponsored terrorist attacks, there was an immediate and unmissable 'enough is enough' air. Few could tell quite why this attack had proven so uniquely numbing—India had suffered bloodier attacks at Pakistan's hands before. Was it the terrifying nature of the attack? Was it a country that had already raised the bar on punitive responses with its strikes inside PoK in 2016? Did looming elections play a role as critics would later allege? Whatever it was, twelve days later, India would cross a new red line with a historic air attack on a terror facility of the JeM inside Pakistan.

The Balakot operation has already attained mythical status even as India and Pakistan fight an uninterrupted stream of claims and counterclaims. While that has always been the nature of the beast between the two countries, the operation itself remains classified in India, with the barest of details ever emerging officially. And that means there is a good chance that the names of over a dozen of those fighter pilots who were assigned the historic and enormously dangerous task of flying into Pakistani airspace in darkness, will never be known with certainty. The prologue to this book carries an account of the Balakot air strikes based on conversations with many of those involved, but who cannot be named. From radar operators, mission support pilots and planners, their work will linger, possibly forever, in the shadows.

Like them, there are hundreds on land, in the air and at sea, who cannot be named because of the nature of their feats.

In the course of writing *India's Most Fearless 1* and *2,* if there's one unshakeable truth that we have come upon, it is this: if you look hard enough, every soldier has a shattering back story. But, like in the case of Maj. Sashidharan, does it take their deaths for such stories to bubble to the surface?

Jim Morrison of The Doors may have been right when he said:

> Death makes angels of us all
> And gives us wings
> Where we had shoulders
> Smooth as raven's claws.

('A Feast of Friends', The Doors)

But it is equally true that the stories of our soldiers are there to be told if only someone were willing to ask. As in the first book of this series, many of the heroes in *India's Most Fearless 2* are no longer alive. But as you will hopefully discover as you turn the pages, death was only a final flourish in lives lived with constant heroism.

What this journey started out as was two guys who've spent their entire careers as journalists listening to stories of military valour deciding it was time to begin documenting them. Not out of any lofty sense of responsibility, but simply because these are stupendous stories that everyone needs to hear. Since the first *India's Most Fearless* was published in 2017, a common response we receive from readers is, thanks for the inspiration. Any thanks, we always tell them, is due only to the heroes, their comrades and their families. We haven't for a moment chosen to tell these stories with the intention to inspire. We have done so because they were amazing stories for us personally. If inspiration is the inevitable effect of these stories, then thanks is due only to the men we've written about.

The book you hold in your hands takes forward the legacy we didn't imagine would take so powerful a root when the first *India's Most Fearless* was released in 2017. Neither of us fathomed that stories of military heroes would be read by so many people in so many languages and with a thirst for more. The book you are (hopefully) about to read is the result of thousands of messages from readers urging us to write about more heroes and their amazing feats.

When we announced this book early in 2019, one reader sent us a message saying, 'The India's Most Fearless series can never end, because India will never run out of heroes.' This is the truth, but a disturbing one. It has never escaped our minds for a moment that it is India's uniquely difficult security atmosphere that creates opportunities for military heroism—the Balakot strikes will serve as a numbing reminder to a generation of Indians that wasn't born when India and Pakistan last locked horns in Kargil. And in many ways, it does mean there will be a steady flow of acts of courage from our frontlines and disturbed areas. But not for a moment is *India's Most Fearless* a romanticization of conflict. If there's one thing we've learnt in telling these stories, it is the silent and humble trust of every soldier that India will not send them into combat unless absolutely necessary. That their heroism comes at an enormous premium. And that the country would rather its soldiers were safe, than forced into a situation where they have to decide between life and death.

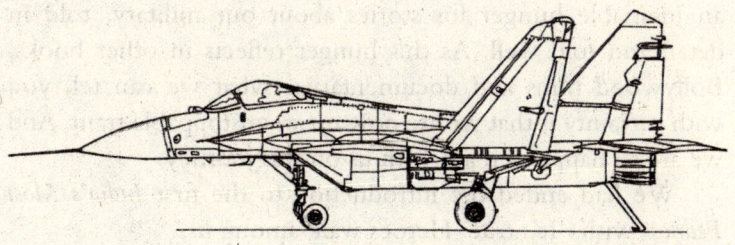

MiG-29K

While the stories in this book seek to keep the spotlight on individual heroism, it doesn't exclude the many hard questions that must rightly be asked about military operations. Questions of leadership and decision-making, of tactics and training, of spirit and initiative, all of which have a bearing on why the heroes needed to face their ultimate dilemma in the first place. Let no one tell you that these questions don't bear asking. If there's one thing our military heroes deserve, it is an unending stream of questions about the circumstances surrounding their operations. The stories that follow answer some of these questions and raise some more. But through them all, as you will see, the core of courage stands unshaken.

If the first book in the series made the front pages of newspapers for the first and only personal account of the 2016 surgical strikes into PoK, the book in your hands features several operations that took place in the aftermath of that momentous mission. Missions that throw important light on India's security after that daring revenge strike by the Indian Army's Special Forces squads, including the 26 February 2019 air strikes in Balakot, Pakistan.

As in the first book, we are also privileged to have been able to feature heroes who went beyond the call of duty to save the lives of others. Stories that still give us sleepless nights, and that we hope will keep you awake too. If there's one thing that *India's Most Fearless* has showed us, it is that there's an insatiable hunger for stories about our military, told in detail and told well. As this hunger reflects in other books, Bollywood films and documentaries, what we can tell you with certainty is that we are now on an unstoppable train. And we are so happy you are with us on this journey.

We had ended the introduction to the first *India's Most Fearless* with, 'It's true. Heroes walk among us.'

It won't ever stop being true.

1

'Killed, Maybe, but Never Caught'

Major Mohit Sharma

Undisclosed location near Shopian, Jammu and Kashmir
March 2004

'Something's not right,' Abu Torara whispered to the man slouched on a cot next to him. A pair of early summer evening sunbeams streamed into the room from a half-open window in their small hideout not far from Shopian, just over 50 km south of Srinagar.

Abu Sabzar drew deeply on a cigarette, exhaled through his nostrils, roughly scratched his beard and turned to look at Torara, who was on his feet, leaning against the wall. A pair of AK-47 assault rifles lay at the foot of the cot. Torara was looking straight ahead of him at the tiny doorway that led to the next room—a small balcony-cum-kitchen that opened out into the woods. Emanating from that direction was the sound of boiling water, the aroma of kahwa, the frothy pour of liquid into glass tumblers and their clink as they were placed on a tray.

'You want to talk to him some more?' Sabzar asked, stubbing out his cigarette on the windowsill next to him. Torara said nothing. A few seconds later, bearing a steel plate with glasses of tea, Iftikhar Bhatt stepped through the tiny doorway and into the room.

Six feet two inches tall, with hair down to his shoulders and most of his face covered with a bushy beard that flowed down his neck, Bhatt wore a stony expression as he stepped forward to offer the other two terrorists their tea. His own

3

rifle was slung from his neck, resting at his side. After they had picked up their glasses, Bhatt picked up his own and sat down at the edge of the cot, silent, staring straight ahead.

Minutes passed as the three men sipped from their steaming glasses. Then, Torara stepped forward and spoke.

'Iftikhar, I'm going to ask you only once,' Torara said, still sipping his tea, placing the other hand on Bhatt's knee. 'Who are you?'

Bhatt said nothing, his face rigid, unmoved, his hand still bringing the tea up to his lips. He had met the two terrorists two weeks earlier in a village near Shopian. They had never seen him before and he said very little apart from telling them the village he was from. A few days later, he opened up a little more, speaking about how his brother had been killed in an encounter three years ago. Another young man, they thought, looking for revenge, looking for work with a militant outfit, both for a livelihood as well as for closure. At the end of a full week, he spoke his first full sentences, telling them he wanted their help with an attack on an Army checkpoint. He showed them hand-drawn maps depicting the movement of Army patrols along a little-known hill trail, research that suggested this young, bearded man of few words had already begun reconnaissance, the most crucial groundwork for a successful attack on security forces.

Torara and Sabzar were moderately impressed. Bhatt, clearly in his twenties, though the beard hid much of his youthfulness, had demonstrated the motivation to take matters into his own hands—half the battle in the process of radicalization. Tall and well-built, there was no doubt he could be useful in the rough, dark life of a militant in Kashmir. Over the following week, the two Hizbul men questioned Bhatt, presenting him with situations and asking him what he would do. Bhatt's answer would remain the same, 'I need your support, I want to learn.'

Torara and Sabzar were no ordinary terrorists. Both had gained a reputation for leading a highly effective recruitment campaign in south Kashmir. If Bhatt wanted to pick up a gun and get started, these were the men to get in touch with. The men weren't surprised that Bhatt knew who they were.

At the end of two weeks, Torara and Sabzar told Bhatt that they would help with his proposed attack on the Army's foot patrol north of Shopian, but that they needed to disappear for a few days, coordinate the logistics and finer points. Bhatt said he would not return to his village without completing his mission, with or without them. So they took him along to their hideout, where they now sat sipping hot tea.

The attack plan had been detailed and fleshed out. A consignment of grenades would arrive that night. Bhatt would be joined by three Hizbul men, who had been summoned from another village and would show up the following morning. They would then proceed in the evening to launch the attack, with the intention of killing as many of the soldiers as possible as they trudged through a short trough in the trail.

But Torara was having second thoughts. Something didn't seem to fit. Squatting before Bhatt, he asked again.

'Who are you?'

Bhatt, who had been circumspect and soft-spoken thus far, placed his tumbler down on the ground with a splash. Rising to his feet, he took the rifle from around his neck and dropped it on the ground with a clatter. Then, looking from Torara to Sabzar, he spoke, his voice quivering. 'If you have any doubts about me, kill me,' he said, his voice raised to its highest. 'You cannot do this if you don't trust me. So you have no choice but to kill me now.'

Torara rose to his feet, looking at Bhatt closely. And then, just as he turned to Sabzar, perhaps to ask what to do next, Bhatt pulled out a concealed 9-mm pistol and shot both

the terrorists in the head. Sabzar slouched back into the cot.
Torara was thrown against the wall, blood splattering against
the white as he crumpled to the ground. Bhatt fired two more
bullets, to be sure.

As the swirl of gun smoke cleared, Bhatt sat down on
the cot, picked up the tumbler he had set down earlier and
drained the tea. Then he waited for the sun to set before he
could walk, in the darkness, back to where he had come from.
And when he reached there, he would, for the first time in a
fortnight, be able to use his real name: Maj. Mohit Sharma, of
the Army's 1 Para Special Forces.

* * *

Back at his field base before dawn the following day, another
officer in the unit would quip at breakfast, 'You know, Mohit,
with that look you've got there, you'll probably end up getting
captured or killed by the Army itself while you're on your
next covert mission.'

Mohit had replied, 'Killed, maybe. But I'll never get caught.'

Two months later, the twenty-six-year-old officer took
leave, his first in over a year since he had joined the elite
Special Forces unit, to visit his parents in the sprawling
Delhi suburb of Ghaziabad. His brother, Madhur, was at the
railway station to receive him, but when the train pulled in,
the Major was nowhere to be seen. Perplexed, his brother
searched the compartments, wondering if Mohit had missed
his train or got off at the wrong station. It was then that he
realized that Mohit was standing right in front of him, just a
few feet away, silent and completely unrecognizable. He was
wearing a Kashmiri phiran, and his long hair and beard now
covered everything but his eyes. He had simply stood there,
wondering if his brother would recognize him. Madhur was

clueless till Mohit finally shouted his name, the beard parting to reveal a toothy grin.

The young officer had just finished twelve months of living his greatest dream—to be a commando in the Special Forces. At the dining table that evening, his mother, Sushila, a Delhi Jal Board employee, anxiously pressed him to describe what his work was like in Kashmir. Maj. Mohit smiled mischievously and said that he had made the right decision to reject the path his parents had chosen for him.

Nine years earlier, in 1995, fresh out of Delhi Public School, Ghaziabad, Mohit had been persuaded by his parents to pursue an engineering degree. Older than Mohit by a year, Madhur had already been admitted to an engineering college. Like many middle-class families, their parents hoped that engineering degrees would allow both to start a manufacturing business of some kind to take care of the family and settle into a 'stable' life near the national capital.

Strangely, though, for a family that couldn't have had less to do with the armed forces, both their sons wanted to join the military. Very eager to become a fighter pilot, Madhur had made three failed attempts at cracking the Services Selection Board examination, finally giving up and deciding to stick with his engineering course. Mohit, who was now at a crossroads, agreed to go with his parents' wishes. As he said goodbye to them and hugged his brother, he whispered something into Madhur's ear.

'Don't worry, I'll serve for both of us.'

Then the seventeen-year-old hopped on to a train that took him to Amravati in Maharashtra, and then on a bus to Shegaon, where he joined the Shri Sant Gajanan Maharaj College of Engineering.

'He took admission, but from the start, it was clear he did not plan to stay,' says Madhur. 'He had taken the NDA (National Defence Academy) entrance exam before leaving.

When the results came, my parents tried to hide them from him, hoping he would just continue with engineering. But Mohit got suspicious and called the UPSC (Union Public Service Commission) himself to find out if his name was on the list. Without telling any of us, he left his hostel keys with his friends and got on a train from Amravati to Bhopal, which was his centre. There, he took the final test, cleared it and was summoned to Delhi for his medical. That's when he called our parents, announcing that he was coming back home, bag and baggage, and that he would not be returning to his college.'

Unnerved by the sudden turn of events, Sushila was comforted by Mohit's father, Rajender Sharma. They decided they would persuade Mohit to see sense, return to college and finish his degree—to forget about the military and focus on his studies instead. What they hadn't accounted for was just how energized and focused the successful crossing of the first few entrance hurdles had made their younger son. When he arrived home, he would respectfully raise his hand and tell his parents to save their breath if they intended to try changing his mind. The Sharmas had never seen their son this way. It worried them, but they quickly realized that any attempt to talk him out of his chosen path was futile.

With a month left for his medical tests, Mohit marshalled his mother's help to surmount a difficult immediate hurdle—he was 6 kg underweight to make the cut. A diet that included litres of milk, high protein foods and a dozen bananas a day was employed to help Mohit pack on a hefty 8 kg in just four weeks.

Clean-shaven and wiry compared to the hairy, hulking man he would become a few years later, Mohit entered the hallowed portals of the NDA in Pune, quickly proving how committed he was to his journey into the Army. Emerging as one of the best all-round cadets with trophies for excellence in boxing, swimming and horse-riding, Mohit finished his three years at the NDA and then graduated from the Indian Military Academy (IMA) in Dehradun in 1999.

While he was at the IMA, India would fight the Kargil War with Pakistan, an intense fifty-day skirmish that put the two countries on edge. Mohit, along with every other cadet at the Academy, would crowd into common rooms to watch the evening news bulletins carrying reports from TV journalists in Kargil and Dras. These reports would shock, depress, but ultimately galvanize the entire batch of cadets, infusing in them an even greater urgency to get out into the field and serve.

Years earlier, when Madhur had asked him which arm of the military he wanted to join, Mohit had fired back without a pause, 'Infantry. What else is there?'

Watching dispatches containing interviews of soldiers and young officers in Kargil, Mohit found the strongest affirmation yet that the path he had chosen was a worthy one, even if it had meant disobeying his parents. Electrified and angry at Pakistan, like each one of his course mates at the time, the young cadet from Ghaziabad awaited his turn.

On 11 December 1999, Lieutenant Mohit Sharma was commissioned into the 5th Battalion of the Madras Regiment, one of the Army's oldest infantry regiments, which dated back to the eighteenth century and which had the famous war cry '*Veera Madrasi, adi kollu, adi kollu* (Brave Madrasi, hit and kill, hit and kill)!'

That war cry couldn't have been more appropriate. During celebrations and revelry on the night of his commissioning, when Mohit was asked by a course mate where he was hoping to be sent, he replied, 'Anywhere I can hit and kill terrorists.'

'Get yourself into the Special Forces, then,' his course mate said. 'If that's what you're looking for, that's the knife-edge.'

Mohit clinked glasses with him and said, 'That's the plan. I'll get there.'

He would need to wait only five months. In mid-2000, Mohit's unit was sent to the Poonch–Rajouri sector of Jammu and Kashmir to operate under the 38 Rashtriya Rifles (RR) counter-insurgency force. Thrown into one of the most challenging sectors along the LoC, he would frequently operate alongside officers and soldiers from the Army's Para Special Forces units. On foot patrols and reconnaissance missions, from high-altitude cordon-and-search missions to full-blown terror hunts, Lt Mohit would take to his duties with great eagerness. That eagerness would see him peel away the layers of counter-insurgency operations to discover the terrifying complexities he was being entrusted with.

Almost exactly two years into the Army, with over half of that time spent in Jammu and Kashmir, a piece of news exploded on the morning of 13 December 2001. Military intelligence reports streamed in from Delhi as five terrorists from the JeM and LeT breached security at India's Parliament House. They were stopped in their tracks only after a tense and extended firefight that killed nine Indians—six Delhi Police personnel, two Parliament Security personnel and a gardener. The terrifying assault, coming as it did just three months after the 11 September attacks in the United States, was the most insidious and shocking provocation from Pakistan's notorious state-sponsored terror instruments. It was, in effect, a call to war.

In Jammu and Kashmir, where Lt Mohit and thousands of Army personnel were deployed, an already volatile situation had just been violently escalated. The attack on Parliament triggered a massive five-month standoff between India and Pakistan, with enormous mobilization of troops, missiles, artillery, tanks and other weaponry to the LoC and the international border, the largest both countries had seen since they had gone nuclear three years earlier. India code-named the mobilization 'Operation Parakram'.

On the threshold of war, Lt Mohit and his unit were compelled to significantly crank up the tempo of their operations. The Poonch–Rajouri area was seeing major exchanges of fire across the LoC. The hostilities were also aiding the movement of terrorist infiltrators across and through these sectors, resulting in a significant increase in the need for patrolling and search operations.

On one such cordon-and-search operation (CASO) on the Poonch–Mendhar road in early 2002, as Mohit and a Special Forces Major took a break, Mohit asked the officer if he had a chance with the Special Forces. The officer's reply was all the validation he needed at the time.

'*Bhai, dekh* (Brother, look), I've seen absolute beasts of men come and fail to make the cut. And I've seen seemingly unimpressive guys come and nail it better than anyone. So what I'm saying is, there is only one way to find out.'

Weeks later, Mohit would receive his first recognition, a Chief of Army Staff commendation from the Army chief, General Sundararajan Padmanabhan, for leading a counter-insurgency operation in the Rajouri area. But the fight against terrorists was hardly being won. For every small victory, there was a body blow waiting around the corner.

A horrific reminder of how the flow of terror across the LoC hadn't been stopped by the massing of forces came on 14 May 2002. Three Pakistani terrorists infiltrated across the LoC, boarded a bus in Jammu's Vijay Pura, then went on to massacre Armymen and their families at their living quarters in Kaluchak. The dead included ten children (the youngest was four) and eight women, with a total of thirty-one killed. The attack would have a devastating effect on Mohit. On a phone call to a course mate in Chandigarh four days after the attack, he would say he had totally lost the ability to sleep, asking how the terrorists could bring themselves to fire at toddlers.

Then Prime Minister Atal Bihari Vajpayee would describe it as 'a most brutal and inhuman carnage'. If bad blood defined relations between India and Pakistan after Kargil and the Parliament attack, the Kaluchak massacre stirred India into a frenzy of disbelief and fury.

Operation Parakram would see military exchanges along the length of the frontier, but enormous restraint from India, diplomacy and a great measure of global pressure helped ease tensions, resulting in a gradual pullback of troops starting June 2002.

A year later, in June 2003, after being rejected in his first attempt due to an illness, Mohit, now promoted to Captain, was welcomed into the 1 Para Special Forces.

Brigadier Vinod Kumar Nambiar, then a younger officer with the unit and later its Commanding Officer (CO), remembers the time clearly.

'Mohit had come for Special Forces probation to 1 Para. He didn't get selected initially because he was unwell. And when we turned him away, he said he would be back. Normally, people give up. Probation *breaks* you. But Mohit recovered his strength and came back to the same unit again. So his feeling for 1 Para was very strong. He had the humility to come back despite being rejected in his first attempt. Any normal person would hesitate a bit about going back to the same unit for a probation attempt. He could have volunteered for another battalion, to avoid being rejected twice by the same people. To me, that was the determination aspect of it. Not afraid of failures and not afraid to come back to the same challenge. It also showed his love and affinity for 1 Para "*ki kuch bhi ho jaaye* (whatever happens), I will come back to the same unit."'

Getting to wear the maroon beret and *balidaan* ('sacrifice') badge for the first time would be a day of solemn reflection for Mohit, once the celebrations at his unit had died down. On a phone call home, his mother, Sushila, who had become used to worrying every day ever since her son was deployed in J&K, would now demand to know why he had felt the need to move into even more dangerous territory with the Special Forces. At that moment, Mohit had news he knew would calm his mother down—he was to report shortly to Chandimandir, the headquarters of the Army's Western Command on the outskirts of Chandigarh, to take charge as a Special Forces team leader. He was right. Sushila Sharma did exhale, but knew it was only a matter of time before he was sent back to Kashmir.

'You have finished your responsibility in Kashmir, Mohit,' his mother said. 'Why don't you seek a posting elsewhere now?'

In Chandimandir, Mohit met and fell in love with Captain Rishma Sareen, an officer with the Army Service Corps (ASC). Months into their relationship, in early 2004, Mohit would be summoned back to Kashmir. It was at this time that he would spend weeks planning a dangerous, deep-cover mission to kill the two dreaded Hizbul Mujahideen recruiters, Abu Torara and Abu Sabzar, near Shopian. The hair-raising mission would win him a Sena Medal, with a deliberately vague citation crafted to mask the true nature of what he had managed to achieve. The mission instantly became legendary.

'He did share that story with me a few weeks later, when he came to see me,' says Rishma, now a Lieutenant Colonel posted with an Army unit in Haryana. 'The story shocked me—it would have scared anyone. I asked him to please be careful. I mean, I knew for sure that he was very good at what he did. I knew he planned everything very well. He was very meticulous by nature. He was professionally very sound. So I didn't panic, but I did ask him to be careful.'

On 19 November 2004, Mohit and Rishma were married in Chandigarh. Two months later, with less than two years of Special Forces experience under his belt, Mohit was dispatched to Belagavi (previously Belgaum) in northern Karnataka for two years to serve as an instructor at the prestigious Commando Training Wing of the Infantry School there, followed by a two-year stint in Nahan in the hills of Himachal Pradesh, home base of the 1 Para unit and the Indian Army's Special Forces Training School (SFTS), the largest elite warfare training institute in the country. While there, he would drive every week to see his wife, who was posted at the Army base in Patiala, a little over two hours away by road. Mohit's parents had welcomed the idea of their son working at the two training schools. He had won two gallantry awards in J&K—surely he had nothing more to prove?

The training in Nahan was pointed in one direction, and one direction only—which his parents knew but dreaded. In October 2008, Mohit, now a Major, was summoned back to Kashmir, arriving on the threshold of one of the fiercest winters in decades.

Over the next five months, through steadily worsening weather, Maj. Mohit and his team of Special Forces men would prowl the countryside in the forbidding snow-blown stretches of northern Kashmir, operating alongside the RR as he had at the start of his career.

In spring 2009, when the snow had reluctantly begun to melt, Maj. Mohit and his team were at their base in Kupwara. It was the afternoon of 20 March. His buddy,[1] Havildar Rajeev Kumar, then twenty-two years old and holding the rank of Paratrooper, remembers it like it was yesterday.

'It's been more than ten years, but there are some things that you can never forget throughout your life,' says Havildar Rajeev. 'I remember the sequence of events very clearly. We were being briefed by our team leader, Maj. Mohit, at our

---

[1] The buddy system, which dates back to World War II, places officers and soldiers into pairs to enhance efficiency, bonding and lethality in combat. The system is based on the theory that camaraderie will amplify safety and combat effectiveness.

base in the Kupwara sector about an anti-terror operation to
be conducted in Bangus Valley.[2] Mohit Sir had a solid network
of informants and he had received some intelligence about
terrorist activity in that area. We were deployed in Kupwara
with 7 Sector, RR. While Maj. Sir briefed us, he got a call
from the Sector Commander. We knew it was an important
call because he cut the briefing short. The Sector Commander
told Mohit Sir that a terror squad had crossed the LoC and was
hiding in the dense Haphruda forest. Our team leader told us
that the Bangus operation was no longer on and we would all
be proceeding towards Haphruda instead.'

The team knew immediately that something big was
unfolding at Haphruda. Every man in that briefing room was
familiar with the tone and tenor of Maj. Mohit's voice. This was
clearly something urgent, something that could not wait. The
men didn't have all the details yet, but they knew they had been
placed on quick reaction alert, which meant they were poised to
leave their base within minutes of being asked to do so.

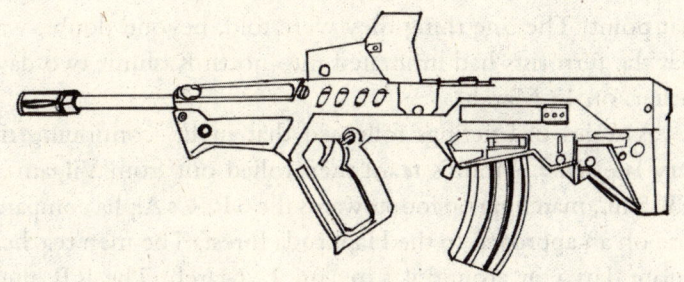

Tavor TAR-21 assault rifle

---

[2] Bangus Valley in Kupwara district is a stunningly picturesque sub-
valley at 10,000 feet, rich in animal and plant life and sprawling sun-
kissed meadows in the mountains. But for the security situation,
this would count as an effortlessly beautiful tourist destination.
Efforts are now on to build more infrastructure there.

'Within a few minutes of that call from the Sector Commander, we left our base at 3 p.m. and headed towards the headquarters of the 6 RR Battalion based at Vilgam,' says Havildar Rajeev. 'We always stay ready for operations and can launch within minutes. It was dinnertime, about 9 p.m., when we reached Vilgam. There were twenty-five of us, including Maj. Mohit.'

After reaching the 6 RR base, Maj. Mohit quickly met the CO of 6 RR, receiving a briefing on the infiltration and the operation he now needed to launch and lead. He then communicated the full picture to his team, including a situation report from the forest. An hour earlier, a team from 6 RR had spotted some terrorists in the Haphruda forest and fired at them—for the second time that day. The terrorists had managed to give their attackers the slip again and had ventured deeper into the darkness of Haphruda.

The reason Maj. Mohit's team had been called in was clear. It was to hunt down the terrorists that had escaped the 6 RR net. It was unclear how many terrorists there were in the group, and details about their movements remained sketchy at that point. The one thing they were told, beyond doubt, was that the terrorists had infiltrated into north Kashmir two days earlier, on 18 March.

A series of briefings followed that night, continuing till very late. Maj. Mohit's team then rolled out from Vilgam at 2.30 a.m., marching on foot towards the 6 RR's Alpha company base on an approach to the Haphruda forest. The men reached before dawn, at around 4 a.m., on 21 March. The RR men there knew that the Special Forces team was coming and were waiting for them. In the shadow of the looming Haphruda forest, Maj. Mohit met the 6 RR Alpha company commander, also an Army Major.

The 6 RR squad that had fired at the terrorists was from the Alpha company. Maj. Mohit and his men asked an officer and two men from the Alpha company to take the Special Forces team to the point where they had fired at the terrorists, which was nearly 700 m inside the forest. He then told the 6 RR men to return to base, saying the operation would be conducted from that point by the Special Forces team. Maj. Mohit's team checked their weapons again before trudging deeper into Haphruda.

Another soldier from Maj. Mohit's team, Naik Hajari Lal Gurjar, holding the rank of Paratrooper at the time, recalls how the next few hours played out.

'The sun had risen and was out in the skies, but the Haphruda forest is so dense that sunlight doesn't reach the ground in many of its parts,' says Naik Hajari Lal. 'It was a deviously thick forest, offering a large number of hiding places to terrorists. It was around 8 a.m. and it seemed like it was the middle of the night. A few minutes later, we noticed something. It was our first clue. We could see footmarks in the snow. It had to be the terrorists. Mohit Sir asked us to be very careful, watch every step we took, and scan every nook and cranny around us.'

'*Woh jungle itna bada hai ki kuch pata nahin woh kahan chhupe ho sakte the* (The forest was so vast that there was no way to tell where they could have been hiding),' says Havildar Rajeev. 'The ground was covered in a sheet of snow at least two-and-a-half feet deep. The entire squad halted as soon as we saw the footprints.'

Maj. Mohit's team had been split into three squads by this time, with eight men each. One man from the team had been asked to remain at the 6 RR Alpha company base to coordinate with the 1 Para Battalion headquarters in Srinagar, in case additional forces were needed.

One squad was led by Maj. Mohit himself, along with his buddy, Havildar Rajeev. The second squad was led by Naib

Subedar Uttam Chand with his buddy, Naik Hajari Lal, and the third was led by Havildar Rakesh. The second squad, led by Naib Subedar Uttam Chand, was ordered to deploy along an elevated ridge inside the forest, so it could keep an eye on the other two squads moving forward on lower ground and give them warning or cover, if required. The two squads, led by Maj. Mohit and Havildar Rakesh, followed the footsteps in the snow, moving forward as noiselessly as possible.

'The terrorists were moving very cleverly, most likely in single file,' says Havildar Rajeev. 'It appeared that the terrorists following the lead terrorist were deliberately stepping right into his footmarks and moving forward, so it was difficult to even guess how many terrorists we were hunting that day. A small thing, but a pointer to how well they were trained to evade us.'

Separated by a short distance, the two squads on lower ground trudged forward, crossing a ridge after a few minutes and halting. Maj. Mohit ordered the third squad to move quickly and hold that ridge, and then continued to move forward soundlessly. Every soldier in the forest that day remembers how silent it was that morning.

The men then hit a patch where the snow had melted, washing away further footprints. The trail they had been following ended abruptly. The men stopped, looking carefully in every direction for where the footprints could have moved to next.

'At that point, we got a call over the radio from the three men from 6 RR's Alpha company, who had earlier been asked by Mohit Sir to return before the Special Forces could take over the operation,' remembers Havildar Rajeev. 'All of us had radios, and all of us could listen to the conversation. The Major from 6 RR said that they had spotted footprints in the snow in another location inside Haphruda. So we linked up with the three men from 6 RR. Now we were two squads of sixteen men plus the three RR men, including their company commander. The third squad continued to move on higher ground to cover us. We again started following the footprints. But once again, we hit a dead end. There was no snow there, there was a small *nallah*[3] and the footprints just vanished.'

Just as the squads were deliberating about which direction to head in, there was a breakthrough.

The team's scout, Paratrooper Netra Singh, who was marching ahead of Maj. Mohit's squad, froze in his tracks and relayed a message over the radio that he had spotted two terrorists. He quickly added that the terrorists had spotted him too at virtually the same moment.

And then it began.

'Less than a second after Netra Singh radioed us about the two terrorists, our squad came under heavy automatic fire,' says Havildar Rajeev. 'Netra took a bullet straight to his forehead and collapsed to the ground. He was killed instantly, the first casualty of the operation. Nothing could be done. It was a headshot.'

In seconds, the intensity of the attack escalated, with bullets flying in from multiple directions now.

'It is hard to explain the intensity of that incoming fire,' says Havildar Rajeev. 'Bullets were flying all around us. They seemed to be coming from all directions. We all fell flat on our stomachs and took whatever cover we could. Someone took cover behind a tree, someone took cover behind a boulder. *Jo kuch bhi humein mila, humne ussi ke peeche cover le liya* (We took cover behind whatever we could find).'

---

[3] Nallahs are mini tributaries of rivers, more like small streams, common across Kashmir. They are commonly used by infiltrators as pathways for undetected movement.

It was 11 a.m., and it was clear that the terrorists were spread out ahead of the squads. The bullets were flying in from the front. Maj. Mohit ordered the men to engage the terrorists by approaching them from the right. The moment a pair of soldiers tried to crawl in that direction to take position, they were pushed back with heavy firing from the right. It was now abundantly clear that they weren't up against just one small group of terrorists.

An attempt to engage the terrorists from the left was met with a similar burst of pre-emptive fire. The terrorists knew exactly where the squads were. And the firing was becoming more accurate.

'*Jis tarah se hum par firing ho rahi thi, aisa lag raha tha ki terrorists ne humein gher liya hai* (The way we were coming under heavy fire, it seemed the terrorists had surrounded us),' says Havildar Rajeev, who was, at that time, crouched behind a tree a few metres away from his team leader. 'The pattern of firing made it clear that the terrorists had spread out in a semicircle, in a C-shape, ahead of us and that's why it appeared that bullets were coming in from all sides. They had formed an arc around 50–60 m away and all of them had AK-47s. So their guns were very effective at that range. They were terrorists trained military-style. It was obvious, given the way they brought us under fire.'

Then Maj. Mohit radioed Havildar Rakesh, who was leading the second ground squad, and told him to head left, try to get up on a ridge and engage the terrorists. The moment Havildar Rakesh headed in that direction with his squad, the terrorists started firing at them. His squad split into four buddy pairs and fired back at the terrorists with their Israeli Tavor TAR-21 assault rifles.

'In front of my eyes, Rakesh was hit by a burst of bullets in his thigh,' says Havildar Rajeev. 'His leg seemed to have been split open. He immediately removed his combat patka (the scarf that Special Forces men wear) and administered himself first aid to prevent excessive blood loss. And right through this he kept firing at the terrorists with one hand.'

With one man dead and another badly injured, Maj. Mohit quickly realized that their vulnerability had gone up several notches. Quickly assessing the situation, he decided that the squads needed to retreat a short distance to find protective cover, so they had a moment away from the line of fire to plan their counter-attack.

'The terrorists clearly had the better position,' says Havildar Rajeev. 'I asked Mohit Sir to move back and take cover, but he kept firing and said he would stay right there and engage the terrorists so that other squad members could move back and take cover. I said, "*Sahab, aap pehle cover lo. Main inko sambhalta hoon* (Sahab, you take cover first. I will handle them)." But Mohit Sir said, "*Main nahin, tum jao. Main last mein aaoonga* (Not me, you take cover. I'll take cover in the end)."'

Havildar Rakesh's squad continued to return fire from halfway up the ridge. The squad on higher ground, led by Naik Subedar Uttam Chand, hadn't yet linked up with the two ground squads at that point. As Maj. Mohit provided cover fire to the squads pulling back, a bullet tore into his left arm. Havildar Rajeev watched as his team leader quickly tied his patka around the wound to prevent blood loss, and then resumed firing non-stop at the terrorists.

'I called out to him, saying, "Sir, pull back, let me check your wound." He replied, "Nothing has happened to me. I have only been shot in the arm, Rajeev. You keep firing and moving back till you have good cover. Don't worry about me. Don't stop firing."'

The sun was high in the sky by this time, but none of the men could see it.

'The idea was to get in the nallah and get cover as soon as possible,' says Havildar Rajeev. *'Netra Singh shaheed ho gaye the, Rakesh Sir ko kaafi goliyan lagi thi aur Mohit Sir bhi ghayal the* (Netra Singh was dead, Rakesh and Maj. Mohit were injured).'

Still firing from the front, Maj. Mohit ordered a soldier on his squad, Lance Naik Subhash Singh, to fire at the terrorists with his multi-grenade launcher (MGL). But just as the soldier positioned himself to fire the grenades, a bullet tore through his elbow, incapacitating him. There was no way he could effectively operate the weapon.

'Mohit Sir saw Subhash getting hit, ordered him to move back and took his MGL,' says Havildar Rajeev, who, by this time, had taken cover and was still firing at the terrorists. 'Mohit Sir then started firing the MGL at the terrorists. By now, his entire squad had taken cover. Mohit Sir fired six grenades from the MGL in the direction from where the maximum fire was coming at us. After that bombardment, the firing from the terrorists slowed down considerably. *Woh shaant ho gaye the. Firing bahut kamm ho gayi thi.* We didn't know it at the time, but the bombardment had killed four terrorists.'

As the smoke from the grenades cleared, Maj. Mohit crept back towards the rest of his men to finally take cover alongside them. Just as he was about to reach them, a bullet smashed into the left of his chest. His bulletproof vest provided protection in the front and back, but the sides remained vulnerable. The bullet tore right through him. He staggered a little, but stayed on his feet. Havildar Rajeev rushed to his team leader.

'I am fine, Rajeev, these are not serious wounds,' Maj. Mohit told his buddy. 'Keep firing at them. We cannot let them get away today.'

By this time, the squad deployed on higher ground reached the site of the firefight. Naik Hajari Lal, who was on that squad, saw a clearly injured Maj. Mohit leaning against a tree and firing at the terrorists without pause.

'Mohit Sir kept telling my squad *ki terrorists ko bhagne nahi dena. Upar se fire karo inn par* (that don't let the terrorists escape. Fire at them from your height). The only thought in his mind was that the terrorists should not escape. It appeared to us that there were around eight to ten terrorists, given the volume of incoming fire.'

Maj. Mohit had turned visibly pale from blood loss. But he didn't stop firing. And over the roar of the crossfire, the men could hear their team leader on the radio, his voice short of breath now, '*Firing rukni nahi chahiye. Hum issi ke liye SF* [Special Forces] *mein hain aur issi ke liye train karte hain* (Don't stop firing. We joined the SF [Special Forces] for this and this is what we train for day in and day out).'

Havildar Rakesh's squad then got their rocket launchers out and started firing at the terrorists, giving the soldiers a chance to climb higher up and assume what they hoped would be an advantageous position. Maj. Mohit had stopped firing by this time. He was sitting with his back against the tree he had taken cover behind.

'*Buri tarah woh ghayal ho gaye the* (He was grievously injured),' Havildar Rajeev says. '*Poore squad ko unhone wahan se nikala tha* (In that condition, he had helped the entire squad take cover).'

The situation was grim.

'Four commandos in our squad had been hit,' remembers Havildar Rajeev. 'Netra was no more, the man behind him was also injured, Subhash had been hit and Mohit Sir had taken bullets. Despite being so gravely injured, he kept firing for as long as he could, kept telling us not to leave cover, and keep engaging the terrorists. He just refused to stop.'

Minutes later, Havildar Rakesh would be killed in the crossfire. Attempts by the other soldiers to pull him to safety failed, because the terrorists were using his body as a trap—whenever somebody tried to crawl towards him, the terrorists would start firing to cause more casualties.

Maj. Mohit had gone completely still, but his eyes were open. Firing from the terrorists prevented the soldiers from reaching him.

'He kept saying, "*Meri injury normal hai* (My injury is normal)." He was only concerned about two things that day: that we should not take casualties, and that the terrorists should not get away. *Woh akhiri dum tak bas yehi bolte rahe, "Main theek hoon. Baakiyon ko dekho"* (He kept saying till the end that he was fine and asked us to take care of the other injured commandos). He ensured that every man on the squad had taken cover before he did.'

At 4 p.m., Havildar Rajeev managed to reach his team leader, still seated with his back to the tree, clutching his weapon. Maj. Mohit had bled out and wasn't breathing any longer. The magazine in his rifle was empty.

Along with Maj. Mohit and three from his squad, a total of eight Special Forces men were killed in the Haphruda operation, which continued for four more days, till 25 March. A total of twelve terrorists would be killed in the operation.

'He kept guiding us throughout that firefight. We didn't even know the exact extent of his injuries. He was just not bothered about his safety,' says Naik Hajari Lal. 'By late night, multiple squads from our unit had come down to Haphruda forest from different locations and cordoned off the entire area. Choppers were flying above us, looking for any patch where they could land to evacuate the injured commandos to hospital. But there was no place for the choppers to land in the dense forest.'

Havildar Rajeev and another soldier carried their team leader's body back to the 6 RR Alpha Company base.

'His body was still warm. I was hoping he would survive,' Havildar Rajeev says. 'It was not possible to think of him as dead. His back was on my back and the other man was holding his legs. There was no time for a stretcher. There was no time to check vital signs. His eyes were closed and he was unconscious. We carried him for over a kilometre. I kept thinking throughout that Mohit Sir had guided and protected us non-stop till he fell unconscious. Even after taking so many bullets, he was only concerned about us. *Unko humari life ki zyada chinta thi* (He was more worried about our lives). When I was carrying him, I just wanted to take him to the chopper as quickly as possible. He was taken straight to 92 Base Hospital in Srinagar. We returned to join our squad in the forest. We learnt only the next morning that Mohit Sir did not make it. Carrying Mohit Sir that day to the helicopter is the heaviest burden I have ever carried.'

Maj. Mohit's CO, Colonel Vinod Kumar Nambiar, had arrived and was monitoring the situation from an adjoining area.

'Mohit laid the foundation for the success of that five-day operation,' says Col. Nambiar. 'He had killed four terrorists himself with the MGL. Two or three more terrorists were also killed by his squad. Because of Mohit, the terrorists were unable to break away and run from the firefight. And by that time, we had positioned multiple teams. It was his decision to stay engaged even after so many injuries, because of which our boys were able to finish them off.'

Naik Hajari Lal says, 'Our squad was more furious than sad. We had to avenge Mohit Sir's death. If we were alive, it was because of him. In an intense firefight, it is difficult to remember or recall every detail, but his last order was that the terrorists should not escape. Those were his last words. And that was our only focus. I was with him till his last moments. He knew, perhaps, that it was his last firefight. But he kept motivating us till the very end. We kept thinking that Sa'ab is OK.'

'Had Mohit Sir not guided us and led us the way he did that day, perhaps I too would have returned home in a coffin,' says Havildar Rajeev, his eyes welling up. 'I was only twenty-two then. *Aakhiri saans tak unhone humein lead kiya ek sherdil commander ki tarah* (He led us till his last breath, just as a braveheart commander should).'

His CO cannot forget the reports he heard from the other soldiers after the operation.

'When three of his boys have died, he has also been hit, and he is sticking around, leading his men from the front and continuing to fight, it means a lot and it raises the morale of others; it motivates them. And him sticking around despite his injuries was precisely what motivated the other squads there.'

All twelve terrorists killed that day were Pakistani nationals from the LeT. The firefight with an elite Indian Army squad proved just how well-trained and armed the terrorists were. They had scale maps of the whole area on them—some better than the ones used by the security forces. They even had route maps.

'We lost eight men that day, some of the best trained men in the Indian Army. That can happen in battle. But how could it happen to someone like us, the 1 Para?' Col. Nambiar remembers wondering at the time. 'We had a sense of professional arrogance. How could we take such casualties? We are good, we are well-trained, we have worked hard, we have prepared for such scenarios, we have conducted some remarkable operations in Kashmir in the past. We knew we were very good. Against that background, when something like this happens, it was like, "How could this happen to us?"

I still think of it. I play it back in my mind very often. The terrorists were on a height. They had some advantage. I think they had a sharpshooter in their squad, someone with the ability of a sniper. That person took accurate shots. While the other terrorists were firing, this person was taking his time and picking his targets. Invariably, most casualties happened when they were trying to retrieve injured commandos. I think it was this guy who caused the maximum damage. I play the whole thing in my head over and over even now.'

When a message was transmitted to the 1 Para base camp informing them that Maj. Mohit had been hit and was sitting motionless, saying nothing, an officer at the camp remembers replying with disbelief.

'It was not possible to imagine him lying motionless,' he says. 'I kept asking the jawan to make sure, and then make doubly sure. How could Mohit be gone? But they replied, he isn't moving and he isn't saying a word. It took us a long time to accept this.'

At 1 a.m., just under 1000 km away in Ghaziabad, the phone rang at the Sharma residence. Maj. Mohit's father answered the call.

'Two officers were at the other end of the line, saying they wanted to visit us,' says Rajender Sharma. 'I woke Mohit's mother and shared this with her. She found it strange that Mohit's friends were coming over this late and that Mohit had not informed us. She felt there was something amiss. Then we called his wife, Rishma, who was in Patiala at the time. His colleague Maj. Bhaskar Tomar took the call, but didn't tell us, despite knowing what had happened. Half an hour later, the two officers arrived at our house. They shared the news with us eventually.'

Sushila Sharma lost consciousness when the two officers finally said the words. In Patiala, Mohit's wife, Maj. Rishma, had also fainted when she was told.

'The news was broken to me on the night of 21 March by his unit officers, accompanied by my formation officers,' says Maj. Rishma. 'They had come to my house. I went blank. I couldn't believe what I had heard. I fell unconscious. I was taken to the intensive care unit at the base hospital and was there all night. I was taken in an ambulance to Ghaziabad the next day, 22 March.'

Every 21 March, Maj. Mohit's unit observes a two-minute silence in his memory and of those who died with him.

'Everyone in the unit knows about that operation,' says Havildar Rajeev. 'He never made us feel he was our senior. He treated us as equals. In fact, that's the culture of 1 Para, officers and soldiers intermingle a lot. Eat together, live together, stay together, fight together.'

A decade since the operation, there isn't a Special Forces soldier who doesn't know about the Haphruda operation or the covert strike by Iftiqar Bhatt.

Lt Col Vikas Dhuria, Second-in-Command of the 1 Para Special Forces, says, 'I first met Mohit Sir when I was doing my probation with 1 Para Special Forces in 2004. People outside the unit may not know about it, but all of us in 1 Para Special Forces had heard about what he had done in Shopian, masquerading as a terrorist and killing two of them. It used to be a dream for us to conduct such an operation. It involved a lot of risk. Few can pull off something like that. I actually cannot think of anyone who could do so except Mohit.'

Maj. Shantanu Sinha, twenty-eight years old and currently adjutant of 1 Para (having joined the unit in 2012, three years after Mohit's death), says, 'The first thing that anyone joining 1 Para learns of is Maj. Mohit's gallant actions. One of the tasks for probationers is to find out about the major operations of the unit. His actions still motivate everyone in the unit.'

'In many incidents, you have a person suddenly becoming brave in the heat of the moment. In Mohit's case, it wasn't in the heat of the moment. It was repeated—he was brave each and every time. I will say it was in his grain,' says his CO at the time, Col. Nambiar. 'His bravery was there for all to see, and not just in the battlefield. It was visible during training and in other aspects also. That moral courage was remarkable and that physical courage was also remarkable.'

Speaking of the Iftikhar Bhatt operation, Naik Hajari Lal says, 'He could have easily died in that operation. We would not have been able to even find his body. But he still went in. *Bahut bada jigra tha unka. Itna dum rakhte the woh* (He had a big heart. He had a lot of courage). Nothing mattered more to Mohit Sir than operations. The moment he would get a lead or an intelligence input, he would be ready to launch an operation. *Sher ki tarah toot kar padte the ki chalo abhi, operation karte hain* (Like a lion, he would pounce at the opportunity and say, let's go, let's launch an operation). He believed in launching operations quickly. He was in Kashmir only to conduct operations. *Operation ke maamle mein shayad koi nahin hoga unke jaisa* (There might not be many like him when it comes to operations). He knew the area very well and had an excellent network of informants. *Agar source ka phone aaya toh Rishma ma'am ka phone woh rakh denge* (He would choose to speak to his source over Rishma ma'am).'

Maj. Mohit's buddy, Havildar Rajeev, says that the officer's story is incomplete without a mention of his guitar. 'Guitar *hamesha saath rakhte the. Unke haath mein ya toh Tavor hoti thi, ya guitar* (He would either be holding a Tavor in his hands or a guitar),' says Havildar Rajeev.

The men of 1 Para Special Forces also remember Maj. Mohit for a particularly morbid sense of humour. Naik Hajari Lal recalls how he had completed a medical course several months before the 2009 operation.

'*Mohit Sa'ab asked me kya grading aayi hai.* I said, "*Sa'ab, grading toh B hai.*" He said, "*B toh theek hai par agar mujhe goli lagegi toh mujhe bacha lega na* (B grading is fine, but if I get shot, you will be able to save me, won't you)?"'

At 6.45 a.m. on 23 March, Maj. Mohit's body reached Ghaziabad, where his inconsolable mother, wife and father received him amid crowds from the locality and several more who had arrived to pay their respects. Later that morning, the officer's mortal remains were transported to the Brar Square crematorium in Delhi Cantonment, where, in the presence of senior officers from the Special Forces, his pyre was lit by his brother, Madhur, who had given up his plans to move abroad as a software engineer when Mohit had decided to join the Army.

A decade later, the wounds of their loss are still fresh for Mohit's parents.

'I can't hold back my tears when I talk about my Mohit,' says his mother, Sushila. 'He was such a wonderful and caring son. He left us ten years ago, but we feel his presence in every corner of our home. We have not stopped crying. All parents love their children. But the qualities he had, he was different. I find it very difficult to talk about Mohit even today. We are his parents and you can understand how we feel, having lost our loving boy at such a young age. We have lost everything. He had tremendous respect for us and he would stop associating with anyone if they had a bad word to say about his parents. There can be no one like Mohit. Yes, he made the country proud and served with honour and bravery, but nothing can heal the loss of a child. For a mother, the void left by the death of her child can never be filled. A son gone will never come back.'

Sushila Sharma looks at pictures of Mohit as a boy, and one of how his face had lit up when his parents had brought him a Casio keyboard from Nepal, a gift to encourage his love for music.

Mohit's father, Rajender Sharma, a retired employee of the Punjab National Bank, says, 'Our life has come to a standstill after Mohit left us. Nothing can fill the void he has left behind. We miss our son every moment of our lives. It's a wound that will never heal. My wife and I often talk about his childhood and go through our old photo albums. Mohit would always say that if he lived, he would rise to the rank of General, but if he was killed in action, he would earn the top-most gallantry award.'

On 15 August 2009, Maj. Rishma would receive her husband's posthumous Ashok Chakra, India's highest peacetime gallantry award, from Pratibha Patil, the then President. Rishma had last spoken to her husband the morning he died.

'It was a short call. He never told me about operations,' Maj. Rishma says. 'He would never disclose those things. He only said he would not be available on the phone for some time, and I understood he was out there doing what he liked most— conducting an operation. He was to come home on leave in a week's time, around the end of March. He had delayed his own leave as he wanted his team's officers and men to be able to visit their homes on Holi. He was very concerned about them. He used to consider them at par with family. He postponed his own leave. We had bought a house in Noida and were planning to take possession during Mohit's leave. These ten years have definitely been difficult. And every single day, I have thought of Mohit and missed him. I miss the songs he would sing to me with his guitar. His two favourites to sing to me were "*Pal pal dil ke paas*" and "*Pyar humein kis mod pe le aaya*". The love we had for each other gives me immense strength. He was the best. There can be nobody else like him. When you have been with the best, you just can't think of anything else. I miss him. He was a happy guy. The rage and aggression would surface only when he was conducting operations.'

'"*Pyar humein kis mod pe le aaya*" was Mohit's signature song,' says Maj. Shantanu Sinha of his unit. 'We still play it sometimes in his memory at our parties.'

Col. Nambiar, now a Brigadier and a decorated commando himself, says the Special Forces will never stop talking about Maj. Mohit Sharma.

'Loads of his friends keep meeting up and they come to me. We miss him. It is not about mourning his death, his sacrifice, it's about celebrating his life and his legacy. *Yeh nahin ki aankhon se aansoo nikal rahe hain, inspiration waali baat hai* (It's not that we shed tears, it's more to do with being inspired by his bravery). We stare into the eyes of death every day. It's like, Mohit has done such a great job, so how can we not do better and make him proud?'

Maj. Mohit's name and legacy have been preserved in many ways. Apart from the infinite recounting of his exploits within the Special Forces, a road and a metro station were named after him in Ghaziabad, and his family has established a trust that gives out scholarships and holds medical camps in his name each year.

'We in 1 Para rejoice that he walked this earth and he was one of us,' says Col. Surendra Singh Rajpurohit, current CO of 1 Para. 'Mohit will forever inspire 1 Para and the Indian Army. Men come and men go. Being remembered the way Mohit is is the ultimate honour for a soldier.'

Before he set out for the Shopian operation in March 2004, a soldier remembers hearing another officer call Mohit by his name. Mohit turned around, his bushy beard and hair covered in a checked scarf.

'*Mohit nahin, Sir, Iftikhar.*'

2

'He Avenged Them, Didn't He?'

Corporal Jyoti Prakash Nirala

The familiar tone of an incoming WhatsApp video call interrupted proceedings inside the darkened operations room at a secret Garud Special Forces base in Kashmir's Manasbal. The call was from three-year-old Jigyasa. In one corner of the room, a commando pulled out the phone from his pocket, looked at it furtively for a moment and then cancelled the call. Five hundred kilometres away in Chandigarh, the child would try to reach her father three more times over the next few minutes before giving up. She had to—he had switched his phone off.

There was no way Corporal Jyoti Prakash Nirala could have taken the call. He thanked technology every day for allowing him to video chat with his baby daughter, but that day, on 17 November 2017, Jigyasa had chosen to dial him when he was halfway through the most sensitive mission briefing of his life. The sort of briefing where there was no option to excuse himself even for a moment. He would speak to her later, after the briefing was done.

The dimly lit operations room, like most the Army used, had maps covering nearly all its walls. At one end stood Sqn Ldr Rajiv Chauhan, thirty-six, the oldest man in the room and CO of the IAF Special Forces unit, 617 Garud Flight. Facing him were eighteen commandos, each man listening intently as their boss shared a piece of fresh intelligence with them. Intelligence the unit had been awaiting for weeks.

'I've just received word that Osama Jungi and Mehmud Bhai have been tracked to a house in Chandargeer. It's a reliable input from one of our known guys. He has said we can launch a hunt immediately,' Sqn Ldr Rajiv told the men.

He didn't need to say the names twice. Every man in the room knew that Osama Jungi was special—if the blood that flowed through him was any measure, he was terror royalty. Jungi was closely related to India's most wanted Pakistani terrorist, Hafiz Muhammad Saeed, founder of the LeT and its front organization, the Jamaat-ud-Dawa. Known by the alias 'Ubaid' in Kashmir, Jungi was the son of Hafiz Saeed's brother-in-law, Abdul Rehman Makki, the LeT's second-in-command, and nephew of the LeT commander, Zakiur Rehman Lakhvi, a key Pakistani terror boss who helped organize the 26/11 attacks in Mumbai. Ajmal Amir Kasab, the lone terrorist captured alive in the Mumbai attacks, would later refer to Lakhvi during an interrogation as 'Chacha Zaki', a term of familiarity that established Lakhvi's role in the operation. For Osama Jungi, Lakhvi was an uncle related by blood.

The other man named in the briefing wasn't of high terror pedigree, but no less notorious—Mehmud Bhai was the LeT's north Kashmir commander. In classified lists of wanted terrorists in Kashmir, both names were high up on the first page.

Corporal Jyoti knew what every man in the room was thinking. Actually, for over a month, they had thought of little else.

Only thirty-seven days earlier, two men from their unit, Sergeant Khairnar Milind Kishor and Corporal Nilesh Kumar Nayan, had painstakingly tracked down and killed two senior LeT terrorist commanders in a brutal, brief eight-minute pre-dawn operation at Bandipora's Rakh Hajin village. It was an operation that would have tested the IAF in a never-before role, and its Garud commando force had only recently been deployed in counter-terror operations. At 4.40 a.m. on 11 October 2017, while Sergeant Milind fought and killed one of the surrounded terrorists at close range, even after being shot multiple times, his buddy, Corporal Nilesh, in an act of indescribable courage, took direct fire from the terrorists just so he could provide cover to his track leader. He killed the second terrorist before succumbing to his injuries.

The killing of these terrorists, Abu Bakar and Nassuralla Mir, had come at the heaviest price imaginable for the Garuds—both soldiers returned to their villages the next day in flag-draped coffins. Both would later be posthumously awarded the country's third-highest peacetime gallantry award, the Shaurya Chakra. Milind would be decorated for displaying 'bravery of the highest order in leading the attack', while Nilesh would be commended for having displayed 'extreme valour and the highest order of camaraderie with total disregard to personal safety'. Their partnership and sacrifice would serve as a reminder of what it meant to be 'blood brothers', a phrase that's just a cliche outside the military.

Since then, not a moment had passed when the men of 617 Garud Flight hadn't wanted revenge on the LeT.

'The operation that we are about to launch could be one of the biggest we Garuds have undertaken in Kashmir,' the CO told his men. 'Milind and Nilesh are gone because of these people. It's time to make them fear the Garuds and show them what we are capable of.'

No detail was spared during that hour-long briefing. Every patch of the layout of Chandargeer, the village the two terrorists were said to be in—the number of houses, count of inhabitants and the approach that the squad would take—were all laid out on a series of projected images on a screen. Not that the commandos were not familiar with Chandargeer—it had been on their radar since the 11 October operation, which involved not just the Garuds, but also the Army's 9 Para Special Forces and two units of the counter-insurgency force, the RR.

The unit had actually been on the hunt for Osama Jungi and Mehmud Bhai long before the fatal Rakh Hajin operation. Both terrorists had managed to give the Garuds and their counterparts in the Army's Special Forces units the slip on a number of occasions, reinforcing the superiority of the LeT's formidable on-ground intelligence and its local network, enforced as much with ideology and radicalization as with threats of violence to families. The deaths of their two comrades had energized the Air Force unit's hunt. When their CO revealed that the terrorists' locations had finally been reliably tracked to Chandargeer, every man in that room was thinking of payback.

Corporal Jyoti certainly was. A week after the loss of Milind and Nilesh, he would proceed on leave for Diwali to see his family in Chandigarh. Over three days, he spoke of virtually nothing else, remembers his wife, Sushma Nand.

'Jyoti surprised us by landing up at home on Diwali, 19 October,' says Sushma. 'I immediately sensed that he was a changed man, shaken by the loss. He showed us pictures of Milind and Nilesh and talked endlessly about them. He was in a very emotional state and couldn't hold back his tears. He wanted to get even, but he was just in deep mourning. They lived and worked like brothers.'

Nudged by his wife, Corporal Jyoti would change the subject and put on a happy face when Jigyasa accosted him for stories, always with an endless barrage of questions about his work. They had named her prophetically—'Jigyasa' means curiosity, inquisitiveness.

'I could tell that it was very difficult for him to come to terms with the loss,' says Sushma. 'It was Diwali and we were spending happy times with our extended family. But Jyoti was a broken man. I would find him awake in the middle of the night, in tears. He would tell me that nothing was going to console him except getting even. What could I say to that?'

A month later, Jigyasa would wonder why her father wasn't accepting her incessant WhatsApp video calls. Her mother assured her that her father was probably busy and would probably call later that evening.

Walking back to his barracks from the operations room, Corporal Jyoti switched his phone back on, immediately tapping WhatsApp open and placing a video call to his wife's phone. Jigyasa answered in a second.

'Papa, aap ghar kab aaoge? Mujhe nayi kahani kab sunaoge (Papa, when will you come home? When will you tell me a new story)?' the little girl asked straightaway.

'Main jaldi aaonga, beta. Mujhe subah chaar baje uthna hai. Aap ko bhi ab so jaana chahiye (I will come soon. I have to wake up at 4 a.m. It's your bedtime too),' said Corporal Jyoti, convincing his daughter after several minutes to hand the phone to her mother.

'He sounded tired but alert,' says Sushma. 'He said he had an early start. We spoke for barely two minutes and then said goodnight.'

Corporal Jyoti and the eighteen men of 617 Garud Flight barely managed three hours of sleep that night. Before first light the following day, they were at the high-security base of 13 RR.

Attached to the Indian Army's 13 RR, the Garuds had arrived in Kashmir three months earlier for a six-month tour of duty, the first such IAF ground deployment in a counter-terror role in the Kashmir Valley in more than a decade. Corporal Jyoti had volunteered to join the Garud Commando Force in 2006 when he was twenty years old and just a year into service in the Air Force, enlisting from his village of Badladih in Bihar's Rohtas district.

The Garuds had been raised in 2004 as a specialized force to protect airfields, IAF bases and sensitive establishments, though their role and spectrum of responsibilities has widened and evolved since then. Garud units have begun training for offensive strikes, much like the Army's Para Special Forces, been deployed on UN Peacekeeping missions abroad, assisted in humanitarian operations and, as of 2017, with the unit in question, been inserted into the Kashmir Valley for counter-terror operations. Corporal Jyoti had been handpicked by Squadron Leader Chauhan.

More detailed briefings followed at the 13 RR base in Manasbal, and it was decided that two combat teams would take part in the daytime operation to hunt down Osama Jungi and Mehmud Bhai. An eleven-man Garud squad was formed to be led by Sqn Ldr Rajiv, and another eleven-member 13 RR team was to be commanded by an Army Major.

As it happened, the 18 November operation that was about to unfold had tangible links to the previous month's encounter in Rakh Hajin, in which Milind and Nilesh had died after killing Abu Bakar and Nassuralla Mir. Seized mobile phones, documents and intercepted phone calls of terrorists injured in that firefight had all pointed in one direction—Chandargeer in Bandipora.

Two rickety civilian trucks trundled out of the 13 RR base that afternoon, moving at an unhurried pace towards Chandargeer. The first truck, covered with yellow tarpaulin, carried the eleven Garud commandos. Their CO sat in the front passenger seat wearing a phiran, a traditional loose Kashmiri garment, to conceal his camouflage battle fatigues. All the men sported beards of varying thickness.

Corporal Jyoti had been quietly pensive since the day began. In the truck, he spoke, '*Aaj Milind aur Nilesh ka badla lena hai* (We have to avenge the deaths of Milind and Nilesh today),' he whispered to Sergeant Sandeep Kumar, the commando sitting next to him. '*Bahut intezaar kiya hai. Jitne zyada terrorists milen, utna hi accha hai* (We have waited too long for this. The more terrorists we find, the better).'

Sergeant Sandeep nodded, the barrel of his fully-loaded, standard issue, Israel-made Tavor TAR-21 assault rifle pointing towards the truck's floor. Corporal Jyoti was holding the deadliest weapon the squad was carrying that day, an Israel-made Negev light machine gun (LMG), with an attached ammunition belt that held 150 rounds.

In the second truck was the RR squad. The two trucks kept a distance of 100 m between them on the 30-minute drive to Chandargeer. It was crucial that they maintained some distance between them because the village sat on a knoll, and any movement could be spotted from 2 km away. The trucks were handpicked to minimize suspicion, since such vehicles regularly plied the roads in the area, including in Chandargeer, hauling timber from sawmills.

The Garud squad had done its homework well before their CO had received the intelligence input that had set them on course for Chandargeer. A few houses in the village had been under surveillance by the Garuds for several days, with reconnaissance patrols sent out regularly to collect information about terrorist movement. The patrolling teams had taken photographs and videos to acquaint the squad with the likely target area.

'Dressed in civvies, my men had managed to gather intelligence both from vehicles and on foot,' says Sqn Ldr Rajiv. 'The last recce was on 14 November. My men had walked the nooks and crannies of Chandargeer, which has around 200 houses and 1500-odd inhabitants. When our ground contact phoned me with intelligence of the whereabouts of Osama Jungi and Mehmud Bhai, it basically confirmed what we had suspected.'

Not for a second doubting the reliability and trustworthiness of his source, Sqn Ldr Rajiv still decided to corroborate the information he had received. Counter-terrorism specialists in the Kashmir Valley know that nothing is of greater significance than accurate and detailed intelligence. And given that the input was coincidentally about the very terrorists the unit had been hunting, the CO needed to be absolutely sure that he wasn't leading his men into a death trap.

The effort to cross-check the intelligence input paid off in a critical way. Shortly before the squads left the base in their trucks, a fresh intelligence input landed, indicating that Osama Jungi and Mehmud Bhai were not alone in the target area— at least four other LeT terrorists had joined them there. The encounter would now be against six highly trained terrorists, instead of two. While the intelligence was useful in confirming the presence of the two senior LeT commanders, it also made the men realize that the operation ahead would be dramatically more difficult than they had anticipated.

When the Garud squad had piled into their truck, their CO had exhorted them with, 'We have a chance to eliminate the top LeT leadership in one go. It doesn't get better than this. And this time, we are doing this without taking any casualties. *Hum badhiya operation karenge* (We will conduct a good operation).'

The teams were to be launched from the base at 1 p.m., but their departure was rescheduled as the squad leaders factored in the possibility of the terrorists going to the village mosque for afternoon prayers.

'We thought, let the afternoon prayers be done, let them have their lunch and allow them some rest. That's when we will strike,' says Sqn Ldr Rajiv.

The trucks rumbled past a gently rising and falling landscape studded with poplar trees, the serene Bandipora countryside belying the dangerous work on the path ahead. A commando in the truck quipped, 'Bandipora is famous for its three As—*a'lim* (knowledge), *adab* (good habits) and *aab* (water). But we are more interested in a different 'A' today—Category A++ terrorists!'

Conversation remained sparse on the 30-minute drive to Chandargeer. If any man spoke, it was about the mission that was about to begin.

'*Jitne zyada terrorists milein, utna hi accha hai* (The more terrorists, the better),' said another commando in the truck, Corporal Devendra Mehta, echoing his buddy Corporal Jyoti's words.

Arriving at Chandargeer at 3.27 p.m., the trucks rumbled slowly past a few houses in the village before rolling to a halt a short distance from where the terrorists were believed to be hiding. The two teams quietly climbed out of their trucks and approached a cluster of six houses that had to be cordoned off to prevent anyone inside from getting away.

Sqn Ldr Rajiv and the 13 RR Army Major quickly organized their twenty-two men into a wide cordon around the cluster of homes. Corporal Jyoti and Corporal Devendra were positioned at a spot that their CO felt would be the route most likely to be used by any terrorist looking to escape. Just how many terrorists were hiding in that cluster would be revealed minutes later, but with the cordon, all escape routes were now effectively blocked.

'We wasted no time and immediately took our positions,' says Sqn Ldr Rajiv. 'These positions concealed us from the terrorists' possible line of sight and provided us cover from any incoming fire. The chances of getting hit are the highest when a cordon is being laid. But once you have settled into your positions, you are likely to be relatively safe.'

The CO had scanned the target site as best he could before the cordon was laid. An effective cordon demands that soldiers have the best possible view of what they've cordoned off, with as much of the target area visible as possible, while still providing protection from outbound fire. It's a delicate, difficult balance that goes way beyond simply surrounding a house. A cordon that relies too much on providing cover to the troops deprives them of a view of the target, and exponentially increases the chances of the target's escape. Too little cover makes troops vulnerable to terrorists, who get to fire accurately from the protected confines of a house.

'It was critical for me to make sure that my Garuds could acquire targets with ease,' says Sqn Ldr Rajiv. 'I wanted my men to have the best field of fire to take the terrorists out quickly. If the terrorists were to take the escape route I thought they would, they would most certainly run into Nirala and Mehta.'

The cordon established, the men lay in wait. The two squads had moved stealthily to take position around the cluster of homes, but none of the men was depending very much on the element of surprise. It was almost certain that if the terrorists were indeed inside one of the houses, they were likely to have been tipped off by their contacts in the village. The LeT's human intelligence network remains

without par among foreign terrorist organizations operating
in Kashmir.

The men didn't have to wait long. The first rounds
were fired less than sixty seconds after the cordon was laid.
And they came from an AK-47 inside one of the houses.
As those first bullets came flying, an ironic sigh of relief
passed through the cordon. It was confirmed now that the
intelligence they had received was accurate. The terrorists
knew they had been surrounded and, as expected, had
decided to put up a fight.

The 13 RR team was tightening its cordon at around 3.40
p.m. when a figure emerged in a flash from the rear door of
one of the houses, lunged towards the soldiers with his assault
rifle and directed a fully automatic spray of ammunition at
them. In seconds, a Garud commando cut him down with a
hail of return fire. He crumpled to the ground, motionless.
How many more terrorists were holed up in the cluster was
still unclear at this point. But if their intelligence was 100 per
cent accurate, there were at least five more inside. The cordon
slowly tightened.

Ninety seconds after the first terrorist was gunned down,
five men leapt out of the same house, each one of them firing
their weapons at the Garud and RR men, who were barely
20 m away. Two of the terrorists were also firing from under-
barrel grenade launchers (UBGLs) attached to their rifles. So
in addition to the spray of bullets came the deadly explosion of
grenades, with shrapnel flying in every direction.

The bullets flew inches above their heads, making a
distinct *crack* sound. Corporal Jyoti and his buddy had a split-
second to dive for cover as one of the grenades exploded
dangerously close to them. The sound of gunfire now filled
the air in Chandargeer.

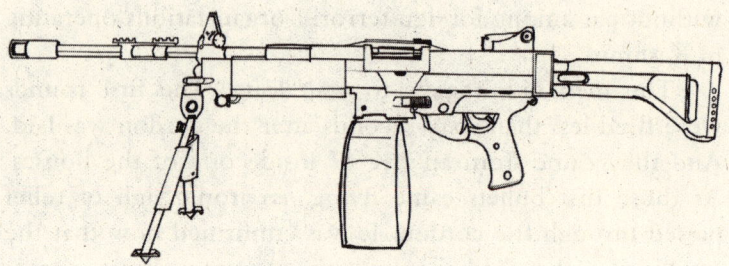

Negev light machine gun

The five terrorists quickly grouped together and began moving in a single file towards the position held by Corporal Jyoti, their fire focused in every direction so they could approach unchallenged. Jyoti and Corporal Devendra were the first commandos to engage the advancing terrorists in close-range combat because of where they were stationed. Their CO had been right—the terrorists had chosen precisely the path he thought they would to make a break and escape.

From his position, Corporal Jyoti watched the terrorists, now just over 10 m away and advancing. In moments, they would be within breathing distance. How many times had he heard of situations like these, when small groups of terrorists had used heavy and indiscriminate fire to fight their way out of a cordon and escape? Milind and Nilesh flashed through his mind as he wondered if the opportunity for revenge was seconds away from slipping out of his hands. He turned to look at his buddy Garud. Their eyes met for a moment. And then, Corporal Devendra watched as his fellow Garud did something entirely unexpected.

Corporal Jyoti suddenly sprang up, abandoning the safety of his cover. He had decided to head straight into the advancing line of terrorists.

'I remember Jyoti's eyes just before he stepped forward,' says Corporal Devendra. 'They were blazing with fury. He had made up his mind. Nobody could have stopped him.'

Corporal Jyoti charged at the terrorists with his LMG. This was a weapon capable of firing a stunning 150 rounds per minute. Expending the 150 rounds the belt was holding that Saturday afternoon would take mere seconds once the trigger was pulled.

'What are you doing, Jyoti?' Corporal Devendra screamed. 'Don't lose your cover! Come back!'

But, as his buddy had correctly observed, Corporal Jyoti had made up his mind. With the terrorists in his direct line of sight and a perfect field of fire, the commando tightened his hold on the handle of his machine gun and rained down multiple 5.56 mm rounds at them with a single trigger squeeze. A few metres to his right was his CO.

'Tck tck tck tck tck tck—all I could hear was the rat-a-tat of his LMG fire,' says Sqn Ldr Rajiv. '*Woh fire karta jaa raha tha. Bas karta jaa raha tha. Woh nazara kuch aur tha* (He just kept firing non-stop. That sight was something else).'

The nearest terrorist in the advancing group was unmistakably Osama Jungi. The men had seen his picture in a woollen cap and sleeveless olive-green T-shirt, smiling at the camera. Now, he wore a dark grey phiran over an olive-green garment. Wrapped around his neck was a checked black and grey scarf. The nephew of the 26/11 mastermind, Zakiur Rehman Lakhvi, was firing in short bursts from his AK-47 as he stepped towards Corporal Jyoti.

Refusing to move from the line of fire, fearing he would lose his chance to cut them down, Corporal Jyoti returned fire, pumping a hail of shots directly into Jungi's chest, sending him crashing to the ground. It was the sixth minute of the firefight, and the Garuds had got the man who had given the security forces the slip at least a dozen times before.

Watching Jungi collapse, the four other terrorists stopped in their tracks, now vulnerable and exposed. Corporal Jyoti could have retreated at this point to his position of cover. He decided not to, most likely because he believed that if he stepped away now, the remaining terrorists would find an opportunity to escape.

Corporal Jyoti had expended half his ammunition in the initial burst, but he still had enough left. Corporal Devendra watched in silence as he saw his buddy step towards the remaining terrorists.

'Cover me,' he shouted, advancing to close the few metres that separated him from the terrorists. As the crossfire erupted again, he got a fleeting glimpse of one of the terrorists. Once again, a flash of familiarity—it was Mehmud Bhai. Unlike Jungi, though, the LeT's north Kashmir commander was in Army-like battle fatigues, a common ruse used by terrorists to evade capture.

Stumbling over Jungi's bullet-ridden body lying in front of him, Mehmud opened fire at Corporal Jyoti, his bullets missing their target by bare inches. Before the terrorist could regain his balance and fire another burst, the Garud commando let loose a spray of machine gun rounds straight into him, throwing him off his feet.

Less than three minutes apart, the names of two of the most wanted terrorists in Kashmir had just been struck off the hit list at Chandargeer. But the operation was far from over.

Half of the six-man terror squad had been eliminated. Two of the remaining three terrorists behind Mehmud Bhai had suffered grave injuries in Corporal Jyoti's relentless machine gunfire. Unable to hold out against the advancing commandos and running out of ammunition, the three scampered for cover into a ditch a few metres to their right. They were now out of sight.

Once again, Corporal Devendra called out, pleading with Corporal Jyoti to wait until his buddy could reach him. It was clear that there was no stopping Corporal Jyoti, though. He turned around momentarily, the faintest smile on his face, which was glistening with sweat.

'He said nothing, he just looked at me. And I understood. He had avenged Milind and Nilesh. Jyoti had made his peace even before it happened,' says Corporal Devendra.

Sqn Ldr Rajiv screamed from the right, asking Corporal Jyoti to slow down, to wait until his buddy could reach him. The commando stepped forward towards the ditch, his machine gun blazing. It was then that one of the terrorists popped up for a fleeting second to fire a burst straight at the advancing Garud.

One of the bullets hit Corporal Jyoti in the head. As he fell to the ground, the machine gun in his arms kept firing, his finger still squeezed around the trigger.

'I remember the moment so clearly,' says Corporal Devendra. 'The LMG was still firing when Jyoti fell. I can never forget that.'

On that grassy patch under a winter sun, Corporal Jyoti breathed his last.

'I saw it with my own eyes,' says Sqn Ldr Rajiv, who, by this time, had advanced and called for Corporal Jyoti's buddy to join him in the front. 'That sort of courage is almost impossible. I had never seen anything like that in my life. And perhaps, I never will. Few men can match that kind of grit. I am not exaggerating when I say it was an honour for every man in my squad to fight alongside Jyoti that day.'

'Pata nahin uske dimaag mein kya chal raha tha us din (I don't know what was going on in his head that day),' says Sqn Ldr Rajiv. 'The mission had become an obsession for him. He was completely transformed in those minutes. He didn't feel the need for self-preservation. He had heard a call from above, maybe for Milind and Nilesh.'

The operation was not over. The three terrorists in the ditch were still firing at the Garuds, emboldened now after felling the machine-gunner who had torn Osama Jungi and Mehmud Bhai to shreds. Their ammunition running out, they were now desperate to take out a few more soldiers before the inevitable.

Dodging direct fire, Corporal Devendra dashed forward from his position to drag away Corporal Jyoti. As he moved, he remembers hoping his buddy was somehow still alive. Corporal Jyoti's eyes were still open, still blazing, he remembers. Dragging him to a point outside the arc of the terrorists' fire, Corporal Devendra looked down at his buddy, reaching down to close his eyes. He set down his TAR-21 rifle next to Corporal Jyoti's body and grabbed his buddy's machine gun. There were no more than ten rounds left, a single brief burst of ammunition. The weapon pointed, Corporal Devendra rushed towards the ditch at high speed, spraying the last remaining bullets in the ammunition belt straight at the terrorists with a scream.

At that same moment, Sqn Ldr Rajiv aimed heavy fire at the terrorists and lobbed hand grenades into the ditch. There was no way the three could have survived that final onslaught. Twelve minutes after the first shots were fired by Osama Jungi, the encounter was over.

Twelve minutes.

In the two-hour mopping-up operation that followed, the Garuds followed standard operating procedure, taking headshots of each of the fallen terrorists from close range to make sure they were dead. Two of the three dead men in the ditch were LeT commanders Abu Qital and Abu Zargam.

'I keep thinking what could have happened had the encounter stretched on,' says Sqn Ldr Rajiv. 'It was because of Nirala's otherworldly courage that we were able to finish

off the terror squad. He saved many lives by refusing to back
down. I watched him throughout that operation. There was a
total absence of fear. He was fully at peace with what he was
doing.'

With the encounter complete, the men loaded Corporal
Jyoti's body into one of the trucks and returned to base. Six
senior terrorists, including a family member of the LeT's
leadership, had been eliminated that afternoon, a stupendous
feat for a unit so new to counter-terror operations. But there
would be no celebration. The men were in mourning. The
CO of 617 Garud Flight knew the loss would never leave his
mind. It was the second big blow to the Garuds in just over
a month.

'An operation is good if you don't suffer any casualties.
We killed six terrorists, but the bottom line is we lost Jyoti. I
will have to live with it. It wasn't a victory for us. A man I had
trained for years was lost on my watch,' he says.

The 18 November operation will go down in the Garuds'
short history as their finest hour yet. A little over two months
later, the government announced India's highest peacetime
gallantry award posthumously for thirty-one-year-old Corporal
Jyoti, the IAF's first Ashok Chakra in combat. Only two other
IAF men had been decorated with the top honour before him:
transport pilot Flt Lt Suhas Biswas in 1952 and Sqn Ldr Rakesh
Sharma (later Wing Commander) in 1984. Biswas was decorated
with the Ashok Chakra for averting a mid-air disaster by belly-
landing his burning de Havilland Devon aircraft, which had
the Indian Army's top leadership on board, while Rakesh was
awarded for being the first Indian in space on board the Russian
spacecraft Soyuz T-11.

When Sqn Ldr Rajiv made the dreaded phone call to
Corporal Jyoti's wife, Sushma paused for a few moments.
Then, faltering, she spoke, 'He avenged Milind and Nilesh,

didn't he? Then Jyoti has gone happy,' she said, before breaking down.

The commando's flag-draped casket would be flown that evening to Chandigarh, where a weeping but calm Sushma would salute his remains. Jigyasa would be kept away from the airfield. The next morning, she and her mother would board a flight with Corporal Jyoti's coffin to Bihar, where the commando's inconsolable mother and silent, stoic father waited with crowds of mourners.

On Republic Day 2018, in a pink fur-lined jacket, Jigyasa would be held aloft by her two grandfathers as Corporal Jyoti's wife and mother, Malti Devi, were escorted to the central dais to receive Corporal Jyoti's posthumous Ashok Chakra from President Ram Nath Kovind. As the two women were escorted away, cameras would zoom in to find that the President had broken down, wiping away tears with a handkerchief.

Sushma, now thirty-four, will never forget a moment of that cold January morning. 'I was trying hard to hold back my tears, but eventually broke down,' she says. 'I kept thinking that Jyoti should have been there on the dais to receive his medal. The country is proud of him, but imagine our joy had he been alive to receive the medal himself,' says Sushma, herself the daughter of a retired Army Subedar.

Married for six years, she remembers how life had changed completely in the month after Sergeant Milind and Corporal Nilesh were lost in combat.

'He would always tell me to be independent, but after Milind and Nilesh were killed, something just snapped in Jyoti,' says Sushma. 'He was permanently distracted and restless. He loved us and gave us time every single day. But I could tell that his mind was constantly troubled by their deaths. He was looking for peace. Even in that state he would reassure me that things were not as bad as I was imagining.'

Sushma remembers the day the two commandos were killed in October. Hours later, her husband had video-called her, fully aware that the news on television would have terrified the families of Garuds spread across the country.

'We were all panic-stricken as we didn't know what was going on in Kashmir,' says Sushma. 'The Garuds were launching operations every day. I received a video call from him on the night Milind and Nilesh died. He said, "Look at me. I am absolutely fine. Stop panicking unnecessarily." But he was saying that only to calm us down. He himself was shattered, though he had no choice but to be strong.'

If there was one thing that Sushma and Jigyasa looked forward to, it was those video calls. The WhatsApp video call tone became the sweetest sound in a barracks room in Manasbal and in a small Air Force quarter in Chandigarh.

Corporal Jyoti's call on the night of 17 November was his last. Rebuffing his daughter's incessant calls earlier, he had remembered to call back before turning in.

'How was I to know that I would never hear his voice again?' says Sushma. 'I so wish we had spoken a bit longer that night. He said his Kashmir deployment would end on 17 January 2018 and he would be back with us. *Main bahut khush thi ki bas doh mahine ki baat hai* (I was very happy that it was a matter of only two months).'

Sushma remembers how ecstatic her husband had been when he was deployed to Kashmir for the first time early in 2017. His dream to do something real for the country had come true, she says. In their final call, Corporal Jyoti made no mention of the big operation planned for the next day at Chandargeer.

Hours after the Republic Day ceremony, Sushma and Corporal Jyoti's father, Tej Narayan Nirala, would remember him live on a television news channel.

'*Itihas racha hai mere pati ne. Mujhe aur desh ko garv hai in par. Woh hamesha mujhe bolte the ki desh ke liye kuchh karna hai. Bahut bahadur the mere pati* (My husband made history. The country and I are proud of him. He always told me he wanted to do something for India. He was incredibly courageous),' Sushma would tell the channel, fighting back tears.

'Nobody can alleviate our pain. But we are immeasurably proud of Jyoti. He truly believed in country above all else,' the commando's father had said.

Corporal Jyoti's Ashok Chakra citation says he demonstrated 'exceptional battle craft' as he positioned himself near the terrorist hideout, cutting off all possible escape routes. Laying such an ambush at close quarters, the citation says, demanded exceptional courage and professional acumen.

It added: 'While the detachment was lying in wait, six terrorists rushed out, shooting and lobbing grenades at the Garuds. Corporal Jyoti, disregarding personal safety and displaying indomitable courage, retaliated with lethal fire and gunned down two Category "A" terrorists and injured two others. In this violent exchange of fire, Corporal Jyoti was hit by a volley of small arms fire. Despite being critically injured, the Corporal continued retaliatory fire. Subsequently, he succumbed to fatal gunshot wounds received in the fierce encounter, which resulted in the killing of all six dreaded terrorists.'

The commando's buddy, Corporal Devendra, and CO, Sqn Ldr Rajiv, would also be decorated with gallantry awards for their actions during the Chandargeer firefight, receiving the country's third-highest peacetime gallantry award, the Shaurya Chakra, and a Vayu Sena Medal for gallantry, respectively.

The Shaurya Chakra for Corporal Devendra recognized his fearless role in the operation, his choosing to disregard his own safety to remove the body of his fallen buddy. Terrorist Abu Qital had fallen to a bullet fired by Corporal Devendra during the firefight. 'To counter other advancing terrorists, he readjusted his arc of fire and provided cover to the LMG man, fully aware of the risk of being exposed to automatic gunfire. Disregarding personal safety, he displayed indomitable courage while assisting in the evacuation of his buddy, Corporal Jyoti, who was critically injured in the gunfight,' reads Corporal Devendra's citation. It also mentions the heroic final act, when the Corporal picked up his fallen comrade's machine gun and charged at the remaining terrorists.

Sqn Ldr Rajiv's citation credits him for being directly responsible for the killing of Abu Zargam, a Category 'A' terrorist and a key Lashkar frontman in north Kashmir. It says the officer exhibited 'indomitable courage and admirable leadership' during the 'intensely fought close quarter battle.'

On Air Force Day on 8 October 2018, the 617 Garud Flight was awarded the Air Chief's citation for outstanding performance in counter-terrorism operations in Kashmir.

'I am the Jyoti Prakash of the family now,' says Sushma, who lives with Corporal Jyoti's parents and sisters in the IAF quarters allotted to them in Chandigarh. 'I have to take care of everyone the way he would have.'

It was Corporal Jyoti's dream to see Jigyasa join the IAF, perhaps as a doctor. He would fondly address her as 'Dr Jigyasa Kumari', hoping to instil ambition in the little girl.

'*Jigyasa ko doctor banna padega har keemat par*,' says Sushma, smiling. '*Unka yeh sapna toh poora karna hai* (Jigyasa will have to become a doctor at any cost. This dream of his has to be fulfilled).'

Sushma herself aspires to join the IAF as a short-service commissioned officer. She is above the age one needs to be to pursue a career in the service, but the military is known to relax rules for widows of men killed in the line of duty.

Jigyasa, fortunately too young to fully understand where her father is, often tells her friends that he has taken his trolley bag and gone away on duty for a long time.

'She is the spitting image of him,' Sushma says. 'She knows the truth, but she's too young to process it. I will tell her everything when she's older and has her father's strength. She already does, in many ways. I pray that no one has to face such a situation. It is very difficult to live without him. We miss him every second of our lives. He surprised us by coming home on Diwali in 2017. Even on Diwali in 2018, I was hoping he might surprise us again by some miracle. It's that hard to believe that he's gone.'

'*Agar kissi baat ki khushi hai toh bas yeh ki unhone badla le liya* (If I am happy about one thing, it's that he avenged the death of his comrades). I know Jyoti must be smiling from above hearing me say this,' she says.

Hoping to ease her daughter into the truth, Sushma has begun telling Jigyasa a tale of how her father left the world to meet his two close friends, Milind and Nilesh. Jigyasa listens rapt, with her usual barrage of questions.

Sushma doesn't know yet how to finish the story.

3

'Fire when You Can See Their Faces'

Lieutenant Navdeep Singh

Gurez, Jammu and Kashmir
18 August 2011

'*Woh kya kar rahe hain? Dikhai de raha hai?* (What are they doing? Can you see anything?)'

Fifteen minutes before midnight, on 19 August 2011, the soldier focused his night-vision device, the circular view it captured glowing a techno-green, his hushed whisper sharp with disbelief, '*Kuchh inflate kar rahe hain* (They're inflating something)!'

Standing on the west bank of a bend in the Kishenganga River, barely a kilometre from the LoC in north Kashmir's Gurez sector, both men were looking in precisely the same direction, their battery-charged monoculars whining softly as the lenses shifted in the tube to focus on a point across the river.

'Is this possible? Are you able to see clearly?'

'Zooming now. Looks like a boat. A dinghy.'

The picture was as sharp as it could possibly be. A dozen men stood huddled around an inflatable boat on the Kishenganga's east bank. Two of the men had just pushed the dinghy into the river. In pairs, they began to clamber aboard.

'That's definitely a boat,' said the first soldier, lowering the night-vision device and staring straight out into the darkness across the 80-m-wide swell of the river.

'Call Rana. *Now!*'

Rana, about 800 m away, was the operational headquarters of the 15 Maratha Light Infantry and sat in a small clearing alongside the tiny Kanzalwan village. In the small mess on site, two officers—the senior-most and the junior-most in the unit—had just finished dinner and were going over the following day's patrolling plan when the surveillance unit deployed on the Kishenganga's west bank called in to raise the alarm.

Thirty minutes earlier, Lt Navdeep Singh, twenty-six years old and barely five months into the Army—the baby of the unit, really—had grumbled to his CO, Col. Girish Upadhya.

'Sir, what is the point of going on ambushes every day when we are unable to make contact?' Lt Navdeep asked gloomily, spooning the mess staple of chicken curry and rice into his mouth. 'I'll tell you frankly, Sir, I can't wait for some big group of infiltrators to show up.'

Col. Girish had chosen to keep Lt Navdeep with him at the tactical headquarters, keen that the 'baby' learn the ropes while also learning how to keep his emotions in check. That March, Lt Navdeep had graduated from the Officers Training Academy in Chennai and been commissioned into the Army Ordnance Corps, a combat logistics arm that supplies the Army with weapons, ammunition and clothes. He had been posted with the 15 Maratha Light Infantry for his mandatory three-year tenure with an infantry fighting unit, a regimen that gives officers of every arm an initial burst of ground experience that stays with them, no matter where they go. Deploying an officer fresh out of the academy in the Gurez sector along the LoC, an infiltration hotspot, was the very definition of dropping a man in the deep end.

Col. Girish knew he needed to employ a careful mix of indulgence and firmness to handle the young officer who was hunched over his food.

'Navdeep, wait, relax,' said the CO. 'Don't worry. You will get your chance. *Yeh Gurez sector hai* (This is the Gurez sector). It's only a matter of time.'

Returning to his barracks at 11.15 p.m., Lt Navdeep dialled his girlfriend back home in Gurdaspur, Punjab, a town just 10 km from the international border with Pakistan. He was on the phone for only a few minutes when the first call from the surveillance team came in.

The 11.45 p.m. call from the surveillance unit was actually the second call that night. It had first raised the alarm 15 minutes before, at 11.30 p.m. Lt Navdeep quickly ended the call with his girlfriend and ran out of his barracks, back to the operations room.

In Gurdaspur, Lt Navdeep's girlfriend sighed and put her phone away. She couldn't really complain—he had told her that if he ever disconnected abruptly, it meant that the boss was calling him. Lt Navdeep had gone a step further with his mother, telling her that if he was unreachable for four or five days or more, she needn't worry, since his missions normally lasted that long. Her husband a thirty-year Army veteran, Jagtinder Kaur would wonder *when* the endless anxiety would finally end. Lt Navdeep's father, Subedar Joginder Singh, would calm her by gently admonishing her, telling her that the worry in her voice shouldn't distract their son from his work.

In the operations room, the CO was waiting for Lt Navdeep.

'In that first call, the surveillance boys had reported some suspicious movement near the Kishenganga River,' says Col. Girish. 'The team reported seeing three or four infiltrators. This was not uncommon, since we had already anticipated the route any Pakistani infiltrators were likely to take. But crossing the river is not easy. It has a very fast current.'

Only the previous day, a team from Rana had tried to cross the river using ropes, but had to give up halfway and return because of how cold the water was and how aggressive the current. So when the second call from the surveillance unit came in, informing Rana about the inflatable boat launched into the water, there was disbelief.

Lt Navdeep wasted no time, asking to be sent out immediately to lead an ambush team.

'At 11.30 p.m., our guy confirmed movement,' says Col. Girish. 'I immediately told him to keep the infiltrators in his sights using night-vision and his hand-held thermal imager (HHTI). We already had a few ambush parties, with eight men each, scattered in that area as part of our regular anti-infiltration deployment. And within minutes of receiving the input, we sent out more ambush parties to cover the likely infiltration routes. I asked Navdeep to lead one of these ambush teams.'

The gloom dissipated in seconds. Lt Navdeep quickly gathered his team of seven soldiers, picked up his AK-47 assault rifle and departed from the Rana headquarters.

'Navdeep had sensed that this was his operation,' says Col. Girish. 'All that sulking about having to wait endlessly for an encounter was washed away in seconds. I had never seen him so electrified as he left the base with his team.'

Lt Navdeep's ambush team positioned itself near a bend in the river, about 500 m upstream from the surveillance unit that had detected the infiltrators. There were three more ambush parties along that section of the river, scattered between Lt Navdeep and the surveillance unit. The ambush party next to Lt Navdeep's was led by Naib Subedar Mengare Shankar Ganpati, and sat across a small nallah that branched off from the Kishenganga River to run through Kanzalwan town. Two other ambush parties closed in to cover every patch of vulnerable ground between Lt Navdeep's position and the surveillance team downstream.

Once they were deployed and ready, the radios of the ambush teams crackled, delivering another message from the surveillance team.

'Counting fourteen or fifteen men with weapons and backpacks,' came the alert.

'I received the message in the Rana operations room too, as I was monitoring every move,' says Col. Girish. 'This was a big number being reported. The surveillance unit requested permission to fire at the infiltrators using their LMG. Their targets were across the river diagonally, and about 700 m distant.'

The team was denied permission to use the LMG.

'*Aur paas aane do unko. Jitna paas aa sakte hain utna aane do. Jab unki aankhon mein dekh sakte ho, tab hi engage karna* (Let them come closer. As close as they possibly can. When you can look in their eyes, then you open fire),' Col. Girish said over the radio to the surveillance team.

He had asked the team not to use the LMG because he knew that the chances of hitting the infiltrators at that distance in the dark were low.

'At best, the team would have managed to bring down only one or two guys, and the rest might have escaped and gone back. Even if you engage at 100 m at night, it is very difficult to get kills,' he says.

The call that came at 11.45 p.m. about the boats suddenly changed everything. Never before had infiltrators tried to cross the river in a boat. There was a bridge less than 100 m upstream, which was used by locals to cross the river to cut wood or graze their animals. But the bridge was manned by Army soldiers. So it was impossible for infiltrators to use it without a fierce firefight first.

Around midnight, the infiltrators had begun crossing the Kishenganga in their dinghy. The surveillance team watched as, repeatedly, teams of four men would climb into the boat, with two of them at the oars. The boat would drop two terrorists to the west bank of the river before returning to collect the next batch. This continued until a dozen infiltrators had been transported to the side of the river where the Army ambush teams lay in wait.

A soldier in one of the ambush parties remembers the scene that played out over the next few minutes, starlight painting the darkness with a faint milkiness, made possible by how high above sea level Gurez sector is—*8000 feet.*

'As we were watching, within a few minutes, they all crossed the river and started moving in the direction of Lt Navdeep and Naib Subedar Ganpati's ambush parties,' the soldier says. 'They were slowly approaching the nallah where these two ambush teams were stationed. Naib Subedar Ganpati's team was one side of the nallah, and 25 m away was Lt Navdeep's party on the other side, closer to Rana. There was a small nallah behind Lt Navdeep's position too, so his party was sandwiched between two nallahs.'

From the Rana base operations room, where he was receiving a stream of real-time inputs from the surveillance team, Col. Girish got on the radio with Lt Navdeep and Naib Subedar Ganpati, who were leading the two ambush parties closest to the unit base, a distance of about 500 m.

'Navdeep, Ganpati, here is what you will do—try and engage the infiltrators when they reach a point between both your parties along the riverbank,' Col. Girish said over the radio. 'From there, both of your positions can bring the group under combined fire from two directions and ensure sure-shot kills.'

The CO needed to stay at the base to provide crucial command and control to the unfolding operation.

'Navdeep, stay calm and wait till they are between your two parties,' Col. Girish called in. 'Once trapped there, there will be no escape for them. But wait till they get there. Under no circumstances should you fire early.'

The two ambush parties were positioned behind sangars,[1] their weapons ready and waiting. Their CO at Rana, half a kilometre away, had given them broad guidance on what to do next, but he knew that the final call could only really be taken by his men on the ground.

'I was depending on the surveillance guy for the latest inputs,' says Col. Girish. 'He allowed them to come as close to the Navdeep–Ganpati point as possible. The standing order was to wait till they were very close, then take the call and open fire.'

Seconds later, the group of terrorist infiltrators appeared in front of Naib Subedar Ganpati's ambush party. By this time, Lt Navdeep could see the group too.

The infiltrators were walking in a tactical single file, their weapons raised and ready. They were taking no chances either. The high-altitude Gurez sector comprises a scattering of villages that are largely friendly to the Army, and therefore provide no safe havens or stop-over points for terrorists crossing the LoC and making their way into the Kashmir Valley. There is a steady flow of infiltrators in this sector, but those who manage to sneak in successfully never stay too long in Gurez, using it only as a transit route before disappearing into the hinterlands of Bandipora and onward to the Kashmir Valley.

'Ganpati, hold fire,' Lt Navdeep called into his radio. '*Koi fire nahi karega* until my orders.'

From behind his sangar, Lt Navdeep counted each terrorist as they all stepped into the range of his weapon just 10 m in front

---

[1] A temporary breast-high fortification constructed with stones and sandbags. The term is understood to have been used first by the British Indian Army in the nineteenth century.

of him. The number of terrorists on foot was now clear—there were nine of them. Some of them had scarves wrapped around their heads. Others didn't. All of them had rucksacks and assault rifles. Finger on trigger, every man in the two ambush parties held his breath.

Eight metres.

'This young officer was demonstrating an amazing measure of resolve in allowing the terrorists to come close enough to finish them,' says a soldier from Naib Subedar Ganpati's ambush party. 'It was hard to imagine he had joined the Army just five months before. Every word he spoke was with confidence. He was sure of the order he was giving. There was no hesitation in his voice.'

'*Sa'ab, ab fire karte hain* (Let's fire now),' came a whisper from Lt Navdeep's left. It was his buddy soldier, Sepoy Vijay Gajre, a jawan who had joined the Army only the year before. He was, in effect, the other baby of the unit.

Lt Navdeep signalled to him to wait.

'*Aur paas aane do, Vijay. Aur thoda paas. Unke chehre dikhne chahiye* (Let them come closer, Vijay. A little closer. We should be able to see their faces),' Lt Navdeep said.

Five metres.

Lt Navdeep looked to his buddy for a moment, nodding. Every one of the infiltrators had stopped at a position between the two ambush parties and they were now just 5 m away. A few minutes past midnight, the young officer gave the order to open fire.

The sequence was thus: the infiltrators had been spotted at 11.30 p.m. The ambush teams had been deployed by 11.45 p.m. And at 12.03 a.m., the first bullets flew.

'The terrorists were barely 5 m from Navdeep when he ordered the men to open up their weapons at them,' says Col. Girish, who heard the first shots fired over the radio, but could also hear them echo from the site half a kilometre away. 'The terrorists were around 20 m from Naib Subedar Ganpati's ambush team. Once Navdeep opened fire, everyone began firing simultaneously.'

The decision to wait till the last moment had paid off. The first hail of bullets from the two ambush parties instantly killed eight of the infiltrators. A sniper from an ambush party further downstream shot and killed three more terrorists who were still in the dinghy that had brought them. Across the river, the surveillance team spotted three more infiltrators break into a run back to the LoC once the firing began.

Of the nine terrorists ambushed by Lt Navdeep and Naib Subedar Ganpati's men, one had sustained a gunshot wound but was still alive. He picked himself up and crouched between two small boulders on the riverbank. From that position, he began to fire at Lt Navdeep's team.

'He had taken such a position that our team could not fire directly at him; only Navdeep's could,' says the soldier on Naib Subedar Ganpati's team.

Lt Navdeep kept his squad's fire focused on the space around and between the boulders, pinning down the last terrorist. As the firing continued, the terrorist lobbed a grenade from behind his cover towards Lt Navdeep's position.

The grenade smashed into the sangar Lt Navdeep and his team were using for cover and exploded, sending shrapnel flying everywhere. The men dived for cover, but a splinter hit Lt Navdeep's buddy soldier, Sepoy Vijay, throwing him off his feet, a wound torn into his shoulder.

'*Vijay, tum theek ho* (Are you okay)?' Lt Navdeep screamed over the gunfire, crawling up to his buddy as the six other men in the ambush party continued to fire at the last terrorist behind the boulders.

'*Sa'ab, laga hai par chhota ghaav hai* (I'm hurt, but not badly),' Sepoy Vijay said. '*Main theek hoon* (I'm fine).'

'*Tum neeche raho, Vijay, main sambhal loonga* (Stay down, Vijay, I'll take care of this),' Lt Navdeep said, as he pulled his buddy soldier closer to the sangar. Sepoy Vijay slouched, with his back against the fortification, bleeding profusely. He looked up at the young officer, who had got to his knees and begun firing again.

'*Sa'ab, sambhal ke* (Be careful),' Sepoy Vijay said. 'Give me a few minutes, I will pick up my weapon again.'

'*Neeche raho* (Stay where you are),' Lt Navdeep said. '*Yeh khatam hone wala hai* (This is about to end).'

Lt Navdeep raised his head a few inches to get a better look at precisely where the last terrorist was—whether he had changed his position from behind the boulders while still firing. At that precise moment, one bullet came flying in, grazed the edge of Lt Navdeep's bulletproof patka and went straight through his head. Just as he was hit, Lt Navdeep squeezed the trigger of his own AK-47, sending a burst of ammunition straight into the face of the last terrorist. Five metres apart, both fell in their positions at the same time.

A few seconds passed and the guns fell silent. It had been just 5 minutes since the first bullets were fired. Back at Rana, Col. Girish received a radio message from a Havildar in Lt Navdeep's squad.

'*Navdeep sa'ab ko goli lagi hai* (Navdeep Sir has been hit),' he told the CO.

'*Goli kahan lagi hai* (Where has he been hit)?'

'*Sa'ab, sar par* (In the head, Sir).'

Col. Girish told the Havildar not to worry, and immediately sent a column of troops from Rana to remove Lt Navdeep and his buddy from the encounter site. Naib Subedar Ganpati had also suffered a splinter injury in the grenade attack. An ambulance was summoned. Fifteen minutes later, at Rana, a doctor examined Lt Navdeep. He still had a pulse when he was moved from the banks of the Kishenganga.

At 12.30 a.m., a doctor at the base pronounced him dead.

All the ambush squads remained in their positions till sunrise, as the surveillance team had alerted them to the possibility of more infiltrators lurking in the vicinity. In the clean-up operation that morning, the bodies of twelve terrorists were recovered from two different sites—the banks of the Kishenganga near Lt Navdeep's post, and the dinghy on the banks of the river about 200 m downstream, where they had crossed.

At 9 a.m. on 20 August, one of the search parties reported seeing a trail of blood leading into a large meadow to the west of Kanzalwan village. This suggested that a certain number of terrorists had survived the ambush and escaped with their lives. The meadow led to a hilly, forested stretch and on to two more small villages, Bagtore and Taarbal, the final settlements before the LoC.

'We launched a search operation again but couldn't find anyone,' says Col. Girish. 'But I didn't move the boys from the ambush sites for the next two days. They were being fed on site. My gut feeling was that if there were some injured terrorists, they would try to return to the other side of the LoC rather than try to go deeper into our area. The area had to be sanitized.'

His suspicions proved correct. Two days later, on 22 August, one of the ambush parties that was patrolling near the LoC fence spotted a terrorist crouched behind a pine tree. In a brief firefight, the thirteenth terrorist was shot in the head by Havildar Zore Bapu Bhagoji. The terrorist had a gunshot wound on his hand from the firefight two days ago.

'The terrorist had torn off a piece of his shirt and tied it around his hand to prevent blood loss. His hand was swollen.

He was firing with one hand,' says a soldier who was part of that search team.

Lt Navdeep had fallen after killing four terrorists that night, his decision to wait until the final moment ensuring that most of the infiltration group was eliminated in the first few seconds—crucial to the success of the operation.

The dead terrorists had plenty in their rucksacks to sustain them for a long and potentially damaging operation. They were carrying a large load of paranthas, dates, anti-venom ampoules and morphine, with each man also hauling ten ammunition magazines, grenades and military-grade night sights. It was enough material to last them a full week without having to seek local shelter or support of any kind.

Hours after he was pronounced dead, Lt Navdeep's body was flown 500 km to his home in Gurdaspur, accompanied by another officer of the unit and a senior soldier.

As the flag-draped casket arrived at the family home, Lt Navdeep's father, Subedar Joginder Singh, stepped out, his eyes dry.

'*Mera beta lada na? Achhe se lada! Kitne aatankwadi maare usne?* (My boy fought, didn't he? He fought well! How many terrorists did he kill)?' he asked the Army personnel who had accompanied the body.

Subedar Joginder Singh, who retired as an Honorary Captain from the Army's Corps of Engineers, was overjoyed when his son, unenthused by life after a hotel management degree and an MBA, had decided to join the Army. He would be the third generation from the family to put on the olive-greens.

'*Main toh kehta hoon ki Navdeep ne apni duty bahut hi acchi tarah nibhayi hai* (Navdeep performed his duty very well),' says the officer's father. 'But as parents, we are completely shattered. When you lose your twenty-six-year-old son, your world comes to an end. Nothing can be more painful than the loss of a child. *Sab kucch khatam ho jaata hai. Duniya ujjad jaati hai* (Everything ends. One's world becomes barren).'

Subedar Joginder Singh and Jagtinder Kaur have two other children, a son who works in Chandigarh, and a daughter who recently got married to an Army Major. Lt Navdeep's mother has only recently managed to compose herself. Her last conversation with him was days before his operation.

'I spoke to Navdeep for the last time three days before we lost him,' his mother says. 'He told me that the phone network in that area was bad, and that if he was out of reach, we should not worry. We had plans to get him married. He was supposed to come home on leave in November 2011 and we were hoping to get him engaged then. But he said he wanted to be in the field for at least two or three years before marriage. He said he would marry when his wife could join him where he was posted. All those plans were wiped out in a second.'

Lt Navdeep was commissioned into the Army in March 2011.

'He completed other courses, but his real dream was always to become an Army officer,' says Subedar Joginder. 'Had Navdeep been alive, he would have been a Major now. His course-mates are Majors. They meet us. And we think *aaj agar humara beta zinda hota toh woh bhi Major hota* (if our son were alive today, he too would have been a Major). Going by what he achieved in just five months in uniform, I feel he had the capability to rise to the rank of General. The Army has thousands of officers, but Navdeep *jaise kam hote hain* (there are few like Navdeep). I am not saying this because he was my son. Ask anyone in the Army who knew him. *Yeh hamara loss toh hai hi par Army ke liye bhi ek bada loss hai* (Navdeep's death is not only our personal loss but also the Army's).'

Thirteen years his senior at the time of the operation, Col. Girish says he still finds it difficult to think of Lt Navdeep without his heart swelling, both with sorrow and with pride.

'I used to consider Navdeep a kid brother,' says Col. Girish, posted to the Integrated Defence Staff in Delhi at the time of writing this. 'He was a good, solid boy. Losing him is a deep personal loss. I am forty-six now. The kind of *josh* (enthusiasm) he had is hard to describe. He used to go out for ambush missions and return very late at night. Sometimes, if I wanted to leave the base early, I would wait, thinking, let Navdeep sleep a bit longer. And if I left early without telling him, he would somehow find out and catch up with me as early as possible. He was a tough-as-nails soldier.'

Before their final dinner together, the CO teased Lt Navdeep, asking him of what use his hotel management degree was if he couldn't cook them a delicious snack. Lt Navdeep had disappeared into the mess kitchen and rustled up a few plates of paneer tikka.

'He was talking to his girlfriend on the phone when I summoned him for the ambush,' Col. Girish says. 'Who knows how many things were left unsaid? This always plays on my mind. And his parents' too. When I met Navdeep for the first time, I had a hunch that this kid would do something big.'

Col. Girish would know. A two-time recipient of the Sena Medal for gallantry, he would be decorated with a Vishisht Seva Medal for his command and control leadership during the Gurez encounter.

Naib Subedar Ganpati would receive a Shaurya Chakra, while Havildar Bhagoji, who killed the last terrorist, would receive a Sena Medal for gallantry.

Given his youth and astonishing grit and leadership on the ground, the Army had no hesitation in recommending Lt Navdeep for the Ashok Chakra, India's highest peacetime award for gallantry.

On Republic Day 2012, as tears streamed down Jagtinder Kaur's face on the pavilion, Lt Navdeep's father was escorted to the President's dais to receive his son's posthumous Ashok Chakra.

'His father is a brave man, he didn't break down, and accepted the award like a soldier,' remembers Col. Girish. 'It was my life's proudest moment, but I kept thinking Navdeep should have been there to receive this honour. He would have been amused by all the attention. He had a very strong mind, but he was also a kid.'

'His mother and I miss him a great deal,' says Subedar Joginder. 'There's pride and there's sorrow, both, in equal measure. I can't say that there's more pride and less sorrow. *Navdeep ka khayal dil mein hamesha rehta hai* (We constantly think of Navdeep). One room in our home is dedicated to Navdeep, his Ashok Chakra, the citation, his uniform, his boots, his photographs. We often sit in that room and talk about our boy and his short life.'

In Gurdaspur, a ceremonial gate was constructed in his memory, and a local college stadium renamed in his honour. His birthday, 8 June, is celebrated every year at the gate, where his parents set up a *chabeel* and a langar, ceremonial stands with food and sweetened water. On his death anniversary, 20 August, his parents organize a memorial function at the college stadium.

'There's no better life than life in the Army,' says Subedar Joginder. 'If I were to be born again, I would like to join the Army again. And I am sure Navdeep would have said the same had you asked him that question.'

In mid-2011, two months after he was commissioned into the Army, Lt Navdeep and the 15 Maratha Light Infantry moved from Kanpur to Khrew, in Jammu and Kashmir's Pulwama, for pre-induction training at the 15 Corps Battle

School, a curriculum designed to toughen up troops before the demanding nature of high-altitude operations at the LoC. It was the first time Col. Girish saw Lt Navdeep come into his own.

'We had five or six young officers and Navdeep was the youngest,' says Col. Girish. 'Seeing his physical fitness, agility and level of motivation, I put him in charge of a Ghatak platoon. Navdeep was brilliant during the training phase in Khrew. Other battalions were also there for training, and some competitions were held. Navdeep's Ghatak platoon stood first in many. He demonstrated excellent soldierly and leadership qualities in Khrew. He was gelling very well with the troops. They had also started liking him. He came across as a tough guy who understood the nuances of operations very quickly. He could take crucial decisions swiftly. He would demonstrate these qualities just three months later, in a life-and-death situation. Try and think about that for a moment.'

At Khrew, Lt Navdeep would be restless to be deployed at Gurez, calling his girlfriend and parents frequently to tell them how much he was longing to be at the LoC.

'Once he was posted in Gurez, Navdeep didn't waste a single moment; he simply hit the ground running,' remembers another officer from the 15 Maratha Light Infantry. 'He quickly dived into a routine of extensive area familiarization. He became obsessed with understanding every inch of the area, every peak, every nallah, every patch of jungle. By August, he had analysed all previous operations in that area, the likeliest infiltration routes, how better to plan the next mission. The CO would take him around to all the places, and it became clear that Navdeep was picking up the basics very quickly. In his final moments, he showed just what could be done with training and dedication.'

A few months after the August operation, the CO of 15 Maratha Light Infantry invited Lt Navdeep's father to visit the unit in Gurez, to see for himself the place where his son had fallen fighting.

'When Subedar Joginder arrived, I took him to the sangar from where Navdeep fired his last bullet,' says Col. Girish. 'It was a very emotional moment for both me and Navdeep's father. He bent down, dug his hand into the earth and grabbed a fistful of soil from the place where his son had fallen. I can't describe how moving that sight was. There are absolutely no words. It can only be experienced. I think he could feel Navdeep's presence there. I saw that look on his face. *Unhone uss mitti ko maathe se lagaya* (He touched the soil to his forehead).'

Lt Navdeep's father had looked up at his son's CO, a fist filled with the soil stretched out in front of him.

'*Mere bete ka khoon iss mitti par gira tha* (My son's blood fell on this soil),' he said. '*Main ek muthi uss mitti ki Gurez se laya. Uss muthi bhar mitti ki koi keemat nahi hai. Woh mitti mere liye Waheguru se kam nahin hai* (I brought a fistful of that soil from Gurez. It is priceless for me; It is no less than god for me).'

The soil sits in a bottle now in Lt Navdeep's room.

4

'I've Been Ready since the Day I Was Born'

Major David Manlun

Greater Noida
January 2009

Nobody had seen David Manlun dance the way he did that night. Channelling his inner Salman Khan, the young Manipuri ripped off his T-shirt to bounce to the blaring beat of *Oh Oh Jaane Jaana*, mouthing the lyrics almost completely wrong, but with euphoric abandon. Plastic glasses filled with Old Monk rum were passed around the crowd of friends, suffusing one end of the hostel block in Delhi's outskirts with its unmistakable aroma. They wouldn't miss this for the world. They had all agreed that if there was one night they needed to be together, this was it.

Twenty-four years old, bare-chested and playing an air guitar with his eyes scrunched shut, the young man at the centre of the revelry had just achieved something he had failed at twice before and had dreamed about since his days as an NCC cadet. David Manlun had made it to the Indian Army.

He had been biding his time for a year at the Army Institute of Management and Technology in the sprawling Greater Noida suburb of Delhi, filling his days with football, friends and all the heady amusement afforded by student life in India's capital. But with the weekly partying, few of his friends ever got to see David's other side. His, always-on cheery manner concealed a simmering frustration, an unremitting yearning to join the military.

His father, Manlun Khamzalam, had been a junior commissioned officer in the Army, a Subedar. In 2008, through phone calls and text messages from Shillong, 2000 km away, he and David's mother, Mannuamniang Manlun, had kept tabs, with a mixture of pride and parental concern, as David seemed unwilling to let go of the Army dream. They hoped that two failed attempts hadn't broken their son's confidence and spirit. They had really only seen the cheerful, mild-mannered boy they had raised, and prayed that he stayed strong. But far from Shillong, fuelled by the turbulent freedom of life away from home, David's determination had only intensified with each letter of rejection he received.

'I'll never forget the party that night he made it to the Army,' says Rajni Rangra, a classmate and friend of David. 'Every one of us there was very happy for him. None of us had seen him as full of joy as he was that night. And for a guy like David, that's saying something.'

Days later, when he packed his bags to leave for the Officers Training Academy (OTA) on the outskirts of Chennai, Rajni knew she would miss him deeply.

'David knew I loved bike rides in the winter,' she says. 'He would borrow someone else's motorcycle and take me for a ride at 8.45 p.m. at 100 kmph, delivering me back to the hostel in 10 minutes, right before the gates shut. He found joy in taking pains to make his friends happy.'

When they had celebrated his admission to the Army, the alcohol-fuelled Salman-style air guitar had given way to David's real one—a black acoustic guitar. He had acquired it in Delhi and propped it up in his room. It quickly became one of the many things David was popular on campus for. He made sure the guitar followed him wherever he went thereafter.

Sad to be leaving his friends but ecstatic at the prospect of what lay ahead, David began life at the OTA, which trains officers for the Army's short-service commission. A year later, in March 2010, his parents travelled to Chennai to watch their son complete his training and get commissioned into the 1st Battalion of the Naga Regiment, a unit that had cut its teeth in the 1971 war just a year after it was raised.

'All through his training, David was restless,' a course-mate who was commissioned alongside him at the academy, and is still serving, remembers. 'David wanted just one thing—to put on fatigues and get out there. He was hungry for that life. And he could not have got a better first posting.'

That first posting, with 1 Naga, was in Naugam in north Kashmir. Receiving orders to move to the location, which is not far from India's LoC with PoK, David had called his father to give him the news.

'God bless you, be careful,' Khamzalam told him. 'Give it your best and make us all proud, son. But be careful.'

In Naugam, Lt David Manlun threw himself into the daily whirlwind of counter-insurgency and counter-terror operations. Throwing himself into the role that had played out in his mind for years, the young officer would volunteer to lead a non-stop series of operations. He would give up opportunities for leave so he wouldn't miss the chance to be part of missions. In the words of another officer in the unit at the time, young Lt David was now fully in combat mode. When he phoned home every few days, he was aware his father was intimately familiar with the trails and forests where he now stalked militants—Khamzalam had served in Naugam years ago with 35 RR,[1] a unit affiliated with the Army's Assam Regiment, of which he was a member.

---

[1] For a counter-terror operation in Naugam in 2016, Havildar Hangpan Dada of 35 RR would be posthumously decorated with the country's highest peacetime award, the Ashok Chakra. An account of his mission is in *India's Most Fearless 1*.

After nearly five years in Naugam and in regimental training centres in Bakloh in Himachal Pradesh, David received word that he would be heading closer home—a posting to the 164 Infantry Battalion of the Territorial Army[2] in Nagaland.

On his way to the North-east, he stopped at home in Shillong for a quick break. His mother had insisted, since David had postponed leave several times earlier to stay with his unit in north Kashmir. Khamzalam and Mannuamniang Manlun spent those days with a young man who had been transformed by his five years in the Army. More serious and disciplined than before, his grimness lifted only in the company of loved ones.

'He was very satisfied with Army life,' says Khamzalam. 'After Kashmir, he was headed to Nagaland, which is another extremely challenging place to operate against outfits like ULFA (United Liberation Front of Assam) and NSCN(K).[3] I advised David again, stay strong, but please be careful.'

---

[2] Part of the Indian Army, the Territorial Army serves as a second line of defence, drawing its stock from civilians with elements from the regular Army. The 164 Infantry Battalion draws troops and officers from the Naga Regiment and is headquartered in Zakhama, Nagaland. As a 'home and hearth' battalion, it is intended to keep local youth from joining separatist terror outfits, while also directly operating against those outfits.

[3] The S.S. Khaplang faction of the National Socialist Council of Nagaland (NSCN[K]) is a banned terror outfit operating across states in the North-east. In 2015, the group unilaterally abrogated a fourteen-year ceasefire, going on to mount major attacks across Nagaland and Manipur. A June 2015 cross-border operation by the Indian Army Special Forces to destroy NSCN(K) camps, as revenge for an ambush a few days before in Manipur, is detailed in *India's Most Fearless 1*.

The ULFA, a separatist terror outfit founded in 1979, has been banned by the Indian government since 1990, and has mounted attacks, big and small, nearly non-stop, since the eighties. With training camps in the border forests of Myanmar and with proven support from China, the terror group has managed to remain a violent presence in the North-east's turbulent narrative since its inception. A common cause had led the ULFA to forge ties with the NSCN(K), with intelligence pointing to a long list of coordinated logistics that help both organizations. The ULFA has recently re-energized itself, riding on an exploding controversy over India's Citizenship (Amendment) Bill, which it sees as a threat to the indigenous people of Assam. The place that David was headed to was the backyard of both terror groups.

Mokokchung district has a long border with Assam to the east and north. On 5 December 2014, bags and guitar in hand, David reported to the headquarters at Zakhama. With a glowing record of leading counter-insurgency operations in Naugam, the unit's CO, Col. K.V.K. Prakash, immediately dispatched David out into the field to command a company of infantry soldiers at Mokokchung.

'This man had two distinct sides to him,' remembers the young officer quoted above, who served with him in Naugam and was also posted to Nagaland. 'In leisure, he was all about fun and frolic with his music and guitar. But during the lead-up to operations, you could not meet a more serious and focused guy than David.'

A common refrain directed at David by his comrades was to 'grow up', an affirmation of his boundless energy. But fuelled by it, over the next two years, during which he was promoted to the rank of Major, David would lead a series of crucial operations against terrorists in the troubled area that was now his responsibility.

These included two frenetic chases in the River Belt Colony and Dhobinala areas of Nagaland's Dimapur, where three NSCN(K) and two NSCN(R) terrorists were captured alive with a large quantity of arms and ammunition. Another NSCN(K) terrorist was intercepted with bomb-making equipment in Zunheboto, while a foreign terrorist was arrested with a bunch of extortion notes in Namsa village in Tizit, David's backyard.

'He was a Manipuri from Meghalaya operating in Nagaland,' says the officer who served with him. 'And in the North-east, where state and tribe affiliations are sometimes drawn in blood, David was operating at the intersection of all these fault lines. It was infinitely more challenging than his stint in Naugam. Here, terrorists believe in hit-and-run tactics, and operate large-scale extortion rackets to terrorize local communities. So David's own identity was very much in the mix.'

In June 2015, David, like the rest of the Army, had been emotionally shaken by the NSCN(K) ambush of an Army convoy in his native Manipur's Chandel district, in which eighteen soldiers were massacred. Those serving with him remember David wishing that he could participate in the cross-border strike that a unit of the Para Special Forces mounted deep inside Myanmar as an act of revenge on 4 June. Except, David and his unit had their hands full in their own area—they knew there would be an escalation in NSCN(K) and ULFA activity following the raid inside Myanmar.

After ten months spent in operations, David took a few days off to see his parents in Shillong. During that visit, he would take one of his happiest photographs, one that would be splashed across the media less than two years later. Standing on the terrace of a house, the photograph showed David and his two brothers, Jimmy and Siampu, captured while jumping in the air in total glee. 'Three *paagal* brothers,' a friend would comment when David made the photograph his profile picture on Facebook—one that remains till today.

But David knew that visits home were going to increasingly become a luxury on the path he had chosen. Back from leave, he wasted no time in jumping right back into work. There was never a shortage of intelligence inputs about the movement of militants and terrorists—the real work was to judge which of those alerts would actually lead to results. Separatist groups had learnt the fine art of jamming intelligence networks with false alarms in an effort to fatigue the alertness and energy of units like David's. But he knew that even a single show of weakness would embolden groups like NSCN(K) and ULFA to step up the audacity of their operations. And that meant a direct threat to the youth in the area, the fodder needed by these groups to fuel their activities.

Apart from institutionalized outreach methods that included medical camps, vocational training and career counselling sessions, David tried to use football to win over local youth and divert their minds from the lure of militancy. It wasn't difficult for David—if there was one thing he prized nearly as much as Army life, it was the beautiful game. All through his time at the Army Public School and St Anthony's College in Shillong, David had been obsessed with football. When he moved to Delhi, football followed him.

'I can't forget the way David would guide the whole team and take full responsibility,' says Sagar Pande, David's classmate at the management institute. 'He was an amazing centre forward and used to make some of the best passes I've ever seen on the ground. He made sure that every player on the ground was contributing to the game as per his potential.'

A major fan of international football, but even more obsessively, a follower of the North-east's football clubs, David's

Facebook page stands testimony to just how closely he tracked even the smallest games. Only weeks after he took position as company commander in Tizit, he began organizing football tournaments, drawing local youth from surrounding villages. The games would be fiercely competitive and sometimes even turn violent. But David didn't mind. He wanted the youth to be emotionally invested in anything but militancy.

His CO, Col. Prakash, had known immediately that David was special. Unusually motivated and with an action-oriented ethic, he had proven, in just two years of operations in Nagaland, how young officers with comparatively little experience in the area could lead with both lethality and empathy. As the arrests of NSCN terrorists piled up, David's energy levels seemed to permeate his company, transforming it into a highly energized unit in one of the most challenging conflict-ridden areas in the country. In August 2016, the results delivered by a troop team under his leadership in Dimapur won David the Chief of Army Staff's commendation from the Army Chief at the time, Gen. Dalbir Singh, a man who had served as Eastern Commander and personally recognized the worth of the young officer's difficult work.

The operation itself had become legendary in the unit. David and an officer from the Army's Para Special Forces had chased a highly prized commander of the NSCN(K) in broad daylight in Dimapur, overpowering him and capturing him alive. The captive turned out to be a major source of intelligence on the location and movement of terrorist logistics from Myanmar, across Nagaland and into Assam.

In Shillong, David's parents were proud of the award. And Khamzalam knew that it meant his son had truly thrown himself into his work. He sent David a text message that evening: 'Keep making us and your unit proud, son, but take care of yourself and get enough rest.'

The award scarcely interrupted David's work, coming as it did in the middle of an operation near the Myanmar border. He had been moved from Mokokchung to Mon district, at Nagaland's northern tip, with Assam to the west and north, Arunachal Pradesh to the north-east and, most significantly, an international border with Myanmar to the east.

David was excited about the move—a patch of international border in his area of responsibility provided an even greater canvas for combat. But for the young officer, the move up the chain of responsibility also served as a reminder that he had less than a year to go with the 164 Infantry Battalion. And given that he had served back-to-back in two operational areas, it was almost certain he would be sent next to a peace posting, effectively a desk job, for a few years before he could be circled back into active missions. The prospect disturbed him greatly.

'He was simply unwilling to accept a staff posting,' says the officer who served with him. 'He had got it in his head that he needed to stay in active operations at all costs. One of the avenues available to him was the National Security Guard (NSG).[4] So he put up his name. I remember him telling me, "Bro, no way I can do staff posting, too boring."'

Four months after the award, leadership changed at the unit and a new CO, Col. K.K. Mishra, replaced Col. Prakash.

'The moment I met David, I knew he was a maverick,' says Col. Mishra. 'I was impressed by his energy and focus, but was also concerned, right from the start, that these high

---

[4] The NSG is a Special Forces unit under India's Home Ministry. The Special Action Group of the NSG draws its forces from the Army, and is primarily a counter-terrorist force with specialization in counter-terror operations in built-up areas, anti-hijack operations, hostage rescue and bomb disposal missions.

motivation levels should not lead him to harm. I always had this at the back of my mind. *Always*. I needed to make sure that I could harness that energy, but without endangering him and the other boys.'

Embracing the challenge of his new position near Mon district's Tizit village, David set about cultivating new sources of intelligence in the ever-shifting landscape. His fears proved to be true—both the NSCN(K) and the ULFA(I) stepped up activities to recuperate and re-arm following the blistering Myanmar raid of 2015, and clearly had a point to prove. Challenged at nearly every step, they were increasingly desperate to score a major attack on the Army and other security forces deployed against them.

On 4 June 2017, David had picked up the buzz that a group of ULFA(I) terrorists had infiltrated from Myanmar, and were likely to move towards Assam with a large quantity of weapons and ammunition. The buzz was typically vague, with no actionable information. As he always did, David tried to build on the intelligence and flesh out an action plan.

On the evening of 6 June 2017, David was at his base, WhatsApping friends and family. He sent a message to his friend, Richa, in Shillong, telling her he would be back home in a week for a break. To his father, Khamzalam, he asked that he pray that the NSG plan worked out.

'I told him, everything will work out,' Khamzalam says. 'You just focus on your work and stay alert.'

At 8.30 p.m., David's phone buzzed. It was a local contact he had cultivated near the Assam border who was calling with information about the movement of suspected ULFA(I) terrorists in the hilly Lapa Lempong area of Mon. Unlike the many vague inputs that came in daily, this particular piece of intelligence was more specific than anything David had heard before. It not only specified the number of terrorists, but also where they could be intercepted and the direction they were heading in.

'David had been doggedly pursuing that input for three days,' says Col. Mishra. 'It was clear to him that the terror cadres were attempting to cross into Assam. The exact time when they would cross Tizit was not known. That's when he got that call, informing him that the terrorists had commandeered two autorickshaws and were moving towards Tizit to cross over to Lapa Lempong. He was sitting with Para officers. His men were already on standby when he got the call.'

David assembled two groups of men in two Gypsies, one with his own company and another with men from the 12 Para Special Forces. At 9.05 p.m., the two vehicles crept out of Tizit base and sped towards a suspension bridge near Lapa Lempong to establish a mobile check post (MCP). The function of the MCP was to intercept and challenge the two autorickshaws that were expected to pass that way on the Lapa Lempong-Lunglam-Oting road.

Three minutes after 10 p.m., with the MCP established and the men waiting, the two autorickshaws emerged through the darkness from Lapa Lempong village, moving in the direction of Oting. On being signalled to stop, the two autos swerved away and accelerated up a nearby hill, a highly suspicious action that confirmed, if nothing else, that those in the autos were up to no good. Given the intelligence, it was all that David needed to drop everything and give chase.

'Move! Move! Move!' David screamed, diving back into the front passenger seat of his Gypsy, AK-47 armed and ready, bursting out of the location in pursuit of the two autos up the winding hill road.

The second Gypsy with the Para unit followed 200 m behind. He knew he could have fired at the autos, but the smallest chance that those in the autos weren't terrorists stopped him from doing so. Killing civilians accidentally would have destroyed over two years of painstaking work and the many hearts won. And as a non-Naga in Nagaland, he knew such an incident had the potential to spiral into a nightmare for the Army and the people. His weapon aimed and ready, David leaned forward in his seat, watching the two autos race through the darkness.

'This is a very narrow mountain road, so there's no question of overtaking,' says Col. Mishra. 'They could not open fire either, because the last thing David would have wanted was a case of mistaken identity. So he kept pursuing the autos at a distance of 25 m. Then, at one of the blind turns up the hill, the trailing auto halted while the one ahead sped away. Through the darkness, David and his team saw at least three people jump out of the auto and run to the right, up the hill behind some rocks. And almost immediately, these men began firing at David's Gypsy.'

Immediately jumping out of the moving vehicle from the left, David ordered the driver to duck and crawl out of the vehicle, screaming to the six soldiers in the passenger benches to get out and take cover behind the vehicle. Kneeling on the ground with the passenger door as a shield, David fired back at the three terrorists.

'With fire coming from the darkness, David could have gone down the hill to protect himself, but he did not,' says Col. Mishra. 'He ordered his men to a safe spot behind the vehicle and away from the line of fire, while he stood at the door returning fire. It was a very fierce firefight.'

The second Gypsy, carrying the Para Special Forces men led by their commander, Capt. Nitesh Kumar, had pulled up seconds later straight in the line of fire, with bullets flying through the vehicle. The Para soldiers immediately emerged from their vehicle to join the firefight, but a hail of bullets hit three soldiers just as they jumped out of their Gypsy, critically injuring them. A fourth man, Paratrooper Manchu, crawled towards the three injured men in an attempt to pull them to safety.

David screamed again at the soldiers to get out of the vehicle and take cover as quickly as possible. As he did so, a bullet from one of the terrorists tore through the car door and went straight through David's chest. The terrorists followed this quickly with a grenade hurled between the two Gypsies, the shrapnel hitting David in the head and grievously injuring Paratrooper Manchu in both the eyes and his shoulder. Blinded by the injuries, Paratrooper Manchu still pushed himself forward to pull his three injured comrades to safety behind the vehicle. Shaken by the head injury but standing his ground, David turned back to scream once again, telling the soldiers in both Gypsies not to emerge from behind the vehicles, and to stay away from the line of fire.

'If David had not fired back at the terrorists and instructed his men to take cover, the entire party would have been eliminated in seconds,' says Col. Mishra. 'A grenade splinter hit his head, and it was followed by another bullet hitting him in the arm, but he remained standing and firing. He was bleeding out, but he kept firing at the terrorists.'

After minutes of non-stop firing, David realized the engagement was useless unless he got a clear shot of the terrorists. The three men on the hill had every advantage. They were standing on higher ground, had rocks to hide behind, and were raining their bullets down from three weapons, as against David's solo counter-fire from below. Looking back at the three injured Para Special Forces men, David looked down at the wound in his chest. He knew the blood loss meant he might pass out at any moment, so if there was anything he could do, it needed to be right then.

'David signalled to us to provide him covering fire, but we did not understand why,' says a soldier from the second Gypsy.

As the soldiers emerged from their positions to fire back at the terrorists, David got on to his stomach and slowly crawled out from behind his Gypsy, and in the darkness, snaked his way, bleeding, towards the terrorists still firing from the hill. Reaching a spot 10 m from the terrorists, David then used the last of his energy to explode out from in front of the rocks and kill the three terrorists at point–blank range. Then, with a roar that echoed down the hill, he collapsed there, unconscious from blood loss.

'The entire operation took just 5 minutes from start to finish,' says one of the soldiers who provided covering fire to David as he crawled towards the terrorists for the final attack. 'The entire party was saved by this act by David. Initially, it was only David who could fire, because there was nobody else in a position to fire back.'

At 11 p.m., Col. Mishra got a call informing him about the operation that had just taken place and that Maj. David had been injured in the chase.

'David did the very best he could have in the circumstances. My priority at midnight was to send a backup party to the hill and bring down David and the other injured men.'

When the truck arrived at the encounter site, three men from David's unit carried him back to the Gypsies, administering emergency aid. The young officer was unconscious, but he still had a pulse.

'Initially, we were told that the first bullet on the chest is fatal, but after some time, we came to know that David had a pulse. But the damage was severe. It was futile,' says Col. Mishra.

On the way down the hill in the truck, Maj. David succumbed to his three injuries.

His men, still on the hill, secured the position, confirming by dawn that the three men David had killed were indeed ULFA(I) terrorists. The terror group itself would confirm that their names were Bipul Asom, Santosh Asom and Phanindra Asom. The leader of the group was found to be a notorious cadre who was wanted for causing serial blasts in Assam on Republic Day six months earlier. The autorickshaw they were in carried a large quantity of bomb-making material and weapons, along with documents that indicated a widespread network of extortion. The men David had crawled bleeding towards, and eliminated, were part of the cutting edge of ULFA(I)'s terrorist operations.

At 5 a.m. the following morning, Col. Mishra made the dreaded phone call to David's father in Shillong.

'I woke up to the call telling me my son was no more,' says Khamzalam. 'He said, "*Bahut sorry, sa'ab,* but this is the news I have to bring to you." What could I say? When a man does his work honestly to protect his country, this can happen.'

'David may have been impulsive, but he took the correct decision, and acted in the best way he could to save his team and finish the operation. Even in his final mission, his act of total selflessness saved over a dozen men,' says Col. Mishra, who would go on to recommend the young officer for a posthumous Kirti Chakra, the country's second-highest peacetime gallantry award. 'It was a very well-planned operation but it happened quite suddenly. There was no other way of doing it. Some people may say that chasing two autorickshaws in Gypsies and having a firefight like this within a space of 25 m seems more like a police operation than an Army one. But I visited the site and saw how things played out. There was no other way it could have been done.'

After the men under his charge had had a chance to say goodbye, David's body was airlifted to Shillong the following morning. In a truck, accompanied by a full ceremonial guard, the flag-draped coffin would snake its way through Shillong's roads to the Happy Valley area where his parents lived, where a wailing Mannuamniang, helped up by two relatives, would welcome her son home for the last time. Her husband, in contrast, would be unshakeably stoic, calm, even smiling, as friends, family and a stream of officers lined up to offer their condolences.

'I was proud, but I also felt guilty when he died,' says Khamzalam. 'The Army was his duty and it was a dangerous life. It is my good fortune that God gave me a son for good work, and then God took him back. My son fought very bravely. If more men are like him, this country will have peace.'

'My son has passed away, but I'm with my son always, and he is with me,' says Mannuamniang. 'I'm proud of my son. He died a hero.'

Holding herself together with enormous strength under the hot sun, David's mother would take the microphone and tell the crowd, 'I know it is quite warm today for all of you sitting outside like this. But for David's sake and ours, I would request you all to bear it for a while since it is my son's last journey.'

David's sister, Melody, married to an Army officer and settled in Delhi, would be unable to arrive home in time for David's last rites at the Assam Regimental Centre and his burial in the cemetery there. Posting a picture of him that day on Facebook, she would write, 'My hero. You made us proud. You are coming home with the highest honour, wrapped in the tricolour.'

A large number of friends, including those he had spoken to the previous night, would show up for his funeral. The roads leading up to Happy Valley would be lined with mourners that morning, with Armymen and friends taking the microphone to pay rich tributes to David for hours.

'We miss him, but what can we do? He has gone for our country. Gave his life for India,' says Khamzalam. 'Rest in peace, David. I will also be gone some day. But going like this is very good. Such people are good. He is accepted by God. I am happy and peaceful in my mind. I don't worry about his soul. I appreciate his bravery. And I am proud of the son God has given me, and then taken away from me. I don't worry, I am happy. I would like to congratulate my son for completing his duty. I would like to think that God believed my son was too fine a person to be kept on this earth and therefore took him back.'

In Delhi, a group of David's classmates from the management institute received the news with disbelief. They would spend an evening recounting their favourite memories of the young upstart from Manipur who wouldn't sit down for a moment.

'He's generally unforgettable, but there's one small thing I will remember him for, above all else,' says his classmate Sagar Pande. 'Some guys and I would get late getting back to the hostel and would miss dinner. David would always make sure to keep a few extra plates of food in his room. So after 10 p.m., when we returned, we would go straight to his room. And we never once asked him to do this for us. He just did it.'

Another classmate, Parneet Hira, a public relations expert in Delhi, says, 'I couldn't believe he was no more. It's hard to think of him as anything but a bundle of joy. He was never gloomy. I remember how he was on top of the world when he made it to the Army. It was all he ever wanted. His life was the Army, apart from his football obsession, dancing with his shirt off and those secret drinking sessions. Women loved him. Everyone loved him.'

Col. Mishra would travel to Shillong with his wife a few days later to meet David's family, a customary visit for every CO. Condoling the loss of a son and an officer together is a bond few outside the Army can fully appreciate.

'With the heaviest of hearts, I shared with his parents that David had made it to the NSG, and I was to share the news with him just two days after his final operation,' Col. Mishra says. 'Had he not made this sacrifice, by that time, he would have been on his way to Manesar near Delhi to begin his training and probation with the NSG. He was moving out.'

Khamzalam had smiled at the news.

'David would have been very happy to hear he had made it to the NSG,' his father says. 'The last thing he wanted was to sit at a desk in his unit. He would have had his fill of operations with the NSG. He could talk of nothing else in his last six months.'

David leaves behind a legacy of loyal sources that continue to help his unit hunt terrorists in Nagaland. His football-driven connect with civilians in the area is being carried forward despite his loss. Four days after his death, the North East United Football Club (NEUFC) paid tribute to David, hailing his 'outstanding act of bravery'.

'No doubt, football was in his blood, and he had a large female fan following too,' remembers Col. Mishra. 'When he was company commander in Mokokchung, he cultivated many sources who were women. He had this idea that women are the most reliable sources of intelligence. It helped that he was a charmer and they loved him. I know David did not like to be bound by rules and regulations and methods.'

During his final days, David had experimented with a new haircut, shaving one full side of his head. His CO had laughed, asking what he was hoping to achieve with the new look. 'It may give the impression that I'm not serious, Sir,' he said with a wink. 'It will be good if others think that.'

In the months after his passing, David's identity would cause complications in the family's wish to build a house. As Manipuris—and not Khasis—the law forbids them from buying land in Meghalaya despite having lived there for decades. The Army unveiled a memorial bust of David in Shillong, but his parents were prohibited from owning land on account of ethno-political fault lines.

'For all practical purposes, David was from Shillong, and he was a local hero,' says Col. Mishra. 'But instead of accepting him fully as a hero, the community fault lines come into play. The Army tried to help the family get clearance to buy land either in Meghalaya or Manipur. After hitting many walls, the family has now purchased a plot of land in Guwahati, where they are building a house.'

On 27 March 2018, Khamzalam and Mannuamniang travelled to Delhi to receive their son's posthumous Kirti Chakra from the President of India. His citation would declare that the young officer had displayed 'conspicuous personal bravery and leadership of the highest order' in his final mission.

On his final night in Delhi before leaving to join the Army, David had been drinking with a friend, who asked him, 'So you think you're ready for this? It's not fun and games, mate.'

David bottoms-upped his drink, narrowed his eyes into a grinning grimace and replied, 'I've been ready since the day I was born.'

5

'Get to the Upper Decks, Don't Come Back'

Lieutenant Commander Kapish Muwal and Lieutenant Manoranjan Kumar

Mumbai
14 August 2013

The explosion lit up the sky over Colaba in south Mumbai. Those who didn't hear the sickening blast saw it from miles away. A massive ball of angry flames rose into the air, followed by a fountain of projectile explosions that could be mistaken for celebratory fireworks by those observing it from far away. The blast settled into a roaring blaze that would send up an ominous cone of orange light into the sky, as if a portal to hell had opened up on the ground. Not until later that night of 14 August 2013 would Mumbai know what that blast really was.

It was the *INS Sindhurakshak*.

The Indian Navy attack submarine had suffered a terrifying accident in its berth at the naval dockyard, which opened out into the Arabian Sea. As the minutes passed, horrific details of the disaster began to emerge. Eighteen personnel were inside the submarine at the time of the catastrophic blast, which mutilated its double hull and sent out a thudding shockwave that shook the other ships in the cramped dockyard that night. Forty-eight hours would pass before rescue personnel could enter the carcass of the *Sindhurakshak*. As the Navy had feared, none of the eighteen men on board survived. Most had perished in the blast, which was later found to have been caused by a mishandled torpedo and a series of tragic lapses.

For the Indian Navy, the tragedy reverberated in many directions. With eighteen personnel gone, it was the single biggest peacetime loss in its history, an unspeakable tragedy above all else. The eighteen personnel were all submariners, some of the hardiest and best qualified men in service, trained to function in the most difficult and dangerous conditions imaginable. The destruction of a submarine that wasn't at war was a crushing blow—the Indian Navy had already been wrestling with a drastic depletion in its submarine fleet, and was desperate to keep the small number in service as functional as possible for the enormous responsibility that rested on its shoulders in the Indian Ocean.

Preparing to conduct the most difficult accident investigation in its history, a shaken Navy ordered a safety stand-down, a short period of pause where all non-essential sailing would cease and the entire gamut of service safety procedures would be revised across naval bases, air stations, ships and, especially, submarines. It was to serve as a powerful refresher and reminder of just why those rules had been written in the first place, and how horribly wrong things could go if even a single rule was given a pass.

Marine commando divers would inspect the shattered *Sindhurakshak* in its berth, using it for months as a training wreck for salvage operations. The submarine had returned only the previous year after a twenty-four-month overhaul and upgrade at western Russia's Zvezdochka shipyard on the White Sea, where it had been fitted with new sensor systems, communications gear, safety rigs and modern Klub-S anti-ship and land attack cruise missiles. To the commandos swimming among the wreck, the sight below the waterline was devastating. A mangled, twisted hull breached at several points, a far cry from the silent, deadly hunter of the deep it was built to be.

That August night in 2013, two young Navy officers had heard the explosion from their quarters in the Colaba naval area. Word quickly spread about what had happened. Lt Cdr Kapish Muwal and Lt Manoranjan Kumar, both submariners themselves, made for the naval dockyard—they personally knew most of those who were on board the *Sindhurakshak*— but found that the area had been cordoned off and secured for safety reasons. There was every chance that there would be more explosions, since there was no guarantee that all the armament and ordnance on board the submarine had detonated. Hurriedly heading to the naval mess, they found a large group of officers glued to a television screen. News channels had begun beaming live footage of the smouldering orange blaze over the naval dockyard and amateur mobile phone footage of the explosions from earlier.

Nobody said a word as the TV anchor described the hellish images. As he watched, Lt Cdr Kapish's mobile phone rang. It was his father, a retired naval officer, calling from Delhi. Over the next 15 minutes, every submariner in the officers' mess would receive a phone call from a loved one. Lt Manoranjan's father, a retired Subedar from the Army, called from Jamshedpur. Every one of the callers would thank their gods when their son or brother or father picked up the phone.

'I'm okay, Dad, but they are saying eighteen people were inside *Sindhurakshak*,' Lt Cdr Kapish told his father. 'There is no information yet. We are waiting. We can't go anywhere near that place.'

As Lt Manoranjan's mother, Rukmani Devi, came on the line, he asked her not to worry, promising he would keep them posted.

For eighteen other families spread across the country, phone calls to loved ones would go unanswered. Not until the next day would a full list of those who were inside the

Sindhurakshak become available. And it would take another day for the Navy to announce that its worst fears were true. The *Sindhurakshak* had entombed a part of her crew—there were no survivors.

For the two young officers, the tragedy hit even closer home. They were both crew on another submarine, the *INS Sindhuratna*, a sister vessel to the ill-fated *Sindhurakshak*. The two vessels were among ten Kilo-class attack submarines built by Russia's Sevmash shipyard and delivered to India between 1986 and 2000. The *Sindhuratna* was nearly a decade older than the *Sindhurakshak*, and it wasn't sailing for the time being— it had been dry-docked in Mumbai for crucial maintenance procedures and wouldn't be ready to sail for at least another six months.

Lt Cdr Kapish had joined the crew of *Sindhuratna* in August 2011, with Lt Manoranjan joining five months later, in January 2012. Both were electrical officers tasked with overseeing the huge quantity of electrical equipment on board—notably, the large battery pits that provide part of a diesel-electric submarine's propulsion.

With their submarine out of action, the two young officers would be consumed by the storm of intrigue erupting over what could have caused the *Sindhurakshak* disaster.

It was all that submariners would talk about for months. And since the Navy operated ten submarines of the *Sindhurakshak*'s kind, crew members of the other nine submarines—including the two young officers from *Sindhuratna*—were drawn, at some level or the other, into the investigation and its implications. Was there something wrong with the submarine? Could something similar happen to its sister submarines? Were standard operating procedures on the Kilo-class vessels faulty? Were maintenance procedures introducing new, undiscovered risks? Did the new systems added during extensive overhauls in Russia hamper the safety regime on board? These and countless more questions would overwhelm the daily lives of the Navy's small submarine arm in the aftermath of the *Sindhurakshak* catastrophe.

In the military, there is barely any time or luxury to mourn lost comrades. But if the comfort of routine served as a salve to the wounded submarine arm, it would, at the time, be unaware that the next tragedy was only six months away.

In January 2014, Lt Cdr Kapish visited his parents in Delhi to celebrate his twenty-eighth birthday. A day before, on 18 January, the Navy's Kilo-class submarine fleet would have a major scare. The lead submarine of the pack, *INS Sindhughosh*, touched the seabed while entering the naval dockyard in Mumbai, stranded by a combination of low tide and lack of desilting and dredging work in the harbour approach. While the incident was a serious lapse that sent alarm bells ringing, the submarine itself suffered no damage and nobody on board was hurt. What the incident did was remind the submarine arm again how delicate their operations were.

'Why do you need to work on these submarines?' Lt Cdr Kapish's younger brother, Ashish, would ask him during that break.

'Somebody has to do it, and I enjoy it,' Lt Cdr Kapish had said. 'It's difficult, but I've chosen this.'

His mother, Dayawati Singh, who tried not to let her worry show, put all her energy into trying to persuade her son to think about getting married. Before his birthday on 19 January, she had even managed to get him to meet prospective brides.

The officer remained non-committal, playing along for his parents' sake and hoping it would at least alleviate some of their worry. He knew there was little he could really say to take their minds off the horrors of *Sindhurakshak*, but he did have news he felt would comfort them. His three-year tenure with the *Sindhuratna* would, after all, end in August that year, and he had been told that he would be sent by the Navy to the Indian Institute of Technology (IIT), Bombay, to do an MTech degree in electrical engineering. His parents were exultant. His father, Cdr Ishwar Singh, a naval veteran himself, knew this meant the Indian Navy had seen promise in their son, and was investing in upgrading his skills with a prestigious degree.

It was gratifying for his parents for another reason. Kapish, a high-ranking student in school, had got admission to Delhi's prestigious St Stephen's College for physics, but had chosen to leave after only six months to take up an engineering degree, since his real intention was to join the Navy. He had worked very hard to get into one of India's finest colleges, so his parents wondered if he was certain about the path he was choosing. Kapish would prove just how committed he was when he was adjudged best cadet at the Naval Academy, and awarded the Sword of Honour by the Chief of Naval Staff. Backing him for an MTech degree now was an enormous show of faith by the Navy.

Feeling recharged after a week with his family, Lt Cdr Kapish returned to Mumbai, ready to dive back into his work.

The *Sindhuratna* was nearing the end of her refit and would be ready to sail by mid-February. With sailing duties back on the horizon, Lt Manoranjan also took a short break to visit his parents in Jamshedpur.

As an electrical watchkeeping officer on the submarine, Lt Manoranjan was living a childhood dream. In class V at the Army Public School in Bareilly, where his Armyman father was posted at the time, Manoranjan had come home after watching a military demonstration on his campus on the occasion of Republic Day. He had told his parents that there was nothing he wanted more than to wear a uniform. His father, Subedar

Navin Kumar, had given the usual advice: study hard, pay attention in school, do well, and you can choose whatever you want to be. Manoranjan would top his class in senior school, train with the Navy's engineering and electrical establishments and join the Indian Navy in 2009. Three years later, he would join the crew of the *Sindhuratna*.

Back in Mumbai, the two young officers, along with over ninety other men who comprised the *Sindhuratna*'s complement, waited for their submarine to be lowered back into the water. The 2300-ton hunk of metal had been out of action for months, and the crew couldn't wait to get back inside and stretch its legs out in the Arabian Sea.

When the day finally arrived for *Sindhuratna* to be put back into the sea from her dry dock, the anxious crew assembled at Mumbai's naval dockyard. One thousand four hundred kilometres away, another submariner, an alumnus of the *Sindhuratna*, was also closely tracking the events as they unfolded.

Commodore Ravi Dhingra had begun his underwater career on the *Sindhuratna*, doing his entire initial training on it. Many of the younger officers who were part of the crew in February 2014 were personnel he had trained. And since his duties at the Naval Headquarters were directly related to the submarine fleet, he was keeping daily tabs on what was going on.

'*Sindhuratna* had been in refit for some time and when the submarine comes out of refit, officers and sailors have not bonded for months as a full sailing unit,' says Commodore Ravi. 'To get back into action mode, to be able to handle any kind of requirement or emergency, the crew goes through something called a "work-up". It begins with a harbour phase and then out at sea. The crew essentially carries out various drills involving simulated emergencies like a fire or smoke in a compartment. They practise safety procedures like de-energizing compartments and restoring the delicate balance to the submarine so all the other functions work fine.'

On 19 February, with Capt. Sandip Sinha in command, Lt Cdr Kapish and Lt Manoranjan embarked the *Sindhuratna* with the rest of her crew for its first sail out to sea since the refit. For the next five days, the submarine would conduct a series of manoeuvres and trials, returning to its dock on 24 February. According to established procedure, the crew of the *Sindhuratna* needed to demonstrate that the post-refit 'work-up' had been satisfactorily completed. For this, the Western Naval Command's senior-most submariner, the Commodore Commanding Submarines (COMCOS), Commodore S.R. Kapoor, would embark and sail with them for what the Navy calls a Task-II examination,[1] a crucial step before the submarine is cleared for operations.

At 7.30 p.m. on 25 February, before setting sail, Lt Cdr Kapish called his parents. They didn't pick up. He called again. Maybe they were busy, he thought. So he sent them a text message saying he would be out at sea and unreachable for the next few days, and they should not worry.

The crew of the *Sindhuratna* was upbeat that evening. They were finally ready to prove they had held together professionally and were ready to handle the submarine in all respects.

---

[1] The layers of 'work-up' are indicative of how delicate and difficult submarine operations are. The myriad procedures that must align for successful and safe operations require constant checks and balances, since even small deviations or violations can mean disaster in the deep sea. In December 2018, the Indian Navy commissioned its first deep-submergence rescue vehicle (DSRV), a mini submarine designed to rescue the crew of submarines in distress.

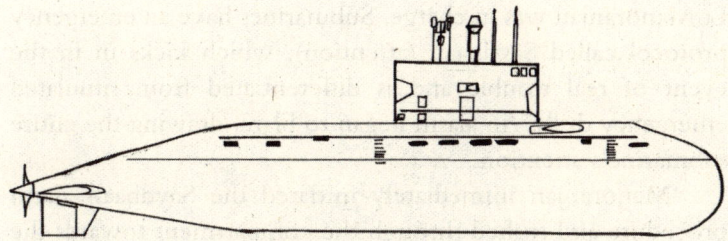

Kilo-class attack submarine

'The purpose of the Task-II was to practically demonstrate what kind of training level the crew had achieved,' says Commodore Ravi. 'So on the evening of 25 February, with the COMCOS on board, the *Sindhuratna* sailed out. During the evening, when they were leaving Mumbai harbour, and also that night, simulated emergencies were being given to the submarine's crew. That's how it works.'

In submarines, the crews work in shifts, since the vessel's stations can never be left unmanned. The night passed as *Sindhuratna* sailed, at a depth of 40 m, 110 km out into the Arabian Sea. As one part of the crew climbed into their 'slot in the wall' cabins—a signature of the highly cramped interiors of the Kilo-class design—the other part manned the submarine, sailing it further out for what would be a long day of tests the following day.

'Everything was calm that night,' remembers a sailor who was in Compartment 3 of the *Sindhuratna*. 'Lt Manoranjan was on duty as the electrical officer-in-charge in Compartment 3, and Lt Cdr Kapish, the deputy electrical officer (DLO), was at his post in Compartment 5.'

The quiet hum of the submarine, as it coursed through the water 40 m deep, was about to be dramatically interrupted.

At 5.30 the following morning, as examination drills began again, smoke was reported from Compartment 3 where

Lt Manoranjan was in charge. Submarines have an emergency protocol called Savdhaan (attention), which kicks in in the event of real trouble and is differentiated from simulated emergency drills. An alarm began to blare, drawing the entire submarine's attention.

'Manoranjan immediately initiated the Savdhaan alarm procedure and rushed through the compartment towards the source of the smoke,' says a sailor who was present at the time. 'Compartment 3 has a lot of electrical equipment, so he knew he had to move quickly before things got worse. Compartments in a Kilo are very cramped, so a fire can quickly spiral out of control. Manoranjan wasted no time.'

Within minutes, Lt Cdr Kapish had jogged through the narrow corridor of the submarine, arriving in Compartment 3 as the smoke was getting thicker.

'The smoke that was coming out had possibly heated some of the panelling running behind the equipment, and there was a short circuit in the cables, which began to burn, leading to even denser black smoke emanating in Compartment 3,' says Commodore Ravi.

Finding his way through the smoke, Lt Cdr Kapish found Lt Manoranjan hunting for the source of the smoke and fire so he could try and contain it as quickly as possible. The first thing both officers did was get the other men in Compartment 3 out, just as the smoke became unbearably thick.

'They forced us to leave the compartment, so we moved to an upper deck,' says the sailor who was in Compartment 3. 'As I exited, I could see them through the smoke. Manoranjan was using the communication console to speak to the CO to tell him that the situation was very serious.'

It was critical, actually. Compartment 3 housed a portion of *Sindhuratna*'s electric batteries and sat on top of another compartment that contained more batteries. A fire in a compartment with batteries could be catastrophic.

'The presence of batteries in that compartment made this a very tricky situation,' says Commodore Ravi. 'When the batteries charge, they give out hydrogen gas in small quantities, which usually get burned off. But if its quantity is not controlled, hydrogen is a highly explosive gas, and with a fire nearby, it could lead to a catastrophic situation. By virtue of being electrical officers, Kapish and Manoranjan would have known that, and therefore were trying to identify the seat of the fire, while at the same time getting everyone else out of the compartment.'

Receiving another call from Manoranjan about the worsening situation, CO Capt. Sinha took a call to take *Sindhuratna* to the surface.

'There was a major advantage in breaking the surface at that point,' says Commodore Ravi. 'With the submarine stationary and above water, there was a better chance of the smoke escaping from open hatches.'

Lt Cdr Kapish and Lt Manoranjan pushed eleven sailors out of Compartment 3, despite every one of them volunteering to stay and help. The sailors remember that Compartment III was nearly impossible to be in when they were pushed out by the two young officers, who refused to leave.

'*Sambhaal lenge* (We will manage), just get to the upper deck and wait for us,' Lt Cdr Kapish told the sailors as he and Lt Manoranjan got them out of Compartment 3.

'There is a certain sense of responsibility that officers have, and being officers, I guess they felt they were duty-bound to remain within the compartment and fix the problem, while getting everyone else out,' says Commodore Ravi.

Both the officers battled the smoke and continued to try locating the source of the fire, seemingly unconcerned about the harm the smoke was doing to them with each breath. As the minutes passed, the deadly smoke got thicker.

'The last anyone saw them, Kapish and Manoranjan were valiantly fighting the fire and smoke,' says Commodore Ravi. 'At some point, they may have possibly lost consciousness and by that time, Compartment 3 had already been sealed from the other side. Communication from the compartment had also stopped. The two boys had ensured that everything flammable that could allow the fire to spread had been removed from the compartment along with the eleven sailors.'

For all their efforts, the two officers couldn't put out the fire—the material that was burning was more smouldering than fully ablaze—but by removing other equipment, they had ensured that the fire didn't spread.

Just before the *Sindhuratna* surfaced, the CO descended the decks of his submarine for a first-hand inspection of the situation, but was forced to return to the upper decks after inhaling the dangerous smoke. Once the submarine surfaced, he got some of the crew members to put on breathing apparatuses so they could go back down to check on the two officers who were no longer answering calls on the submarine communication system.

Four separate rescue parties were sent down to Compartment 3, but were forced to retreat. The approach to the compartment had become extremely hot, and it was no longer possible for the sailors to go anywhere near it. And with the submarine's electrical mains turned off as a precautionary measure at the surface, the rescue teams were walking in total darkness with thick black smoke engulfing them.

'We couldn't see our hands if we held them out in front of our faces,' says a sailor who was sent down with one of the rescue parties. 'Kapish and Manoranjan had pushed us out of that compartment and had refused to allow us to stay. And they refused to come out till the job was done. If you had seen and felt that smoke, you would not have believed that someone had voluntarily remained in that compartment to fight it.'

With repeated attempts to reach Compartment 3 proving dangerous and futile, and the heat and smoke only increasing, the CO of *Sindhuratna* was faced with a terrible dilemma.

'It had now come to a point where the CO had to look to the safety of the rest of the men on board and the submarine itself,' says Commodore Ravi. 'It must have been the most difficult decision to make, since the two young officers were not accounted for, and there was no absolute clarity on their condition. It must have been a very distressing and difficult decision.'

As a last resort, 4 hours after the smoke was first detected, Capt. Sinha ordered the freon gas fire suppression system to be activated. The action would completely seal the submarine's compartments, suck out all the oxygen and pump freon gas at high pressure to quell the smouldering fire. After the freon gas was administered, Compartment 3 remained sealed as a precaution.

The *Sindhuratna* had surfaced, but it was in distress. The CO authorized a message to be sent out to the Western Naval Command headquarters, requesting emergency assistance. Within the hour, a Sea King helicopter arrived over the submarine, flying back with seven sailors who had taken seriously ill after inhaling the noxious fumes of Compartment 3. The helicopter would fly them straight to the INS Asvini naval hospital on Mumbai's southern tip. A naval fast attack craft (FAC) was also diverted towards *Sindhuratna*, picking up more of those affected by the smoke.

The submarine was now practically a floating shell, all its systems powered down as a safety precaution. A naval Sukanya-class patrol vessel arrived on the scene to tow the submarine back to Mumbai. In Delhi, the *Sindhuratna* incident would shake the highest levels of the Indian Navy. Taking moral responsibility for the incident, coming as it did just six months after the *Sindhurakshak* tragedy, then Chief of Naval Staff, an anti-submarine warfare specialist, Admiral Devendra Kumar Joshi, resigned, becoming the first Indian Navy chief to do so. The Navy's vice chief at the time, Vice Admiral Robin Dhowan, would take charge as interim chief with immediate effect, becoming Chief of Naval Staff two months later.

Limping home with its compartments still sealed, the *Sindhuratna* would be back at the Mumbai naval dockyard the following day, on 27 February.

Once docked, the submarine was evacuated. The compartments were then unsealed and ventilated with high-pressure systems to drive out any residual gas or smoke. Later that evening, a team of naval personnel finally descended the decks to Compartment 3.

'The team opened Compartment 3 and immediately saw both their bodies,' says Commodore Ravi. 'Kapish was found near the electrical equipment. Manoranjan was found near the possible source of the fire. It looked like they had been fighting to remedy the situation to the point when they passed out. We will never get to know their side of the story. What we do know is they fought till the end, not for a moment thinking of giving up. They pushed eleven sailors out from that compartment, but by securing that area and ensuring the fire did not reach the battery pits, they essentially saved all ninety-two people on board that day. They saved the submarine.'

Thirty-six hours after the probable time of their deaths, the families of both the officers in Delhi and Jamshedpur were notified. Still in profound shock, an Indian Navy board of inquiry would immediately begin investigations. To lead the difficult probe, the Navy would hand-pick an officer, Rear Admiral Soonil Bhokare, who, in 1988, was part of the first crew of the *Sindhuratna*, and the senior-most officer to have served on board.

In the national media, a controversy would explode over suggestions that the *Sindhuratna*'s batteries were dangerously old, and that government red tape had slowed the replacement of critical safety equipment. With a national election around the corner, the tragedy would briefly showcase the political animosities between the government of the day and the opposition, which had reached fever pitch. Within the military, there would be a call for accountability all the way up to the Ministry of Defence.

As it turned out, the batteries on board *Sindhuratna* were not to blame for the fire, but a failure in a regeneration unit in Compartment 3.

'The batteries being old was certainly an issue, but this particular fire wasn't because of the batteries,' says Commodore Ravi. 'What happens is when a submarine is submerged, there is a steady build-up of carbon dioxide (CO_2) inside it. Now, if you're in a situation where you cannot surface for operational reasons, the regeneration compound (RC) essentially absorbs the CO_2. This process gives out heat and also smoke if it comes in contact with seawater. It's possible that some seawater entered the submarine, causing the smoke and heat. These RC boxes are fitted very close to submarine cables. One of the boxes may have caught fire or the heat might have been so great that the cables associated with it gave way, leading to a short circuit.'

By 4 March, the Navy would release information through a press release to battle allegations about expired batteries, saying, 'The batteries presently installed on *Sindhuratna* have till date completed about 113 cycles, as against 200 cycles available for exploitation. The batteries which were being exploited by *Sindhuratna* at the time of the incident were [therefore] operationally in–date.'[2]

Commodore Ravi, who has commanded the *INS Sindhughosh*, says, 'While the batteries were not the culprit, there was always a threat. The older the battery, the higher the percentage of hydrogen discharge. That's a real threat that Kapish and Manoranjan were aware of.'

The *Sindhuratna* remained in harbour for two months, and an extensive exercise was conducted to re–cable and rebuild many of the interiors that had been destroyed by smoke and flame. The submarine remains in service.

On 15 August 2014, the *Sindhuratna*'s two young electrical officers were decorated with a posthumous Shaurya Chakra, India's third–highest peacetime gallantry award. Their citations would acknowledge what they had truly done.

Lt Cdr Kapish's Shaurya Chakra citation concludes with, 'The officer sacrificed his life keeping the safety of the submarine and personnel above his own. His act of courage and bravery was beyond the call of duty and thus in keeping with the highest traditions of the Indian Navy and the time-honoured military adage "Service before Self".'

Lt Manoranjan's citation says, 'The officer laid down his life keeping with his responsibility as the compartment officer, safety of the submarine and personnel above his own. His singular act of courage and bravery resulted in the damage being contained, casualties being minimized and extensive structural damage to the submarine being averted. His valour and dedication is in keeping with the highest traditions of the Indian Navy and ethos of the officer corp.'

---

[2] https://www.indiannavy.nic.in/content/accident-onboard-ins-sindhuratna

For the two families, though, life stands terribly still.

'I cannot imagine what it must have been like inside the submarine, and I am haunted by it,' says Lt Cdr Kapish's father, naval veteran Cdr Ishwar Singh. 'He was a great boy and made us proud at every stage of his life. Wherever he has gone, whatever he has done. He fought till the very end like a true soldier. He did whatever he could to save the lives of his fellow sailors, the men under his charge. Everybody doesn't get a chance like this—to face such a threat and prove their valour. And when he confronted that situation, he faced it with immense courage and died a brave man. He was a true hero.'

Hours after the gallantry award was announced, Lt Manoranjan's mother, Rukmani Devi, would speak to journalists through tears of sorrow and rage.

'What is the use of this award if the submarine wasn't safe to be in?' she would ask. 'If the government wants to know what I want, then let me say I want my boy back. Can they give me my boy back?'

Time barely heals such wounds. Today, Lt Manoranjan's father, Subedar Navin Kumar Chaudhary, says, '*Jab tak saans chalegi, woh ghaav toh rahega humare dil mein* (As long as we are alive, the wounds will remain fresh). But our heart swells with pride knowing he saved so many lives. We miss him every day and we are sure that the men whose lives he saved that day also think about our boy and miss him.'

Lt Manoranjan's younger brother, Sumant, was also headed into the armed forces. He had cleared the written examination for NDA and was to appear for the Service Selection Board (SSB) interview when the tragedy on board the *Sindhuratna* occurred. His shattered parents barred him from going for that interview, refusing to allow him to join the Army.

Lt Manoranjan had been cleared for a promotion to the rank of Lieutenant Commander at the time of his death.

In the most terrible twist of fate, tragedy would return to Lt Cdr Kapish's family four months after they lost their son. They would lose their second son, Ashish, to a heart attack. Their third son, Manish, is twenty-six.

'Kapish was a very precious child,' says his father. 'He was born in 1986, seven years after we lost two infant sons. Kapish was very special and we had taken good care of him. Whatever he has done, he has brought glory to us. Life has been very hard. Now I am left with only one child. We can't do anything. We have to bear it.'

On 26 February every year, as two families hold small memorial events for their sons, the crew of the *Sindhuratna* doesn't forget either. Over the course of the day, when their duties allow them a spare moment, every man descends in turns to Compartment 3.

There, they stand for a minute before two tiny framed pictures of two officers, who saved an entire crew while sacrificing their own lives.

6

'There Are More Terrorists Inside, Sir!'

Captain Pawan Kumar

At 2.20 in the afternoon of 20 February 2016, Capt. Pawan Kumar sat back in a metal chair at a secret temporary base in the freezing wilderness outside Shopian in south Kashmir. He had arrived an hour before in a jeep with five men from his unit of ten months, the Army's 10 Para Special Forces. Dressed in a black hoodie, black combat trousers and an olive bomber jacket, he was thoughtful.

He had turned twenty-three a month ago, but his dishevelled beard and exhaustion made him look older. As he sat there awaiting orders for a new mission, he pulled out his iPhone from the pocket of his trousers and began scrolling through a news feed he had made it a habit to glance through at least once a day. That afternoon, his phone's screen threw up picture after picture from two places that meant something to him.

The first were images of groups of angry men setting fire to vehicles and smashing property near Capt. Pawan's home in Haryana's Jind district. The violence was part of a protest by Haryana's dominant Jat community, which had mobilized aggressively and with increasing violence to demand reservation in government jobs. Capt. Pawan squinted at the images—many of the places looked familiar. He had grown up there, but had left home to join the military after school.

The second set of images were from Delhi's Jawaharlal Nehru University (JNU), where the student protests that had begun ten days earlier had intensified into a high-pitched controversy. Some students had held a high-decibel protest to mark the third death anniversary of Mohammed Afzal Guru, an Indian teacher from Kashmir who had been convicted and hanged for aiding a Pakistani terror attack on India's Parliament in 2001. The protests had exploded into a national controversy in the media and the political arena after some among the protesters seemed to call for 'freedom' from India and a 'breakup' of the country. The campus looked familiar, though Capt. Pawan had only been there twice. As a graduate of NDA, he had received a degree from JNU like all cadets who become commissioned officers. It was therefore, in effect, his alma mater.

He clicked the phone shut and glanced at his watch, looking out through the small window to the left of where he sat. A slow breeze rustled a row of chinar trees near the perimeter of the camp. Then, he tapped his phone on again, raised his right arm and took a selfie. But he wasn't looking into the camera. His head bent, his expression was of bored gloominess, staring at the floor.

He looked at the picture for a moment, then posted it to his Facebook profile. When the app prompted him for a caption, Capt. Pawan thought for a moment. Then he keyed in:

'Kisi ko reservation chahiye toh kisi ko azadi,
Bhai humme kuchh nahi chahiye bhai,
Bas apni razai'
(Some want reservation, some want freedom,
Man, I want nothing,
But for my blanket)

He looked at the post for a moment, smiled to himself and clicked the phone off. Sitting back in the chair, he closed his eyes for the first time that day. An afternoon siesta was the furthest thing from his mind, but a few minutes of stillness were welcome. When they arrived at Shopian earlier that day, he and his men knew they had essentially been placed on hunting duty. The villages that dotted the landscape beyond the forest that hid his temporary base were an uninterrupted hotbed of local militancy, armed and funded from across the LoC by terror groups based in Pakistan. Any moment, Capt. Pawan expected to hear either from the superiors in his unit or from the J&K Police Special Operations Group (SOG), teams of which prowled the villages of the area sniffing out terror hideouts and weapon dumps.

The wave of tiredness that swept over Capt. Pawan on that cold afternoon was hardly surprising. He had been deployed in Jammu and Kashmir barely six months earlier, in August 2015, and had been on his feet nearly the entire time. In that chair, lulled by the stillness of his camp, Capt. Pawan slipped into a deep sleep, the kind afforded by Kashmir's unmistakably pristine air.

Two hours later, at the 10 Para Special Force main base in Awantipora 40 km away, Capt. Pawan's team leader, Maj. Tushar Singh Tomar, received a call from the Army's 15 Corps Headquarters in Srinagar. Maj. Tushar was used to receiving calls summoning him at short notice for anti-terror operations—it was his bread and butter, after all. But the brief he received over the phone from Srinagar made it clear that this was going to be a nightmarish mission even by Special Forces standards.

At 4.20 p.m., a convoy of buses transporting over 500 CRPF personnel from Jammu had been ambushed by a small group of terrorists in Pampore, a suburb of Srinagar on River

Jhelum's east bank that was famous for its saffron fields. The number of terrorists was unknown at the time, but three of them had taken position outside the main gate of the state government's Entrepreneurship Development Institute (JKEDI), a plush four-storeyed building that stood out for its modern construction and carefully manicured lawns. Two CRPF men, Head Constable Bhola Prasad Singh and Constable R.K. Raina, had been killed and several injured in the initial ambush. But when soldiers who formed part of the convoy's armed escort spilled out of their buses to return fire, the terrorists rapidly shifted position. All the while firing, the three men scaled the compound wall, dashed across the wide lawn and disappeared into the JKEDI building.

It was a Saturday. As the CRPF soldiers took defensive positions, they hoped the building would be shut for the weekend. But it soon became clear that the institute had over 100 Kashmiri civilians inside, attending special weekend lectures on entrepreneurship. And three terrorists had just entered the building.

'Bloody nightmare,' Maj. Tushar found himself whispering to himself as he disconnected the call. Gathering his thoughts, a few moments later he dialled Capt. Pawan in Shopian, asking him and his squad to drop everything and move immediately towards Pampore.

'We got the information that the JKEDI building was surrounded, but that the forces on the ground had no way in,' says Tushar, who is now a Lieutenant Colonel and second-in-command of the 10 Para Special Forces.

As Team Leader, Maj. Tushar's 'boys' were split into squads deployed across the Kashmir valley. He chose to summon two squads immediately to Pampore—Capt. Pawan's from Shopian, and the squad under another young officer deployed near Kulgam further to the south.

Capt. Pawan and his Team Leader agreed to rendezvous at a common meeting point near Pampore before heading into the encounter area. A cold late winter rain had begun to fall as Maj. Tushar left his unit base in Awantipora, the kind that soldiers in the Valley know well. A steady, unrelenting drizzle that drives the cold into your bones.

'When I departed from Awantipora, Pawan called me saying he was stuck—it had been raining hard in his area, and his Gypsy had got stranded in a nallah that had overflowed on to the road,' says Tushar.

Cancelling the rendezvous, Maj. Tushar headed straight to where Capt. Pawan was stuck.

'When I reached, I see this guy had taken off his shoes, rolled up his pants, and with his feet submerged in freezing sub-zero temperature water, he was pushing the Gypsy with all his strength,' says Tushar.

The Major got down from his jeep and helped the squad push the Gypsy out of the slush and back on to the road. Capt. Pawan was soaked to the skin. His Team Leader suggested they head to the nearest temporary camp so the young officer could change and catch a breath before they moved towards Pampore. But Capt. Pawan refused.

'We'll lose time if we don't move now, Sir,' he told his superior officer. 'Let's move immediately.'

It wasn't the first time Capt. Pawan had demonstrated an unusual youthful keenness to be deployed in an operation. Originally commissioned into 7 Dogra, the new officer yearned to be part of the Special Forces. In April 2015, he successfully made the switch and was inducted into the 10 Para unit. With his maroon beret and *balidaan* badge in place, he immediately requested his CO to send him to Kashmir. He wanted to waste no time getting into the thick of action.

Arriving in Kashmir four months after he joined the 10 Para, he was sent on his first mission on 3 October in Pulwama, an operation in which two terrorists were killed. It would be the first of a nearly uninterrupted series of missions the young officer would throw himself into.

'He would be the first to volunteer for every operation,' says Tushar. 'He was doing a lot of operations—almost every single day he was out on an op.'

As the rain showed no signs of stopping, the two officers and their men got back into their vehicles and made for Pampore, arriving just as the sun was setting.

'We arrived in Pampore and had to dodge a crowd of stone-pelters near the encounter site,' says Tushar. Stone-pelting crowds had become par for the course during anti-terror operations in the Kashmir Valley, and this was no surprise.

A month earlier, Capt. Pawan had been struck in the face by a stone following an encounter in Pulwama. A knot of stone-pelters had attacked the Special Forces men as they embarked on a combing operation looking for additional terrorists. The incident had revealed much about Capt. Pawan to his superiors.

'Pawan had taken three days' leave on his birthday (15 January, which happens to also be Army Day) to go to Ferozepur to participate in a reunion of his original unit, the 7 Dogra,' says Tushar. 'Thereafter, he reported to the Valley on 18 January and was deployed the very next day on another encounter, once again in Pulwama. On the way back, there was a lot of stone-pelting. He was hit by a stone on his face and broke three front teeth. I was with him, but he never told me. And he kept on walking. Later, I found out through another boy, who informed me that Pawan was bleeding. We gave him some first aid, which he accepted reluctantly. He was totally calm about his broken three teeth, with the nerves exposed. He refused an injection. We had to force him to see a doctor in Awantipora and then later send him for a check-up to the base hospital in Srinagar. When he was there, they recommended that he take sick leave and rest for a few days. He agreed to visit the doctor once a week, but refused outright to take leave.'

Carefully navigating through the crowd of stone-pelters near Pampore, Maj. Tushar and Capt. Pawan arrived on the scene to the intermittent sound of assault rifle fire and grenade blasts. Tushar had been right about this being a nightmare.

Stepping out of their vehicles, the two officers quickly found the immediate responders at the site. Apart from the CRPF, an RR unit had been deployed around the JKEDI building, cutting off any chance of escape for the three terrorists who were now confirmed to be inside the building. A team from the J&K Police SOG was also present.

The men quickly assessed what was becoming a devastatingly difficult situation. With information confirming that over 100 civilians were inside, soldiers on the outside couldn't fire indiscriminately at the building for fear of casualties. And grenades were out of the question. The three terrorists, who had made their way into the building from the main entrance on the ground floor, had no such constraints. Switching among themselves and moving cunningly between rooms on the top floor that provided them the perfect vantage point, they fired intermittently at the forces below, keeping them on the defensive and effectively pinning them down.

During a brief lull in the firing, one team of CRPF men managed to approach and enter the building. What they didn't know was that the terrorists had seen them and quietly crept to the ground floor to stop them. Watching their every move, the terrorists flung a grenade at the CRPF men as soon as they entered the lobby, injuring four and triggering the first heavy firefight indoors. One terrorist was hit in the leg in the exchange, forcing him and his partners to return to their positions in the rooms on the top floor. The injured CRPF men were carefully extracted from the building. With the terrorists now effectively restricted to the top floor, they resumed firing at the forces below, even as the injured men were being pulled out on stretchers. The situation had got worse.

What this exchange had done, however, was provide a diversion so that additional CRPF units could approach other entrances of the building and extract the civilians. Throughout that night, civilians would be pulled out in small numbers, often two or three at a time, to ensure the terrorists didn't blow the whole building up.

Capt. Pawan observed the building carefully. Smoke rose from a blaze at the lobby level, and a facade of glass panes on the building's front was completely shattered. The darkness was lit up periodically by flashes of gunfire from the top floors. It was clear to him that the terrorists had the advantage. And with some civilians still inside the building, there was every possibility that the encounter could turn into a hostage situation.

It had been decided that the next offensive action would be at first light the following morning. But Capt. Pawan hated the idea. He took Maj. Tushar aside.

'We can't let the night go, Sir,' Capt. Pawan told him. 'I'm proceeding for a recce of the building. We can finish this.'

Maj. Tushar considered it for a moment. It was a highly dangerous proposition. Approaching the building unseen was virtually impossible.

'Every minute we give them, they can plant more traps and extend the stand-off. There will be more casualties. Let

me go in now and do a recce. What else are we here for, Sir?'
Capt. Pawan insisted.

The young Captain had a point. The Special Forces had
been summoned specifically because of the high possibility of a
night fight in an enclosed space. If approaching the building was
difficult under the cover of darkness, it would be impossible in
daylight. And Capt. Pawan was right—the terrorists appeared
well-trained and armed to stretch the encounter out as long as
possible. It was important to do some damage, any damage,
while it was still dark. Maj. Tushar relented.

Capt. Pawan gathered the five men of his squad and held
a quick briefing in the cover provided by a large armoured
vehicle, with bullets raining down from the top floors of the
building. As they spoke, the men began to dress themselves
with the equipment they would need for the recce. First, they
put on their ballistic vests and gloves. Each man then strapped
an American-built AN/PVS-14 night vision device to his
helmet using a harness—a crucial piece of optical gear that
would allow them to hunt in pitch darkness. Finally, each man
checked his main weapon—a Colt M4A1 carbine along with
its ammunition magazines.

The plan was shared with the CRPF, RR and SOG.
With backup forces in position around the perimeter of the
compound, Capt. Pawan and his men crept over a low wall
and headed towards the building. Through his night vision
device, Capt. Pawan could make out the building clearly,
along with the fire still burning in the lobby area that appeared
to him as a bright green blaze.

At the perimeter, Maj. Tushar had put on his battle gear
too, ready to rush in as part of a second wave. In touch with
Capt. Pawan through a combat communications earpiece, he
watched through his own night vision device as the young
officer's squad made its way carefully around the building.

'There was speculative firing from many of the windows on the top floor. But I could tell Pawan was very calm,' says Tushar. 'He took his time. He walked around the entire building, carefully studying the situation. Then we worked out a plan. He informed me that he planned to take it floor by floor, starting with the top.'

Entering the building from the front lobby was out of the question. The wide atrium would give the terrorists an easy vantage point to pick out the approaching commandos and fire straight at them. Capt. Pawan had found a better approach—through an emergency fire staircase at the back of the building.

Calling for cover fire on the front of the building to keep the terrorists focused on that area, Capt. Pawan and his men carefully slipped to the back and began ascending the stairway to the fourth floor. With reconnaissance complete, the plan was a standard drill—to clear the building floor by floor from the top downward. Every few seconds, Capt. Pawan would whisper into his mouthpiece to tell Maj. Tushar what he saw and what the status of his approach was.

'He sounded very confident as he and his five boys went up the stairs,' says Tushar. 'There was a restaurant on the rooftop. Adjoining it were a number of rooms that had to be cleared. Pawan kept me informed at every step.'

When Capt. Pawan's squad was halfway up the stairs, the firing from the top floors stopped. He quickly received word from Maj. Tushar to proceed with caution as there was no way to confirm that the attention of the terrorists remained diverted to the front. Capt. Pawan noted the warning and proceeded up the stairs even more slowly, listening at every step for any audio clues that might give away the location of the terrorists.

Ascending the final set of stairs, the six commandos emerged onto a small landing with a door that led to the top floor. Capt. Pawan waited, straining his ears to check if he could hear anything at all from beyond the door. There was silence. It was totally dark by this time, but Capt. Pawan and his squad trudged forward through the door, seeing a world painted shades of green by their night vision devices, their weapons at shoulder height and ready. They knew they were a whisper away from a firefight.

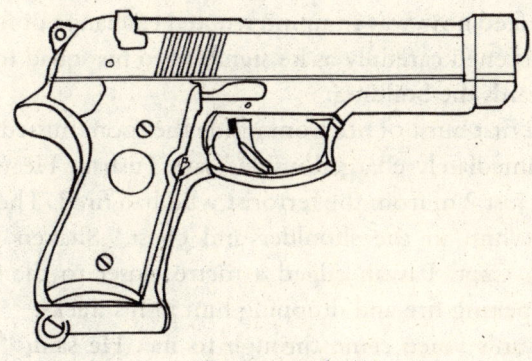

Beretta pistol

'Entering floor now,' Capt. Pawan whispered, as the scout from his squad opened the door and the six men entered the floor into a narrow passageway. At the end of the passage were a pair of doors to rooms that looked out over the side of the building. The men would no longer have the luxury of silence, as they now needed to begin clearing the rooms one by one. With the element of stealth gone, they would be infinitely more vulnerable. There was no other way to get the operation going.

In silence, Capt. Pawan signalled to his scout to break down the door.

'The scout tried to break open the door but couldn't, so Capt. Pawan rushed forward and kicked down the door quickly. As soon as he broke the door down, there was a burst of fire from inside the room,' says Tushar, who could hear the firing through his earpiece in real time. Waiting anxiously for an update, Maj. Tushar gathered his men and asked them to prepare to approach the building for backup.

Capt. Pawan did not have the time to report his every move any longer. He had just broken down the door to the room in which the terrorists had taken refuge. The only audio from his feed now was an uninterrupted exchange of fire. Maj. Tushar listened carefully as he signalled to his squad to follow him towards the building.

'The first burst of fire from inside the room missed Pawan, so he immediately charged inside,' says Tushar. 'He was now standing just 2 m from the terrorist who had fired. The second burst hit him in the shoulder and chest.' Shaken but still standing, Capt. Pawan edged a metre closer to the terrorist before opening fire and dropping him in his tracks.

'Pawan's voice came through to me. He said, "*Lagi hai* (I've been hit)," but continued,' Tushar says.

What the men didn't know was that one of the bullets had pierced through Capt. Pawan's chest and ripped away a piece of his heart. Bleeding profusely, he continued to edge forward as he had seen the other two terrorists retreat into another room.

'Two other boys from the squad continued to fire alongside Pawan—they knew they had only one opening to fire through,' says Tushar. 'For the first time, it became clear what the number of terrorists was. Pawan had dropped one. There were two more.'

Maj. Tushar asked the young officer to move back and let the other squad members go forward, but Capt. Pawan said he now had a vantage point from which to proceed. As he edged forward, a third hail of fire came out of the dark recesses of the room, hitting him again. This time, the two lead scouts of the team, Sabarmal Baji and Nayak Singh, quickly pulled him out of the room. As two members of the squad stood in the passageway firing into the room, Capt. Pawan was carried carefully down the stairs by three men, even as he protested weakly.

As they reached the ground floor, Maj. Tushar's team was preparing to ascend the stairs.

'He was bleeding heavily. Pawan was tall and very well-built for his age. It took three of his men to carry him down the stairs,' Tushar says. 'I couldn't believe it, seeing Pawan like that. He had incredible physical strength.'

As Cadet Sergeant Major of his squadron at NDA, Pawan had displayed unusual endurance and might—course-mates remember him carrying other cadets during cross country training and never betraying more than passing tiredness.

Now dizzy from blood loss, Capt. Pawan stammered to his team leader, '*Bande hain andar, Sir, aur bande hain* (There are more inside, Sir, there are more).'

Capt. Pawan slipped in and out of consciousness as Maj. Tushar quickly took a briefing from the other two men of the squad about the situation on the top floor.

'The terrorist had missed him with the first burst of fire, so this guy had charged inside,' says Tushar. 'It was an intensely close combat engagement. Just about 2 m. He was hit but he continued. It was just an unfortunate moment when he was shot—the range was too close; otherwise, Pawan could have dodged it. There was no room, it was impossible. What he did was beyond brave.'

Apart from his bravery, Capt. Pawan had also given the operation its first foothold into the building, a crucial tactical step upon which the remainder of the operation could be planned and executed. The foothold drastically reduced the advantage held thus far by the terrorists by virtue of their position, and effectively began a countdown to their end. For the first time since the encounter began, the terrorists were on the defensive.

'Any average guy would have died on the spot, or wouldn't have been able to stay awake,' says Tushar. 'Pawan wanted to continue even in that state, but his body wasn't allowing it. I could see it was very difficult for him to realize he could not continue the fight, even as he oozed blood.'

An ambulance was summoned to take Capt. Pawan to the 92 Base Hospital a few kilometres away in Srinagar city. When he was stretchered off the encounter site, he was still conscious, still stammering to his Team Leader.

'*Chaloo kar do, Sir, bande andar hain. Ek-do aur hain andar. Pehla banda iss side par hai aur do bande andar nikal gaye hain* (Start the operation, Sir. There are a couple more inside. The first is on this side and two more have escaped within),' Capt. Pawan said before being loaded into the ambulance and driven away.

'One bullet wouldn't have pulled him down. Even two or three rounds wouldn't have stopped him. Unfortunately, one of those bullets ripped off a part of his heart. It was a very bad hit. If it hadn't hit his heart, he would have continued fighting,' says Tushar. 'He was unbelievably strong.'

An hour later, just before dawn on 21 February, Capt. Pawan, twenty-three years old, succumbed to his injuries at the Srinagar base hospital.

Back in Pampore, the foothold he provided into the besieged building would be crucial to an operation that stretched over 48 hours. A terrible reminder of the deadly difficulty of the operation would come the following day, when two more Special Forces men—Capt. Tushar Mahajan and Lance Naik Om Prakash from the 9 Para Special Forces unit—would lose their lives in gun battles within the shattered, blazing building. In the days that followed, questions would explode over whether tactics had failed, or whether the men had taken unnecessary risks in their pursuit of the terrorists. But leaders in the unit would vouch for their men—they had been in the best possible position to take the decisions they did. And what Capt. Pawan had done was, quite literally, open the door to the end of the operation.

'Traditionally, building interventions are done in the daytime. In Pawan's case, they stormed the building at 2 a.m.,' says Tushar. 'He was extremely calm and confident. Try and picture operating in pitch-black darkness.'

The second thing Capt. Pawan had done was cross what is known as the 'fatal funnel'. In close quarter combat, the fatal funnel is described as the cone-shaped path leading from the entry where the assaulter (Capt. Pawan) is most vulnerable to the defenders (the terrorists) inside the room. Once an assaulter enters, defenders inside the room do everything they can to keep the assaulter inside the fatal funnel, where bullets are most likely to find him. It was in the fatal funnel that Capt. Pawan had killed the first terrorist. And it was here that bullets had found him.

'Pawan had conquered the fatal funnel with his life,' says Tushar. 'The first door you enter leaves you with the maximum probability of being hit. After that, you are inside. The terrorists know it. You know it. Earlier, there was concrete between you. Now there's nothing. They no longer have an advantage. It's only a matter of time.'

'We've lost Pawan,' a voice at the other end of the line from Srinagar said. Maj. Tushar had been ready for the news, but somewhere, he had still believed the young officer would pull through. It was then that Maj. Tushar was informed that Capt. Pawan had been shot through the heart. Anything less, and he would have survived.

'I spent that night in disbelief, more than anything else,' Tushar says. 'It wasn't believable that Pawan was gone. He was a supremely strong, intelligent soldier. He had gone through heavy stone pelting and calmly walked away. He took a leadership call to break down a door, knowing those first bullets could have his name on them. He held his nerve no matter what. And he was an incredibly disciplined fighter.'

Tomar refers to how Capt. Pawan unfailingly carried two weapons into operation—his M4A1 carbine, of course, but always backed up by a Beretta pistol.

'This is basic protocol for us in the Special Forces, but people tend to forget during operations,' says Tushar. 'Pawan also made it a point to keep two ammunition magazines strapped together. And he always changed magazines before the first one was empty. These are basic techniques, but people forget. Pawan didn't. He never made those mistakes.'

Nearly 48 hours after the stand-off began, the operation was finally brought to a close on 22 February with the confirmed killing of all three Pakistani terrorists. Intelligence agencies would later reveal that the three men likely infiltrated weeks before across the LoC in Handwara in north Kashmir, before slipping into the Valley along the Jhelum.

On 22 February, wreaths would be laid on a wooden casket containing Capt. Pawan's body at the Army's Badami Bagh Cantonment in Srinagar. Maj. Tushar and the other men from 10 Para Special Forces would be unable to attend, because they were still deployed at the Pampore encounter site.

With the Jat quota agitations still raging across Haryana, including in areas surrounding Capt. Pawan's native Jind district, it became clear to the Army that transporting the young officer's remains by road from anywhere could meet with roadblocks and unexpected trouble. In a rare appeal, the Army officially called for the people of Haryana to extend their full support in ensuring that the body of the soldier could be sent back to his village. To play it safe, however, the casket containing Capt. Pawan's body was flown from Srinagar to Pathankot in an Army Dhruv helicopter. At the base, the Army held another farewell ceremony for the young officer. From Pathankot, the helicopter airlifted the casket straight to his native Badhana village in Jind. From the air, the crew of that helicopter would have glimpsed any crowds that had gathered.

By this time, Capt. Pawan's Facebook post from two days earlier had been discovered by journalists, his words amplifying the irony of the situation to the fullest. On social media, he may have playfully dismissed the agitation by members of his community. His antipathy towards the mobilization actually ran deeper.

'He loved his iPhone and MacBook. Remember how young he was,' says Tushar. 'These gadgets were his steady connection to the world. He wasn't just active on social media, but was also very vocal about his views. He would openly say Haryanvis don't require reservations, and that this was all politics. He hailed the people of Haryana as brave, strong, as winners in sports arenas and beyond. He was like, "*Yeh sab bakwaas hai, Sir* (This is all rubbish, Sir)."

'I couldn't reach [Jind] for the last rites because of operational duties. Most of our unit did go from wherever they were. Our CO was there. The Jat agitation was quite bad, and tempers were on edge. But news had also spread by this time about Pawan, especially in and around Jind. Many of the road blockades simply disappeared when they heard about Pawan. I hear there was a 3-km long line of people leading into his village, wailing and crying. In Haryana, people do care about their soldiers. Crowds arrived from all over for the funeral in his village.'

As deafening chants of '*Pawan Kumar amar rahe* (Long live Pawan Kumar)' broke through the sounds of mourning, marigold petals came showering down as the casket was carried by four Special Forces men into the forecourt of the officer's small home. Stoic and solemn amid the wailing and the slogans, Capt. Pawan's father knelt next to the casket, while relatives pulled the cover off.

Inside, wrapped in white, was Rajbir Singh's only child. His eyes were closed and his face wore a gentle calm. His dishevelled hair and beard were hidden by the folds of the cloth that now draped him. His mother would be brought out of the house to see him one last time. Inconsolable and reluctant, she would touch Pawan's cheek before collapsing near the casket.

'I had one child. I gave him to the Army. To the nation. No father could be prouder,' Rajbir Singh would tell the journalists who waited to hear from Capt. Pawan's father. The comment, made with unusual composure, would stun the country and echo across the media through that day.

'He was an only child. He spoke to his parents once every few days. He knew they worried about him, but they would never let him know that,' Tushar says. 'He always wanted to be in the Army, but wasn't sure which area. He went through NDA and IMA. Initially, he wanted to join the Ordnance Corps. Then he decided to join the Infantry. One of his instructors was from the Dogra regiment, so he was commissioned into 7 Dogra in December 2013. Later, he learnt about the Special Forces.'

Unable to extricate himself from operations at Pampore and another anti-terror operation a few days later in Pulwama, Maj. Tushar would finally get the time to visit Jind on 4 March, two weeks after Capt. Pawan's death.

'I have never seen a braver family. Their pride is not an act. His father isn't lying about the pride he feels,' says Tushar. 'Their sorrow runs perfectly parallel to how proud they are of their son's sacrifice. It is hard to describe.'

'We had many plans for Pawan,' says his father. 'He was in love with a girl whom we had met and liked. We were looking at a possible marriage in 2017. Had Pawan been around, I would have perhaps become a grandfather by now. *Par zindagi ki raftaar mein sab sapne peeche reh gaye* (But the pace of life overtook all our dreams). He had plans for us too. Pawan would often say that we will sell off the Jind house and buy one in Panchkula. That's a far better place, he would say. *Hum baap–beta ne bahut planning ki thi. Par kismet ne apni planning ki thi* (We father–son had planned a lot of things. But destiny had its own plans).'

Six months after his death, a day before the Independence Day in 2016, the government announced that Capt. Pawan would be decorated with the Shaurya Chakra by the Indian President. Capt. Tushar and Lance Naik Om Prakash too were decorated with the Shaurya Chakra for their sacrifice.

Capt. Pawan's parents speak about him every day, to each other and to anyone who visits them.

'We will miss him till our deaths,' says his father. 'He was our only child. We rummage through old photo albums to revisit the happier times of our lives. We have a cupboard in the house and all of Pawan's stuff is in it. His Shaurya Chakra, his uniform and his boots—everything. Sometimes we just take his clothes out of the cupboard and hold them.'

A year into the Army, Capt. Pawan had chosen to use his first spell of leave not to go home, but to bike up on his Bullet motorcycle to Khardung La in Ladakh, the world's highest motorable pass.

Tushar, now second-in-command of the 10 Para Special Forces, says he will never forget the young commando, but believes he will hang on to one story more than any other to remember Capt. Pawan. It was late one night in December 2015, in a snow-blown patch of wilderness south of Srinagar.

'We had been deployed there for an operation. But the intelligence turned out to be faulty, so there was no encounter and no sign of any terrorists,' says Tushar. 'The weather was very bad, and it was snowing very heavily. My legs were nearly frozen and I suggested we return to base. But Pawan had a smile on his face. He said, "*Sir, thodi der baithte hain, aayenge aayenge* (Sir, let's sit here a while, they'll come)." He persuaded me to sit there the whole night in the snow and brutal cold. My brain knew that no terrorists would come. But Pawan insisted we wait. I don't think he expected any terrorists either. I think he just liked being out there.'

7

'I Rust when I Rest'

Major Satish Dahiya

Jaipur
14 February 2017

'*Aap ghar se bahar mat jana, aaj kucch aane wala hai* (Don't leave home, something is going to arrive today).'

Sujata Chowdhry wondered for a moment why her husband was speaking in an unusually hushed voice. He was calling from the headquarters of his Army counter-insurgency unit in north Kashmir's Kupwara, where voices were kept low as a rule. But it was clear he was making an extra effort to be discreet.

Sujata understood. Maybe Maj. Satish Dahiya's men were within earshot and he was embarrassed to be speaking to his wife on Valentine's Day. Yes, that must be it.

She had taunted him the day before, after he had had a package of children's towels delivered to their home in Jaipur for Priyasha, their daughter who would turn two in a few months. The towels were better suited for an infant, but Sujata had smiled and kept them, despite her husband's repeated pleas that she have them exchanged. Later that night, she sent him a message on WhatsApp: 'You know what day it is tomorrow? You've sent presents for your daughter. What about your wife?'

Sujata didn't receive a reply that night, but was woken early the next morning by her husband's hushed instructions not to leave home.

'After that, he disconnected the phone,' says Sujata. 'I began the day, got our daughter ready for school, dropped her off and returned home to get ready for the day. That's when Satish called again, asking where I was. He again asked me to stay home and to expect a parcel at noon. Like his first call in the morning, he talked for barely 30 seconds before hanging up. He kept saying, '*Bahar hoon kaam pe* (I'm out on duty).'

Shortly before noon, Sujata drove to Priyasha's nursery school near their home. On their way back, Maj. Satish called for the third time.

'He asked me why I'd left the house, and what if the parcel arrived while I was gone,' says Sujata. 'After reminding him that there was nobody else to collect our child from school, I teased him, saying I hoped the parcel was in fact for me, and not for someone else. Satish was anxious but taunted me back, saying it looked like he would be doing nothing else that day but coordinating this parcel business. He hung up again quickly.'

After Priyasha had been fed and put down for an afternoon nap, Sujata settled down to wait for the parcel, half expecting her husband to call back every hour for updates.

But Maj. Satish would call next only at 5.22 p.m. This time, it wasn't from his own mobile phone, but a number she recognized as being from his unit. And this time, he didn't ask about the parcel. He quickly informed her that he was leaving for an operation, before hanging up abruptly.

This wasn't unusual. Maj. Satish was an officer with the RR, the chief counter-insurgency and anti-terror force in Jammu and Kashmir. The 30th Battalion of the RR, which the thirty-two-year-old officer was a part of, operated in Handwara, one of the most militant-infested zones near the LoC in north Kashmir. This was a restive militancy hotspot that was regularly fed by a stream of Pakistan-trained terrorist infiltrators who stole across the frontier fence to launch attacks.

Far from unusual, dropping everything to dash to an operation was, in fact, part of the day's work for the RR. But it had taken Sujata a while to get used to it.

Maj. Satish was already on the road when he had called Sujata. He was headed with his company of soldiers to a tiny village called Hajin Kralgund, not far from Handwara town and only a few kilometres from a vast swathe of some of the thickest forests in Kashmir, west of the Jhelum. The officer had barely been able to call his wife as he and his men rumbled down the highway towards their destination.

For the previous five hours, from his position on the fringes of Kupwara, about 40 km away from the battalion headquarters, Maj. Satish had been in constant touch with Col. Rajiv Saharan, the CO of 30 RR.

'Satish was absolutely clear when he called me,' says Col. Rajiv. 'He had cultivated a very good source in the local community since he had come to us in 2015. That source turned out to be highly reliable and had tremendous faith in Satish. Generally, people are not open to providing information because they fear that sooner or later, their identity will be revealed and it will be game over for them at the hands of the terrorists. But this particular source had tremendous faith in Satish and said, "*Sahab, main aapke liye karoonga* (I will do it for you)". Finding such a source is difficult and dangerous. We have a large number of sources that are not reliable because they might be double agents. But Satish had 100 per cent faith in this guy. On 14 February, the source had confirmed to Satish the whereabouts of certain hardcore Pakistani terrorists whom we, along with several other units, had been hunting for weeks, but without success. Finally, there was specific information about their location. And there was no doubt in Satish's mind that it was reliable information about the group's movements and location.'

The group, as it turned out, was from one of the most specialized terrorist units that Pakistan had sent into India that year—the so-called 'Afzal Guru Squad' of the JeM. These weren't ragtag infiltrators, but men with fearsome commando-style training that armed them with both physical endurance and the sort of tactical combat training required to engage ably with any military force hunting them. Making the fight even more complicated, the terror squad was believed to be using YSMS—a crafty trick in which smartphones were paired with high frequency radio sets to send out SMS text messages that were nearly impossible to intercept. Terror groups had begun using this technology in early 2015, but there was still no credible way of tapping these conversations.

These were men, in other words, who could put up a real fight in total stealth. If no terrorists were to be taken lightly, the men of the Afzal Guru Squad were to be taken least lightly of all. And it was a group of these men that Maj. Satish's source confirmed was hiding in Hajin Kralgund.

'Satish vouched for the tip-off, and requested permission to proceed towards Hajin Kralgund,' says Col. Rajiv. 'The input was very specific. Three to four terrorists were hiding in two separate houses, one on the fringe of the village and the other somewhere in the centre.'

With Maj. Satish and his men on the way, Col. Rajiv phoned Ghulam Jeelani, Handwara's Senior Superintendent of Police (SSP), an officer well-regarded for his commitment to going after terrorists. Militants had flung a grenade at Jeelani's Srinagar home late at night only two months earlier, though neither he nor his family were at home at the time.

Col. Rajiv's calls to Jeelani went unanswered. He gathered from sources that the Jammu and Kashmir police was inaugurating a women's cell in Handwara on Valentine's

Day, and the SSP was probably busy at the function. Twenty minutes later, though, Jeelani returned the Army officer's call.

'Jeelani finally called me back to confirm that he had also received an identical input about the same terrorists,' says Col. Rajiv. 'I told him my men were already on their way to Hajin Kralgund and requested him to join the operation with his SOG. Jeelani was very enterprising and committed. He sent a huge contingent of his men to join the operation.'

In his Maruti Gypsy, Maj. Satish went over the plan that would unfold once they reached the village. With him was a quadcopter drone that would be deployed high over the village to provide them with a bird's-eye view. He remained connected over a secure audio line to his CO. Even if he wanted to call Sujata to check on that parcel, he simply didn't have the time at that point.

Back in Jaipur, Sujata was growing restless about the promised package delivery. Why was it so late if her husband had said it would arrive no later than noon? She switched on the television and flipped channels distractedly for a while before turning it off. She paced about the house, wondering if she should get dinner started, but decided to wait for the package or a call from her husband. Their flat in Jaipur had been rented only two months earlier, when it became clear that Maj. Satish would be stationed in that city after he completed his Kashmir posting in three weeks' time.

'I didn't want to make dinner,' says Sujata. 'I was getting impatient. The only thought that I had in my mind was, why did he lie about a parcel being delivered? And why wasn't he answering his phone?'

Sujata couldn't wait for his Jaipur tenure to begin. Maj. Satish had been commissioned into the ASC in 2009, in the logistical supply wing, which oversees the enormous task of procurement and distribution of food, rations, fuel and other items nationwide to 1.2 million personnel. A few months into service, he was deployed on a three-year posting to Kashmir with the 1st Battalion of RR operating in Nowgam. Returning home to his village in Haryana's Mahendragarh district for a short break a year into the posting, he had met Sujata.

'It was an arranged marriage, but we took our time,' she says. 'When I first met him, I was shocked. He looked scruffy, dishevelled and very different from the clean, handsome man in the photo my parents had shown me. I think Satish noticed my shock. I felt bad later when I thought of this. When he visited me again two months later, he looked just like he did in the photograph—clean-shaven, sharp and tidy. I was over the moon. My heart said yes.'

They were engaged in 2011 and married on 17 February 2012, nearly two years after they first met. Still deployed in a sensitive area, Maj. Satish would get very little time with his new wife, returning to Kashmir within days of his wedding, where he immersed himself in counter-insurgency operations. On 3 July 2012, he would be part of an operation that would later win him a commendation from the Chief of Army Staff.

'He called me during that operation,' says Sujata. 'I could hear the sound of bullets being fired, but didn't know what they were. I ended the call and told my father that the call had been disconnected because there was a lot of background noise. My father understood what I was talking about. He followed the news and later told me that three militants had been killed in Nowgam. Later that night, Satish called and said he had been injured, but was okay. Apparently, a bullet had also missed his head by a few inches. I didn't know what to think.'

Sujata remembered hearing from her husband's friends about an operation he had been involved in before they met, when a young Lt Satish had bravely stormed a location despite being warned to be careful, and had killed a militant. They had told her he was crazy to be so brave. He had calmed her down when, agitated, she asked him about the incident. He did what he had to do, he told her.

For the young officer, an honour from the Army Chief was all the affirmation he needed to be sure that fighting insurgency was what he really wanted to do in the Army. When he had the time for longer conversations with his wife, she listened quietly, noncommittally, as he described his work and achievements. Unfamiliar with an Armyman's work, Sujata would have trouble understanding and navigating the first few years of their relationship.

'After we got engaged, we didn't talk much because he was deployed in that difficult area, Nowgam in Kashmir,' says Sujata. 'I thought he didn't want to talk to me, and was busy with other things. Then, when I asked around, it became clear that his unit was in a place where there was virtually no connectivity. I would call the headquarters, and if I was lucky, I would be patched through for a two-minute call with Satish. That was very precious time. On some occasions, he would walk for an hour up a hill to a spot where he could get a mobile signal and then call me for a few minutes. That one year he was there was a very hard time. He was gentle with me. But I understood how difficult Army life is.'

In 2015, when her husband was deployed once again with the RR for a two-year tenure in north Kashmir, Sujata was understandably apprehensive. She was now expecting their child and hadn't fully settled into life as a military spouse. And while leaving his pregnant wife for a posting was difficult, his return to Kashmir was a calling Maj. Satish had been waiting to fulfil. For three years as a young Lieutenant in Nowgam, and through stints in Nagaland thereafter, he had yearned for an opportunity for some actual combat against foreign terrorists entering Kashmir.

On 30 March 2015, Maj. Satish arrived in Kupwara to join 30 RR. The CO was anticipating his arrival.

'Satish was joining us after operating in a sector adjacent to ours,' says Col. Rajiv. 'So he was quite clued in as far as the modus operandi of terrorists was concerned. It was apparent to me from the start that this was a bold, brave young man, capable of giving us solid dividends in the fight against terror and militancy.'

With an impressive record of counter-insurgency operations, Maj. Satish was given charge of a 'jungle' company. Troops of 30 RR are divided into four 'companies'— two companies stay focused on road-opening tasks and providing escort protection to convoys travelling on National Highway 701, which connects Srinagar to Baramulla and Kupwara. The other two companies are 'jungle companies', deployed in the rural hinterland in thick forests.

In 2016, a year into the posting, Maj. Satish went home for a short break to see Sujata and their infant daughter. And just when he was about to return to Kashmir, a shooting pain in his abdomen forced him to delay his departure. Diagnosed with acute appendicitis, he had no choice but to report to the Army hospital in Delhi.

'It was quite bad, and we gave him thirty days of sick leave initially,' says Col. Rajiv. 'Once you have surgery, you can't be part of active operations for about six months. And the general practice is to send such officers back to their parent unit to make place for a physically fit officer, so that operations don't suffer. But as the CO, I was sure I didn't want to lose Satish. I wasn't sure how much time he would take to recover. But I was sure this young man had the strength to deliver—if not today, then tomorrow.

Since I couldn't send him out for operations, I placed him as adjutant in the unit.'

As adjutant, Maj. Satish was, in effect, a staff officer to the Colonel. And even though he ached to be declared fit for operations, he used this time to advise the unit's company commanders. Crucially, he also spent time cultivating his best source, a local Kashmiri villager who would prove crucial in the months ahead. The time away from field missions also brought with it a special privilege—the opportunity to invite his young family to visit his unit and see for themselves why he was hardly ever free to talk.

'I had a chance to meet his wife and daughter when they came,' says Col. Rajiv. 'I found Sujata to be a very confident and supportive partner to Satish. She had many questions and was curious about all aspects of the tough job her husband was doing. And you could see she was very proud of him.'

The visit was brief. A disturbed area like Kupwara is, after all, usually out of bounds for families of Army personnel. But Maj. Satish had taken the time to introduce Sujata to his men—the men in whose hands he placed his life as a matter of routine, and whose lives he took responsibility for. The tough RR soldiers she met were an enormous source of reassurance. When his family left, Maj. Satish dived right back into his work.

'Satish stayed highly motivated despite the health setback,' says Col. Rajiv. 'I knew he was disappointed, but he still worked round the clock. Whenever I went on operations, he would always be awake, and when I returned, be up and ready to work with me. Frankly, I don't know when he rested.'

When Maj. Satish was finally declared fit to return to combat in late 2016 and was deployed to command one of 30 RR's jungle companies in Behak Harvet, he placed a cardboard placard above the entrance of his bunker with five words, scrawled with a red marker, that appeared to answer his CO's question: *I RUST WHEN I REST.*

If anyone thought it was a cheesy line, Maj. Satish and his company very nearly had to live by it. His company had to keep a lookout for militant and terrorist activity over a huge area—over 23 sq km and nearly thirty villages—with a large part of it covered in thick forest. From the moment he took charge, inputs began blaring in about the arrival of terrorists from the JeM's Afzal Guru Squad. Intercepts and whispers from sources had revealed that four terrorists were on the prowl. The names they seemed to go by were Saad, Baaz, Maavia and Darda.

'This was clearly a fidayeen squad,' says Col. Rajiv. 'They were highly trained terrorists. They had already ambushed Army convoys. They were also instrumental in carrying out attacks on a police station, inflicting casualties there and elsewhere. They had survived and escaped multiple police and Army ambushes too.'

Adding insult to injury, intelligence intercepts showed one of the terrorists bragging about how to break Indian Army cordons: '*Fauj ka cordon kaise break karte hai, main bataunga* (Let me tell you how to break an Army cordon).'

The frustrating game of cat-and-mouse had also become something of a prestige issue by now. And on Valentine's Day 2017, as his vehicle sped towards Hajin Kralgund, Maj. Satish prayed his source hadn't made an error.

'Satish was after these guys, religiously working day in and day out,' says Col. Rajiv. 'He spent every waking hour taking out parties, laying ambushes or cultivating more sources. When his source finally gave him that specific piece of data, it basically confirmed weeks of work he had done. Finally, he had something he could act on.'

A month previously, in January, on a short visit to Jaipur for a medical check-up and to see his family, Maj. Satish, who rarely went into specific details of operations, told Sujata about the terrorists he was hunting.

'Satish had been after these militants for two months,' Sujata says. 'He started to tell me about the ambushes they laid for the terrorists, but that they were managing to escape. I would always ask why he didn't shoot them or attack them. But Satish would say just one thing—if anything happens to my men, even a single man, I will never be able to face their families. "*Mere bande first hain mere liye* (For me, my men come first)."'

When he returned to Kashmir, he wouldn't call home for days. And when he did call Sujata, it would be for a few minutes late at night.

'It was only on Valentine's Day that he called so many times,' she says.

A few hundred metres before Hajin Kralgund, Maj. Satish stopped his vehicle near the side of the road. On his orders, one of his men got out of the Gypsy and pulled out a cardboard box from the back. Inside the box was a quadcopter drone that they powered on and launched. With a low, whining hum, the drone rose into the air, climbing steadily until the village came into view. Fitted with a camera calibrated for low-light missions, the drone streamed live images of the village to a briefcase-sized monitor operated by Maj. Satish's men, which, in turn, beamed the images to the 30 RR headquarters where Col. Rajiv sat in the unit's small operations room, keeping real-time watch on where his men were headed.

In Jaipur, Sujata continued to dial her husband and wait for the promised package delivery.

'I called my mother to tell her that Satish hadn't called and that he wasn't answering his phone,' she says. 'My mother asked me not to worry and to have dinner. She reminded me that there was nothing unusual about Satish not taking calls.'

Maj. Satish's mobile phone was with him, but it was silent—and there was no time to answer. As they arrived on foot at the periphery of the village, he had planned that his company would split into two groups and approach it from two sides. At every step, the CO, armed with the live drone video feed from the operation site, was kept updated.

'Initially, we placed a bigger cordon around the village,' says Col. Rajiv, explaining the plan as it unfolded. 'Then the plan was for Satish to go with his team and lay a tighter cordon around the house at the edge of the village.'

Hajin Kralgund was one of the thirty villages that fell under Maj. Satish's area of responsibility, and he knew it well. He and his men had made it their job to be aware of the layout, the owners of most houses and the numbers of family members. But even with all that data, a terror encounter in a built-up space with unarmed civilians couldn't be more unpredictable. Sure, the drone provided a highly valuable live feed of the intended encounter site. But another pair of eyes was truly indispensable. A pair of eyes that was on the ground, deep inside the village, secretly watching from a vantage point. This was Maj. Satish's source. In an inexplicable show of faith, he was risking his own life and the lives of his family members by agreeing to provide a live commentary of what he saw. In human intelligence terms, nothing could be more valuable than this.

'While Satish and his men proceeded to form the tight inner cordon, I took the responsibility to coordinate the outer cordon with another team of my men,' says Col. Rajiv. 'The inner cordon had two parties—the one led by Satish laid its cordon around the suspected house in the middle of the village. A Junior Commissioned Officer (JCO) was tasked with leading the second party to cordon the house on the edge of the village.'

Hajin Kralgund was a small village, but it still had a large number of women and children residents, many in houses just metres away from the two houses that had been identified as the terrorists' hideouts. The Army was hoping to complete the operation as quickly and cleanly as possible, with the least amount of bother to residents. Both Col. Rajiv and Maj. Satish hoped that the intelligence checked out and that they wouldn't need to search other houses in the village, because if they did, it would greatly amplify the risk of the mission. And given that the terrorists would have a number of hiding places within the village, a more delicate and dangerous operation couldn't be imagined in the circumstances.

'Satish voiced his confidence again in the input as he and his men walked into the village,' says Col. Rajiv. 'After overseeing the cordon laid at the farther house, he proceeded towards the middle of the village to lay his team's cordon. So far, the input had been that both houses had two terrorists each. But just before Satish and his men reached the inner house, his source came on the line.'

'*Sahab, donon bahar wale ghar ki taraf bhag gaye hain* (Both terrorists have left the middle house and are on their way to the house on the edge of the village),' said the source.

Maj. Satish quickly passed the information to his CO. Col. Rajiv looked closely at the drone's video feed, squinting to see if he could make out anything at all.

'Now we knew for sure that the two terrorists from the inner house were headed to the outer house,' says Col. Rajiv. 'But we didn't have a fix on whether there were two terrorists already in the outer house, or whether they had already left. When Satish called me with an update, I advised him to take the party and join the close cordon of the outer house. By this time, SSP Jeelani had also arrived at my headquarters and was monitoring the operation with me. He was in touch with his sizeable team at the encounter site, who were part of the cordons we had laid.'

Maj. Satish's source had been quite accurate. The two terrorists had indeed made their way from the centrally located house to the one on the edge of the village. At the latter house, where two men had been hiding, only one terrorist remained—the second had broken the cordon and escaped into the forests. Now regrouped, the three Afzal Guru Squad members took positions inside the house.

The element of surprise Maj. Satish and his men had hoped against hope for was gone. This was made abundantly clear by a hail of AK-47 fire from the first floor of the house that greeted them upon arrival.

The house sat in a small bowl-shaped piece of land surrounded by hills, with forested slopes on three sides and a stream beyond its front gate. Inside the compound, the ground undulated in small dune-like folds.

In his pocket, Maj. Satish's mobile phone continued to ring silently. A thousand kilometres away, Sujata's restlessness grew. The sun had set and there was still no sign of the promised delivery. She wondered if that was why she was particularly anxious that evening. Her restlessness was turning into full-fledged fear.

By this time, Maj. Satish had reached the front gate of the house with his buddy soldier. He had sent some of his men to its back to reinforce the rear cordon, in case the terrorists tried to make a break for the hills.

'Welcomed with heavy fire, Satish and his men rapidly took position and returned fire,' says Col. Rajiv. 'With the cordon now as tight as it could possibly get, the three terrorists inside the house realized that their only means of escape was to fight their way out. Otherwise, it would only be a matter of time till they ran out of ammunition.'

For six minutes, the firing from the house stopped and silence descended. Maj. Satish and his buddy used the opportunity to enter the compound through the front gate and take cover behind a small mound of earth. At that precise moment, the three terrorists burst out through the front door, firing indiscriminately. Maj. Satish immediately let out a burst of gunfire, hitting the first terrorist in the head and sending him crashing to the ground. The two others immediately leapt behind a clump of earth to take cover.

What Maj. Satish didn't let his men know immediately was that a bullet had found him too. The officer was wearing his bulletproof vest, but the round had pierced him in the gap between two plates and had ripped through a major artery. Minutes later, when his buddy soldier noticed the Major's blood-drenched side, he insisted they pull back so the injured officer could get medical help. Maj. Satish refused, reassuring the soldier that he wasn't badly injured, and that he couldn't leave at that moment.

The firefight was now a tense, pitched battle. Two terrorists, with a mound for cover, were firing at Maj. Satish and his buddy, who had their own knoll protecting them. Nearly ten minutes later, the terrorists began to fling grenades towards the periphery of the compound, where the other men of the inner cordon stood firing. It became clear that the terrorists were well-armed and had plenty of ammunition and grenades to draw out the encounter. And Maj. Satish knew that the longer they stretched it, the greater the chances were of the terrorists causing casualties among the cordon and finding a way through it to escape into the thickly forested hills of Handwara, which surround the village.

In the crossfire, Maj. Satish heard muffled screams from behind him, outside the front gate. Three soldiers from his company, who had been providing covering fire from a few metres away, had been hit by grenade splinters and were seriously injured. Maj. Satish quickly ordered all three men to be pulled out of the firefight, but chose to remain where he was. By this time, he had lost a great deal of blood and had begun to feel dizzy. He made one final call to his CO, informing him that he was going to mount a flank attack on the two terrorists, in which he would crawl out from his position and creep up on the terrorists from the side where they had no cover. Col. Rajiv, by this time, had arrived at the encounter site with more men from 30 RR, deploying them up on the hills surrounding the house.

'Satish insisted he wanted to get closer and finish them off,' says Col. Rajiv. 'He was the operational commander on the ground and best knew the situation. His assessment was that if he waited any longer, the terrorists would definitely kill some of his men. And I knew that was totally unacceptable to him.'

Unknown to Maj. Satish, the three soldiers injured by the terrorists' grenades would tragically succumb to their injuries. It was when the three men—Paratrooper Dharmendra Kumar, twenty-six, Rifleman Ravi Kumar, thirty-three, and Gunner Astosh Kumar, twenty-four—were hit that Maj. Satish decided to finish the encounter himself.

The CO knew it was a delicate situation. He was speaking to an injured officer in the middle of a firefight. He asked Maj. Satish if he wished to pull back. But the officer requested that he be allowed to stay and finish the operation. Maj. Satish's buddy soldier, who had been ordered to retreat towards the front gate of the house, received a call from the CO asking about Maj. Satish's condition.

'*Buri tarah lagi hai, Sir, par khatam karke hi aayenge* (He's grievously injured, Sir, but he'll pull back only when he's finished them),' the soldier said.

Despite the heavy bleeding, Maj. Satish crawled out from his secure position on his elbows, making his way from the side towards the clump that protected the two remaining terrorists, who continued to direct their fire to the front. Taking aim, his bullet killed one. The other let out a burst of fire, narrowly missing Maj. Satish, and bolted towards the hills, only to be felled minutes later by outer cordon troops waiting for precisely such an eventuality.

With the firefight at an end, two soldiers rushed forward to check on Maj. Satish, who had, by this time, slumped in his position and was in and out of consciousness. They carried the Major out of the front gate, where a Tata Sumo vehicle had been arranged to transport the injured officer to a makeshift helipad that had been coordinated minutes after Maj. Satish had been shot.

His eyes still half-open, it seemed as if Maj. Satish wanted to speak, but didn't have the strength. Nobody had the heart to tell the Major that three soldiers had lost their lives in the encounter. He blacked out as the car sped down the highway towards the helipad. A Dhruv helicopter carried him straight towards the 92 Base Hospital in Srinagar, a facility renowned for being able to repair any man with a pulse. But in that helicopter, shortly after 8 p.m., high above the mountains and forests of Kashmir he had come to love and know so well, Maj. Satish drew his final breath.

In his pocket, his silent mobile phone continued to ring.

At 8.15 p.m., Sujata received a call from a lady officer in the Army asking her how she was. The call was followed by a flurry of other calls from Maj. Satish's colleagues in the Army. Now, breathless with anxiety, Sujata asked one of the callers, a course-mate of Maj. Satish, why everyone was calling her. After a moment's pause, the course-mate informed her that her husband had been shot, but that he was recovering in hospital, that there was nothing to worry about. Sujata hung up immediately, dialling the CO of 30 RR. At this time, Col. Rajiv was still at Hajin Kralgund and couldn't answer the phone.

'I lit a diya in my small mandir,' says Sujata. 'My mind and heart were restless and I knew that something was wrong and that everybody was hiding it from me. Priyasha came to me, wondering why I was praying at that time. I usually pray only in the morning. Then the doorbell rang for the first time that day. I ran to the door.'

A delivery man stood outside with a bouquet and a large red package. Sujata grabbed both, took them to the dining table and set them down. She paused for a moment. Then she carefully opened it. Inside was a heart-shaped cake and candles. She stared at it for a moment, then picked up her phone to try her husband's number again. There was no answer.

'I called every single person I had come to know in the Army over the last two years,' says Sujata, 'literally pleading with them to make me speak to Satish somehow. Most of them told me he was fine and that I should come to Srinagar. I still knew nothing. It was when my father arrived at 10.30 p.m. that my heart sank.'

But even Sujata's father, despite knowing what had happened, would say nothing specific to her, instead calming his daughter and telling her that everything would be all right. He also ensured that the television stayed off. News of Maj. Satish's death had begun breaking on social media shortly before 9 p.m. and on television after 10 p.m.

The next to arrive at the house was another Major from 30 RR who was on leave in Jaipur at the time. He told Sujata he was there to help in any way, and if necessary, would take her to Srinagar to visit her husband.

'Nobody was telling me anything clearly,' Sujata says. 'When I pressed my father, he said the Army had told him that Satish was under observation. Everybody was too scared to tell me the truth. I wasn't even admitting it to myself. My heart was praying that these comforting words were true, that Satish was only injured and that he would return my call soon.'

By midnight, with neighbours and Army personnel at her home but still no clear answers to her questions, Sujata begged her father to take her to Delhi so she could catch a flight to Srinagar the first thing next morning.

'I was on the road at night and tried the CO's number again,' says Sujata. 'A voice at the other end, which I did not recognize, told me not to come to Srinagar. They had obviously heard that I was headed to Delhi to take the next flight out. Then the CO came on the line. His voice was breaking with emotion, and he finally told me that Satish was gone, and that I have to take care of myself and Priyasha now.'

Sujata's father directed the cab driver to change course and head to Narnaul in Haryana, the village where Maj. Satish's parents lived.

'After that, I don't remember anything,' says Sujata. 'I went totally blank. Priyasha was on my lap. I wasn't crying. I was just in a daze.'

Sujata had carried nothing with her but the bouquet of flowers and the Valentine's Day cake.

In Narnaul the next morning, huge crowds had gathered on the road leading up to the home of the Dahiya family. Several officers and soldiers from the Army were there too. That evening, amid loud chants in his honour, Maj. Satish's flag-draped casket was brought home.

Surrounded by mourners and friends from the Army, Sujata's calm finally shattered. In front of television cameras that had streamed into the Dahiya house that evening, she wept uncontrollably. The sound of her wails would be broadcast across the country. Priyasha, too young to fully understand what was going on, would be carried by an uncle to light her father's pyre not far from their home.

'Priyasha lives in the belief that her father is alive somewhere,' Sujata says. 'She asks me sometimes when her Papa will come. I have a few audio and video clips of Satish, which I make her watch or hear every now and then, so she feels she has actually spoken to Satish. I order things online and say her father has sent gifts for her. She has begun to insist now on meeting her father. She is very young. I have to take it slow. Maybe, in a few years, she will understand.'

Sujata left Jaipur weeks later to move to Delhi, where she now lives. She hopes to begin a job soon.

'I take out Satish's uniform every few days and look at it,' she says. 'It has the place marked where he was hit by the bullet. It is washed and cleaned and kept in my cupboard.'

Two days after the funeral, on their fifth wedding anniversary, Sujata would learn from a travel agent that her husband had booked a four-day holiday in April for them at the Taj Vivanta in Goa, a place Sujata had longed to visit but hadn't had the chance. The holiday was to coincide with Priyasha's second birthday. Growing up in landlocked towns, the two had often spoken about a holiday at the seaside.

Meanwhile, search operations for other members of the Afzal Guru Squad continued for nearly two weeks after the Hajin Kralgund operation. It was only twelve days later that Col. Rajiv would get the time to visit Maj. Satish's family in Narnaul.

'His father is a very bold personality,' says Col. Rajiv. 'And Satish was a very good son. The way he looked after his parents was truly commendable. He understood his responsibility to them very well. Whenever he went home, he would find out if there was anything he could do to help the village. A street in the village and a nearby college are now rightly named after him.'

'When I think about the Hajin encounter, a part of me wonders what would have happened if Satish had survived,' says a Lieutenant Colonel from 30 RR who was part of the mission. 'He was obsessively concerned about the welfare of his men. I wonder whether he would have been able to digest the news that three soldiers were lost in the operation. It was the most unacceptable thing to him. That's why his men worshipped him.'

Weeks before the Hajin operation, Col. Rajiv had visited Maj. Satish's highly secured jungle post, a base he had meticulously protected with five layers of fencing.

'I told him we were both at the fag end of our tenures with 30 RR,' says Col. Rajiv. 'I told him it was okay if I wrapped up my tenure without eliminating those three terrorists, but Satish Dahiya would never be okay if he moved on without finishing the job, this hardcore group that had inflicted so many casualties on our security forces. Satish said, "I promise you, I won't leave the *paltan* till I eliminate them."'

Months after the operation, Maj. Satish's source in the Hajin Kralgund encounter resurfaced after having gone underground for safety. In a message sent to another officer designated to replace Maj. Satish in the event of the end of his tenure—or worse—he said Kashmir would miss Maj. Satish.

'He wanted to play his small role in ending the militancy here,' the source said. 'I hope Kashmir will not forget that.'

Maj. Mohit Sharma in uniform and a field disguise

Maj. Mohit at the Academy with his family

Maj. Mohit's bust at a heritage hall
built in his memory at the 1 Para
Special Forces base in Nahan,
Himachal Pradesh

Maj. Mohit's wife,
Maj. Rishma, receiving
her husband's posthumous
Ashoka Chakra

Corporal Jyoti Nirala in full battle gear

Corporal Jyoti with his wife, Sushma, and daughter, Jigyasa

Corporal Jyoti with a group of Garud commandos

The last service photo of
Corporal Jyoti

Corporal Jyoti's wife, Sushma, receiving his posthumous Ashoka Chakra

Lt Navdeep Singh shortly after entering service

Lt Navdeep with his parents at his passing-out parade

Lt Navdeep in Gurez, Jammu and Kashmir

Lt Navdeep celebrating during a break from operations

Maj. David Manlun's
service photo

Maj. David with two Kalashnikovs

Maj. David with his entire family
during a summer break

The *3 Idiots* photo of
Maj. David and his brothers

Maj. David's parents receiving
his posthumous Kirti Chakra

Lt Cdr Kapish Muwal on a break between duties

Lt Manoranjan after combat
dive training

Service photos of
Lt Cdr Kapish and Lt Manoranjan

Service photo of
Capt. Pawan Kumar

Capt. Pawan leaving home to head
to Jammu and Kashmir

One of Capt. Pawan's last photos in Jammu and Kashmir

Maj. Satish Dahiya

Maj. Satish with his wife, Sujata, and daughter, Priyasha

Sujata and Maj. Satish's mother receiving his posthumous Shaurya Chakra

Lt Cdr Firdaus Mogal

Lt Cdr Firdaus with his wife, Kerzin, and son, Yashaan

Lt Cdr Firdaus in his bunk on board a submarine

Capt. P. Rajkumar with a Sea King helicopter

People picked up during the record-breaking rescue flight

Capt. Rajkumar with his flying crew

Maj. Rishi with his AK-47
during an operation

Maj. Rishi with his wife, Maj. Anupama,
an officer with the Military Nursing Service

Maj. Rishi still keeps
a part of his face covered

Flt Lt Gunadnya Kharche

Flt Lt Gunadnya and crew after landing

The moment Flt Lt Gunadnya's An-32 touched down on one wheel

Flt Lt Gunadnya and his wife, Shruti, at Wellington with their daughters

Flt Lt Gunadnya and Shruti's favourite pastime in Wellington

Capt. Pradeep Arya near the China border in Ladakh

Capt. Pradeep with his wife, Deepa,
and their daughters, Dhriti and Dhiksha

Maj. Preetam Kunwar with his family after receiving the Kirti Chakra

Maj. Preetam

Maj. Preetam with his wife, Megha

Sqn Ldr Ajit Vasane with a MiG-29 in Jamnagar

Sqn Ldr Ajit with his wife, Sqn Ldr Rajeev Kaur, and
their son, Rishan

8

'Climb over Me,
Get to the Submarine!'

Lieutenant Commander Firdaus Mogal

Arabian Sea, 220 km off the Mumbai coast
30 August 2010, 6.55 a.m.

'THREE MEN OVERBOARD, SIR!'

Seventy feet from where he stood watch on the periscope tower, Lt Cdr Firdaus Mogal had just witnessed three sailors on the submarine's back being violently flung several feet out into the churning Arabian Sea.

The sun had just risen, and as the submarine violently bobbed and pitched in the swell, it was clear to the Executive Officer (XO in military parlance) that the three heads in the water were drifting helplessly away—and fast. He knew a minute wasted would mean the possibility of losing those men. Using his walkie-talkie, he quickly relayed what had happened to the control room. A reply crackled back in. It was the submarine's CO, informing Lt Cdr Firdaus that two sailors were being sent up to go after the three men who had been thrown overboard.

'There's no time, I'm going after them,' Lt Cdr Firdaus shouted down the shaft to the submarine's bridge. As he did so, he saw two combat divers hastily clambering up the ladder from the control room to join him. As they heaved themselves up through the hatch, the three men felt a familiar wave of vibrations course through the submarine. Their CO had just switched the submarine's engines slowly back on to stand by for the rescue.

And just as the three leapt off the back of their submarine into the heaving swell, Lt Cdr Firdous knew he was about to enter a realm he loved and felt most comfortable in.

Mumbai naval dockyard
Fifteen hours earlier

'All set and good to go, Sir,' Lt Cdr Firdaus said, as he welcomed the submarine's CO on board at the jetty. Cdr Gangupomu Murali, the last to walk the gangplank leading into the submarine, returned the younger officer's salute. With the full crew on board, the submarine was ready to depart. The submarine skipper and his second-in-command would be leading a crew of forty on a mission to practise action in a situation that all submariners see in their worst nightmares.

The 211-feet-long and twenty-four-year-old *INS Shankush* dived beneath the surface shortly after sailing out of Mumbai's harbour. The German-built Type 209, one of four attack submarines that the Indira Gandhi government had ordered in 1981, headed straight out into the Arabian Sea for a combat war game that seemed straight out of a Cold War-era film.

INS Shankush was all set to play the quarry in a deadly game of cat-and-mouse, as part of an annual exercise with the French Navy code named 'Varuna'. From the first week of August onwards, the submarine would test the fearsome tracking and detection capabilities of two warships—the Indian Navy's *INS Brahmaputra* and the French Navy's *FNS Dupleix*—both purpose-built to hunt and destroy submarines. The crew of *Shankush*, on its part, would need to dodge the two marauding warships using a combination of silence and evasive electronic techniques. The ships wouldn't be dropping real weapons at the submarine, and *INS Shankush* wouldn't be letting loose any of its deadly torpedoes. But the idea of a war game like Varuna was that the men on board all three vessels performed as if they were in a real war. And pretending isn't difficult. The sound of sensor 'pings' and the distinct tone that blares when a ship's sonar 'locks' on to a submarine, indicating to the crew of the submarine that it has been detected, would make a submariner's skin crawl in any circumstance.

Even the skin of a seasoned submariner.

Lt Cdr Firdaus had opted for the submarine arm of the Indian Navy. A fully voluntary division that permits entry only to sailors and officers who make a certain cut, the submarine arm, like the military Special Forces, requires additional conditioning, both physical and psychological, given the consistently isolated and menacing circumstances that submariners work in as part of their routine duties at sea. It isn't often that military cadets are clear about what they want to specialize in, even if they've chosen between the Army, the Navy and the Air Force. It's rarer still for anyone to join a military academy with the singular intention of choosing to work in submarines.

Lt Cdr Firdaus, on the other hand, had been clear from the day he entered NDA, Pune, that he wanted to be a submariner, spending his years there becoming a masterful swimmer, diver and a prodigious young authority on the history of submarining. When a course-mate asked him why he was so stubborn about joining the submarine arm, cadet Firdaus had regarded him and asked with mock horror, 'How can you want to do anything else?'

When *INS Shankush* sank below the waves and accelerated out into the Arabian Sea, it was just another day on the job for the young officer. But as his body got used to the familiar, gentle hum of the submarine's diesel-electric motors and the warm glow of the lamps that lined the single thin corridor that ran from end to end, he never forgot for even a moment that he was living a real dream.

Lt Cdr Firdaus and his CO had final discussions with the crew on what lay ahead. As XO, Lt Cdr Firdaus was second-in-command. As *INS Shankush* cruised 50 feet below the sea's surface, he walked the length of the submarine that night, speaking with the crew and conducting a list of final checks. It would be past midnight when the XO returned to his cabin for a few hours of sleep before the next day's drills.

Churning the depths with a low roar, *INS Shankush*'s large rear propeller would push it far out into the Arabian Sea overnight, into a section of the ocean where the submarine would spend the following day, 30 August, practising a few manoeuvres before the arrival of the Indian and French warships.

Back on shore in Mumbai, Kerzin Mogal lay awake. As a submariner's wife, she was used to being cut off from her husband for days at a time. Dinner a few hours earlier with Firdaus and their twenty-month-old son, Yashaan, had been quiet. She wouldn't be seeing him for at least the next ten days. After five years of marriage, words seemed almost unnecessary to fill the silence that hung over a now familiar mood just before departure. Seeing her husband off at the door with a hug, Kerzin put their son to bed and retired to her room with a book. Two hours later, making an involuntary mental note that Firdaus's submarine had probably just slipped beneath the waves, Kerzin fell asleep.

'I remember that it was raining heavily the Sunday that he left,' she says. 'I'm usually a late riser. But the next morning, despite falling asleep quite late, I woke up early and I was very restless. I had no idea why.'

Unable to go back to sleep, Kerzin got out of bed, deciding to start the day. She went about getting ready, picked out a red top for work—she always wore red on Mondays. Then she made another mental note. Firdaus and his submarine were probably far out at sea by now.

By this time, *INS Shankush* had sailed nearly 200 km away from Mumbai. As Kerzin sipped her morning coffee, she couldn't possibly have known that all wasn't well on board her husband's submarine.

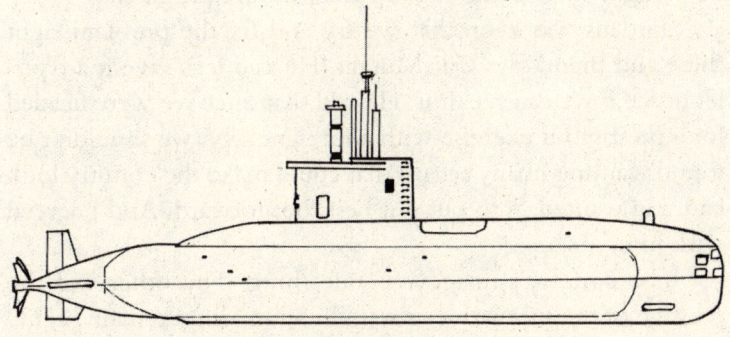

Type 209 attack submarine

There was a problem. A serious one, detected at dawn. An exhaust valve that expelled toxic by-products of the submarine's electric batteries had failed, causing a leak inside *INS Shankush*, a glitch that could be deadly to the crew if left unrepaired. The valve, one of a pair that remains underwater even at periscope depth, wasn't doing its very important job. Cdr Murali wasted no time, quickly ordering the crew to take the submarine to the surface.

Within minutes, the 1800-tonne submarine climbed through the water before gently breaking the surface.

'When we surfaced, we noticed that the sea was quite rough,' Cdr Murali says. 'I knew we had to fix the valve right there and then. I took Mogal with me to the top of the submarine to assess the situation.'

The two men ascended the ladder that led to a hatch at the top of the conning tower, the dorsal fin-like structure on a submarine's back through which the periscope, communications antennae and other sensors jut out of the water, even as the submarine itself remains submerged. It was clear to both the officers that the repair of the exhaust valve would be a stiff challenge in those sea conditions. Cdr Murali asked his XO what he thought about returning to harbour to fix the glitch.

'Firdaus was keen that we try and fix the problem right there and then,' says Cdr Murali. 'He said let's give it a try— let us see if we can repair it. He said that since we were headed for a prestigious exercise with a foreign navy, we shouldn't be found wanting in any regard that could make the country look bad. He wanted us to put our best foot forward. And I agreed with him.'

The country's image was one thing. The other was the prestige of the submarine arm itself. Submarines remain highly stealthy when submerged, but are immediately detectable once they surface. In the restive Arabian Sea, where both China and Pakistan attempt to keep constant tabs on the movement of Indian submarines, sailing the submarine back to base would be a strategic embarrassment. It would be an admission that an Indian submarine, under the watchful eyes of satellites, couldn't complete an exercise with a foreign navy.

Cdr Murali summoned his Engineering Officer, the man trained to oversee the fixing of any technical trouble on a submarine at sea or in port. The three men quickly discussed the operation, deciding that they would try to complete the repair in no more than 20 minutes. The Engineering Officer called three sailors and led them down the outside ladder of the conning tower and onto the submarine's back half. The men were strapped to safety lines as they descended the sides to fix the faulty valve. But to actually conduct the repair, they needed to remove the safety straps since the valve was situated within the submarine's casing. About fifteen minutes into the repair operation, the men were preparing to emerge and strap themselves back into their safety lines, but they didn't see what was roaring towards them.

'You cannot predict waves. This one was 15 feet high and came out of nowhere. Three men were flung overboard,' says Cdr Murali who, by this time, had descended to the bridge, the submarine's control room, to monitor the repair operation while his XO kept visual tabs from the tower.

'Firdaus immediately reported that three men were in the water, with the fourth man not visible,' says Cdr Murali. 'And without a pause, he asked me for permission to go after them himself since there was no time. He said the submarine was stationary, with its engines off, and therefore, he could safely venture after them.'

But something had caught Lt Cdr Firdaus's eye—the fourth man, who had also been flung overboard, was spotted hanging by the side of the submarine, holding on precariously to his safety line. The sailor had injured himself badly and was bleeding from a deep cut on his leg. Lt Cdr Firdaus immediately climbed down the conning tower and rushed forward across the submarine's casing. Reaching down while fighting a ferociously pitching sea, he heaved with all his strength to pull the injured sailor up. Supporting the sailor, since he couldn't walk, Lt Cdr Firdaus rushed him back to the conning tower and had him lowered in for treatment.

'The doctor on board said he needed to be evacuated because he was bleeding very heavily,' Cdr Murali remembers. 'The sea was rough, and it had also started raining quite heavily by then. I knew it would be impossible to send him back to base anytime soon. I held his hand and told him to hold on. I was now doing three things—handling the submarine for the rescue, dealing with my injured sailor, and reassuring the other men that all was in control.'

With one man rescued, Lt Cdr Firdaus cast his gaze away from the submarine to the three men bobbing in the water. Joined by two combat divers, he climbed down onto the back of the submarine and jogged to the far end, as the entire vessel rolled and pitched in the heaving sea. It quickly became clear that the three sailors who had been flung overboard had now drifted too far from the submarine to be thrown a line. And that's when Lt Cdr Firdaus decided to dive into the sea in an attempt to reach the sailors who, by then, had drifted over 100 m from the submarine. The two combat divers jumped into the water after him.

At the bridge, Cdr Murali was faced with a difficult decision. The engines of *INS Shankush* had been turned off for the valve repair operation. And with men overboard, the engines were kept switched off because of the very real possibility of the drifting sailors being sucked into the powerful propeller.

'There were now six of my men in the sea, including Mogal,' says Cdr Murali. 'I waited for the men to drift a certain distance away. Then I switched on the engines to low power, carefully manoeuvring the submarine around in their direction to begin the recovery.'

Two more sailors had been ordered onto the submarine's back to help pull the six men on board. As the submarine edged towards Lt Cdr Firdaus and the two combat divers with him, he signalled to the submarine to proceed forward and rescue the other three men first, who by this time had drifted to over 200 m away.

The submarine gurgled past Lt Cdr Firdaus and the two divers towards the three men. The sea held steady as they were carefully pulled back on to the submarine with some difficulty.

Cdr Murali now had three more men to bring back on board. The wind had picked up and all three had begun to drift similarly. The men were adept swimmers, capable of handling themselves well in rough water, but as the minutes wore on, they would tire. It was imperative that they be brought back on board quickly. As the submarine edged back towards them, the divers gestured to Lt Cdr Firdaus to prepare to pull himself back up on board. But he refused, insisting that he would ascend the rope only after the two sailors were safely back on board.

Back in Mumbai, Kerzin Mogal had just arrived at work.

'I was still extremely restless when I entered the office,' Kerzin says. 'I remember telling a colleague of mine that I thought that day was going to be a holiday. I missed most of my calls that morning and kept mostly to myself. And I had no idea why. I found it hard to focus or concentrate on anything. Something was off. I felt it.'

From the bridge of *INS Shankush*, Cdr Murali edged closer to his XO and the two combat divers in the water.

'Finally, I manoeuvred the submarine close to the three,' says Cdr Murali. 'Mogal told the two sailors to go and board the submarine first and that he would come last. They had been in the water for nearly twenty minutes. All three were tired by this time.'

Realizing that the two combat divers were drifting again despite being highly qualified swimmers, Lt Cdr Firdaus grabbed the rope from the submarine and pushed himself towards the drifting men, offering to be a human bridge of sorts. He shouted to the two men to clamber over him and get to the submarine as quickly as possible. Once again, the two divers pleaded with him to climb on board first. They would manage, they said. But the XO knew they hadn't a chance of being pulled back on board, given how rapidly they were now drifting.

'Do it now! We have no time. Climb over me, get to the submarine,' Lt Cdr Firdaus screamed through the spray.

Finally, they obeyed—they knew it was their only chance of reaching the submarine. Holding on to the XO's shoulders, the two divers pulled themselves into the submarine.

Now, only Lt Cdr Firdaus remained in the water. He began to pull himself up. At the precise moment that he was about to haul himself out of the sea, the swell caused *INS Shankush* to roll violently, hitting the officer in the head and throwing him back into the sea in a splash of blood. It was a devastating knock from a 1800-tonne hunk of metal.

'He was knocked out,' says Cdr Murali. 'I tried to manoeuvre and get him on board, but he was unconscious. One officer and two sailors volunteered and asked me to bring the submarine next to him, saying that they would somehow pick him up. It was clear that someone would have to physically pull him out as he was not conscious. The three sailors who had been rescued by the submarine initially had no choice but to get into the water to pull him out.'

For a few harrowing minutes, the CO wondered about the unrelenting irony of the situation at hand. The unconscious officer was drifting rapidly now in the sea churned by the wind. Over the minutes, he drifted nearly 300 m from the submarine. The three men who went in after him were similarly set adrift in the swell. Even if the submarine was manoeuvred carefully, it would take thirty more minutes to recover Lt Cdr Firdaus and the three men.

The situation was now critical and Cdr Murali knew it was time to call for help. From the bridge, an emergency request was sent to the *INS Shikra* helicopter base in Mumbai, calling for a chopper to be dispatched immediately.

Back on board the submarine's deck, the sailors were ordered a few minutes later to stand by as a Chetak helicopter was arriving to pick up the injured XO. Over a rolling and pitching sea, the helicopter roared in 30 minutes later through the early morning haze and flew back with Lt Cdr Firdaus strapped to a stretcher, still unconscious. Two sailors had been precariously lowered into the churning sea to pick up the officer. The remaining three men were picked up by the submarine.

From the bridge, Cdr Murali conferred with Commodore Mohit Gupta, his boss and the then Mumbai-based COMCOS in the western fleet. Receiving authorization, Cdr Murali turned INS Shankush around and headed straight back to base.

With INS Shankush making an exit from the Varuna exercise, the Indian Navy had ordered another submarine to replace it in the Arabian Sea. The two submarines likely crossed each other that day.

'I was there to receive Mogal at the INHS Asvini naval hospital,' says Commodore Mohit, now a Rear Admiral at the Indian Navy Headquarters. 'When he was brought in, he was still alive. I stood outside the Intensive Care Unit and waited. This was a tough-as-nails officer who had just done something unbelievably brave. I was certain he would make it.'

The first phone call Kerzin Mogal took that morning was from a young Indian Navy officer who worked with her husband. She knew she couldn't ignore a call from the Navy.

'I was informed that Firdaus had been injured and that I should rush to INHS Asvini as quickly as possible,' says Kerzin. 'I wasn't told precisely what had happened, only that I get there as soon as possible. I was in the suburbs and it would have taken two hours to reach the hospital in south Mumbai. I called my mother, who worked in the

Fort area, and asked her to rush there. My son was with my in-laws.'

With a colleague, Kerzin speeded towards the hospital. A series of phone calls to Lt Cdr Firdaus's colleagues in the Navy revealed nothing further.

'There was a lot of anxiety. And a lot of negative thoughts,' says Kerzin. 'The hospital wasn't telling us what was really going on. I guess they needed me to be there. I know their work is dangerous but I had never actually imagined this happening. It's always smooth sailing with these guys. They are such brave souls. They don't tell their wives exactly what happens.'

On the eighth floor of the INHS Asvini naval hospital, a team of doctors battled to revive the injured officer. But scans had revealed a grievous skull fracture. Nobody could have survived such a blow to the head.

A little over an hour after he was flown in, Lt Cdr Firdaus Mogal was pronounced dead.

'As I was crossing Haji Ali, I called Firdaus's colleagues again, yelling and telling them not to lie to me about the situation,' says Kerzin. 'My mother had reached the hospital by then and I was told that she had fainted. I was rushing as fast as I could. Something in my heart told me that when I reached, I would not be seeing Firdaus.'

'Despite the best efforts of everyone, we lost this brave soldier,' says Commodore Mohit. 'His CO knows. His men know. His family knows. And I know—that he did not need to dive into the water to rescue those men. As the XO, he did not need to. That was his super-humanity.'

INS Shankush was headed towards Mumbai at full speed, but had to slow down. A few weeks earlier, a cargo vessel named MSC Chitra, coming from Mumbai's Jawaharlal Nehru Port Trust (JNPT), had collided with another merchant vessel, MV Khalijia-III, about 9 km off Mumbai, spilling oil and strewing its massive containers over a sizeable area of the sea. This was a navigational hazard, forcing the submarine to approach slowly and with extra care.

'Mogal and the others were picked up around 9.15 a.m.,' says Cdr Murali. 'Then we started back to harbour, reaching around 7 p.m. There were people on the jetty. They informed us that Mogal had not made it. That's when I knew I had lost my brave XO.'

Cdr Murali and the others from INS Shankush rushed straight to the naval hospital.

'I wanted to see the body. And I wanted to meet Mrs Mogal and tell her personally what had happened,' says Cdr Murali. 'But the body had been taken home. They wished to hold the funeral two days later. That night, I gathered my entire crew and spoke to them. I told them they had done a brave job that day, and that we had lost a brave brother officer. It was a tragedy.'

But what compelled Lt Cdr Firdaus Mogal to jump into the sea that August morning?

'Imagine how he saw those three sailors struggling—he instantly jumped. It was a flash of leadership. No rational man in his wisdom would have jumped,' says Commodore Mohit. 'It is the eternal spirit of duty that made him jump. He was physically pushing those sailors up, some on his shoulders. But unfortunately, he couldn't save himself. It takes a huge amount of effort to be the one at the top. Even the best of people couldn't do it, it was extraordinary.'

Two days later, Cdr Murali finally got a chance to see the remains of his XO. And to meet his wife, Kerzin.

'I couldn't speak because of the lump in my throat. I tried my best but I couldn't save him. He was brave beyond measure,' says Cdr Murali. 'I remember admonishing the other sailors who were with Mogal, telling them they should have pushed Mogal back in first. They said, "Sir, we tried

that, but Mogal was adamant that he would go only after us."
I told his wife that he could have saved himself by going in
first, perhaps. But he was more concerned about the safety of
his men.'

Lt Cdr Firdaus and Kerzin's son, Yashaan, was barely two
when his father died. Young enough, fortunately, not to know
what was going on, and safely in the arms of a tightly knit
Parsee family. Now ten, he cannot stop talking about a father
he cannot possibly remember, but has heard much about.

'Many people told me I shouldn't be talking to my son
about what happened. But I always felt it was his right to
know what his father did,' says Kerzin. 'Today, Yashaan speaks
about Firdaus very proudly. Even when he was four, he would
narrate in school or to friends how brave his dad was and what
he did. That's how I always wanted him to be. I didn't want
him to be left out because with a father like Firdaus, he has a
legacy. Firdaus hasn't simply left us. We all have to match up
to how brave this soul was.'

In India's small but highly respected submarine arm,
Firdaus Mogal is a name few will ever forget.

'He's definitely considered a superhero,' says Commodore
Mohit. 'You can ask anyone in the submarine arm about the
name Mogal—his memory has not faded. It's been nearly a
decade. We still remember him like it was yesterday.'

To honour his memory, the Navy has named a submarine
simulator complex, marriage accommodation blocks and an
electrical training school in the officer's name. But as with all
military personnel, families and friends of heroes are still the
only ones left to mourn their passing.

'I still tell everybody that if the submarine hadn't suddenly
rolled and hit Firdaus on the head, my husband would have
come back just injured. He would have healed and been with
us,' says Kerzin. 'I can't sit through the stories that are told

of him, and I've read every single one that has been written. They shatter me every single time. He was something else. Firdaus was an angel. He wasn't a human being.'

In the years since his passing, Kerzin Mogal has had time to explore every aspect of her husband's motivation that fateful day. And even though she mourns his loss, she has discovered a strange comfort in what happened.

'I keep wondering how Firdaus would have felt if one of the men had died and he had survived. He would never have been the same again,' says Kerzin. 'He would never have forgiven himself. I've lost my partner, my love, the father of my child. But if anyone else had died in this incident, Firdaus would have been miserable for the rest of his life. Nothing would ever have been the same again for him.'

Lt Cdr Firdaus's family and comrades in the Navy remember him as an officer who unstintingly stood up for the men who served under him.

'"My men are my world," he would say. Everybody will be good to seniors. Real honour is in how you are with your men,' says Kerzin.

Yashaan Mogal is an avid reader. He even reads the Indian Navy journals his mother has collected that contain details of the incident in which his father died.

'Many of Yashaan's teachers say it's amazing how this child doesn't need counselling,' says Kerzin. 'He's a proud and happy kid. He loves talking about his father. When I see and hear that, I feel I can rest easy.'

Kerzin still feels very much a part of the military community in Mumbai. She gets frequent calls from the Navy asking her if she's all right, or if there's anything the service can help her with. She often attends Navy events when she's invited. The one thing she usually declines, though, are the evening functions and dinners.

'I avoid going because I see men in uniform and it's hard for me not to see Firdaus there,' says Kerzin. 'We would go to these together.'

In January 2011, six months after the incident at sea, when Kerzin Mogal arrived in Delhi to receive her husband's posthumous Shaurya Chakra gallantry decoration, a senior Naval officer accosted her at the Rashtrapati Bhavan ceremony.

'He said your husband need not have gone into the water,' says Kerzin. 'His uniform did not tell him to do that. I remember telling this gentleman that he did not know Firdaus. If he were to be in the same situation again, he would go back into the water again and save the six people that he did. Some people are meant to be heroes. It's what I've always said to console and heal myself. Saving those men was his motive. Many of us live without motive. Firdaus just saw it in black and white. He faced no dilemma. He had absolute clarity about what he had to do. I keep saying this—I know in my heart he would have done it again.'

The Shaurya Chakra citation now adorns a wall in Kerzin's living room. It reads:

Lieutenant Commander Firdaus Darabshah Mogal displayed exceptional courage, unmatched show of fearless valour in the face of death and made the supreme sacrifice in saving the lives of six men.

Initially sensitive to questions, Kerzin is now used to them. 'Somebody once asked, when he jumped into the water, did he not think about his child and his wife? And I say, no,' she says.

'I know Firdaus. He was thinking of nothing but his men drowning.'

9

'Just Tell Me when to Begin, Sir'

Captain Pradeep Shoury Arya

Bengaluru
10 May 2017

'Promise me one thing—that you will be alive to receive your medals. And that I won't have to do it on your behalf. Promise me, Pradeep.'

Deepa Arya chuckled nervously. At the other end of the phone line, there was silence. But she could tell he was smiling. It was the sort of verbal ambush Capt. Pradeep Shoury Arya had only been half prepared for in the three years since he began doing what he did.

It was their thirteenth wedding anniversary that day, 10 May 2017. He had promised to celebrate it with Deepa at their Bengaluru home. But his flight out of Mumbai a week ago had taken him in the opposite direction.

Capt. Pradeep disconnected the call from a Special Forces camp in Uri in Jammu and Kashmir's Baramulla area. Contemplating Deepa's words, Capt. Pradeep walked back to the cramped dormitory he shared with three other officers. Unlike them, he knew he didn't fully belong here. For the Captain wasn't an Army officer at all; he was from the Indian Revenue Service (IRS), a comfortably settled tax bureaucrat who had made a career choice that most who knew him would blame on a moment of insanity.

Eight days earlier, on a sultry Mumbai afternoon at his Ballard Estate office, Pradeep Shoury Arya had placed piles of tax papers aside for a quick lunch of his favourite grilled fish. Work didn't stop, though. As he ate, he pored over an International Monetary Fund (IMF) report on money laundering and terror financing, a subject he had grown increasingly obsessed with. Three minutes into lunch, his cell phone rang. It was a call he couldn't dodge.

'Pack your bags, buddy, and take the first flight out of Mumbai. Details after you reach the location,' said the caller. Capt. Pradeep instinctively straightened in his seat, placing his fork with a bit of fish still on it down on the plate with a gentle clang.

The caller was the CO of a Para Special Forces unit operating in Jammu and Kashmir's volatile Uri sector near the LoC—a place where nineteen Army soldiers had been killed in a devastating terror attack that had shaken the entire country eight months before. It was an attack that had sparked the devastating surgical strikes of 29 September 2016, which left thirty-six terrorists and two Pakistani Army personnel dead in one of India's most brutal revenge missions till date.

'On my way immediately, Sir. Jai Hind,' Capt. Pradeep said. If details were to be provided later, it meant there was no room for questions or clarifications now. The only thing to do was to move, and fast.

Except that he was a taxman.

As an Additional Commissioner of Income Tax for International Taxation, Capt. Pradeep couldn't just get up and leave whenever he got a call. He was a Captain in India's Territorial Army, but he still needed leave from his boss in the Income Tax department when the Army needed him. Reserve troops from the Territorial Army, which Capt. Pradeep had joined in 2009, come from all walks of life and are required to operate alongside regular soldiers for at least two months each year, though they are expected to answer calls at any reasonable time. Capt. Pradeep was a reservist attached to a Special Forces unit in the Kashmir Valley. If his full-time work garb involved shirts, ironed trousers and formal shoes, his other job allowed him to wear the coveted maroon beret of India's most lethal soldiers. Capt. Pradeep would jump into combat fatigues in minutes if he could.

But first, he needed leave.

Capt. Pradeep quickly tossed the IMF report he was reading into a drawer. The report was central to his growing expertise in how terror networks raise and move money, and how that financing could be disrupted. The officer's talent for detail in the murky world of terror financing had been recognized a few months earlier, in January 2017, when he was awarded the Chief of Army Staff's commendation for a useful data report he had generated tracking fundraising methods employed to fuel terrorism and anti-state activities in Jammu and Kashmir. The report he had been reading had confirmed many of his suspicions, which he would further study. But the phone call from his CO meant he would have to read it another time.

He moved quickly. A flight to Delhi had to be booked, a promise to be with his wife in Bengaluru for their thirteenth wedding anniversary had to be broken. And a long overdue evening with friends at the popular Leopold Café in Colaba had to be called off. The really tricky part would be seeking leave from his boss, Charanjeet Gulati, the Commissioner of Income Tax. His reasons were solid—a tour of duty in Kashmir—but as Capt. Pradeep walked towards his boss's cabin, he knew he would still have to make a persuasive argument. He wasn't going to be pleased at all, Capt. Pradeep thought to himself as he knocked on the cabin door.

'For heaven's sake, are you saying you are going to be away for a couple of months again? This is utterly unbelievable,' the Income Tax Commissioner spluttered, staring at Capt. Pradeep in disbelief.

'Sir, I am sure we can find a solution,' Capt. Pradeep said quietly, hoping the conversation would end soon so he could go home to pack.

'If you have to go to Kashmir so often, then who will work for the department? Why don't you take a long break and do what you have to do, Arya?' his boss snapped.

'Sir, the call was from the CO himself. I have to reach Uri as soon as possible. You know how things are in Kashmir,' Capt. Pradeep said, a touch of apology in his voice. His boss's irritation wasn't unreasonable, considering Capt. Pradeep had been spending more time with the Special Forces unit than the minimum required under the rules for an Army reservist. Instead of the compulsory two months, he had been out on missions for four to six months in each of the previous two years.

His boss sighed.

'And when do you intend to leave?'

'Tomorrow. As soon as possible, Sir.'

The drive from Ballard Estate to Capt. Pradeep's Malabar Hill apartment overlooking the Arabian Sea took 20 minutes that evening, giving him enough time to make the required phone calls to his family in Bengaluru and the friends who would be nursing hangovers the following morning.

After an early dinner, Capt. Pradeep called it a night. Staring at the ceiling in the dark, he wondered about the reason he had been summoned. He thought of his Special Forces comrades in Kashmir, men for whom a full night's sleep was a rare gift. However, he knew this was likely to be his last predictable night for the foreseeable future and soon fell asleep.

On an early morning flight to Delhi the next day, Capt. Pradeep was typically restless, his mind crowded with thoughts of the new 'adventure' that lay ahead.

'My CO hadn't dropped a single clue. I had no idea what I had been summoned for. And I was restless to get there and find out,' says Capt. Pradeep.

His CO, an Army Colonel, had become something of a cult figure after the September 2016 surgical strikes on terror camps in PoK.[1] He had received a Yudh Seva Medal for the operation that would blast its way into popular culture in the following months and become both a strategic and political buzzword.

The Special Forces unit under the Colonel's charge, to which Capt. Pradeep is attached, played a central role in the blistering covert action behind enemy lines, a mission that drew attention to India's hardened political resolve in the face of repeated terror provocations from Islamabad, as well as its military capability to strike terror havens on the hostile territory beyond the LoC.

Catching a connecting flight from Delhi to Srinagar, Capt. Pradeep settled into his seat for a 40-minute nap. He was certain sleep would be elusive once he landed and made his way to the base he had been summoned to.

'Normally, I would be called to the unit once a month for about a week. But this time, I was told I should be prepared to stay back for a couple of months. That was straightaway unusual,' says Capt. Pradeep. What he was certain about was that he was headed to a site of action.

Darkness had enveloped Uri by the time he reached the quiet and secluded base, where he was greeted by familiar fragrances, a faint breeze whispering through the chinar trees and some of the Army's most elite fighters with long hair and

---

[1] The first and only official account of the 2016 surgical strikes is in the first part of *India's Most Fearless 1*, published by Penguin Random House India in 2017.

full beards, not uncommon among Para commandos operating in Kashmir.

'What took you so long? We were expecting you earlier,' said an officer, high-fiving Capt. Pradeep as he stepped out of his jeep.

'I stopped at Khunmoh [near Srinagar] to meet the boys from 106. You know I always do that,' said Capt. Pradeep, referring to his parent unit, the 106 Infantry Battalion Territorial Army (Para).

'Well, the CO should be here any moment. He's coming from Udhampur,' said the officer.

'I am dying to meet him. Let's stick around till "Tiger" comes,' Capt. Pradeep said.

The Army's Northern Command, headquartered in Udhampur, serves as the nerve centre for the command and control of all Army operations along the LoC with Pakistan. COs from Special Forces and other units are frequently summoned to Udhampur to brief seniors on missions or plans.

Minutes later, the CO's vehicle screeched to a halt inside the base and the muscular six-foot-tall officer stepped out. Respected highly by his men—his credentials burnished even more after the 2016 surgical strikes—the Colonel cut a menacing figure in his battle fatigues, enveloped by an unmistakable aura of authority.

'Good to have you back with us,' the Colonel said, flashing Capt. Pradeep a grin. 'You are curious, aren't you? You were always looking for that great adventure and there's one coming your way.' And without another word, the Colonel walked into the camp quarters.

Capt. Pradeep knew there was no use pushing to know more. He would be told about his mission when the CO was ready, and not a moment earlier. Fortunately, he only needed to wait till dinner to find out why he had been summoned at such short notice, and why he had been asked to stay back almost indefinitely.

At dinner that night at the camp mess, the CO finally got down to briefing Capt. Pradeep and the others.

Intelligence had been gathered that week pointing to unusual activity across the LoC, with military-style terrorist infiltration squads biding their time to slip into Kashmir. The information was apparently solid and required a careful but urgent follow-up with a full-fledged action plan. And Capt. Pradeep was being put in charge.

Capt. Pradeep remembered the jarring images of death and havoc at the Uri Army base from eight months earlier. The base had been struck by a four-man infiltration squad from across the LoC in a nightmarish assault that left nineteen soldiers dead. The pre-dawn strike at the base, rimmed by verdant hills, had taken place in this Special Forces unit's own backyard, and the stinging memories had scarcely faded.

Following India's revenge attacks a few days later, Pakistan was keen to keep the terror pot simmering despite a heightened alert along the LoC.

Capt. Pradeep's CO had told him that the broader intelligence picture pointed to a strong possibility of brazen attempts by terror cells to infiltrate Kashmir in the course of the next few days or weeks.

'The CO gave me a clear mandate to establish a robust intelligence network involving a variety of elements, and simultaneously, prepare my squad to ambush the infiltrators before they could set foot on our soil,' he says.

It was a simple enough job to describe. Capt. Pradeep needed to corroborate intelligence, establish an intelligence network of his own to fine-tune the information and then act upon it with all the force at his disposal. But nothing about it was remotely elementary. Terrorists, aided by the Pakistan

Army and Inter-Services Intelligence, frequently altered plans and locations for infiltration, leaving Indian units guessing till the last moment—sometimes also being caught off guard, with devastating consequences.

As he headed towards the barracks later that night, the single thought that gripped Capt. Pradeep was how a few hours had taken him from a grilled fish lunch in upscale Mumbai to the prospect of a fight to the death in one of the world's most dangerous conflict zones. He contemplated Kashmir's steady and obvious slide back into chaos, with the chances of normalcy dispiritingly remote. The mission at hand for him was but a dot in a larger constellation of threats. But a dot missed, he remembered someone telling him, could make all the difference.

It was early May and the border state was wrestling with what was turning out to be an especially grim year, a tumultuous phase that Pakistani Army-backed terrorists were seeking to exploit to the fullest. Still smarting from the humiliation and damage of the previous year's surgical strikes, Pakistan's Army was ratcheting up efforts to stir up trouble in Kashmir by dispatching commando-trained terror graduates from the camps, running with its full support and funding, in PoK.

'They are up to their old tricks again. But there is no way we will let them succeed,' thought Capt. Pradeep, as he settled into his bunk for the night.

There was nothing remarkable about the barracks at the Special Forces base except its occupants, and the special weapons their dangerous missions afforded them. Capt. Pradeep shared his dormitory with three other officers, and the small attached toilet was the only luxury. It was a far cry from living in one of Mumbai's most exclusive neighbourhoods. He was equally at ease in either world, but secretly loved the austere barracks and the company of these fellow officers he had worked with since he had been attached to the unit in 2015. Their respect for him was inflected with a laughing curiosity—why would a comfortable tax babu living the good life want to be in the Special Forces?

Everyone knew the story, though.

It all began a few years ago with a conversation between the Army's Lt Gen. Subrata Saha, then heading the Srinagar-based 15 Corps, and a senior IRS officer at the prestigious National Defence College in Delhi, where military and civilian officers are groomed for higher leadership positions.

Lt Gen. Subrata, a scholarly three-star General, had just finished delivering a talk on civil–military cooperation when the senior bureaucrat introduced himself to the Corps Commander.

'General, do you know that one of our IRS officers is serving in the Territorial Army? Maybe you can meet him and find a way to put his expertise to the best use,' he said. Lt Gen. Subrata was pleasantly surprised—he didn't know of any bureaucrat who was a military reservist. He made a quick mental note, took Capt. Pradeep's details and a few days later, summoned him to the Badami Bagh Cantonment in Srinagar for a meeting.

When he arrived, the General cut straight to the chase: 'Arya, why don't you consider conducting a study on money laundering and terror funding, and how we can combat these activities? It can be very useful in the context of Kashmir.'

'Just tell me when to begin, Sir,' Capt. Pradeep said.

With the formalities completed a few weeks later, Capt. Pradeep moved to Srinagar for the new assignment. And after Lt Gen. Subrata was transferred to Delhi to take over as one of the Army's two deputy chiefs, the Captain was attached to the Special Forces unit for the next phase of his study, and to also help provide intelligence on terror groups. When he heard about Capt. Pradeep's expertise, the unit's CO grabbed the opportunity with both hands. This man could be very useful, he thought.

'I had had enough of the academic bit and longed to be part of actual operations that the unit was carrying out, the rough-and-tumble of the Special Forces. The peril, the action-packed adventures enticed me,' Capt. Pradeep says.

A notoriously gruelling three-month-long training course followed, which Pradeep describes as the ultimate test of endurance. Upon its completion, he was declared ready to join the ranks of the Special Forces as a Territorial Army officer and be counted among those who would go into combat with them. A higher vote of confidence was unheard of. The Special Forces are not known to outsource. As an outsider, Capt. Pradeep had been welcomed into a unit that normally drew its strength from a nightmarishly difficult training. The tax bureaucrat had proved worthy of the entrance examination. And now he was in.

'One of the critical lessons I learnt during training was that all limits are in our heads. The trainers were ruthless, but they did an amazing job of preparing me to operate at my maximum potential,' says Capt. Pradeep, the training regimen having sculpted his already wiry 5-foot-9-inch frame into raw sinew.

Kashmir had not been far from Capt. Pradeep's mind ever since he had signed up for the Territorial Army in 2009. Regardless of where his work took him, he made sure he kept a finger on the troubled state's pulse. Entries in his personal diary capture how Kashmir was beset with problems at the time, with a sharp escalation in unrest after Hizbul Mujahideen commander Burhan Wani's killing in July 2016, and the flare-up of tensions ahead of the by-elections to the Srinagar and Anantnag Lok Sabha constituencies in April 2017. The diary, his constant companion, details other distressing symptoms of a situation spinning out of control—the swelling crowds at terrorists' funerals, the fury of protesters hurling stones at security men, the surge in anti-India protests, the disturbing images of teenage students clashing with the police, and worse.

'There is no way to sugarcoat the truth that Kashmir was absolutely shaken by a destructive spiral of violence,' says Capt. Pradeep, flipping through the diary.

It was against this backdrop of heightened turbulence in the Kashmir Valley that the operation assigned to Pradeep by his CO assumed greater significance. And he didn't have to be told twice what to do next.

Vital to the effective conduct of army operations is the enormously difficult business of gathering and interpreting intelligence, which requires unsparing effort, skill and patience. The success of a mission depends, unwaveringly, on how good the intelligence is.

The scale of the task before Capt. Pradeep was clear to him from the moment the words left his CO's mouth. And before he hunkered down to sleep that wind-whipped night, Capt. Pradeep wrote a tentative to-do list in the same diary. Activating a carefully cultivated network of informants on either side of the LoC topped the list. He would also need to work with units specializing in diverse intelligence collection disciplines, such as human, technical, communications, electronic and imagery. The different strands would need to come together to form a clearer picture of the mission. The greater the number of strands, the sharper the picture.

Preparing the commando squads for the job ahead was the least of Capt. Pradeep's worries, since Special Forces units train hard every day for some of the Army's most dangerous and secret missions. One of the elements critical to the mission's success would be the difficult task of denying

Pakistani infiltrators the use of well-known routes across the LoC through increased and targeted surveillance. It is no secret that despite the Army's best efforts to hold them off, infiltrators manage to exploit the blind spots produced by the rough terrain to sneak into the state.

And since this was Pakistan the men were dealing with, they weren't just up against terrorist infiltrators, but also Pakistan Army units that actively aided their crossing of the LoC. In the world of Pakistani terror schooling, infiltration operations were the ultimate job placement.

Capt. Pradeep, like all military men operating in Kashmir, was fully aware of the dual face of the adversary—civilian youth honed into dangerous terrorists by a sharp, focused military machinery. And, for the job at hand, Capt. Pradeep knew that the Pakistani Army posts across the LoC in Uri that had steadily provided logistical support to infiltrators could not be allowed to slip under the radar.

'Following through with each element of the broad plan was crucial. We still had a long way ahead of us and had to go about our business keeping a low profile to avoid alerting people across the LoC about Special Forces involvement,' says Capt. Pradeep.

He spent the next three weeks shaping a plan of action, ticking off his to-do list and working to narrow down the enormous amount of information he had collected in a few short days. He knew that this was far from being a 'wham-bam' operation. Success, Capt. Pradeep knew, would rest on a multi-pronged approach that drew on the strengths of key local Army units, especially the electronic warfare detachment, whose primary responsibility was to listen in on terror-related chatter across the LoC. He was assured of their unstinting support.

Before Capt. Pradeep's arrival in Uri, there had been a steady flow of broad, general warnings of a major infiltration

'*Mehmaan aa rahe hain. Bhejne ki tayyarian shuru karein* (The guests are coming. Begin preparations to dispatch them).' The audio intercept of a piece of communication from the Pakistani Army post to a terror launch pad seemed to confirm, at the very least, that an infiltration attempt was afoot.

'Raw data had been distilled into possibly accurate information. The attackers were coming. The likely route they would take was now practically known. It was hard to predict the exact timing, though,' Capt. Pradeep says.

No amount of intelligence can conclusively establish the precise date and time of infiltration activity. The likelihood of plans changing at the last minute is a near certainty, the modus operandi, in fact, of the Pakistani Army. But what seemed certain by this time was that the countdown had begun.

Briefing the CO that evening, like he did every day, Capt. Pradeep presented his recommendations based on an analysis of all the intelligence he had—what was made available to him and what he had additionally gathered. The 'unwanted guests' were likely to try their luck in the last week of May, he said.

At the base, Special Forces squads were busy doing what they do all year round—training for the worst possible situations and the most difficult missions. Capt. Pradeep remembers the training regimen that involved each commando firing hundreds of rounds every day, irrespective of whether a mission was at hand or not. War readiness was a defining requirement. Once a mission had been assigned, there would be no time to train.

Three squads, consisting of six soldiers each, were now training to be a part of Capt. Pradeep's upcoming operation. Two additional squads would provide backup. As mission leader, he was now armed with enough intelligence to brief the squads about how things were likely to unfold and where they would lie in ambush for their Pakistani 'guests'.

attempt by a group of military-trained suicide attackers, aid
by a Pakistani Army Mujahid battalion. He knew that suc
alerts were not uncommon and seldom served as actionabl
intelligence for a successful operation. It was the dispiriting old
maxim of intelligence: only a tiny percentage of all warnings
resulted in anything meaningful at all, let alone a successful
operation.

'It is no exaggeration that we are swamped with a number
of alerts every day. Rarely do one or two inputs out of, say,
ten carry any real intelligence value. But no input is ignored,'
says Capt. Pradeep.

There were plenty of dots that he had to connect in the right
order to generate that elusive big picture of a major infiltration
attempt—the where, the when and, crucially, with what strength.
For days, Capt. Pradeep thoroughly reviewed volumes of fresh
intelligence data and cross-checked it for accuracy with multiple
sources, including his informants on the ground on both sides of
the LoC. He prayed for a breakthrough.

On 10 May, after he called his wife to wish her on their
anniversary and spent some time in the dormitory with the
Special Forces men, Capt. Pradeep returned to the camp's war
room to continue the hunt for intelligence. By noon, it became
clear that the Pakistani Army's 652 Mujahid Battalion, the
'Mountain Tigers' based across the LoC in Chakothi, was very
likely to be the unit that planned to 'administer' the looming
infiltration. Detailed analysis of intercepted communication by
Capt. Pradeep revealed that terrorists belonging to Pakistan's
two most formidable terrorist groups, the LeT and the JeM,
would be launched from the notorious Sugna post to enter Uri
at an appropriate time.

In the war room, Capt. Pradeep also received a crucial
audio report from the electronic warfare unit he had enlisted
for help.

'*Humare paas pukki information hai ki yeh log Sugna post se launch honge. Launch hone ke baad yeh KDK nallah ke raaste se ayenge LoC paar karne ke liye. Hum inko wahin par khatam kar denge* (We have solid information that these people will be launched from the Sugna post. After being launched, they will take the route through KDK nallah to cross the LoC. We will finish them off there),' Capt. Pradeep told his men during a briefing using a sand model of the Uri sector, which contained detailed features of mountains, valleys and the LoC itself. The men knew that the KDK nallah was a notorious blind spot.

The sand model was necessary for such a briefing, even though the Special Forces men knew the area supremely well. This was, in every sense, their backyard. But there could be no mistakes, no assumptions and no short measures before a mission as delicate and dangerous as this one.

'*Abhi yeh clear nahin hai ki yeh fidayeen dasta PoK se kab rawana hoga. Par hum apne kaam par abhi se lag jayenge* (It is not yet clear when the terrorists will leave PoK. But we will begin our work now),' Capt. Pradeep said.

The area was now under intense covert surveillance day and night. With indeterminate intelligence on the date and time of the infiltration, Capt. Pradeep decided to take no chances. On 22 May, he ordered the first of his Special Forces squads to proceed to the likely ambush site along the LoC, near the Indian Army's Chabuk post, to make sure the infiltrators could spring no surprises. The men rolled out from the camp in a jeep that would drop them off near a trail they would trek to get to the LoC. Vehicular sounds needed to be kept to a minimum. Pakistan's own highly evolved intelligence networks could smell a Special Forces man from a mile away.

That morning, Capt. Pradeep had woken up with a stinging pain in his bloodshot eyes. With one squad dispatched to the LoC, and the CO waiting for a morning briefing on his next step, there couldn't have been a worse time to contract conjunctivitis.

'Arya, stay the hell away from me. I don't want to see you within 10 miles of the squads. If you are still harbouring any desire to lead this operation, go see a doctor now,' the CO snapped.

Capt. Pradeep knew he was right. On his way to the field hospital, he felt miserable thinking that the mission may just have ended for him even before it began. The doctor sent him back with a small bag of medicines and much-needed assurance that the infection would go away soon.

'It was a nightmare. I didn't come all the way to Uri to let an eye infection ruin everything,' he says. By that evening, fortunately, the infection had resolved itself.

Preparations for the ambush began in full swing 22 May onwards. The Special Forces squads, fully kitted out with their assault weapons and body armour, were assigned to the operational area in rotation to keep an eye on terrorist movement along the KDK nallah.

'When a Special Forces team waits in ambush, it's not like we just sit in one place. We sit, we move, we sit, we move. It allows us to cover a larger area and hunt better,' Capt. Pradeep says.

Every step had to be taken carefully since the area was peppered with landmines, a legacy from a series of conflicts in 1965, 1971 and even the 1999 Kargil War. It is standard practice for the Army to mark minefields, but these hidden killers are known to drift because of extreme weather conditions, shifting of soil and landslides. Soldiers have blundered into minefields along the LoC, especially in the Baramulla sector, despite mine-risk education to prevent death and injury from these deadly explosive devices.

'A landmine is a weapon of war that treats friends and foes alike. The mines presented a peril to the squads. The men had to proceed with extreme caution,' says an officer from one of Capt. Pradeep's squads.

The next stage of the operation involved planting radio-controlled IEDs along the expected entry routes and testing signal strength to detonate explosives from a distance. Stealth was crucial. Manoeuvres were carried out under the cover of darkness, with the commandos taking every precaution to avoid detection.

Numerous reconnaissance missions to collect intelligence and scout out this particular stretch of the LoC followed. The squads went on night-time patrols, remaining unseen during the day.

'We have a real problem in this sector. Dense foliage across the LoC offers the infiltrators excellent camouflage. We don't enjoy that advantage on our side,' says Capt. Pradeep.

After four days of non-stop alert in the area of expected infiltration, Capt. Pradeep received a late-night phone call on 26 May. It was one of his most trusted informants, a man he had cultivated over two years and someone Capt. Pradeep had been hoping he would hear from. This was a man who operated on both sides of the LoC, but Capt. Pradeep had learnt to trust him.

In a hushed voice, the caller informed Capt. Pradeep that the infiltrators were now being guided by the Pakistani Army's Sugna post and that they would attempt to cross the LoC on the night of 28 or 29 May. The information not only confirmed much of the intelligence Capt. Pradeep and his men had gathered over two weeks, but provided them with that final bit they needed to sharpen their plan of action. A two-day window for an infiltration is a highly specific piece of information, a fragment of data in an otherwise chaotic, unfixable ocean of vague intelligence.

Capt. Pradeep thanked the informant and hung up, immediately ordering four squads to stand by to head to the

Chabuk post the following evening. The post was a 90-minute drive from the Special Forces base.

It was time to move, and fast. Capt. Pradeep made a phone call to his wife, telling her he would be off the grid for the next few days and that he would call her as soon as he was 'free'. Deepa Arya listened in silence, only reminding him of her words from a few days before, 'I'm not collecting your medals.'

As planned, two of the squads would stay at Chabuk for backup and the other two would set out on foot for the last leg to join the squad that had already been positioned in the area on 22 May across the KDK nallah and directly facing Pakistan's Mujahid battalion post on the other side of the LoC. But just when everything seemed to be moving as per plan, the commandos encountered an expected hurdle the next day, on 27 May, which threatened the stealth necessary for their mission.

Sabzar Ahmad Bhat, a Hizbul Mujahideen commander and successor to Burhan Wani, had been killed in a firefight with security forces late the previous night in south Kashmir's restive Tral, a hotbed of Hizbul terror. Bhat's killing had set off a wave of protests across Kashmir.

'Bhat's killing and the ensuing violence was unforeseen. Moving out of the base presented a challenge because of the heightened possibility of being seen. But we somehow managed to reach Chabuk by nightfall on 27 May,' Capt. Pradeep says.

The Special Forces squads were armed with a full array of weapons, including their standard-issue Israeli Tavor TAR-21 assault rifles, Colt M4A1 carbines, Uzi silenced submachine guns, Pulemyot Kalashnikova general-purpose machine guns, C90 disposable rocket launchers, UBGLs and 9-mm pistols. Some commandos carried two pistols, a personal choice.

Each man was armed with six grenades and four magazines of ammunition. They were kitted out for a short, sharp firefight.

From the Chabuk post, large midsummer storm clouds quickly rolled in, raining torrents on the two squads as they departed over slippery ground and up pine-clad slopes towards the LoC.

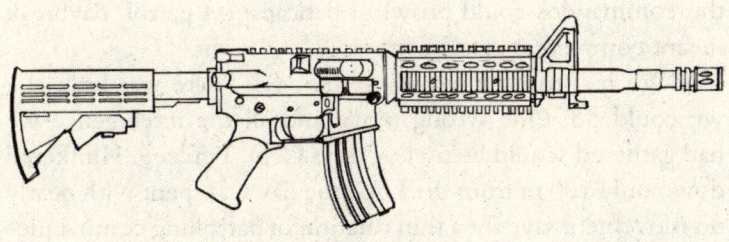

Colt M4A1 carbine

'*Sir, aaj raat ko contact hone ke kitne chances hain?* (What are the chances of establishing contact with the terrorists tonight?),' asked Capt. Pradeep's buddy soldier, a Havildar whom the officer affectionately called Hero.

'*Ho bhi sakta hai, Hero. Jitna jaldi ho utna accha hai* (It's possible. The sooner the better),' said Capt. Pradeep, remembering that his informant had indicated a window that began the following night. The Havildar's question wasn't surprising. Armed and in position, nothing occupies a commando's mind more than the prospect of coming in contact with the enemy. As soon as possible.

The rain had weakened to a drizzle by the time the two squads reached the appointed rendezvous point, where the first squad, sent there five days earlier, was holding fort barely 100 m from the LoC.

'Sir, we haven't spotted any activity yet. It's all quiet here,' one of the commandos reported to Capt. Pradeep as soon as he arrived. The commando was holding an HHTI, a sensor device that provides clear infrared images in complete darkness.

Capt. Pradeep's informant had said the infiltration would happen on 28 or 29 May, but that didn't mean plans from the other side couldn't change. The squads carried out reconnaissance patrols through the night, taking up their pre-assigned positions long before dawn broke over the hills. If the commandos could prowl in darkness on patrol, daybreak meant complete concealment.

'We had to lie low during the day. There was little else we could do. One wrong move and all the intelligence we had gathered would be useless,' says Capt. Pradeep. Hunkered down only 100 m from the LoC, the day was spent with nearly no movement save for a thin rotation of patrolling commandos to keep the proverbial knife edge pointed and ready.

'Nobody likes the day. It's at night that we go to war,' says an officer from the squad deployed in the area that day.

As the sun set on 28 May, Capt. Pradeep gathered his men for a night briefing. The two squads would be on an even sharper state of alert, if that were possible. The third squad had spread out and positioned itself some distance away to cover more area. He felt an inexplicable pang of certainty that his informant was about to be proven devastatingly correct. Two hours later, he saw the confirmation on a thermal imager in the hands of a soldier near him.

At 10.30 p.m., the dark infrared image of an infiltrator entered the frame, his body heat helping draw a sharp picture against the cold, wet surroundings. In a crouched walk, he was moving forward from the expected direction of the KDK nallah. Within moments, the image of a second terrorist became visible on the thermal imager's monitor, followed by a third and then a fourth. Two more terrorists could be seen 200 m behind them in the nallah.

Capt. Pradeep sent up a silent prayer of thanks to his informant. Like every time before, he was dead on.

As the two squads watched in silence, the six infiltrators trudged forward, hoping to cross the LoC undetected that moonless Sunday night. Each terrorist walked about 12 feet behind the other, to reduce the chances of damage if they were ambushed. Minutes later, the first four terrorists crossed the LoC, pausing briefly to gauge their surroundings. On the thermal imagers, the dark figures hesitated for a few minutes, appearing to assess the air for any telltale signs that they were expected. The Indian squads held their breaths. Capt. Pradeep prayed the four infiltrators wouldn't abort their mission and run back, like several infiltrators had done in the past. But after a few minutes, the infiltrators continued onward.

When they were comfortably on 'our turf', Capt. Pradeep signalled to one of his men to detonate the first IED using the radio-controlled receiver. And then another. Two sharp explosions went up near the terrorists, throwing one of them to the ground. The other five, all supremely well-trained, reacted instinctively by splitting themselves into three groups and taking cover in the thick foliage to better engage their hunters.

When the blast had cleared, and the thermal imager was able to redraw its heat picture of the attack site, it became immediately clear that the terrorist injured in the blast wasn't moving. A second terrorist was trying to pull him away to safety while firing aimlessly in the dark in the general direction of the squad, which hadn't fired a single bullet yet. With his assault rifle, Capt. Pradeep took aim and fired a single shot, killing the second terrorist. The injured terrorist remained motionless. It was likely that he was dead too, though the squad wouldn't take any chances.

The remaining four terrorists were firing a steady stream of bullets in the direction of the Special Forces squads, which had by now dispersed into six smaller groups of two men over short distances up the hill, firing back at the intruders from various directions. As Capt. Pradeep returned fire and the terrorists' bullets hissed around him, he realized that Hero had separated from him and was now dangerously vulnerable. With fire returned, the terrorists had a clear idea of where the Indian commandos were. The chances of their bullets finding targets were higher now, and their vantage points in the foliage meant it was far harder for bullets to find them in return.

'I may kill a dozen terrorists, but if I lose even a single soldier under my command, the mission is nothing but a big failure. It's as simple as that,' Capt. Pradeep says.

Rounds from an AK-47 were peppering the ground near Hero as he lay flat on his stomach, firing back. The Havildar was aware of the approaching fusillade, making a move to get out of the way. At that moment, Capt. Pradeep leapt from his position briefly into the line of fire, before diving for cover behind a large fallen tree. There, he flattened himself against the ground. Stepping on one of the hundreds of landmines dotting that stretch would have left him limbless, if not dead. But the safety of his buddy soldier was now his chief concern.

Partially shielded by the fallen tree, Capt. Pradeep concentrated his fire on the two intruders who were targeting Hero with a hail of Kalashnikov bullets from their position of advantage. The incoming 7.62 mm shots missed Hero by a matter of inches, yet the Havildar held his nerve and fired back at the terrorists. There was no way he was backing out of the firefight.

'Hero, cover lo! Sab control mein hai! (Hero, take cover. Everything is under control),' Capt. Pradeep yelled above the sound of gunfire.

The muzzle flash of their AK-47s gave away the positions of the two terrorists. Steadying his breath so he could aim better, Capt. Pradeep pointed his weapon straight into the darkness from which the bullets poured. In a series of short bursts, he unleashed a magazine of ammunition into the foliage. The Kalashnikovs fell silent. Both terrorists had either been killed, or injured enough to stop firing.

Capt. Pradeep leapt once again into the open to pull the Havildar to the safety of cover.

'Every man in my team knows he is being covered by his comrades all the time. He knows he will never find himself alone. This faith is unshakeable and it helps us conduct the toughest of operations,' Capt. Pradeep says.

Twenty minutes from when the first IED was detonated, four terrorists had been either killed or injured. By about 10.50 p.m., the volume of terrorist assault rifle fire had significantly reduced. The next few minutes saw a sporadic exchange of gunfire. The Special Forces squads aimed sustained fire in the direction of the two remaining infiltrators from six different directions.

'The terrorists would fire a burst, and then pause. The pattern was repeated. They were running low on ammunition and were clearly firing only to draw return fire and pinpoint our positions,' Capt. Pradeep says. This was military-style tactical training on display. But Capt. Pradeep noticed something else too.

The spread of the bullets fired by the infiltrators wasn't the typical AK-47 spray. They were using sawed-off Kalashnikovs, with the shortened barrels scattering the bullets wider. The idea was to try and hit a larger area at once, increasing the chances of a bullet finding Indian flesh.

With gunfire now reduced to sporadic pops, the intruders carefully emerged from their hiding positions in the foliage, slipping between rocks down the hill before reaching level ground. Then they started to run back along the trail they had taken towards the LoC, in a clear effort to escape back into Pakistan. As they ran, the two terrorists raised their weapons and opened fire at the Indian commandos. It was clear that the two weren't prepared to fight to the end.

'They wanted to stay alive so that they could return another day. We were not going to let that happen,' Capt. Pradeep said.

Slipping out from their own positions, the dozen Indian commandos trudged down the slopes to the trail and towards the retreating intruders, taking supreme care to duck behind rocks as the Kalashnikovs fired without interruption. As Capt. Pradeep and his men watched, with a six-way line of fire zeroing in on the running terrorists, it became clear that they were now only about 50 m from the LoC. Once they crossed over, the operation would become exponentially more difficult, given that the Pakistani Army was already likely to be on high alert to aid the intruders in any way possible.

Capt. Pradeep shouted to his men, who were close by, that the two intruders had to be stopped at all costs. The teams paused for a few seconds to reorient, slam fresh magazines into their assault rifles, and then took aim as the terrorists sprinted across the final stretch to the LoC. That final burst found the two terrorists. The Indian commandos fired their last shots at 11.15 p.m. No more fire was returned. The last echoes of gunfire and smoke lifted off the mountain as the sounds of the night returned.

But the operation wasn't over yet.

'We had to be certain all the terrorists were dead and more weren't lurking around the corner. We waited till first light to clear up the area. That's standard protocol,' Capt. Pradeep says.

At dawn, the men recovered the bodies of the six infiltrators, along with their AK-47 rifles, dozens of unused magazines, grenades, a satellite phone and Chinese-made pistols—all standard issue kit for Pakistan-sponsored terrorists.

With four commandos still on the lookout for more terrorists, Capt. Pradeep gathered his men for a quick debrief at the site. He congratulated them on a sharp, well-executed mission without casualties or injuries. On their way back to base, he sent a text message to the Colonel announcing the outcome of the operation. After more than three weeks of near-sleepless preparation, the encounter had lasted 45 minutes. Every bit of intelligence gathered by the teams that worked with Capt. Pradeep had proven to be correct. It was a small but ringing success against a machinery that would view the night's proceedings as no more than a temporary setback against an inexhaustible arsenal of human weapons.

But back at the Special Forces camp, celebrations had begun. Capt. Pradeep allowed himself a full bottle of Old Monk rum, his first drink in many years. The CO had also ordered a cake. It was Capt. Pradeep's birthday the following day, 30 May.

In Bengaluru, Deepa got the call she had been waiting for. As their daughters, Dhriti and Dhiksha, slept, Deepa stayed awake—she hadn't closed her eyes with any success for 48 hours. Exhausted but overcome with relief, she listened to her husband matter-of-factly report that he was well and that he had returned to base. Nothing more was discussed.

'The commando missions were completely new to me. I thought I had married an IRS officer,' says Deepa. 'I remember telling him I never signed up for this. He was always very passionate about the uniform, but I had no clue that this was the route he was going to take.'

Being the spouse of an Army officer can be unimaginably hard.

'There are ups and downs, there's pride and there's pain. My heart skips a beat whenever Pradeep calls to tell me that he will be away from his phone for a few days. Yet, I feel proud of his accomplishments and have no regrets. The uniform will always be his first love. The way I look at it is that I should not be the reason for him to not do what he wants to do,' Deepa says.

Eight months after the LoC operation, the government announced India's third-highest peacetime gallantry honour, the Shaurya Chakra, for Capt. Pradeep on the eve of Republic Day in 2018, for prominent acts of courage. His citation for the medal is replete with expressions that describe his actions: 'pre-eminent valour', 'heroic initiative', 'inspirational combat leadership' and 'unmindful of his own safety'.

On 23 April 2018, after President Ram Nath Kovind bestowed the award on Capt. Pradeep, the Rashtrapati Bhavan tweeted, 'He displayed audacity in the face of terrorists' fire and extraordinary valour in risking his life beyond call of duty and eliminated six terrorists.'[2]

At forty-six, Capt. Pradeep, holding a position in the IRS equivalent to a Brigadier's in the Army, was the oldest recipient of a gallantry award at the Rashtrapati Bhavan Durbar Hall that evening. Unlike regular Army officers who walk out of military academies as lieutenants at twenty-one or twenty-two, the Territorial Army allows volunteers to join its ranks till the age of forty-two. Allowing a forty-six-year-old, seconded to a Special Forces unit, to lead a dangerous mission can only be an indicator of the calibre of the officer in question, his reputation, his soldiering skills and commitment.

'Here's a guy, a bureaucrat. A tax official. Who one day packs his bags and goes off to stop terrorists,' says another officer from Capt. Pradeep's unit. 'It's rare to find people outside the uniform who can influence others to get out of their comfort

---

[2] https://bit.ly/2EUxOe8

zones. He could have chosen to be in Mumbai, but here he was, willing to put his life on the line on a godforsaken hill. There has to be something in you if you are ready to take that step.'

He prefers to be addressed as 'Capt. Pradeep', his identity as an Army officer taking precedence over the other hats he wears. His unit comrades don't find this even remotely surprising.

'A bureaucrat sporting combat fatigues is a rarity in India, where the military is neither the first nor an obvious career choice for many. Most of all, the Special Forces are a different species and a prized national asset,' one of them says.

The mission and the Shaurya Chakra have changed many things for the officer. Capt. Pradeep, for all practical purposes an outsider in the Army, is now accorded all the trimmings and respect due to one of their own.

But many things haven't changed at all. Capt. Pradeep continues to divide his time between income tax work and his study of terror financing.

He listens frequently to his favourite Hindi song, which is, '*Main zindagi ka saath nibhata chala gaya, har fikr ko dhuen mein udata chala gaya*'.

And he remains in constant anticipation of that next phone call.

10

'What's Higher than Saving Someone's Life?'

Captain P. Rajkumar

Arabian Sea, off Kerala coast
1 December 2017, 7 p.m.

'Laser beam spotted on port bow, Sir!'

The navigator was screaming the words to the pilots, forcing them to throw the helicopter into a sharp left turn and descend from 200 feet towards the storm-churned blackness of the Arabian Sea. What the five men on board the Indian Navy's SK 528, an old grey Sea King helicopter, had been hunting through the furious, beating rain and swells of Cyclone Ockhi had finally been sighted.

As the helicopter shuddered and carefully lowered itself over a sea churned dangerously into a heaving maelstrom by the rain, the flight commander followed the thin, green, barely visible laser beam. Buffeted by 150 kmph wind speeds, he strained to keep the helicopter in control while maintaining a visual lock on the thin green line. With the laser beam flickering and disappearing for whole seconds into the surging sea surface, he knew that if he lost sight of it, this mission would end in at least one death.

At least.

A few feet at a time, the SK 528 descended into an engulfing darkness, broken only by a cone of light from its search lamp that was rendered virtually useless in the storm. At 50 feet, the flight commander paused for a moment, bringing the helicopter into a tense shuddering hover, the roar of the rotors nearly drowned out by the thrashing rain and wind. The four other men looked at him. They knew his dilemma. Any lower in this weather, and there was a good chance that the smallest in-flight emergency would mean disaster. But seconds later, the helicopter heaved and continued its downward drift, the darkness and rain so overpowering that the crew couldn't see the tips of their helicopter's rotor blades.

The elusive laser beam threatened to disappear as the helicopter descended from 50 feet to 40, then to a dangerous 30 feet. In a final push, the SK 528 came to a hover 20 feet over the sea surface, its swells so rough that they threatened to reach up and hit the helicopter's belly. Vulnerable and dangerously close to a catastrophe, the flight commander knew he had bare minutes. As the cone of light from the helicopter's search lamp scanned the small patch of sea below, he saw it.

'Target spotted!' he breathed into his cockpit talkback. Twenty feet below the trembling helicopter, bobbing dangerously in the middle of the sea, was the very object they were looking for.

Six hours earlier

In his green flying overalls, Capt. P. Rajkumar jogged to the SK 528, the Sea King helicopter that was primed and waiting on the tarmac at *INS Garuda*. The Indian Navy's air base at Kochi was on high alert. The Captain's crew was waiting at the chopper as an unusually dark afternoon sky visibly worsened over their heads. The Navy's INAS 336 squadron, of which the helicopter and crew were a part, is dedicated to anti-submarine warfare operations. On that December afternoon, they had been scrambled to embark on a hunt. It wasn't submarines they were being sent after, but fellow citizens.

By 1 December, Cyclone Ockhi was just hours away from assuming its most fearsome form right off the coast of Kerala. Capt. Rajkumar, a decorated Navy pilot, had flown terrifying helicopter missions in the past, including a daring rescue of scientists in Antarctica three decades earlier. But the next 6 hours would dwarf everything else. As he climbed into the SK 528 that afternoon, strapped in and turned the chopper's twin Rolls-Royce Gnome engines on, he had only the barest sense of what lay ahead.

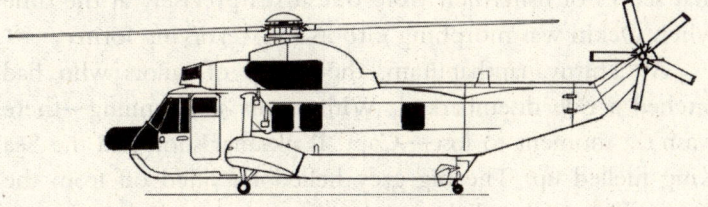

Sea King helicopter

'We're trained for this, but you can never be completely ready for what's in store for you. In the military, we're fatalists. And I'm even more of a fatalist because I'm a Malayali,' Capt. Rajkumar laughs. 'Nobody expected the full force of a cyclone in this side of the country. The meteorological department can cry hoarse as far as the onset of Ockhi is concerned, but on the west coast of India, cyclones are a rare phenomenon. You only hear about them on the east coast. It's only when she hit that it became clear what we were dealing with. And it was not pretty.'

Capt. Rajkumar was flying with co-pilot Lt Cdr Abhijit Garud, navigator and tactical coordinator Lt Cdr Mayoor Chauhan, and two young combat divers, Sumit Raj and Deepak Saini. Also hitching a ride on that 200-km flight to Kerala's capital, Thiruvananthapuram, were a handful of sailors who were required for a separate mission on another aircraft waiting for them there.

Flying south over land, the crew of the SK 528 could easily see the ominous swathe of rain-laden cyclonic cloud on their right, bearing down along the coast. Ockhi had struck two days before and was rapidly whipping itself up into one of the most devastating cyclones of the year. Advisories had been sent out to fishing communities not to venture out to sea and to stay away from the waterline. But a mixture of scepticism—nobody really expected a west coast cyclone to be anything major—and the sobering reality of earning a livelihood meant that scores of fishermen were out at sea precisely at the time when Ockhi was morphing into its full, terrifying form.

At Thiruvananthapuram, the group of sailors who had hitched a ride disembarked. With rotors still running—there wasn't a moment to lose—Capt. Rajkumar soon had the Sea King fuelled up. The big grey helicopter lifted off from the dispersal area and peeled away, heading out directly over the sea.

On a normal day, 2 p.m. would have offered the best flying conditions, with thick sea air giving the helicopter all the lift it needed, and a crisp blue sky offering the crew perfect visibility for an even flight. But on that day, just as Capt. Rajkumar pitched the helicopter forward and speeded out over the Arabian Sea, the crew immediately ran into cyclonic weather with sudden, violent winds and rain that put paid to any visibility they had been blessed with only a few minutes earlier.

'In a typical cyclone like Ockhi, the wind speed picks up, and therefore the sea starts churning and a long swell develops. We knew we were going to face extreme weather. The rain was coming down in sheets, visibility was almost zero,' the Captain remembers.

Very soon, visibility *was* zero. Flying 100 feet over the lashing sea surface, the SK 528 was completely blind. The crew couldn't see more than a few feet beyond the helicopter's windshields and the on-board sensors were tuned to hunting submarines, not finding wrecked fishing boats and survivors in the middle of a maelstrom. Thankfully, the crew had another pair of eyes flying far above them.

Cruising at 1000 feet through the storm clouds was an Indian Navy P-8I, a Boeing 737 aircraft fitted with an arsenal of high-performance sensors and cameras capable of penetrating the worst weather and finding the smallest objects at sea—something the Sea King, flying 900 feet below, could only dream of doing. A small, swivelling camera in a bubble-like container on the P-8I's chin scanned the sea surface in infrared mode, while a team of officers on board scanned the visual feed, hoping to spot a fishing boat in distress.

Or even humans in the water, for that matter.

For 3 hours, the helicopter and the aircraft combed huge sections of the sea roughly 70 km off the Kerala coast. For 3 hours, they found nothing but a roiling sea getting steadily worse.

'The P-8I crew and I knew that with each passing minute, spotting boats or survivors would become increasingly more difficult. But just as we were about to shift our attention to a different part of the sea to look for possible survivors, the P-8I's commander came in through the radio,' says Capt. Rajkumar.

'Boat sighted. Capsized condition. Four survivors visually confirmed,' the P-8I pilot called through the radio, relaying a set of coordinates to the precise location. Daylight was fading fast, but a target had been spotted and the helicopter's crew wasted no time. Still flying at 100 feet, the Sea King banked hard as it changed direction and headed straight for the coordinates of the location where a medium-sized fishing vessel had turned over.

'Unlike thirty or forty years ago, when catamarans and older fishing boats had deep bottom keels that were much more resistant to capsizing, new generation cheaper-to-build fishing boats with their fibreglass hulls are top-heavy, and therefore capsize easily in rough seas,' says Capt. Rajkumar. It was an important nugget of information he had picked up on visits to fishing villages. Now, cruising over a thrashing sea, he knew it meant two things: one, that the time available to rescue the boat in distress was extremely short, and two, spotting them in the water was going to become exponentially more difficult as the minutes passed.

But this needle-in-a-haystack hunt had actually thrown up a target, thanks to the aircraft flying above them. Fully aware of how futile search and rescue operations could be without modern technology, Capt. Rajkumar sent up a silent prayer of thanks for the privilege of precise coordinates to home in on.

The dull grey daylight was fading fast, but as the Sea King swooped in over the coordinates they had locked on to, it was clear that the information supplied by the P-8I was of very high quality. There, bobbing in the sea, was a capsized fishing boat with four persons on top. Fortunately, all four appeared to be in good condition, suggesting that the boat had been out at sea for not more than a day.

'We descended to about 30 feet. The weather was worsening. Our two divers lowered a rescue strop, and we winched the four men up one by one. This was, of course, a time-consuming job and time definitely was at a premium as there was not much daylight remaining,' remembers Capt. Rajkumar.

With the four rescued fishermen on board, the crew of the SK 528 made straight for the coast. By the time the helicopter landed at Thiruvananthapuram with the survivors, it was well past sunset. A difficult mission had been flown with the unlikely bonus of actually finding and rescuing survivors during a cyclone at sea. It was time to switch the engines off and take a break for the night. For one thing, the P-8I that had provided rescue intelligence had flown back to base for the night. And without it, hunting for survivors was virtually impossible. It was a difficult day that had ended with success. There was no reason not to call it a night.

But something nagged at Capt. Rajkumar. And as the SK 528 was refuelled for the second time that day with its rotors still turning, the four other men on board waited for orders.

'My crew understood fully. I did not have to say much. There were no orders or coordinates. We lifted off and headed out to sea for the second time on a hunch and a prayer for those yet to be rescued,' the Captain remembers. 'If there's a very fine line between calculated risk and sheer foolhardiness, we in the military know there's a very fine line between a court martial and a gallantry award.'

In the cabin, combat diver Deepak Saini read the Captain's mind. And then he heard the words.

'We will launch again and search for survivors till the last drop of fuel in our bodies as well as the machine are expended,' the flight commander announced from the cockpit.

'The Captain was determined that we would be angels to someone else that night. He was not ready to give up. And his words energized us afresh,' says thirty-one-year-old Saini, who joined the Navy in 2004 and took the commando diving course in 2010, making him one of a rare breed of sailors tasked with one of the most death-defying mission profiles in the service.

Everything had changed off the coast of Kerala in the hour since the rescue of the four fishermen. The swirling grey of dying daylight had now turned into complete darkness. The only aids the crew had were floodlights and a single controllable spotlight on the helicopter's chin.

'Unfortunately, these lights, when switched on in heavy rain, will only reflect it back to the cockpit, making flying very disorienting. Flying at 200 feet, scanning the night sea in rain, is one of the most challenging tasks out there,' Capt. Rajkumar says. With the sea and sky melting into one endless mass of blackness, the pilots of SK 528 were flying largely with the help of their instruments, most importantly at this time, a radio altimeter that reassured them of their height above the turbulent sea surface.

The SK 528 cruised through the darkness, the bad weather of earlier in the day now working up into an unrelenting viciousness of rain and wind, the endless darkness of the sea and sky only broken by the faint luminescence of waves in violent churn. For a helicopter crew, a more unfavourable set of flying conditions couldn't be imagined.

'We were hunting blind. We had no coordinates. And we couldn't see. This was exponentially more difficult than finding a needle in a haystack,' Capt. Rajkumar now remembers.

The bitter irony of the situation wasn't lost on the crew. They were now searching for something they couldn't see in the first place—and probably wouldn't be able to see even if they flew right over it. But they stayed out over the sea, five men squinting through the darkness in every direction, hoping to see something, anything, combing stretches of inky black ocean on a hunch that there were more people out at sea and waiting to be rescued.

The endless combing of the sea continued for an hour without result. And as the winds picked up, it was clear to the crew that this mission was only getting more challenging. Capt. Rajkumar gathered his thoughts as his co-pilot took control of the helicopter and kept it on course for the search. And that's when a voice with a heavy Chinese accent crackled in on the radio.

The nearly incomprehensible call, inflected with panic, was from a huge 1,00,000-ton Maltese container vessel, the *MV Cosco Beijing*, that had been making its way across the Arabian Sea. The message had come in through Channel 16, the 156.8 megahertz frequency used by merchant vessels. While refuelling at Thiruvananthapuram an hour earlier, Capt. Rajkumar had tuned one of the cockpit's very high frequency (VHF) radios to Channel 16, thereby keeping an ear open for emergency messages from merchant ships. The anguished Chinese voice was distorted and high pitched.

'BOAT! CAPSIZE! FISHING BOAT! CAPSIZE!' came the call through Channel 16 from the *MV Cosco Beijing*. The ship was visible to the helicopter's crew on their radar and had been identified. The call was the first miraculous confirmation to Capt. Rajkumar that his hunch had been correct. He immediately responded to the radio message, confirming that he was headed towards the ship.

Through a series of garbled exchanges deciphered while manoeuvring the Sea King through the cyclone, the crew of SK 528 understood that the *MV Cosco Beijing* had spotted a capsized fishing boat, and was using a laser pointer device to mark its location.

Co-pilot Lt Cdr Garud responded immediately, slowly relaying a message back to the merchant ship. 'This is Indian Navy helicopter on a search-and-rescue mission. Information received on capsized boat. We are heading to your location.'

With the co-pilot in tenuous touch with the merchant ship, Capt. Rajkumar turned the chopper sharply in the vessel's direction.

'We picked up the massive merchant vessel on our radar and speeded towards it. Though the big ship had stopped, it was also rolling and pitching in the churning sea. Its only connection with the fishing boat it had spotted was with the laser pointer. There was no way it could come close to the fishing boat, because that would have been dangerous to any survivor. The safe distance coupled with the heavy rain meant the laser beam wasn't quite reaching its intended target. The huge waves were also constantly hiding the boat. So the man on *MV Cosco Beijing*'s deck was also, in effect, searching for the boat with his laser pointer,' says Capt. Rajkumar.

Arriving within minutes over the huge container ship, the Sea King circled it a few times, its crew hoping desperately to find the tiny laser beam the ship was using to identify the general direction of the capsized boat it had spotted. The sea was so rough, it was very likely that the capsized boat had drifted considerably. But with the Indian Navy helicopter now buzzing over it in a desperate hunt for visual contact, the laser pointer remained switched on.

'Normally, one would not fly this long in such conditions. We had been flying hands-on for 6 hours, without the luxury of autopilot. But in such missions, it's adrenaline that keeps you going, and it's mind over body. In retrospect, you feel you were fatigued, but in the moment, you're not. Your mind is racing, with not a moment to spare,' remembers Capt. Rajkumar.

The flight commander knew it would be much easier to spot the boat from the deck of the ship in these conditions than from a helicopter at 200 feet. The 'slant visibility' at this point was next to nothing over a black cauldron of a sea.

'I had just been circling over the ship for 10–15 minutes, but the laser and the boat were simply invisible to us. We could see absolutely nothing as I manoeuvred the Sea King over the seas around the ship. And it was purely by chance that our navigator saw a flash of the green laser,' he says.

'Laser beam spotted on port bow, Sir!' shouted navigator Lt Cdr Mayoor Chauhan. It was the elusive visual lock the crew had been hunting for. A single laser beam being pointed from the deck of the 1,00,000-ton Maltese container ship, 80 km off the coast of Kerala in the middle of one of 2017's worst cyclones. A single little beam of green light was all that the SK 528 now had to depend on.

'As a submarine hunter, the Sea King has the ability to hover low over the sea surface, but none of those manoeuvres applies in a scenario like this, where you have to slowly descend and depend on excruciatingly delicate hand-eye coordination. There is no autopilot flying here. It's all manual,' Capt. Rajkumar says.

'At such low height, I had to constantly look out and then back at my instruments, because we were flying at very high angles of tilt. In a big helicopter like the Sea King, when you bank in excess of 30 degrees, you lose a lot of your lift power, so you have to compensate with collective power. Piloting becomes much more challenging. So there was no way I could completely ignore the instruments and look for the boat. It was half instrument flying and half visual flying.'

With the laser beam in their sights now, the helicopter swivelled around and headed straight for it, descending slowly. There was no time to lose, but a single hurried step could have meant certain death for the crew of the SK 528.

'Normally, we don't hover lower than 50 feet because there are various criteria that make this a risk, including lift performance and sea spray. In this case, there was no question of staying at 50 feet, because there was sea spray all around anyway. There was no way we could have done this rescue from 50 feet. So we kept descending slowly until we were at 20 feet, which is extremely low over the surface of a swelling sea. It was a constant up-and-down motion for us, because the whole sea would rise and then it would fall. We had to constantly maintain distance between the boat and the aircraft. I knew that if I was not careful with my collective[1] coming up on power, the sea swell could hit the helicopter tail rotor, which was my main worry, and this would have been catastrophic. We had to be extremely sharp while manoeuvring the chopper.'

At 20 feet, a bare whisker in flying terms from the violently undulating sea surface, Capt. Rajkumar finally spotted it.

'It was an unreal moment. By sheer luck, I saw the boat. It had a blue fibreglass hull. And spread across the hull, holding on to a rope for dear life, was a solitary fisherman. From 20 feet above, it was clear that this man wasn't moving,' he remembers.

'I knew if I lost sight of him for even a moment, it would be over. It was extreme, untamed flying, because we were just 20 feet above the boat, which was going all over the place. We didn't know where the natural horizon was. There was the risk of the swell hitting the underbelly of the helicopter or hitting the tail rotor. In daytime, we have a small amount of horizon orientation between the sea and sky. At night, nothing. We could only fly with the reference we had. That was this boat. The boat was moving left, right, up, down, and if I flew looking only at the boat, then I would get disoriented and lose control of the aircraft too. I was looking at the instrument panel for less than a second, then at the boat. And if I looked too long at the instrument panel, I would lose the boat for sure. That's the sort of flying we were doing out there.'

In a shuddering hover at that dangerously low altitude, Capt. Rajkumar collected his thoughts once again. The man on the boat wasn't moving and hadn't responded to the arrival of the helicopter above him. It was clear to the crew that the man

---

[1] The collective lever in a helicopter controls the angle of the main rotor blades, and causes the helicopter to ascend or descend.

was either unconscious or in shock. The horrifying realization also dawned that the man may have been out there for up to three days without water, food or a surface to stand on.

'We realized, much to our dismay, that he was not ready to let go of the rope he had been clutching for at least the last three days. He was a defeated man, both physically and mentally.'

Diver Saini will never forget the moments that followed as he stared down at the boat, with a solitary naked fisherman holding on for life, glaring up with a mixture of yearning and uncertainty.

'High waves and downwash were making the boat drift and making it difficult to hover above it. I suggested to the Captain that we winch down a diver—both of us [divers] were ready to take the risk. The proposal was not approved because of the risk to a diver's safety in the open sea, and that too, on a dark night. The man was staring longingly at us as the last hope of his survival, while above him, discussions were going on about how to pick him up,' Saini remembers.

But how could the crew rouse him?

'There was no question of descending any lower. I was worried about the powerful downwash of our rotors throwing the man off the capsized hull. And if that happened, we would lose him for sure,' Capt. Rajkumar says. His hands on the controls, eyes darting between his cockpit instruments and the man on his doomed boat, the flight commander fought to keep the Sea King in a steady hover.

The first of a series of terrible dilemmas presented itself.

'You think with your mind, you use your logic, backed by knowledge and experience. But there's that other sense which you get. That is often what helps you in a situation where you have to take a decision. If I had used my reasoning and logic, there's no way I would have attempted what happened next.

Because I knew the risks were huge. But I knew we had to. I just knew it,' remembers Capt. Rajkumar.

But before the thought crossed the flight commander's mind, one of the combat divers seated in the helicopter's cabin behind the cockpit shouted.

'*Main jaoonga, Sir* (I will go, Sir)!'

It was Deepak Saini, the younger of the two divers. He already had his black diving suit on. The other diver, Sumit Raj, had prepared the winch and harness. The two young divers were awaiting orders.

'I get emotional talking about it now, because there is disbelief that I even took that decision,' remembers Capt. Rajkumar. 'The most difficult decision for me was to send diver Saini down. It was the only way to attempt any rescue. There was no other way. Sending the diver down was a dilemma for me. He was a young man. I didn't know if he knew the dangers involved. He may have volunteered in his innocence and enthusiasm for the mission. It was very easy for him to step forward. He may not have known the dangers, but I couldn't be excused. I *did* know what those dangers were. It was a gut-wrenching moment for me as I took the decision to send him down. I had to look at the possibility of therefore losing two people at sea—the diver and the fisherman. I knew that if I sent him down, I might never see him again. If I had a small emergency on board—like an engine oil lubricant light coming on, or the rescue hoist not coming up properly, or a vibration developing in the helicopter—there would have been no way to pick up those two people in the water. I would have had to abandon them at sea. I cannot describe how enormous this risk was.'

Capt. Rajkumar hesitated for a moment, turning briefly to look at the two young divers behind him. But that moment of pause, born from a leader's concern for the lives under his charge, dissipated when both co-pilot Garud and navigator Chauhan said taking the risk was the only option they had.

Turning back to his instruments, Capt. Rajkumar wasted no time.

'Roger, lower the diver,' called the flight commander, as he held the SK 528 in its shuddering hover, also issuing instructions to Saini to hold on to the boat after rescuing the fisherman and placing him in the strop harness.

'In that moment, I knew I had to save that fisherman. In that moment, you aren't thinking about yourself. How can you? Here was an opportunity to save a life. Here was my chance to prove all those years of training and money spent to make me a search-and-rescue diver. There was no time for emotion then. It was my job and I did not think for a moment that I had a choice,' remembers Saini.

'I don't know what strength that poor fisherman had left in him as he held on to the rope. The Sea King's downwash can be extremely menacing. At night, if he had slipped off, there would have been no saving him,' says Capt. Rajkumar.

Diver Saini was lowered into the churning sea wearing the rescue harness and with an omni-glow stick so that he could remain visible. He immediately swam up to the boat and grabbed the fisherman, strapped him to the harness and gave the other diver a thumbs-up sign, a signal for him to winch the fisherman up.

'When I swam up to the boat, the fisherman was still refusing to let go. He was totally confused and in fear. In that wild, rough sea, I shouted to him over the sound of the wind and the helicopter rotors above me, trying to calm him down and telling him that we were there to rescue him. He must have thought it was a dream,' Saini says.

Twenty feet above, the Sea King was barely holding steady, thanks only to the flying acumen of the two pilots.

'As I was holding the chopper steady, I was watching Saini as he went about the task of strapping the fisherman to the strop. This young diver, with his entire career ahead of him, had thought nothing of seriously risking his life. As the flight commander, I was immensely nervous about losing him. But I was also immensely proud.'

As the fisherman was winched up, Saini clambered on to the boat, whirling the omni-glow stick in his hand to remain in sight. But a fresh threat had just presented itself.

The winch with the fisherman had begun to swing wildly about halfway up to the waiting helicopter. The other diver was bringing him up carefully, but he had to ensure the fisherman didn't hit his head on the undercarriage. With a measure of difficulty, he was brought aboard the SK 528 without injury.

'We got him on board, removed the rescue harness, made him comfortable,' says Capt. Rajkumar. 'Then we lowered the rescue harness and winched up Saini as well. Thankfully, this wasn't difficult because he was fit and an expert swimmer. It could have been worse if there was an emergency or if we had lost sight of him.'

'I positioned myself on top of the inverted boat, staying calm and remaining positive. As the harness approached the water again, I grasped it and was winched up quickly by diver Sumit Raj. There were huge celebrations in our hearts, but we remained silent. It did not sink in what we had just accomplished,' says Saini.

The rescued fisherman was in shock, delirious and dehydrated.

'He was like a corpse. He was almost dead. More than physically, he was mentally broken. He had given up on life. We could clearly make that out while we rescued him. He had no emotion, he was totally numb. We gave him water and glucose biscuits. He couldn't even eat properly, so we had to do it very carefully because his throat was completely parched. He had been at sea for three days. But had displayed phenomenal strength of mind to not let go and to not give up,' Capt. Rajkumar remembers.

With the all-clear and its doors shut, the SK 528 climbed and turned, speeding immediately towards the coast. Hovering at 20 feet in those weather conditions would have been dangerous even for a few minutes. Capt. Rajkumar and his men had hovered at that sickeningly low altitude for 28 minutes.

Climbing up to 200 feet, the Sea King cruised back to Thiruvananthapuram. On the way, Capt. Rajkumar radioed the *MV Cosco Beijing*, thanking them for the miraculous piece of intelligence and guidance that led to the rescue of one man.

'If the *MV Cosco Beijing* had not spotted the boat, there would be no rescue. Hats off to them. The fisherman wouldn't have had a chance in a million if the Chinese ship hadn't spotted his boat. It was a great humanitarian gesture by them to stop and wait for a rescue.'

Landing a short while later, the fisherman was wheeled away for treatment.

'Funny thing is, we never got the fisherman's name. There was no time. As soon as we landed, he was moved to a hospital for treatment. And frankly, I didn't ask. It never occurred to me. He was a human being and we were doing our job,' says Capt. Rajkumar.

The SK 528's engines had been on for 7 hours with two refuels. The day's mission was now truly over. The cyclone would continue to wreak havoc along the coast for five more days before dissipating. Of a total of 661 missing in Cyclone Ockhi, 261 persons remain missing from the Kerala coast. Official estimates place the damage caused by the cyclone in Kerala alone at over Rs 15,000 crore.

'This was a massive team effort, with contributions from all the men on board. Helicopters are not like fighter aircraft. If I didn't have a competent co-pilot and a smart navigator and brave divers, this wouldn't have been possible,' Capt. Rajkumar says.

What made the seasoned helicopter pilot, trained to take calculated risks, make such an enormous leap of faith?

'The script was probably written by someone who lives up there that this one man had to be rescued. And I had a very small part to play in it. The emotions come in waves once the mission is over. I'm not devoid of them. After landing, I was shivering all over. While flying, you're calm and focused,' he says.

Older and more experienced than the other men who flew with him that day, dangerous flying is scarcely new for Capt. Rajkumar. In 1989, he received his first gallantry decoration, a Nao Sena Medal (NM), for saving four scientists in Antarctica during an Indian Navy expedition.

'I had thought that was the most challenging mission I had ever flown. It was a similar mission in snow-blown white-out conditions. During the Ockhi rescue, I was flying in complete blackout conditions,' says Capt. Rajkumar.

'I can safely say I've come full circle.'

Capt. Rajkumar and the crew of the SK 528 didn't have time for a celebration that night—they were physically and emotionally drained by the mission. And the next morning, they returned to the air for another day of rescue missions over the cyclone-torn Arabian Sea, rescuing eight more fishermen off the Kochi coast. It would be days before they fully took stock of the mission they had flown—India's first-ever helicopter rescue at night at sea.

Saini, whose family in Haryana's Bhiwani live in that curious mixture of anxiety and pride typical of families of military personnel, says the mission was his life's most critical and that he hopes to have the opportunity to do it again.

'What can be a bigger achievement than saving someone's life?'

On Independence Day 2018, the Indian Navy announced that Saini would be decorated with the Nao Sena Medal for gallantry in the mission, his citation recording that he had displayed exemplary presence of mind and bravery in the face of grave danger.

On the same day, the Navy announced that Capt. Rajkumar would be decorated with the Shaurya Chakra, India's third-highest peacetime decoration for gallantry. His citation would note that, 'The bold decision and daring act of the officer enabled saving a human life in extreme conditions and was possible only because of the sheer determination, courage and decision making abilities of the officer,' and that he had displayed 'undeterred commitment to save human life in the most trying conditions accompanied by courage, fortitude and display of valour in the face of danger'.

In Kochi, there literally wasn't a moment to celebrate. In August 2018, Kerala was in the devastating grip of historic floods, a catastrophe that forced the Indian Navy's Southern Naval Command to suspend all training activity and commit all available military assets to rescue and relief operations. On the forefront of this effort, code-named Operation Madad, once again, was Capt. Rajkumar and his trusty Seaking 42B.

On 16 August, the national media struggled to divide its focus between the intensifying disaster in Kerala and the death of former prime minister Atal Bihari Vajpayee. Elaborate state ceremonials for the departed leader diverted the media's attention at a time when the flood situation had become impossible to ignore.

The following day, 17 August, as Capt. Rajkumar and his five-men crew lifted off into the rainy haze over Kochi, little did they know that they were about to break a world record in aviation rescue.

The day started with a gruelling four-hour flying mission in which Capt. Rajkumar and his crew rescued seventeen people from the submerged outskirts of Kochi. After a quick refuel, the helicopter was back in the air to hunt for marooned families scattered across the many devastated colonies beyond the metropolis.

'The weather was dire and marginal for flying with low clouds, poor visibility and rains,' says Capt. Rajkumar. 'After flying for about an hour, we located and winched up fifteen people on board. We were returning to the Kochi base when I sighted someone frantically waving a red flag from a rooftop. I turned the aircraft around in order to investigate. I realized the rooftop of that house was low with tall trees all around the vicinity and it was difficult and dangerous to hover anywhere near.'

Capt. Rajkumar carefully manoeuvred the Sea King close to the roof of the house and winched down a Navy diver. The diver reported back that there were eleven people waiting to be rescued. With fifteen people already on board, in addition to the six crew members, the pilot was faced with a very difficult decision. Daylight was fading, and if Capt. Rajkumar returned to base without rescuing the eleven, they would have had to wait till the following morning. He knew that the rock-steady hover he had managed eight months before out at sea was going to have to return. And this time, with the danger of trees just a few feet away.

'I carefully made the Seaking hover between the tall trees. The winds were gusting and visibility was receding,' says Capt. Rajkumar. 'Winching up one person after the other was a time-consuming exercise and I had to continue maintaining the helicopter's positions in those tough conditions testing the limits of the aircraft.'

As the eleven were winched up from their submerged home, the crew of the helicopter had to carefully rearrange the fifteen who had already been rescued, so everyone could be safely accommodated. The shifting positions made the helicopter's hover much more difficult, the shuddering airframe threatening at any moment to career into the trees.

After many tense minutes, with the eleven safely pulled aboard, Capt. Rajkumar eased the helicopter out from between the trees and peeled away for the Kochi base. He wouldn't know it at the time, but that night, he would be informed that the mission had set a world record for the maximum number of persons rescued in a single helicopter sortie—twenty-six. And this was in a militarized Sea King, not specifically built to carry so many people.

The record, once again, couldn't pose a distraction. Capt. Rajkumar would fly out early the next morning to continue to hunt for marooned persons.

By the time relief operations ended, he had rescued 114.

11

'Half of My Face Was in My Hands'

Major Rishi Rajalekshmy

'Amma, I may not come back. But be proud of whatever comes to you. Promise me.'

Standing outside a house in Hafoo village in south Kashmir's Tral, Maj. Rishi held his mobile phone to his ear with one hand. In the other hand were 15 kg of plastic explosive, wired and ready to be detonated. Some 3500 km away, in Alappuzha, Kerala, it was a wonder that his mother had even answered the phone. It was 7 minutes past midnight, well past her bedtime.

Hearing the sound of assault rifles firing in the background, Rajalekshmy froze. The phone slipped from her hand and crashed to the floor. She stared down at it. She could still hear her son's voice coming from it, asking if she could hear him. After a few seconds, she heard him abruptly say he would call her back later. Then, the call was disconnected. Rajalekshmy hurriedly bent down to pick up the phone, dialling his number. But there was no answer. Maj. Rishi had silenced his phone and tucked it into a zipped pocket in his combat trousers. And with a bomb in his hands, he had stepped carefully into the darkness, picking his way through the debris and into the house in front of him.

In 3 minutes, standing inside that house on a spring night in 2017, the thirty-one-year-old officer's world would be torn apart. He had called his mother because he had been prepared to die that night. But a nightmare like he had never imagined was about to begin, which would have no end in sight.

That Saturday in March 2017 couldn't have started more differently. Maj. Rishi and his company of men from 42 RR were conducting an early morning medical camp for men in Tral, part of a regular humanitarian outreach programme conducted by the army to establish friendships and trust among the local population, whose free will is frequently held hostage by militant and terror groups. The camp that morning was a busy one, with nearly a thousand men with all manner of ailments from villages in Tral lining up for treatment. As a team of Army doctors examined each one of them, giving them injections or little brown paper bags with medicines and tonics, Maj. Rishi chatted with the men in the queue. Most were from villages that fell in the young officer's area of responsibility. He had met some of them before, played with their children, dined in their homes. They even had a special name for him.

Khan, they called him.

As he handed out glass tumblers of steaming hot tea, he looked at the faces in the queue—young and old, some smiling, some haunted, but nearly all relieved in some measure when they saw him. Maj. Rishi of 42 RR couldn't have been farther from his Kerala home town. In the two years since the Malayali had set foot in Tral, the Kashmiri town infamous as a hotbed of militancy and a safe haven for Pakistani terrorists, he had never felt such a deep sense of comfort and belonging.

Tral had been unusually peaceful that year. Just eight months earlier, one of its most notorious natives, the young Hizbul Mujahideen commander Burhan Wani, had been shot in an encounter 60 km south, in Kokernag. His death had sparked widespread protests, stone pelting and Pakistan-backed revenge attacks in several parts of the Valley. But in Wani's own village and the areas surrounding it, where Maj. Rishi's unit operated, other than sporadic protests, a strange peace had prevailed.

As the day wore on, word of the medical camp spread and more men emerged from their villages, persuading the team of doctors to extend their timing so that all the patients could be examined. It was just after 3 p.m. when Maj. Rishi received a call from Col. Neeraj Pandey, who had taken over as the CO of 42 RR a month ago.

'My CO told me that an intelligence input had just come in regarding Aaqib Molvi, a terrorist commander whom we had been tracking for a long time, a man who had managed to escape against all odds,' says Maj. Rishi. 'I was at the medical camp with all my men, so I quickly mobilized a Quick Reaction Team (QRT) and we proceeded towards Hafoo village.'

The officer's team didn't need to go back to its base first, since RR companies almost always travel with everything they need for an encounter or a cordon-and-search operation (CASO). This includes weapons, ammunition, bulletproof vests and helmets, and the ingredients required to assemble IEDs—bombs customized for a desired objective.

Hafiz Muhammad Aaqib alias Aaqib Molvi was no ordinary terrorist. Like Burhan Wani, he was a young Hizbul Mujahideen commander. A native of nearby Awantipora, he had swiftly climbed the ranks of the terror group. His methods and habits in many ways mirrored Wani's. He had embraced avenues afforded by social media to spread propaganda against the Indian state and to exhort Kashmir's young to enlist with terror organizations, especially in the aftermath of Wani's killing. Photographs of him with kohl-lined eyes, carrying a camouflage backpack and an AK-47 rifle, were regular on Hizbul's recruitment posters and digital feeds. But it was his skill in organizing attacks, and his survival and evasion tactics, that had brought him to the notice of Pakistani intelligence. And that's why, as Maj. Rishi and his men speeded towards Hafoo village, they were informed that Aaqib Molvi wasn't alone. A Pakistani JeM terrorist called Saifullah (alias Usama) was with him.

Intelligence agencies had long suspected that Pakistani terror groups were looking to join forces with the local Hizbul Mujahideen to conduct terror attacks in the Kashmir Valley. Information that a Jaish terrorist was with Aaqib confirmed this. The intelligence input suggested that the two had met a few days earlier at Aaqib's home in the village next to Hafoo. When Maj. Rishi and his men approached, the two terrorists deployed a crowd of stone-pelters and protesters, using the cover and distraction to run to a house on the edge of Hafoo village.

'Stone pelting started as soon as we arrived,' says Maj. Rishi. 'Units of the Jammu and Kashmir Police and the CRPF were quickly deployed to keep the pelters at bay so we could focus on the hunt.'

Maj. Rishi knew this was going to be an extremely delicate operation. If the stone-pelting crowds managed to overwhelm him or pin him and his men down, Aaqib Molvi would simply add another successful escape to his already impressive record. Moving quickly, Maj. Rishi and a soldier from his squad darted into a series of houses to make sure Aaqib and his Pakistani accomplice were not hiding in one of them. Clearing one house after another over the next 30 minutes, the officer emerged into the open, frustrated.

The two terrorists could have gone in any direction. At the far end of the village street he stood in, he could see stone-pelters being held off by a police cordon. Closer to where he stood, a small group of men stood outside a house. In that kind of volatile situation, the last thing Maj. Rishi was realistically hoping for was local help of any kind. But as he watched that group, their gaze fixed on him, a single pair of eyes silently turned towards a house about 200 m down the path, in the opposite direction. Maj. Rishi stopped, turning to look at the house the man was staring it. When he turned back, the same pair of eyes was fixed on him. Followed by an almost imperceptible nod.

Immediately summoning his men, Maj. Rishi jogged towards the house that had been silently pointed out, ordering a cordon to be formed around it. If the two terrorists were indeed inside that house, it was almost certain they would be armed. Maj. Rishi and his men took protective cover behind a low boundary wall around the house.

'I shouted out, requesting several times that they come out, saying we didn't want to kill them,' he says. 'At this point, we were still not sure if they were in that house. Either way, I didn't expect them to surrender under any circumstances. That's not how they're trained. But I had to give them that chance. The moment after I made my offer, there was firing from the second floor of the house.'

A fierce firefight commenced, with Aaqib and Usama shifting positions on the second floor and firing at the cordon with AK-47s. Policemen joined the operation as well. The exchange continued in waves for nearly 3 hours—bursts of fire punctuated by minutes of silence, as the terrorists reloaded their weapons or planned their next move. It was a stalemate that would end only if the terrorists ran out of ammunition. The steady firing from the second floor suggested that the house they had chosen was something of a safe house, stocked with rifle magazines—clearly enough to draw out a firefight for hours.

But for how long? In any encounter, the forces do their best to finish in daylight. Darkness brings with it obvious challenges. Most of all, it gives the terrorists an exponentially better chance to escape through the cordon and into the night. No search operation in the darkness thereafter stands a chance of tracking them down. Something had to be done before sunset.

'I called my CO, letting him know that we needed to step up the offensive,' says Maj. Rishi. 'I told him I needed to get an opening into the house. The best way to do that was with an IED. He approved the idea, so I got moving.'

Maj. Rishi asked the explosives expert in his squad, a young soldier, to quickly construct the IED with 15 kg of highly explosive material. Just before last light, at about 6.30 p.m., the officer picked up the IED. Then, under covering fire from two soldiers off to the side, Maj. Rishi held up a bulletproof shield and stepped lightly over the boundary wall and moved towards the house. Following close behind him was his buddy soldier, Lance Naik Avesh Kumar, who fired in short bursts at the top floor till the two reached the front door.

'The two were still firing from above, so we had to move to the side for cover,' says Maj. Rishi. 'I placed the IED right outside the house, near the front. The terrorists stopped firing for a minute. They were clearly changing positions to protect themselves. They must have known what we were trying to do.'

With hand signals, the officer let the soldiers standing around the house know that he was about to detonate the IED. Then he pushed down on the detonator, which was wirelessly connected to the bomb device. The loud thud of an explosion shook the ground and smashed a big hole in its front wall, causing one half of it to come crashing down. A piece of rock knocked loose by the blast came flying straight at Maj. Rishi, hitting him on the side of the head just below his helmet.

'The stone hit a vein, so I began to bleed almost immediately,' says Maj. Rishi. 'My buddy said he would escort me out so I could get medical help and leave the site. I told him the bleeding didn't matter. It was a small wound. I don't remember even feeling it.'

With half the house now a smoking mess of debris, Maj. Rishi and Lance Naik Avesh carefully returned to their positions outside the boundary wall. After a 15-minute pause, firing resumed from the top floor. If the terrorists were jolted by the blast, they were showing no sign of it yet. And since there were clearly two streams of fire still, it was clear that both Aaqib and Usama were alive and well enough to keep the fight on.

'One of my men helped me bandage my head to stop the bleeding,' says Maj. Rishi. 'I remember that we moved a little distance away to do this, because the firing had started again, very aggressively. The IED had given us an opening into the house, but Aaqib and Usama were still holding out. It was time for the next move.'

Darkness had fallen when the squad received word that Maj. Rishi would approach the house again, this time with an armload of Molotov cocktails. These improvised incendiary weapons, made of bottles filled with petrol and a kerosene-soaked cloth as wick, are famous around the world as a weapon of choice for rioters and guerrilla fighters. It is unclear when these simple devices were invented, but the name, in sardonic honour of Soviet foreign minister Vyacheslav Molotov, was a gift from the Finns during the Winter War, a conflict between the Soviet Union and Finland before the Second World War broke out in 1939. Molotov cocktails were widely used by the Finns to attack Soviet tanks when the latter rumbled in to invade. Eighty years later, nearly identical devices had been assembled in that tiny village in south Kashmir.

Maj. Rishi's squad needed to try smoking out the two terrorists by starting a fire, and Molotov cocktails seemed their best option. The ones they made were with half-filled rum bottles and kerosene-soaked rags.

'I stepped into the compound again and tossed one Molotov into the kitchen area from the outside,' says Maj. Rishi. 'There was an immediate flare-up and the fire caught. There was carpeting and wooden panelling inside, so the fire intensified quickly, with a lot of smoke. I flung a few more of the Molotovs deeper inside the house and stepped away. It was now a proper fire. I was hoping this would force the terrorists to jump down or try and escape. That would be our moment to get them.'

But Aaqib and Usama stayed put. The fire raged for 20 minutes, burning large parts of the house, and then died out. The two terrorists were proving to be extremely resilient and were showing signs of having received the sort of combat survival training that was only possible to get from a military force. It is well known that Hizbul Mujahideen terrorists train at Pakistani Army-run camps in PoK, where the curriculum has rapidly evolved from basic hit-and-run tactics to full-fledged commando-style training. It was clear by now that Aaqib and Usama were among the elite in their batches.

As he watched the fire die out, Maj. Rishi knew that another big move had failed to push the mission to a conclusion. And that each passing hour was making it steadily more unacceptable that the encounter end in failure.

'At midnight, I called my CO again,' says Maj. Rishi. 'By this time, he had arrived at the encounter site. I told him I would like to go back in with another IED.'

Col. Neeraj heard out his officer, perhaps wondering what it was that gave Maj. Rishi the nerve to volunteer to venture into the house again with another IED. The bandage on his head was soaked in blood, but the officer didn't seem even remotely tired or weakened from the blood loss. Col. Neeraj asked Maj. Rishi if he was absolutely sure about the move he was proposing. Surely, someone else from the squad could be sent in to plant the device. But Maj. Rishi politely refused, requesting that he be permitted to proceed inside with the second IED. Col. Neeraj relented.

'I don't remember any fatigue. I was pumping adrenaline and very keen to finish the mission,' says Maj. Rishi. 'I'm not saying I'm a hero or anything, but when you're in the middle of an operation with your men, you don't feel pain. I had forgotten about my head injury.'

Maj. Rishi asked the explosives expert to assemble two IEDs with 15 kg of explosives each. When they were ready, he held a quick briefing with his men, telling them that if the explosions didn't kill the terrorists, they would at least have nowhere left to hide—and would therefore have no option but to attempt to break the cordon and run away into the darkness. Under no circumstances should that be allowed to happen, he told them.

The two IEDs were brought to Maj. Rishi. The moment he picked one up, he felt a strange foreboding. Fighting terrorists in a situation like this was always a life-and-death affair, but for the first time that day, Maj. Rishi sensed personal danger. And that's why, even though he knew his mother would be fast asleep at her home in Kerala, he decided to call her.

'Maybe my mother had an intuition too, and that's why she picked up the phone despite it being so late,' says Maj. Rishi. 'I just wanted to hear her voice. She dropped the phone when she heard the firing from the house starting again. I tried to say a few words to calm her, but she couldn't hear me. I put the phone away and picked up the second IED. There was no time to spare. I knew that back home, my mother would spend the whole night praying.'

The firing from the second floor was coming in a furious non-stop barrage now, without the usual pauses. The two terrorists were alternating their fire, so there were no pauses when one of them reloaded with ammunition. They were clearly becoming desperate and it was likely that they were holed up in a corner on the remaining part of the top floor. With the sort of training they had demonstrated that evening, desperation could make them even more dangerous to deal with.

'I went inside with the IEDs in both my hands,' says Maj. Rishi. 'There was a lot of firing and action. Once inside, I used my walkie-talkie to speak with Lance Naik Avesh, whom I had asked to stay behind, right outside the house. He wanted to accompany me inside to plant the IEDs, but I knew I wouldn't be able to live with myself for the rest of my life if anything happened to him.'

When Maj. Rishi had taken charge of his company in early 2015, 42 RR was going through its darkest phase. On 27 January that year, the unit's then CO, Col. Munindra Nath Rai, had died fighting terrorists in Tral. He had led the operation from the front, but was shot in the head as the hiding terrorists burst out of the house in a bid to escape. Hailed for his leadership in fronting the assault, he will be remembered even beyond the Army, for something else. Two days after his death, images would go viral across the country of his inconsolable eleven-year-old daughter, Alka, saluting her father's casket while screaming lines from her father's parent regiment, the Gorkha Rifles, '*Keta 9 GR ko ho ke hoina* (Is this boy from 9 GR or not)?' and joining in the refrain from her father's comrades, '*Ho, ho, ho* (Yes, yes, yes, he is ours, he's our pride!).' To the Army and beyond, this would be telling of how the officer had ensured that even his children imbibed the fearless spirit that the Gorkhas are famous for.

The 42 RR didn't have the time to mourn, but grieving it was when Maj. Rishi arrived. He had spent the first few days promising his men that he wouldn't let them come to harm, no matter what.

'I told my boys I will never send you back in coffins or with broken limbs,' says Maj. Rishi. 'I might die or take a bullet, but even though you are my strength, I will never let a bullet get past me to you. That was my promise. The battalion was tense and grieving. Before the CO's death, two jawans from the unit had also died in operations. The boys were restless but focused. We trained hard together to get through that phase.'

Two years later, the scene was still Tral. And the two terrorists holed up inside that half-destroyed house in Hafoo were probably contemplating a final bid to escape Maj. Rishi's cordon, a full 9 hours after the encounter began.

'I entered the house with two IEDs, crawling in through the huge hole blown by the first IED. I found a spot near the undamaged part of the ground floor to place the IEDs in my hand. Then I crawled back out.'

Once outside, he spoke on his walkie-talkie with his buddy, calling for a 30-second countdown to detonate the IEDs. The countdown began as Maj. Rishi crept away from the house. Two IEDs would be double the intensity of the first. But from a safe distance, with 10 seconds to go for the detonation, Maj. Rishi clicked his walkie-talkie back on and asked the soldier holding the detonator to pause.

'It suddenly struck me that if I placed the IEDs a little deeper inside the house, they would have a more destructive effect. So I told my men to wait, crept back inside and went to pick up the IEDs. As I was about to, I saw a shadow move down the shattered staircase and instantly, there was a flash of fire. I was thrown off my feet and fell to the ground. In that moment, I fired back at the shadow with a long burst of ammunition, hitting him straight and dropping him right there.'

Three bullets hit Maj. Rishi. One blew his helmet right off his head. Two more bullets hit his nose and jaw, ripping a large part of his face off.

'I was thrown several feet by the impact, but I was lucid. I couldn't tell what damage had been caused, but I realized that I couldn't speak and my vision was a little blurred because of the blood. But I couldn't feel any pain. I started crawling out from the house and I remember thinking in those moments about action movies, where you see people function even after getting shot. I'm telling you that it's possible. I didn't feel weak at all. In fact, I don't think I had ever felt stronger.'

Maj. Rishi saw Lance Naik Avesh run towards him as he crawled out, but raised a hand to stop him. The second terrorist hadn't been accounted for yet and was probably still alive on the top floor. With cover fire from his men, the officer exited the house from one side, dropping off a metre-high platform, and then began moving on his elbows towards the outer boundary wall. Halfway back, Lance Naik Avesh and two other soldiers leapt into the compound and pulled him out. He stood up, and the other officers reeled when they saw what was left of his face.

'I couldn't speak, as a large part of my face had come off,' says Maj. Rishi. 'I signalled with my hands to them that the second terrorist was likely still inside, even though the firing had stopped.'

Unknown to the team, the Jaish terrorist, Usama, had been hit in the firefight and was incapacitated on the top floor. He would die shortly thereafter.

Escorted from the encounter site, Maj. Rishi was bundled into a jeep that sped to a helipad nearby. The Army medic who accompanied him couldn't help but stare at Maj. Rishi's injuries. He knew there was nothing he could do with the emergency equipment he had. This needed specialists. The officer's eyes were open but clouded now, and he was sitting upright. But his face was a bleeding, gory mess of flesh and bone. Loaded into a helicopter a few minutes later, he was quickly flown to the 92 Base Hospital in Srinagar.

As the helicopter flew at low altitude towards Srinagar, a soldier from Maj. Rishi's unit phoned the injured officer's wife. Maj. Anupama Rishi, also an Army officer, was posted at the 92 Base Hospital with the Military Nursing Service. She was awake when the call came through, waiting for her husband to check in with her before she called it a night. Hearing that he was en route and badly injured, she got dressed and rushed to the trauma ward to wait for him.

'When I arrived, I saw Anupama waiting with the doctors,' says Maj. Rishi. 'The doctor who examined me went blank when he saw me. I could tell from his face that he knew there was nothing he could do for me in Srinagar. My eyes were closing because of the blood, but I was conscious. The doctor repeatedly asked me, "Are you awake?" I couldn't speak, so I signalled to him with a thumbs-up. I also took his hand to try and communicate that he was looking too worried.'

The doctors, some of the country's finest and trained to bring back men very nearly from the dead, were on edge. The officer's injuries suggested that he should, at the very least, be unconscious. Maj. Rishi was not only conscious, but lucid too. The doctors continued to speak with the Major, keeping him engaged and telling him to relax. They worried that if he became fully aware of the nature of his injuries—they looked far worse than they felt, apparently, at the time—adrenaline would jog his circulatory system and he would lose blood even faster.

After performing emergency procedures that night to contain the blood loss, the 92 Base Hospital was forced to recommend that Maj. Rishi be sent as soon as possible to the Army's premier Research and Referral Hospital in Delhi. Late the next morning, the Major and his wife would fly to Delhi on an IAF An-32.

The same day, the United Jihad Council, the Pakistan Army's purported umbrella outfit for unified command and control of anti-India militant and terror groups active in Jammu and Kashmir, issued a statement mourning the loss of the two terrorists. It said: 'The two slain militants, Hafiz Muhammad Aaqib alias Aaqib Molvi and Saifullah alias Usama are the two shining stars of Jammu Kashmir who will always shine like gems. Their sacrifices will always be remembered.'

Like the crowds that gathered for Burhan Wani, large crowds would gather at the funeral of Aaqib too. The man who had killed him would see the pictures days later, when he was finally allowed to sit up.

'For forty-five days, I could not speak a word,' says Maj. Rishi. 'I could not breathe through my mouth or nose—I had a tube in my throat. All my food intake was also through a tube. In a matter of weeks, I lost 20 kg. I was reduced to skin and bones.'

When Rajalekshmy saw her son at the Army hospital in Delhi a few days after he was admitted, she said nothing. Her daughter-in-law had requested her not to break down in front of her son.

The injury had brought to a violent stop Maj. Rishi's tenure in the Kashmir Valley. But losing most of his face had, in his mind, destroyed something else too.

'Somewhere deep inside, I had also wanted to be a model, and most of my loved ones knew that. Maybe that's why Anupama ordered people not to cry in front of me. She was very strong. We had laughingly discussed how I would seek permission from the Army to do some modelling. But there I was, totally disfigured.'

One of Maj. Rishi's first visitors was the Army Chief, Gen. Bipin Rawat, himself. The General had heard about the young officer's courage—and condition—and wished to meet him in person.

'When the chief came, my eyes were filled with tears, because I was in a situation where my chief was standing in front of me and I could not even get up and salute him,' says Maj. Rishi.

As the General spent time with Maj. Rishi and his family, he was briefed about the officer's story, starting with how Maj. Rishi had left a 'safe' career in civil government service to join the Army so he could literally fight for the country.

'It sounds clichéd these days, I guess, but it's true—I had dreamt of picking up a weapon and fighting for the country since childhood,' says Maj. Rishi, who has had eight reconstructive surgeries since the incident, and is yet to fully recover. 'Whenever the national anthem is played, I get goosebumps. I don't know what that makes me, but I really wanted to protect this land.'

Rishi had studied engineering in college in Kerala and had been employed as an assistant engineer with the Kerala State Electricity Board upon graduating. A year later, he took a special selection board examination to join Air India as a direct entrant. Posted to Mumbai international airport, he would quickly grow restless, his boyhood dreams of combat and weaponry luring him to IMA, Dehradun. For the first time since his studies had ended in Kerala, he finally felt he was on the path to fulfilment. In 2010, he would be commissioned into the Army's Mechanised Infantry, a regiment raised to provide combat mobility to infantry troops, functioning with armoured vehicle units.

Over the next five years, he would serve in the deserts of Rajasthan, and then be dispatched to the Congo to join India's UN Peacekeeping Force. On a month's break in India, he and Anupama got married. When he returned to India after the mission in Congo, he finally had the opportunity to go down the path he had his heart set on—one that led to the Kashmir Valley. He quickly opted for the RR, arriving in Srinagar in early 2015.

'I told my CO, if you are sending me somewhere, please send me to the ground, and not as staff,' says Maj. Rishi. 'I just want to be with my weapon. With men in the field. My mother had told me, if you want to join the Army, then you have to lead men. Even if you are wounded, you have to lead them. Her words are always with me. Maybe that's why I called her the night of the incident.'

Dispatched to Tral to command a company, Maj. Rishi was soon immersed in the daily tensions of counter-insurgency and anti-terror operations, a universe away from the life he had chosen to leave behind.

'I never thought for a moment about my old life. I feel very strongly for this country, and my weapon is an addiction for me. When you hear bullets fly, when a firefight begins, the rush you feel is incomparable. I always look forward to the next operation.'

That is probably why, a year into his duties with 42 RR, his then CO, Col. Vikram Kadyan, would frequently slap his back and say, '*Kabhi toh darr liya kar* (You should be a little afraid sometimes), you should be afraid of something.'

For all the combat he would be immersed in, Maj. Rishi knew he was in one of the most complicated and difficult places in the country. Before he joined the Army, he had watched the Kashmir conflict fester endlessly, its intensity ebbing and flowing. On the inside and dealing with them first-hand, the complexities were amplified in all their terrifying detail. If fighting terrorists was the job he had signed up for, he knew that an even more difficult duty was to win the hearts and trust of the Kashmiri people. He had been sent into a notoriously hostile hotbed of militancy and terror—Tral, the south Kashmir stronghold of the Hizbul Mujahideen, and a place well known for the implacably anti-Army stance of its people.

'I loved Tral from the moment I set foot there,' says Maj. Rishi. 'There is something magical about that place.'

With a stance that was low on aggression and high on cheerful friendliness, Maj. Rishi embarked on patrols with his men in the villages of Tral, taking every opportunity to befriend residents or help them with their everyday problems.

In late 2016, patrolling soon took Maj. Rishi and his men to Tral's Dadasara village, home to Hizbul Mujahideen's most infamous commander at the time. After they had searched Burhan Wani's family home, his father, Muzaffar Ahmad Wani, a high school principal, stepped up to Maj. Rishi.

'His father offered me *Zamzam ka paani*[1] and made me drink it with him,' says Maj. Rishi. 'It was a very humbling

---

[1] The Well of Zamzam is a well inside the Masjid al-Haram in Mecca, Islam's holiest site. Islamic mythology says that the well appeared by way of a divine miracle to bring forth water from God. The well is visited, and its water consumed annually, by millions of pilgrims during the Hajj or Umrah.

experience. We were there to get information about his son. And we spoke to him about it. But relations remained decent.'

Over the next few months, 'Khan', as Maj. Rishi came to be known among local residents, would become that friendly face in the *fauj*, an officer whose Hindi was tellingly inflected with a Malayalam accent, but who refused to give up on his efforts to improve at the language, even learning bits of the local tongue.

'We established a very good rapport with the people of Tral,' says Maj. Rishi. 'Operations continued, but there was never an attack on my camp. When we patrolled, stones were never pelted at us, like they are in certain other locations. We knew there were tensions, but we were succeeding in convincing perhaps a few that we had good intentions and wanted to keep everybody safe.'

In medical camps, like the one he had organized the morning of his fateful encounter, Maj. Rishi would frequently treat children who would line up with a common winter ailment—burns on their hands from spilling boiling tea.

'I would treat the children and make friends with them while extracting a promise that they would not join militancy. These were just friendly chats, but I know that we were making a connection. I was holding those little hands and treating them. I knew the same hands were highly unlikely to pick up a rock against me. These are really good kids. They have a lot of honour.'

In early 2016, a huge protest erupted at a village in Tral. Hundreds of residents from nearby villages joined in to agitate against power outages in the area. This was an area that came under Maj. Rishi's operational responsibility, even if he had no control over the corruption-ridden civil utility supply system.

'I was advised not to go to the protest area because people there were very agitated,' says Maj. Rishi. 'I remember thinking they had every right to be agitated if they had been deprived of electricity for days on end. With my men, I went there right away. When the protesters saw me, an amazing thing happened. They removed the roadblocks, and they were dispersed by some seniors among them. I went up to the elderly men and told them I would not leave until I had solved their problem, whatever it was. When they told me about the electricity nightmare, I called the District Collector and requested him to urgently send an engineer to fix the problem. He arrived a short while later and sorted out the power situation. I needed the people to know that I lived among them and wanted to serve them.'

The friendships established this way would often lead to awkward, uncomfortable incidents. Many of the elderly men that Maj. Rishi had befriended, and whose trust he had earned, were parents of known militants. When intelligence inputs arrived about their presence in the area, Maj. Rishi would be forced to place those friendships aside, but always with dignity.

'They knew I would never trouble them unnecessarily,' says Maj. Rishi. 'I would bother them at home only if there was a specific input. And they understood my compulsions. Sometimes they would voluntarily give me information, confirming that their militant son had visited but had departed to an unknown location. They were very cooperative with me all throughout. We never irritated them or harassed them.'

Restricted to Delhi as a result of his injuries, Maj. Rishi has had eight surgeries so far, each of them nearly 14 hours long. His face has had to be reconstructed with metal prosthetics, with a new jaw fashioned from bone drawn from his leg.

'It takes almost two months to recover from each surgery,' he says. 'By the time I am strong enough to stand up, it's time for the next surgery.'

The officer needs several more surgeries over eighteen months—and even then, it is uncertain if he will be declared fit for combat again.

'Never have I slept for more than 5 hours,' he says. 'From that to this sedentary life, where there is anaesthesia everywhere I look.'

In August 2017, with a black band covering the lower half of his face, Maj. Rishi would receive a Sena Medal (Gallantry) on India's Independence Day. But the future remains a cruel question mark.

'I want to get back to Kashmir,' says Maj. Rishi. 'I need to get back into combat. I will try all my options, even the NSG. I need to get back on my feet, and with a weapon.'

The morning after the encounter, Maj. Rishi's men and seniors received an unusual barrage of calls from residents across villages in Tral. Word had spread about his injuries. Each one of the callers asked if the young officer would survive. One of them, a teenager perhaps, had broken down on the phone.

'*Khan sahab ko wapas bhejo*,' he said, in faltering Hindi. '*Hum unko theek karenge* (Please send Khan Sahab back. We will heal him).'

12

'I Repeat! Fire in My Cockpit!'

Squadron Leader Ajit Bhaskar Vasane

Airspace over Gujarat
10 October 2011

It was the one place that fighter pilots in the area were forbidden from flying anywhere near. If they saw it on the horizon, standard operating procedure made it compulsory for them to swerve their jets away well before they got anywhere within a 2-km radius. Flying closer than that safety margin could mean serious career trouble. But when the site loomed into view that October afternoon in 2011, sitting in his MiG-29 at 10,000 feet, Sqn Ldr Ajit Bhaskar Vasane knew he had only seconds to make a decision.

A decision nobody in the IAF had ever been forced to make.

The out-of-bounds site to the young pilot's left was easily the most dangerous place to fly an aircraft. It was the world's largest crude oil refinery, the Reliance Industries complex outside Jamnagar in Gujarat's Gulf of Kutch. An aircraft crashing anywhere in the sprawling 7500-acre facility could result in a devastating inferno, the likes of which had never been seen in India. And with over a million barrels of petroleum churned out from the site every day, a fiery visitor from the sky had the potential of sending destructive shock waves throughout India's economy as well as the world's volatile oil markets.

Sqn Ldr Ajit's mind was racing. A paralysing choice had just presented itself to him. He needed to get back to base in

Jamnagar as quickly as humanly possible. *Before he lost control of his jet.* But his cockpit maps had confirmed that the shortest flight path back wouldn't just require him to violate the 2-km safety restriction, it would actually take him directly *over* the refinery. If he obeyed protocol and flew just outside the 2-km radius, in the increasingly likely event that he lost control of the jet, there was still every chance that it might helplessly career out of the sky and drift straight into the refinery complex. The third option was to risk himself and the aircraft and fly a long circuitous flight path away from the refinery and back to base.

As he struggled to decide, Sqn Ldr Ajit wondered if the aircraft would hold on. If it would stay in one piece. If he would stay conscious. Because, despite over 600 flight hours logged in a decade of flying fighter aircraft, he had never been strapped into a more hostile cockpit.

Nine minutes earlier, at 3.30 p.m., Sqn Ldr Ajit had roared into the air in his MiG-29 from the Jamnagar Air Force base, the country's westernmost military air station. A second MiG-29 with his wingman, Sqn Ldr Rohit Singh, lifted off seconds later in pursuit. Climbing into a perfectly clear sky, the two pilots from the IAF's 28 Squadron were out to perform a supersonic intercept mission, a simulated confrontation to rehearse how they would respond[1] if a Pakistani military aircraft were to violate Indian airspace.

The two fighter pilots throttled up on their jets to position themselves 80–100 km apart from each other in a patch of

---

[1] Intrusions, while rare, are far from unlikely. In August 1999, shortly after the Kargil war, a Pakistan Navy Atlantique-2 maritime reconnaissance aircraft that repeatedly violated Indian airspace off the Gujarat coast was shot down over the Rann of Kutch by an Indian MiG-21 jet.

designated training airspace at a height of 30,000 feet, where Sqn Ldr Ajit would simulate a hostile air intruder and his wingman would rehearse an aggressive interception, and if necessary, a shoot-down. All electronically, of course. If this were a real-life interception, Sqn Ldr Rohit would likely fire, as a last resort, a Russian Vympel R-73 heat seeking air-to-air missile at the unwelcome intruder. That morning, both MiG-29s in the air were armed with missiles, but there would be no firing—the brutal air drill was purely to hone the flying reflexes of the pilots, and the time-tested standard operating procedures, in the event of a hostile air intrusion.

Back in Jamnagar, a young Flight Lieutenant from the squadron manning the air traffic control cleared the two pilots to begin their manoeuvres. The 28 Squadron, operating a pack of eighteen MiG-29s, is codenamed 'The First Supersonics' for being the first squadron to be equipped with a fighter that could break the sound barrier, the MiG-21, in the late 1960s. The MiG-29 could easily throttle up to over twice the speed of sound, but the two pilots would be keeping their velocities in check. The IAF does not permit supersonic flights over populated areas as the deafening 'boom' that fighter jets produce when they break the sound barrier can shatter glass panes on the ground and cause panic.

Sqn Ldr Ajit pointed his jet west to fly a long, curved loop 120 km from Jamnagar and back towards the mainland to simulate the intrusion. He squinted as he pulled up, bringing the aircraft head-on with the sun, flooding the cockpit with blinding light. The pilot adjusted his oxygen mask and pulled down the integrated tinted visor on his helmet, crucial to protecting the eyes during such flights.

Suddenly, the MiG-29's head-up display (HUD), a pane of glass sitting on top of the aircraft's 'dashboard', which superimposes aircraft instrument readings and mission data onto the pilot's viewpoint, flickered and blanked out. As the name suggests, an HUD allows a pilot to keep their 'head up' without having to lower their gaze at maps or instruments on the cockpit panel. The HUD going blank wasn't, by itself, a catastrophic emergency. But what had caused it was something unheard of. Something that had never been documented or reported before on any MiG-29 anywhere in the world.

At 30,000 feet, Sqn Ldr Ajit called out to the young Flight Lieutenant at Jamnagar base air traffic control.

'Finback 1 reporting fire in the cockpit,' Sqn Ldr Ajit said over the radio, keeping his voice as casual as possible. It was important not to cause panic if there was no need to. Except, if there was ever a cause for panic in the cockpit, this would be it.

The stunned radar controller wasn't sure if he had heard Finback 1, Ajit's radio call sign, correctly.

'Finback 1, did you just say fire? Can you confirm it?'

Sqn Ldr Ajit replied, 'I repeat, there's fire in the cockpit.'

'This is not engine fire?'

'No, I repeat. It's cockpit fire,' said the pilot. 'Turning urgently back towards base. Need guidance and permission to descend.'

'Finback 1, base 08085 [80 degrees and 85 km away]. Let's get you out of there, Sir. Let's get you home.'

Flying 60 km to Sqn Ldr Ajit's left and on the same radio frequency, his wingman, Sqn Ldr Rohit, heard the terse radio exchange. With permission from the ground, he immediately swerved his jet rightward to get closer to his friend, to see if he could help from the outside. As he banked towards Finback 1, he wondered just how he would be able to help at all.

When the HUD had flickered off, Ajit had immediately spotted the reason—a small fire right below it. At first, he thought he was hallucinating. It could be a symptom of hypoxia from decreased oxygen supply owing to a possibly faulty oxygen supply system. The condition can have deadly consequences—a fighter pilot's mind can blank out completely, for instance. They could even begin to imagine things or experience spatial disorientation, rendering them incapable of taking informed decisions in an unforgiving environment.

'A fire in the cockpit? No way,' remembers Sqn Ldr Ajit about that day. 'I was certain that classic hypoxia was messing with my mind. I thought someone had just lit a matchstick in the cockpit and offered me a cigarette.'

The fire had started small, about the size of an index finger, but was quickly growing. Ominously, it was filling the cockpit with thick, black smoke. But the most obvious danger flowed through the MiG-29's shuddering airframe—over 4000 litres of fuel in a series of tanks in the fuselage and wings that could easily ignite if the fire spread. It would take just a small lick of flame in one of those tanks to destroy the aircraft in seconds.

Sqn Ldr Ajit knew that the first thing he needed to do was check his oxygen supply. He could still breathe through the thickening smoke, but not for long. And it was imperative that he rule out hypoxia, so he could get busy dealing with the emergency at hand. The drill had to be followed. He tilted his head to the left to check the oxygen regulator in the side console. There was no malfunction visible. The system appeared to be generating a steady supply of oxygen. So this wasn't hypoxia. This was a real fire in the cockpit, and it was spreading fast.

'Good news, I wasn't hallucinating,' says Sqn Ldr Ajit. 'Bad news, I was now positive I had a proper fire in the cockpit and had to do something about it before it was too late.'

In seconds, the fire spread to the visor attached to the HUD. The thick dark smoke swirling within the cockpit carried with it the odour of burning plastic. Worse, the smoke was depositing a layer of soot on the aircraft's glass canopy, blocking out the pilot's frontal visibility.

'The time lag between spotting the fire and taking recovery action was less than 60 seconds,' says Sqn Ldr Ajit. 'It's a different matter that that one minute seems almost eternal when you're in the air. Ask any fighter pilot. They'll tell you that 60 seconds in an emergency is longer than a year on the ground.'

Guided by the radio controller at Jamnagar, Sqn Ldr Ajit had carefully manoeuvred his jet and was cruising in the direction of the base. Deprived completely of outside visibility, he was now peering through the thick smoke at the clouded cockpit instruments. He would have to fly the fighter plane with a burning cockpit for another nine minutes if he was to make it safely back to the base, where preparations for an emergency landing had begun.

Hearing about the emergency, more officers had rushed to ground control at Jamnagar. But there was little they could do. Sqn Ldr Ajit chuckled over the radio as he updated the team on the ground. Both he and the men on the ground knew the next few minutes could either end in a messy disaster, or give the IAF a new entry for its flight safety manual. Either way, there was no advice, no precedent. And no earlier record of such an emergency that he or the ground team could fall back on. The pilot knew this was entirely on him.

Even with the daunting uncertainties he was faced with, Sqn Ldr Ajit was sure about one thing. That he would not eject from the aircraft. The option had been presented to him from the ground, but he had calmly declined. Yanking the ejection handle would have given him a possibly safe emergency escape, blowing away the aircraft's canopy, the rocket-powered Russian Zvezda K–36 ejection seat with an attached parachute violently firing him out of the MiG-29. But Sqn Ldr Ajit wasn't so sure.

In the next few minutes, Finback 2 arrived alongside the troubled MiG-29. Sqn Ldr Rohit was still not fully aware what his friend was going through in his cockpit. He looked out of his jet at Finback 1, hoping to get a visual confirmation that things were in control. The sight was a shocking one.

'I couldn't see Vasane,' says Sqn Ldr Rohit. 'The cockpit was filled with smoke and the front portion of the windshield had deposits of soot. I don't know how he was sitting inside that cockpit and flying the aircraft.'

He was right. Sqn Ldr Ajit had, by now, been forced to resort to desperate measures to contain the fire and put it out. He had first tried to douse the spreading flames with his hands. The plan didn't work as the strands of molten plastic adhered to his fire-resistant gloves and he could feel his fingers getting singed.

'I swiftly took off my gloves and tried using them as a duster to smother the fire,' says Sqn Ldr Ajit. 'Nothing seemed to be working. The intensity of the fire kept increasing. My hands were being badly burnt.'

The pilot's eyes had begun to sting and water because of the smoke, and the irritation was growing with every passing second. His helmet's anti-glare visor provided no protection against the dense smoke. For the moment, breathing wasn't as much of a problem in the smoke-filled cockpit as he had his oxygen mask on. But visibility was totally gone. All he could see was a blanket of grey before his eyes. The cockpit panel, with its bright screens, was no longer even faintly visible.

'My vision was now blurred and there was a heavy deposit of soot on the windshield,' says Sqn Ldr Ajit. 'It was all covered in black. I was flying totally blind.'

About 40 km from Jamnagar, Sqn Ldr Ajit was advised by the radio controller to descend to about 10,000 feet. Making any abrupt moves with the aircraft was out of the question. The pilot carefully pushed his stick forward, easing the aircraft into a slow descent. If the cockpit fire was eating away at the insides of the aircraft, its structural integrity could be compromised. Fighter aircraft are made of some of the sturdiest materials in the world, but a fire could easily cause the sort of damage that would seriously hamper flying stability. Worse, it could take control completely out of the pilot's hands.

As the pilot weighed his options, the thought of jettisoning the MiG-29's canopy crossed his mind. On one hand, it would immediately help with the smoke situation. But on the other, there was every chance it would make matters much, much worse. For one thing, blowing off the canopy would suddenly bring a roar of noise from the wind—the MiG-29 was still flying at over 600 kmph.

'In that sound blast, I would not be able to communicate with the radar controller or my wingman,' says Sqn Ldr Ajit. 'I would essentially be flying the aircraft on my own. I was tempted, because the smoke had by now nearly blinded me, but I knew I would be deprived of whatever help I was getting from the ground.'

A sudden rush of wind could also fan the fire into a larger blaze and push it to other parts of the aircraft. It was out of the question, the pilot decided. He had no choice but to continue flying the jet with the canopy firmly on for as long as he could bear the now-choking fire and smoke. What he therefore decided was to fly blind, almost fully dependent on his radar controller.

The toxic plastic fumes became thicker and continued to build up inside the cockpit as the fire quickly spread to other parts of the panel, melting them and distorting the instrument frames. It had now become unbearable. Bits of smoke had begun to seep into the pilot's oxygen mask from the gaps, causing him to cough and wheeze through the cloud, forcing him to again attempt to put out the fire with his now-burnt gloves. But the fire was there to stay. Trying to douse it in any manner was futile.

'Picture a fighter cockpit. It's a really small, confined place stuffed with a whole lot of things, including the pilot. And then there was the fire. It could do a lot of bad things and really fast,' Sqn Ldr Ajit says.

It had been seven minutes into the emergency when the radio control piped up again. He wanted to check if the fire had shut off any of the aircraft's vital systems. Even if the flames hadn't eaten into the fuel tanks, they could still destroy the complex electronics that kept the aircraft stable and the systems that kept the engines running. Losing system indicators was bad enough. Losing the systems themselves would have meant Finback 1 possibly falling out of the sky, with no more options left to the pilot.

'I had begun to see the lack of a catastrophic explosion or deviation as a stroke of luck,' says Sqn Ldr Ajit. 'But I was sure that luck wasn't going to last. Fire does not cooperate. It does not wait for you to take a decision. If the fire killed my systems, none of my flying skills would be of any use. With my instruments and visibility gone, I was entering a situation where I would have no clue about how high and how fast I was flying, and what the engine was doing or, for that matter, the other systems required for landing.'

Finback 1's cockpit had become a terrifying workplace for any fighter pilot, irrespective of his training and experience. If one of the MiG-29's two Klimov RD-33 engines failed, there would be no instrument forewarning in the cockpit. It is instruments that routinely tell the pilot about engine power and revolutions per minute (RPM)—information that is vital to the pilot.

'If panic were to set in, this would have been a good time—at this point, I was desperately concerned about the fire depriving me of vital inputs,' says Sqn Ldr Ajit.

Descending to an altitude below 10,000 feet quickly was critical at this juncture. At that height, Sqn Ldr Ajit would have the option of activating a ventilation system to drive the smoke out of the cockpit.

'The air-conditioning system on the plane has options of "normal" and "flood". Additional air flows inside the cockpit when "flood" is selected. I needed to descend below 10,000 feet for this option to work effectively,' says Sqn Ldr Ajit.

Strapped in his seat in the burning cockpit with the sun blazing behind him, Finback 1 continued its descent, with Finback 2 flying above and to his left.

'I could see Vasane's plane below mine, but I didn't know what exactly was going on inside,' says Sqn Ldr Rohit, who was flying Finback 2. 'The MiG-29 technical handbook lists several emergencies such as engine failure, flameout and loss of oil pressure. We know how to deal with those issues. But a cockpit fire was something new. How do you handle a problem that has never been encountered before?'

As the pair of MiG-29 jets cruised towards their Jamnagar base guided by the radar controller, the cockpit fire assumed its worst form thus far. For the first time in the minutes since the emergency began, Sqn Ldr Ajit wondered if ejecting from the jet was the only way to survive. Ejection seats in modern fighters have a success rate of more than 90 per cent, but pilots can end up with broken spines because of the sudden explosive force that rockets them out of the cockpit.

Even as the fumes choked the pilot, activating the ejection sequence was near unthinkable. Because emerging over the horizon just then was the Reliance Industries refinery complex, which is south-west of Jamnagar city. The heat inside the cockpit was now painful and it had become difficult to breathe. If the human instinct to survive overwhelmed everything else, a pilot in such a situation would bail out of his aircraft. Self-preservation, after all, is the most powerful impulse in all human beings. It would matter less at that point that the blazing wreck of his aircraft could glide unstoppably like a missile into the world's largest crude oil refinery to spark what would almost definitely be an inferno of historic proportions.

'The radar controller was continuously giving me the course and distance from the base,' says Sqn Ldr Ajit. 'If I were to draw a line from my position to the base, the flight path ran right above the refineries. I had been operating from the Jamnagar base for nearly three years and I knew I was in the danger zone. You just can't crash your plane there. I am not supposed to violate the 2-km rule under any circumstances. So there was absolutely no question of not adhering to the restriction on a day when my plane could explode any moment. I decided not to eject until it was absolutely necessary. And by this time, I felt I was mere seconds away from that situation.'

Fully aware of the consequences of his aircraft crashing into one of the refineries below, the pilot of Finback 1 took the most difficult decision of his flying career. Informing the radio controller that he was concerned he would soon lose control over his jet, Sqn Ldr Ajit peeled away to the right, putting 10 km between his aircraft and the refinery complex. He had chosen to take a longer route back to the base, adding excruciating minutes to his flight time.

'In my mind, it was the only way to ensure that there wasn't a bigger tragedy on the ground,' says Sqn Ldr Ajit. 'The refineries were my big worry. I kept thinking, what if my aircraft suddenly turned left and I was unable to control it? In a matter of seconds, the plane could have crashed into one of the refineries, causing unimaginable damage. I just wanted to land my aircraft soon.'

Slowly descending to below 10,000 feet, the pilot set the MiG-29's air-conditioning system to 'flood' mode, hoping to drive the thick cloud of smoke out of the cockpit. It didn't help, but with wisps of smoke dragged out of the cockpit, it stopped the smoke build-up from intensifying any further. The aircraft had begun to shudder slightly, indicating possible internal damage. Sqn Ldr Ajit still held off on ejecting—he made a mental note to be fully prepared to make that split-second decision, if required. Fighter pilots don't eject at the first sign of trouble. They are trained to calmly follow drills to avoid casualties on the ground as a priority, even if that means putting their own lives at enormous risk.

Under the guidance of ground control, Finback 1 continued its descent. The Jamnagar base was finally now on the horizon, except that Sqn Ldr Ajit couldn't see it. He and his wingman had flown identical missions several times before and were fully familiar with the ground features that would lead them back to the base they had departed just 15 minutes ago. But with his visibility completely compromised, Sqn Ldr Ajit listened carefully for visual cues from his wingman flying above and behind him now.

Still flying blind as he lowered his landing gear and came in for a final approach, Sqn Ldr Ajit finally spotted the runway. A tiny keyhole-sized gap in the smoke had opened right above his HUD, affording the smallest glimpse of the outside world for the first time since the fire had ruined his visibility. With a rush of confidence, reducing his speed to 280 kmph, Finback 1 descended the final few metres to touch down on the Jamnagar tarmac.

As the aircraft sped down the runway, Sqn Ldr Ajit saw a pair of crash tender trucks positioned alongside the far end, ready to douse the aircraft with high pressure water jets, if necessary. Sqn Ldr Rohit watched from above as the stricken MiG-29 jet deployed its chute and rolled to a halt.

'I breathed a sigh of relief in the cockpit,' says Sqn Ldr Rohit. 'It takes exceptional skill, presence of mind and plenty of guts to fly and land a doomed plane safely. I sent a word of congratulations from the air before circling around and bringing my own jet in.'

The fire in the cockpit was still burning as Finback 1's canopy finally opened, rapidly dissipating the smoke that had collected inside. Quickly unstrapping himself, the pilot climbed out of the aircraft. He had flown in a crippling blanket of smoke for nine minutes, but as Sqn Ldr Ajit stepped out of the aircraft, he found himself reaching into a zipper in his overalls for his pack of cigarettes. Picked up by a jeep, he lit up on the short drive back to the operations room, where he would brief his seniors about an emergency none of them had ever encountered before.

The MiG-29 was put down as 'unserviceable' in the aircraft's logbook. Over the next two hours, Sqn Ldr Ajit would brief his Flight Commander, the CO, the Chief Operations Officer and the Air Commodore commanding the Jamnagar fighter base about those nine minutes. He was the first Indian fighter pilot to encounter what had just happened. And he had managed to not only survive, but bring the aircraft safely back to base. If he had ejected—and nobody would have blamed him had he chosen to—the mystery of the cockpit fire would likely have remained unsolved, throwing into peril the flight safety of the fleet going forward.

By the time he was done with the briefings, the MiG-29's technical crews were waiting to get a low-down on the unprecedented emergency—a crisis that had brought the base to a virtual standstill for two hours.

'I had to explain everything to the technical guys so that they could identify the glitch and fix it,' says Sqn Ldr Ajit. 'It was in their hands now. I told them everything I had experienced. It was up to them now to ensure that the same thing never happened again.'

The one thing the young pilot hadn't done was call his wife, Sqn Ldr Rajeev Kaur, also an Air Force officer posted in Jamnagar. The two had met and fallen in love three years ago, when they were posted at the Adampur fighter base in Punjab. It was their first posting after joining the Air Force as officers. When Sqn Ldr Ajit survived that cockpit fire, they had been married for a year.

Sqn Ldr Rajeev, part of the IAF's accounts branch at the base, had gone to the office of the Jamnagar base commander at around 4 p.m. that evening to get a file cleared. It was there that she heard hushed whispers among the staff that a MiG-29 piloted by her husband had managed to land at the base a few minutes earlier after a critical emergency in the air.

'In the waiting room, the base commander's personal assistant told me about the incident,' says the pilot's wife. 'I asked him for details but he said, "Oh, nothing to worry about." Then my husband's flight commander, Wg Cdr Shekhar Yadav, entered the room. I asked him the same question. All he said was, "Don't worry, nothing serious," and went in to brief the base commander.'

Extremely anxious about what could have possibly happened, she began to dial her husband's phone, fuming that he hadn't called her on landing. After several attempts, he answered. She demanded to know what had happened and why nobody was sharing any details with her.

'*Yaar, kuch nahi hua hai* (Nothing has happened),' Sqn Ldr Ajit said with a chuckle. 'Who told you all this? I am having chai in the squadron. Don't worry. All is well.'

She was enormously relieved that he was back on the ground, but resented being kept out of the loop. At dinner that evening, she probed her husband a little more, perplexed that he wasn't more forthcoming with details of the mid-air drama.

'It was a routine sortie,' her husband had said, taking her hand from across the table. 'I am here having dinner with you. Everything's okay. Can we talk about something else, please?'

Sqn Ldr Rajeev would learn the full story only by chance that weekend at a party hosted by the Jamnagar base commander and his wife to celebrate their wedding anniversary. At the party, the chief operations officer of the base, a Group Captain, asked her if she had taken a look at the aircraft her husband had flown five days earlier.

'You must go and see that aircraft if you haven't already,' the Group Captain said. She had almost forgotten about the incident by this time. 'It was only then that I got to know about the scary emergency. I remember losing my cool with him for keeping me in the dark. I told him this was not done. How could he not tell me? But I realized later that he was dealing with it as well, and didn't want me to worry. But nobody can stop worrying. Fighter pilots can be crazy. Mine definitely is.'

That night, Sqn Ldr Ajit was allowed to sleep only after he recounted, in the minutest detail, what had happened on his flight that October afternoon.

'The briefing I gave her was probably more detailed than the ones I gave to my seniors after landing,' Sqn Ldr Ajit smiles.

The emergency on board Finback 1 had become the talk of the entire flying branch of the IAF that week. If the pilot needed a few days to calm down and recover, he didn't give any such indication. Sqn Ldr Ajit was back in the cockpit of a MiG-29 the day after the incident, out on another training mission.

'I was only doing my job, which is to train every single day through the year. I don't think it was an event to celebrate, or for that matter, to even thank God for keeping me safe. I never thought it was a big deal,' he says.

Sqn Ldr Ajit may not have thought it a big deal, but it was more than that for the IAF. Awkward about the sudden flood of attention he was receiving, the pilot wouldn't know at the time that the calm resilience he had displayed in the air for nine hellish minutes would earn him a Shaurya Chakra, the country's third-highest peacetime gallantry award, ten months later.

Posted out of Jamnagar a few months after the incident, to Tambaram in Tamil Nadu to train as a flying instructor, Sqn Ldr Ajit would learn about the decoration headed his way as he sat down to dinner with his course-mates one night.

'I had completely forgotten about the incident,' he says. 'I had moved on. And here I was, fielding congratulatory calls all night. I must admit it felt good, but I still say I was just doing my job.'

On the eve of Independence Day 2012, the government announced the gallantry decoration for a pilot who had displayed 'nerves of steel and unwavering commitment to the mission assigned to him'. His Shaurya Chakra citation would heap praise on him for handling a critical emergency that was 'neither documented nor had occurred before in a MiG-29 aircraft'.

He would play down the courage he displayed in the air when he spoke to his colleagues and his wife, but the Shaurya Chakra citation was unequivocal in its praise:

'Amid rapidly increasing intensity of fire and at great personal risk, he initiated emergency recovery of the aircraft. With exceptional presence of mind and courage of highest order, he elected to avoid flying over various petrochemical installations in the vicinity of the airfield, even though this prolonged flight endangered his life,' reads the citation. 'Despite limited visibility due to soot deposits on the windshield, Vasane skilfully positioned the MiG-29 jet for an emergency landing that he executed flawlessly. Throughout the flight, he maintained extreme calm and composure, devotion to duty and thorough professionalism in keeping with the highest traditions of the IAF.'

That August of 2012 was special for another reason. Sqn Ldrs Ajit and Rajeev had a baby boy eleven days before the Shaurya Chakra announcement. Little Rishan hasn't heard the full story yet. His mother says she will ensure he does, though, and soon.

Back in Jamnagar, Air Force specialists had finally managed to get to the bottom of the mysterious cockpit fire on Finback 1. A team from the IAF's base repair depot in Maharashtra established that the source of the fire was a short circuit involving a use-and-throw component in the HUD system. The lesson was a major reminder of how every single component on an aircraft—no matter how small or seemingly insignificant—could literally bring it down if it failed to do its job.

The findings were shared with MiG-29 bases across the country. Three months after the Jamnagar incident, in January 2012, a similar fire broke out in a MiG-29 cockpit at the Adampur fighter base while the jet was still on the ground for safety checks. The repeat incident forced the Air Force Headquarters in Delhi to order a comprehensive safety audit of the MiG-29 fleet to fix the problem. No such incident has been reported in the last seven years, but maintenance crews remain alert nevertheless.

In 2007, four years before Sqn Ldr Ajit's close call in the air, the Indian government decided to upgrade its entire fleet of MiG-29s with new engines, radars, weapons and modernized cockpits. The upgrade programme of seventy-eight aircraft is now nearing completion, though the IAF is still fighting a troubling decline in its combat aircraft strength. The IAF remains the only air force in the world still operating old MiG-21 jets, and is hoping to retire them soon with the arrival of the indigenous Light Combat Aircraft Tejas, the Rafale from France and new fighters that it hopes will be built in India.

Even as successive governments try to modernize the Air Force, pilots don't have a choice but to wait for better and safer aircraft in the interim.

Posted as a flying instructor at an air base in the country's east, Sqn Ldr Ajit—now a Wing Commander—frequently flies with trainee pilots, men who've heard about the Jamnagar incident and almost always ask him about it. His reply is brief and almost always the same: 'Fly with all you have.'

13

'Not a Sound until They Enter the Kill Zone'

Major Preetam Singh Kunwar

Near Badori, PoK
23 May 2017, 10 a.m.

The six men stepped out furtively, one by one, separated by a minute each, from a bunker built into the side of a hill. The advancing summer had turned the blanket of ice into rivulets of snowmelt, but the remnants of a harsh winter still clung stubbornly to the terrain. The six men had draped shawls over their phirans to keep warm; they wore camouflage fatigues and combat boots underneath. They stood for 10 minutes on a ledge cut into the mountainside, talking and sipping steaming hot tea from glasses passed around by one of them. Behind them, to their right, rose the towering Badori mountain, 3700 m high, with a couple of Pakistani Army posts that used its height for an unmatched view of the Uri sector. And 5 km ahead was the place the men were headed to next.

The LoC.

A small, rectangular gadget, an eyepiece, sat on a short tripod less than 2 feet high as an Indian Army soldier on his stomach peered through it. He scanned the area before him slowly, focusing the lens until he had the sharpest picture he could get from that distance. On an LCD screen, the six men appeared as dark, fuzzy blobs, flickering as the thermal imager captured their movements, a flourish of warm, dark pixels against the cold white of their surroundings. The soldier lowered the imager and looked straight out over the LoC. If those dark figures really were who he thought they were, then they were at least a month early. The soldier quickly picked up his communications console and sent a coded message to his base, which was then relayed to other Army units operating in the area. What it said was: 'Six men in view near Badori. Five kilometres inside. Suspected infiltration team. Maintaining surveillance.'

The soldier was part of a reconnaissance unit at a forward post in the Uri sector, manned by men from the 4th Battalion of one of the Army's most decorated infantry regiments, the Garhwal Rifles. In the 1962 war with China, 4 Garhwal Rifles fought overwhelmingly large Chinese forces in the North-east, inflicting significant damage and earning themselves a rare battle honour. They would thereafter be known as the Nuranang Battalion, in honour of a fearsome, courageous battle that the men from the unit fought in Nuranang in Arunachal Pradesh. Fifty-five years later, the unit was watching over a critical length of the frontier in Jammu and Kashmir. And at an enormously tense time.

Eight months earlier, Indian Army Special Forces units had crossed[1] the LoC in Uri and neighbouring sectors to strike at terrorist infiltrators. The mission was, in part, an act of revenge against Pakistan-sponsored terror groups for a commando-style attack by the LeT on the Indian Army's Uri Brigade headquarters, in which nineteen soldiers were killed. The revenge mission resulted in thirty-six terrorists and two Pakistani Army personnel killed across four infiltration launch

---

[1] The only first-hand account of the 2016 surgical strikes, by the Major who led them, is in *India's Most Fearless 1*. At the time of the publication of this book, the account is also being produced as a major web series.

pads. Over the next eight months, through a typically cruel winter, the Indian Army nursed no illusions that the surgical strikes would put an end to the flow of terror. It was only a question of *when* the infiltrations would begin again. And the six dark, pixellated figures that flickered on the soldier's thermal imager confirmed that it was finally time.

At his unit's Alpha company base in Uri's Rustom post, thirty-four-year-old Maj. Preetam Singh Kunwar listened carefully to the incoming message. No one from the unit had had any rest for weeks. Ten days ago, intelligence networks had begun to buzz[2] with the possibility of infiltration just over 10 km away, along the LoC in the Chakothi area, where a parallel operation to intercept infiltrators had begun. The sighting of six more infiltrators near the Badori mountain proved one thing— the final week of May 2017 was about to see the first big burst of infiltrations since the 2016 surgical strikes.

'The surveillance detachment had observed five to six people,' Maj. Preetam says. 'They alerted everyone over radio and line communications about these men. Our CO, Col. Samarjit Ray, then asked the same surveillance detachment at the LoC to check if the people were still there and whether the suspicious movement was still on. The jawan had first observed that movement at 10 a.m. on 23 May and reported it to everyone in the area. He was ordered to stay put and keep his sights fixed on those men.'

At noon, another report came in from the surveillance unit at the LoC. The thermal imager was able to paint a slightly more detailed picture than before, but it still couldn't give precise information on what the six men were doing. The soldier reported that it wasn't possible to tell if the men were carrying weapons or equipment, but that they were repeatedly

---

[2] See Chapter 9.

moving in and out of a *dhok*, a mountain hut constructed by local shepherds.

'*Bande uss dhok ke andar-bahar jaa rahe hain* (The men are going in and out of the dhok),' the soldier reported.

At 12.30 p.m., the CO, Col. Samarjit, sent out a communique to the unit's company commanders, including Maj. Preetam at Rustom, ordering them to be on maximum alert. The Rustom post sat on top of a snow-blown mountain, with other posts on spurs on the ridgeline ahead of it.

'Col. Ray told us there was something fishy on the other side,' Maj. Preetam says. 'He told me to gather as much information as I could about what was happening across the LoC.'

The first person Maj. Preetam called was a source across the LoC, asking him to check as quickly as possible who the six men in the shadow of the Badori mountain were, and to get back with any information about their movements. While he waited, the officer also activated his local sources and intelligence units, a network significantly weaker in that area than the one commanded by the Pakistani Army and the terror groups it armed. Asking a source in PoK about the specific presence of a certain group of men *always* ran the risk of alerting the other side and compelling them to change their plans.

An hour later, Maj. Preetam's PoK source called back, telling him that the group of men had crossed the Haji Pir pass, a veritable gateway for infiltrators looking to cross the LoC and make their way into the Kashmir Valley. Half a century earlier, in the 1965 war, Indian forces had captured this very pass, situated between the Poonch and Uri sectors, but had returned it to Pakistan following the Tashkent peace negotiations the following year, a decision that is widely rued even today by Army veterans and strategic thinkers. It is through the Haji Pir pass that the lion's share of Pakistan-sponsored terrorist infiltrators reach the LoC.

'I was told that a group had crossed the pass, but there was no way of knowing what their intent was,' says Maj. Preetam. 'This intelligence input was corroborated by other agencies and my people on the ground on both sides. We didn't need to wait long for another input to come in.'

That night, the surveillance unit at the LoC kept its thermal imagers trained on the location they had been monitoring. For the second time that day, the six figures emerged from their mountain dhok. The moment the new input was relayed back to base, the CO felt all doubts vanish about what was afoot. Immediately, he sent out orders to tighten the entire anti-infiltration grid, which included a sprinkling of forward bases and patrolling teams that would now step up their readiness to the maximum level.

'We set up extra anti-infiltration posts, sent out more patrols and placed more ambush units out there than usual,' says Maj. Preetam. 'There was little sleep that night. Every man in the sector wondered if this was going to be the infiltration that would break the eight-month lull that followed the surgical strikes.'

The following morning, 24 May, CO Col. Ray arrived at the Rustom post. In the operations room, he met Maj. Preetam and another young officer, the commander of the battalion's Ghatak[3] platoon. On a laptop screen, Col. Samarjit showed the two officers a video clip from the thermal imager used by the surveillance unit the previous day.

'The CO asked me what I thought,' says Maj. Preetam. 'I said those people had no business being in that area in the month

---

[3] Every Indian Army infantry battalion has a 'Ghatak platoon', comprising men trained in special operations and reconnaissance. It is often this platoon that leads offensive operations as 'shock troops'.

of May, as infiltration groups usually assemble at their staging areas only after June, when the snow has melted. He agreed. He also informed us that fresh intelligence had come in indicating that the infiltrators could take any of a number of routes to cross the LoC. So all units in the area needed to stay awake at all costs.'

Already in a high-pressure stand-off on a daily basis, in May 2017, the Army units in Uri were being tested to their limits of preparedness. The loud political narrative in Delhi surrounding the surgical strikes had painted a picture of an Army that Pakistan messed with at its own peril. On the ground in Uri, soldiers and officers knew that the swell of public attention meant that it was even more imperative that not a single infiltration attempt be allowed to succeed. The daily dance of life and death at the LoC had become tinged with an element of prestige, fuelled by political pronouncements of how the Army had been given full tactical freedom to avenge any 'misadventures'[4] by Pakistan.

Maj. Preetam decided to leave his base and lead a reconnaissance party to the LoC. Before departing with eight men from his company, he again checked in with his network of sources, both local and in PoK. Everything he heard only confirmed what was, by this time, beyond any real doubt— that the six pixellated figures captured by the Army soldier's thermal imager the previous day were set to be the first terrorist infiltrators since the 2016 surgical strikes.

'I had been the company commander in that area for over eighteen months, and I knew it like the back of my hand, be it the terrain, the likely infiltration routes or even the phases of the moon,' says Maj. Preetam. 'I checked with my men and assessed which areas were vulnerable and needed to be

---

[4] The word 'misadventure' would be used by the Army Chief, Gen. Bipin Rawat, days after taking office in December 2016, and would be repeated several times the following year.

covered better. We had a shortlist of such patches, the nallahs that had to be covered to prevent infiltration. The entire 161 Infantry Brigade (under which the 4 Garhwal Rifles Battalion operated) was on high alert.'

At 3 p.m. on 24 May, Maj. Preetam and his squad rolled out from their post, arriving 4 hours later at a point 500 m from the LoC, immediately spreading out to get a wide view of that stretch of frontier.

'If the infiltrators were on the foothills of Badori, this was one of the routes they were likely to take,' says Maj. Preetam. 'There were two-three nallahs in this area. I had eight other men in my squad. We were fully equipped.'

The squad had arrived armed for a big fight. They carried AK-47 assault rifles, MGLs, UBGLs, night-vision goggles, grenades, Motorola radio sets and enough rations and ready-to-eat meals to sustain them for a few days. As the sun set, they pulled on their night vision goggles, which allowed them to peer across the LoC to a distance of about a kilometre.

'We had positioned ourselves on a spur to cover the maximum area possible,' says Maj. Preetam, who was using his night-vision goggles to keep a close watch on the two nallahs that flowed in across the LoC in the kilometre-long stretch that he and his men were now keeping tabs on. 'We sat there through the night on 24 May, but didn't see a thing.'

Switching between surveillance and quick naps, Maj. Preetam's squad watched the LoC till the sun rose. Early in the morning of 25 May, a radio message came in from the surveillance unit that had first detected the infiltrators two days earlier.

'Six persons detected leaving Badori foothills, moving towards LoC,' the message said. 'Movement being kept under observation to the extent possible.'

The foothills of Badori sat at an elevation that allowed the Indian Army to get a glimpse of the infiltrators. But once they began to move, folds in the mountainous path and patches of thick foliage soon engulfed them. They were last sighted crossing a nallah on their side of the border.

The Ghatak platoon had placed itself a short distance away along the LoC, on another likely infiltration route. Later that morning, Maj. Preetam, still scanning the stretch of LoC in front of him, called his CO back at the Rustom post.

'I asked for permission to move closer to the LoC,' Maj. Preetam says. 'I was now certain that I wasn't going to be able to detect anything from where I stood. I needed to get a better view of the LoC stretch we were monitoring. Col. Samarjit said it's fine, you and your squad can move ahead. But he had one major concern that he warned us about.'

Tracts of land close to the LoC on both sides are infested with landmines, planted over the years by both armies as a deterrent to attempts to cross the LoC.

'Move ahead if you have to,' Col. Samarjit said over his Motorola handset. 'But I'm telling you right now—I don't want any landmine casualties. *Operation shuru hone se pehle hi mine casualty ho jaaye* (Mine casualties even before the operation)—that's not happening. Make sure each of you treads the ground very carefully.'

Maj. Preetam and his men were wearing anti-mine combat boots, reinforced footwear with a multilayered composite armoured sole built to dissipate the blast of a landmine and spread it outwards. But these were heavy and didn't exactly lend themselves to rapid movement.

'The boots affect mobility, no doubt,' says Maj. Preetam. 'They definitely slow you down a bit and are nowhere near as comfortable as our regular combat boots.'

The landmines pose a real threat. The Army has maps of where these mines are, but the shifting of soil has meant that mines might have moved many feet over the years and could be lethal to infantry troops. A landmine had struck during the 2016 surgical strikes—a Lance Naik from the Special Forces suffered a serious leg injury in a landmine blast when his team was returning from across the LoC after the offensive operation. Two months later, in November 2016, a soldier of Maj. Preetam's own battalion had lost his left leg in a mine explosion in that very sector. Maj. Preetam had seen it happen.

'The minefield was an enormous challenge,' says Col. Samarjit. 'I made it clear I didn't want any blood spilt before the operation began. It hadn't even been a year since we nearly lost one of our boys.'

'I cannot forget that sight,' Maj. Preetam says. 'My men were also aware of that incident. It never left our minds. So when the CO kept telling us to beware of the mines, he knew what he was talking about.'

The dilemma that morning was real. Maj. Preetam needed to lead his men closer to the LoC while crossing, quite literally, a minefield. They also needed to do it as quickly as possible. And with only eight men on the squad, he couldn't afford[5] to lose a single man.

Wondering how they could safely cross the mined stretch in the shortest possible time, the solution presented itself in the form of a six-foot log they found lying on the descent from their position.

'My buddy soldier, Lance Naik Sukhpal Singh, saw that log and had a flash of inspiration,' says Maj. Preetam. 'It was

---

[5] No troops are dispensable in any sense, but in mission planning, an expected casualty count is realistically considered to enable pragmatic, objective-oriented decisions.

basically a simple improvisation. He picked it up and suggested that we use it as a bridge to navigate the minefield.'

Lance Naik Sukhpal quickly demonstrated what he had in mind to the rest of the squad.

'We will not put the other end of the plank on the ground but place it on a small boulder or a helmet—like a small makeshift bridge,' the soldier said. 'And then we will walk across it to ensure we aren't putting our feet on the ground.'

The idea was clever, though complicated. But with no other option, Maj. Preetam ordered the squad to begin the breaching operation.

Before they began, the officer placed an arm around Lance Naik Sukhpal's shoulder, reminded of the time they had stared death in the face together. In 2011, while deployed in the Keran sector of the LoC in north Kashmir, Maj. Preetam, then a Captain, had saved young Sepoy Sukhpal's life during an anti-terror operation. The act had only cemented an already close bond.

Early in the morning on 25 May, aided by the log, the Alpha Company of 4 Garhwal Rifles stepped into a minefield.

It would turn out to be a deeply tense operation, with each man on edge, sweating even at that altitude as they placed the log carefully, hoping it wouldn't be on a mine. The log allowed them to reduce the risk of a mine explosion significantly, but didn't eliminate it completely.

'The innovation ensured safety, but it really slowed down our movement,' says a soldier from the squad. 'We were moving ahead at a snail's pace, one man covering six feet at a time. We took almost 5 hours to cover around 350 to 400 m. We were very mine-conscious and took every precaution possible as we approached the LoC. The moment we got close to the LoC, Maj. Preetam started scouting for an area that would give us both good cover and an unrestricted view.'

By noon, having finally crossed the mined stretch, the squad took up its new positions. Only minutes after they had done so, Maj. Preetam's radio crackled.

'Spotted and in sight, two men in the nallah straight ahead,' said a soldier, who was looking through his binoculars from an elevated position.

The two men the soldier had spotted were about 75 m away across the LoC. Maj. Preetam had split his squad into four buddy pairs, each positioned 30 m apart to provide a 90-m breadth of cover.

'The moment he gave me that information, I immediately changed my position and moved closer to him,' says Maj. Preetam. 'This was the first time I was observing likely terrorists from such a close distance. I felt an adrenaline rush. I instructed my squad to hold fire, and not to open fire until I said so. We sat in our positions and observed them. The two men retreated a short distance.'

Maj. Preetam then called his buddy, Lance Naik Sukhpal, and told him that the two of them would be moving about 40 m closer to the LoC, descending from a spur. He ordered the remaining three pairs to remain in their positions.

'We took our chances without the log, since there was very little time, and because there's little possibility of mines sticking to a slope,' Maj. Preetam says. 'From our new position, closer to the LoC, we could see not only the two men who had been spotted a few moments earlier, but also three more. So there was a total of five, and three of them were carrying weapons. It was now clear that this was the same terror squad that had left the Badori foothills the previous night and was now approaching the LoC.'

Maj. Preetam radioed his CO. When Col. Samarjit was informed about the sighting of five likely infiltrators, he ordered Maj. Preetam and Lance Naik Sukhpal to retreat to their earlier position immediately. Gathering a QRT at Rustom, Col. Samarjit immediately departed the post and made for the LoC to join what was clearly turning into a major operation.

'Looks like they plan to take one of the three nallahs to cross the LoC and come to our side, Sir,' Maj. Preetam told his CO. 'I have a hunch about which nallah they will use, but we are watching all three to be safe.'

At 3 p.m., suddenly, the five terrorists retreated and disappeared from view. Col. Samarjit ordered Maj. Preetam's squad to retreat further, to the position they had taken up before crossing the mined area. Using the log once again, the eight men crossed back over the minefield in failing light, reaching their earlier position by nightfall.

'We took positions there and sat throughout the night—it was extremely cold,' says Maj. Preetam. 'Once again, we kept an uninterrupted watch across the LoC.'

The squad had been outdoors in the cold for over 48 hours since they left the Rustom post. At first light the following day, 26 May, Maj. Preetam took five men and used the log to return across the minefield for the third time, towards the LoC. Two men were left behind at the position from where the first two terrorists had been spotted in a nallah the previous day.

The hours rolled by, with no sign of the men who had been spotted 24 hours earlier. Then, at 4 p.m., three figures came into view.

'It was sudden and very clear—we detected three people about to cross the LoC with weapons and rucksacks,' says Maj. Preetam. 'They were barely 40 m away. We decided not to engage them because we were not certain how many of them there were.'

At 40 m, and undetected, Maj. Preetam and his men had a clear shot of the three terrorists and could have finished them right there. But the officer waited. And there was good reason.

Forty metres. Close enough to hear their footfalls splashing in the nallah.

'We could see only three men—but if there were six or eight of them, the remaining guys would retreat the moment firing began,' says Maj. Preetam. 'I wanted to wait until the full group showed itself. Maybe there were six or eight or ten. It was hard to say. If we opened fire at the three we could see, the remaining would easily escape. We didn't want that to happen. We would lose the element of surprise and alert the Pakistani Army post 500 m away. It would have been ready to provide firing support to the infiltrating group, which is their standard operating procedure.'

Once again, Col. Samarjit ordered Maj. Preetam and his men to fall back. The CO and his team had reached a nearby post, taking position to block another infiltration route.

'As we fell back this time, we decided not to use the log as we wanted to retreat quickly and without a moment to lose,' says Maj. Preetam. 'There was a sense of urgency and I decided to take that risk. I didn't have the luxury of wasting 4 or 5 hours. Once we crossed the minefield, we went straight to link up with our CO and his squad. We were now in a place that the infiltrators needed to cross, no matter which route or nallah they chose.'

The three infiltrators who had been spotted crossing one of the nallahs that day had stopped and waited after they crossed, observing the stretch of LoC ahead of them. Col. Samarjit's team had already blocked one of the nallahs.

'I told my CO that they have crossed the nallah, and now their intent is clear,' Maj. Preetam says. 'They are terrorists about to infiltrate. It was still hard to say which route they would take from where they had stopped, as there were two more nallahs ahead, giving them two more possible routes. But we had both covered now. All possible infiltration routes were closely guarded.'

The men of 4 Garhwal Rifles were now simply waiting for the full group to cross the LoC and reveal its true numbers.

'My buddy and I had taken position on one side of a nallah,' says Maj. Preetam. 'Our CO was on the other side. The complete area was covered. At around 6.30 p.m., as the sun was setting, our squad spotted five men. They were walking with weapons in one hand and walking sticks used for hiking in another. They had rucksacks slung on their backs.'

Five men. *Thirty metres away.*

Maj. Preetam radioed his CO. Col. Samarjit immediately told the squads that no one would open fire until Maj. Preetam gave the go-ahead as he was the closest to the approaching terror group.

As Maj. Preetam and Lance Naik Sukhpal fixed their night-vision-aided gaze on the five figures stepping slowly towards the LoC, another terrorist stepped into view behind them.

'So six were visible now,' says Maj. Preetam. 'The sixth guy was far behind. The first five men were walking 5 m behind each other, but the sixth man was about 20 m behind the fifth. I thought that the sixth man was perhaps leading another squad. There was a possibility that we were dealing with a dozen well-equipped and trained terrorists.'

It was perplexing. Why was the sixth terrorist walking so far behind the group? Again, Maj. Preetam told his squad not to open fire till all six men had crossed the second nallah. The men were climbing up a small spur. They would need to descend from it before crossing the second nallah that was covered by Col. Samarjit's squad. The moment they crossed the nallah, they would be in the field of fire of both squads.

They were being lured into a kill zone they had shown no signs of having noticed yet.

Just as the terrorist group entered the second nallah, the sixth man splashed through the water to catch up with the first five, now marching with a distance of 5 m between them. Crossing the second nallah, the terrorists stopped for a moment. And their postures completely changed.

'As we watched, they crossed the nallah and immediately assumed tactical postures,' says Maj. Preetam. 'They were now holding their weapons in an offensive posture, each of them scanning the area carefully, I could tell. Perhaps they knew that they could be ambushed any moment and were prepared to fight back. They knew they were in the danger zone now. They didn't know we were waiting for them, but they had clearly been trained not to let their guard down for even a moment.'

As CO of the unit, Col. Samarjit had his weapon pointed and ready. He didn't have a moment to think of anything else, but couldn't help remembering that this group of infiltrators were the first since the 2016 surgical strikes—it was beyond doubt that they weren't coming in for small kills. They had to be stopped at all costs.

'I remember thinking—these guys are looking to infiltrate, and then they'll recruit a whole lot more people and head straight for the Valley to create mayhem,' says Col. Samarjit. 'We had to stop them, no matter what.'

Maj. Preetam and his buddy soldier were in a position to cover an area between the two nallahs, where both merged into a single large stream. In the dark, with their night-vision goggles strapped on, the squad watched as the six terrorists stepped through the middle nallah into the Indian side, climbing a spur and passing Maj. Preetam and his buddy.

'The moment they crossed me and my buddy, I looked at my watch—it was 7.13 p.m.,' says Maj. Preetam. 'That was the time I radioed all squads to open fire.'

The men had been out there for two days, tracking the terrorists from a stone's throw away, but hadn't fired a single bullet so far. Finally, they had their orders.

'Every man was waiting for this,' says Maj. Preetam, who opened up his AK-47 with a furious burst of fire directly at the six terrorists. 'We had already decided that we would open fire when the first terrorist crossed the first buddy pair. And the other buddy pairs would engage when they saw them. All the infiltrating terrorists were now in the range of our men. They didn't get much reaction time because they were climbing up a spur when we opened fire. But remember, they were very well trained. Like commandos.'

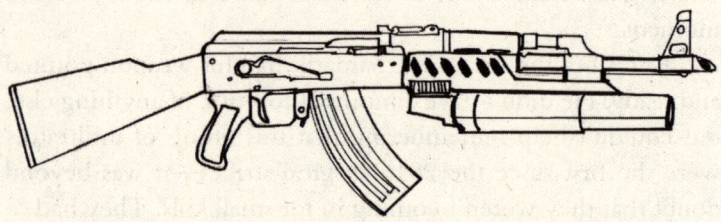

AK-47 with under-barrel grenade launcher

The six terrorists stopped in their tracks, taking cover, with two of them firing back in the darkness. The Indian squads were well concealed, so the two terrorists fired randomly.

Maj. Preetam realized he needed to move if this operation was to end well.

'From the place where I was firing and carrying out surveillance, I didn't have much cover,' he remembers. 'This was from where I was informing the other pairs about the number of terrorists that had passed me and my buddy. The plan was that the moment the crossfire started, I would change my location.'

The lead infiltrator and the man behind him were killed in the opening minutes of the ambush. The third and fourth terrorists scrambled to take cover behind a boulder, disappearing from the line of sight of Maj. Preetam and his buddy soldier. The fifth and sixth terrorists were being fired at by the other pairs. A full-fledged firefight had erupted by this time.

'Sukhpal and I crawled for about 25 m to take position on slightly higher ground to see if we could engage the two terrorists who had hidden behind the boulder,' says Maj. Preetam. 'We took cover behind a tree with a thick trunk, from where we spotted both of them. We could see the muzzle flashes of their rifles as they stuck their weapons out to fire at us.'

As the crossfire intensified, the sixth terrorist was spotted climbing up the hill towards Maj. Preetam and his buddy. He was crawling quietly, without firing, hoping to sneak up on the two and kill them at close quarters.

'*Uss ko dikh gaya tha ki maximum firing kahaan se ho rahi hai* (He could make out where the maximum fire was coming from—our position),' says Maj. Preetam.

The terrorist got to his feet and began running towards the officer and his buddy. Just as Maj. Preetam and the soldier began to move from their position, a grenade fired by one of the terrorists from behind the boulder landed a few metres away.

'I think he must have seen us as we also had to stick out our rifles from behind the tree trunk to fire at them,' says Maj. Preetam. 'The blast was deafening and for a few moments, I could neither hear nor see anything. The grenade exploded very close to us. Splinter *kaanon ke itne paas se gaye ki mujhe unki awaaz aa rahi thi* (The splinters were flying so close to my ears that I could hear them). I screamed out to Sukhpal to see if he was okay. He asked me the same question. We were both okay.'

The sixth terrorist, who had been running towards them, was shot down by another buddy pair who had seen him when he got to his feet. Three terrorists were now dead, with three left. At this point, Maj. Preetam left his cover briefly and stepped closer to the boulder that still hid two terrorists.

'From 15 m away, I lobbed a grenade at them,' says Maj. Preetam. 'They were both killed by that grenade. After throwing the grenade, I was still firing in their direction, and so was Sukhpal. Firing from behind the boulder had stopped totally.'

There was now only one terrorist left, and he was approaching Col. Samarjit's squad. Maj. Preetam quickly radioed his CO to tell him, asking him to throw grenades in his direction. Col. Samarjit's buddy soldier flung a grenade directly at the approaching terrorist, while the other squads fired towards him. The terrorist crumpled to the ground just as he was about to break into a run for the final stretch to the CO's squad position.

All the six terrorists were dead. They had been drawn into a kill zone and eliminated in under 15 minutes. The ambush site had been designed in such a way that there was no real place for the infiltrators to hide for any extended period. They had been, in effect, lured to a place from which there was no escape. From any one point, the terrorists were visible to at least one of the buddy pairs, allowing for two pairs of eyes to be fixed on them no matter where they tried to run or hide. By the time the assault rifles finally fell silent, the sky was overblown with clouds and it had begun to drizzle.

'I radioed my squad to check if everyone was okay,' says Maj. Preetam. 'My CO was doing the same. There were seventeen of us in that operation. My team of eight, which included me, and our CO and his QRT of eight soldiers. That area, in the shape of a bowl, was covered by these seventeen men. We enjoyed the advantage of height. The ambush site was chosen very carefully. And we got the results we wanted.

We were still very alert, because you never know—there might still have been some more people hiding somewhere. We told our surveillance detachment to keep the area, that section of the LoC, under watch to see if more infiltrator squads were coming. All our men were still behind cover.'

The squads then launched quadcopter drones equipped with thermal cameras for an airborne sweep of the area—a final effort to ensure they hadn't missed a terrorist or two. With the all-clear from above, the squads settled down on the mountainside to wait the night out.

At first light on 27 May, the men of 4 Garhwal Rifles began searching the ambush site to retrieve the bodies of the six terrorists. Standard operating procedure needed to be followed in every case.

'First, we took headshots from up close to make sure they were dead,' says Maj. Preetam. 'Terrorists trained by the Pakistani Army frequently use their bodies as booby traps. We use a rope and hook to move the bodies. We throw the hook hoping for it to catch on the body. If not, we go close to the body and anchor it and then pull the body from a distance of about 25 m. If there is a grenade under the body, it will explode.'

The search and retrieval of bodies continued till early in the afternoon. The men looked at each other, and then to their company commander as they completed the 'mopping up' operation, which involved special surveillance by two buddy pairs to ensure that the ambushers weren't ambushed by any more infiltrators.

'As a company commander on the ground, my top priority was that there should be not be a single scratch on any of the men I am leading,' says Maj. Preetam. 'This was a highly dangerous mission, and we needed to stay motivated as a team. Imagine the scout out front. He knows the area is infested with mines. But he follows my word because he trusts me. No mission is possible without the trust of your men. And it was that trust that allowed me to plan the mission in such a way that all of us stayed alive.'

With the mopping up operation complete, the squads finally received word that they could return to Rustom post.

'This was an operation that stood on a razor's edge,' says Col. Samarjit. 'This terrain was tricky. The minefield threat was real. The terrorists were very well trained. And we still managed to ensure there wasn't a single scratch on any of the men. It was a very clean operation, although Preetam and his buddy, Sukhpal, exposed themselves to a lot of risk during close-quarter surveillance of the infiltrators and by frequently relocating their ambush site.'

Back at Rustom post a few hours later, Maj. Preetam, Col. Samarjit and some of the team gathered in the operations room for a debrief about the mission and to take stock of what they had accomplished. If there was a unanimous takeaway, it was fire discipline.

'If you lose the element of surprise, chances are you will end up getting killed,' Maj. Preetam says. 'I made it very clear that no one will open fire till I order them to. Imagine you are observing two or three guys with weapons and they are a few metres away. You know that either they are going to kill you or they will get killed. That is the time you have to be in control of the situation. You cannot suddenly become trigger-happy and fire. Had we done that that day, we would have got only three of them. The remaining would have escaped. It is easy to forget how crucial and indispensable fire discipline is.'

He may have been the one issuing orders to his men to hold their fire, but Maj. Preetam would remember how it was his own momentary lapse of fire discipline that had cost him at least seven terrorists two years earlier, when he operated as part of the Army's 28 Division in Kupwara. An early burst of fire had sent a large group of terrorists scurrying back across the LoC, depriving the squad of a major encounter. The squad had waited 96 hours for the terrorists, but a few bullets fired too early had brought the operation to an end in minutes. A major infiltration had been foiled, but the terrorists had got away alive—certain to attempt another infiltration at a future date. That day, Maj. Preetam had decided he would never make the same mistake again.

The other major lesson from the debrief was the intricate detailing and design of the operation. The large number of moving parts—when and where the terrorists would cross, the number of terrorists, how well they would be armed, and whether they would split up after being challenged, to name a few. Luring them into a zone of no escape was therefore both critical as well as an enormous challenge. And with all of these difficult variables, there were still the minute levels of planning that needed to be built into the operation even as it rolled out. Fighting effectively at night, for instance, was a certainty, and the squads were armed for it. But they needed to take it a step further to face an infiltration squad.

'The moon was in its first quarter at the time, so there was not much moonlight,' Maj. Preetam says. 'It's almost pitch dark and an intense firefight is on. So how do I spot my buddy? We can talk over the Motorola radio sets, but what if I need to know his exact location? We had passive night-vision goggles, which emit an infrared blip when you press a button on the side. Those IR lights can be seen through passive night-vision sets only. You can see a red dot. So the moment I see a red dot, I know it's a man from my squad. This is how we knew who was where.'

There was the additional challenge of targeting the terrorists accurately in the dark.

'We were also carrying laser pointers with us,' explains Maj. Preetam. 'Say, your buddy has spotted some movement, but you haven't—he will direct you to that movement with the laser light. The fifth terrorist was killed with the help of a laser pointer. One of the buddy pairs directed the laser at him, and following the pointer, the CO's buddy lobbed a grenade and killed him. We had kept it simple. No one would laser in the direction of our parties' positions. *Jahan mein laser karunga, wahan bina soche ya time waste kare wahan fire karna hai* (Where I point the laser, just fire in that direction without thinking or wasting time). It was a simple plan. Once the target is painted by the laser, just fire or throw a grenade.'

The six terrorists killed that night were all Pakistani nationals from the LeT. Their rucksacks contained enough rations and ammunition for a major strike—bigger than the September 2016 attack at the Uri Brigade headquarters. The encounter had ended so quickly. The Pakistani Army post that had been expected to wake up and attempt to help the terrorists hadn't fired a single shot. It wasn't clear why, though it was likely that the post had let its guard down and was unable to focus fire before the encounter ended.

Back at the Rustom base that evening, the rest of the battalion was waiting for the return of the squads. Celebrations erupted, but had to be kept brief—the surveillance unit, still at the LoC, had radioed in, alerting the men to the possibility of more infiltration attempts. The alert would prove true the following day, but in a stretch of the LoC over 10 km west of the Rustom post's area of responsibility.

On a visit to the Kashmir Valley five days later, the Army Chief, Gen. Bipin Rawat, took a chopper out to the Rustom post to meet the men who had foiled the first infiltration since the 2016 surgical strikes. Rawat had played a part in the surgical strikes as Vice Chief at the Army Headquarters in Delhi.

On 14 August 2017, the government announced that Maj. Preetam Singh Kunwar would be decorated with the Kirti Chakra, the country's second-highest peacetime gallantry award. He would find out about it only the following day.

A few days earlier, in August, a soldier from the unit, Kuldeep Singh Rawat, had been killed by a Pakistani Army sniper not far from where the 26 May infiltration attempt had taken place. The entire unit had been drawn into the imperative for revenge, an impulse which, at the tactical level, doesn't normally escalate into a larger confrontation.

'We were focused on avenging our boy,' says Maj. Preetam. 'On the night of 14 August, I was out on a patrol in the same area. On 15 August, I began getting congratulatory calls for the Kirti Chakra. It was a good day. On the same day, we sniped three Pakistani Army men. That was a very special Independence Day. It was a doubly joyous day, more so because we were able to take revenge.'

'My boys were so aggressive that they did not allow the Pakistanis to retrieve the bodies,' says Col. Samarjit, describing an anger it is hard to imagine outside the confines of the bonds that functioning in a single Army unit can engender. 'If they tried to retrieve the bodies, my boys would fire at them. They gave up after trying several times. Finally, it happened only after their Director General Military Operations called his Indian counterpart.'

The Kirti Chakra citation would record that Maj. Preetam had shown 'great courage and valour' in the 'highest traditions of the Indian Army'. Like most heroes, the officer plays it all down.

'I don't think I did anything extraordinary,' he says. 'I signed up for this job and did what I am expected to do.'

Col. Samarjit, who was decorated with a Sena Medal for the operation, feels differently.

'That night, Preetam came very close to being awarded an Ashok Chakra, posthumously,' he says. 'That was the kind of risk he took to make sure we were able to kill all the intruders. The terrorists were barely metres away from him. No man in that operation would have been surprised if we had lost Preetam that night. His courage was extraordinary.'

Maj. Preetam's buddy, Lance Naik Sukhpal Singh, and Col. Samarjit's buddy, Havildar Brijendra Lal, received Sena Medals for gallantry in the operation.

Back home in Dehradun, Maj. Preetam's father, Narender Singh, a retired soldier from the Garhwal Scouts, couldn't have been happier about his son's mission. On the evening of 27 May, when Maj. Preetam called his wife, Megha—after being off the grid for three days—to tell her about the operation, she had been overcome with relief and had broken down. Her father had come running, his heart in his mouth. Taking the phone from his weeping daughter's hand, he exhaled when he heard the voice at the other end.

'The day the operation took place, I was very restless,' says Megha. 'I couldn't sleep the whole night. I guess it was intuition that not all was well with Preetam. Maybe he was caught in a difficult situation. When he finally called and told me he was okay, I just couldn't hold on, and I began to cry. My father then hugged me and said, "You should be proud of what your husband has done for the country."'

Megha Kunwar says that ever since her husband has been deployed in Kashmir, she sits in their home's puja room and meditates twice a day, morning and evening.

'I had spoken to Megha before the operation—I told her I had been given some extra responsibilities and I would be unreachable for a few days,' says Maj. Preetam. 'That's what I tell her when I am going out to conduct operations. When I came back on 27 May, I saw around forty missed calls from her and a barrage of WhatsApp messages on my mobile. You can tell your family that everything is going to be alright, but you can never really convince them. As we lie exhausted in our bunks, getting the deepest sleep so we can perform well the next day, our loved ones are sitting far away, having sleepless nights. That thought never leaves our minds in forward areas.'

'Every time he doesn't take my call, I am on the verge of a breakdown,' says Megha. 'One day, one of his unit officers told me, "Ma'am, don't worry if he doesn't call you, but pray *karo ki aapko unit se koi aur kabhi call na kare.*" That day, when he called me after the operation, he had dialled from a different number. I was praying that I hear his voice at the other end. I remember praying fervently till I heard his voice.'

'Of course, I feel pride every single day, but let nobody fool you about our fears,' says Megha. 'The fear I feel as the partner of an Army officer in Kashmir is a constant, daily, hourly fear. It never goes away. It is a constant companion. It is part of the deal.'

Their son, Abhiraj, now four years old, has been told about his father's mission.

'He always asks me, "*Papa, aapne chheh bhooton ko kaise maara* (How did you kill the six ghosts)?"' says Maj. Preetam.

'What struck me most about Kunwar was his unshakeable cool in a very tense situation,' says Col. Samarjit. 'His radio messages from a position just metres away from the terrorists were always conveyed in the calmest manner. There was not a hint of fear or anxiety in his tone. Through the firefight too, he didn't get even slightly jittery. It takes nerves of steel to hold yourself together like that. If he had not, both he and Sukhpal would have been lost. The terrorists were so close to them, perhaps not even 5 m away. He took an unimaginable risk. It was profoundly brave.'

At the time of this being written, the 4 Garhwal Rifles has completed its tenure in Kashmir and is being moved to the decidedly more tropical climes of Port Blair in the Andaman and Nicobar Islands, where there's a far lower chance of snow, and where starlight keeps even the darkest nights bright enough to see. Not surprisingly, though, Maj. Preetam isn't enthusiastic.

'I've got to be honest—I'm not really looking forward to it,' he says. 'I will miss Kashmir. That's where I wish to operate. It's the reason I joined the Army.'

At a ceremony at the Rashtrapati Bhavan where her husband was decorated with the Kirti Chakra in 2017, Megha sat in the audience sending up a prayer.

'I was thanking God from the bottom of my heart that I was attending this function *with* my husband,' she says. 'There were families who were receiving posthumous awards. I couldn't imagine being in that position. And yet, how difficult is it to really imagine? The unthinkable plays in our minds, invades our peace every minute of every day, doesn't it?'

As they walked out of the Rashtrapati Bhavan that evening and waited for their car to pick them up, Megha held Maj. Preetam's hand and whispered, 'Listen, enough. I don't want you to serve in Kashmir any more. Do you hear me? Enough.'

Maj. Preetam smiled and held her close.

14

'You Cannot Sustain Fear of Death'

Flight Lieutenant Gunadnya
Ramesh Kharche

Airspace over Tamil Nadu
6 February 2012

'Would you rather die in a big fireball? Or in a mangled mess? Take your pick.'

K-3060's co-pilot, Sqn Ldr Aditee Bhangaonkar, found herself smiling at the question. She turned to the flight's captain, sitting to her left in the cockpit, his words helping break the tension, even if only for a few moments. He chuckled into his headset, which was the only way the people in the cockpit could hear one another over the roar of the twin Soviet-built turbo propeller engines. Except, like his co-pilot, Flt Lt Gunadnya Ramesh Kharche knew that there was nothing even remotely amusing about their situation.

It was simple, really. The aircraft couldn't land. Not without ending up in either a deadly fireball or a mangled mess, as its captain had joked. And the realization had taken the crew of the Antonov whole minutes to digest.

As Gunadnya's nervous humour hung heavy in the cockpit, he manoeuvred the aircraft for a pass over the air base they had departed from 2 hours earlier. He needed a few moments to process the K-3060's terrifying situation. As he banked the aircraft sharply to the right, the base popped into view below them. The pilots and navigator saw it instantly—a large crowd had gathered on the apron near the air traffic control (ATC) tower. It looked like every soul at the air base had dropped whatever they had been doing that busy Monday morning in February 2012 and gathered near the runway. Right next to the tarmac, the pilots spotted three crash trucks, their lights flashing and engines running.

Two hours had changed everything.

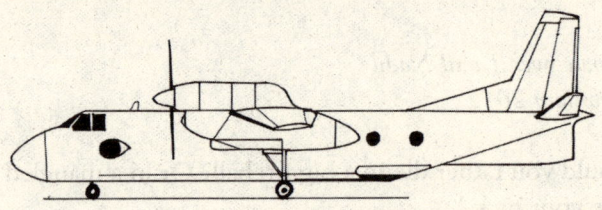

An-32 military transport aircraft

At 11 that morning, as the K–3060 had taken off from the IAF's Sulur air base near the ancient city of Coimbatore in Tamil Nadu, the pilots remembered noting what perfect weather it was for the sort of flying they were about to embark on. The Soviet-era Antonov An-32 transport plane had been its usual reliable self—rugged, rough and reassuringly loud. Apart from co-pilots Flt Lt Gunadnya and Sqn Ldr Aditee, the crew included navigator Sqn Ldr Saumitra Mishra and flight engineer Sergeant Shailendra Singh. As the aircraft roared off the runway, the pilots pulled the 20-tonne airplane into a gentle climb to 1000 feet, levelling out at that altitude. The crew from the 33 Squadron, nicknamed 'Soaring Storks', was on a 'low-level navigation mission' that day, which meant they would stay at a low altitude and practise a series of manoeuvres to test flying capabilities, navigation skills and, quite simply, to keep the complicated and skill-intensive daily business of operating aircraft in shipshape.

'It was a beautiful day. No clouds, bright sun, cool weather. Perfect for flying,' recalls Gunadnya about that February morning in 2012.

Cruising at 1000 feet, the crew had pointed the aircraft south, flying it out over a relatively low-population area, quite a distance away from the Coimbatore metropolitan area, which was to their west and which was Tamil Nadu's second largest city after its state capital, Chennai. Flight paths for military air missions are usually carefully chosen to avoid built-up and densely populated areas. It was especially so that morning, since the crew of K-3060 was flying low.

Flying low wasn't a problem at all. The An-32 has endeared itself to generations of Indian pilots for its forgiving toughness and stability even during risky, low-level flight missions. It is the An-32, after all, that transports the bulk of military personnel and material to forward bases in the north and the north-east, frequently flying through narrow valleys and navigating past lofty ridge lines. Some would say the An-32 is an ungainly sight, with its narrow fuselage and hulking shoulder-like engine pods. But to the men and women who fly them, there are few aircraft more cherished than the An-32.

The K-3060 may have been over two decades old, but it was still a reliable airframe that had given its current and past crews little cause for worry. It was already in line for an elaborate upgrade. Three years earlier, in June 2009, the Indian government had engaged a Ukrainian firm to give 105 IAF An-32s an additional fifteen years of service. This would be done by overhauling each aircraft from the inside out, fitting them with better engines, modern navigation and survival electronics, advanced new sensors and a superior cockpit. The aircraft would also receive better ergonomic and functional seats for the pilots and crew. The plan was to squeeze a valuable decade and a half out of a rugged fleet of planes by making them more modern, safer and easier to fly. The K-3060 was in queue for the big makeover, but that wouldn't stop normal, daily missions like the one it was on that morning.

'We had planned to fly for 3 hours. Everything went perfectly smoothly,' says Flt Lt Gunadnya. 'Until, of course, it was time to return to base.'

Their mission complete, Flt Lt Gunadnya ordered his crew to prepare for the return leg of the journey. The aircraft was eased out of its already low altitude to about 800 feet. It was gently descending as the pilots manoeuvred it into a return flight path towards Sulur. Over the radio, co-pilot Aditee notified air traffic control at Sulur that they were returning to base. With a clear flight path back, it was time to lower the K-3060's wheels. The An-32 has three wheels—two tough main landing wheels that pop out from the engine pods in the wings, and a smaller nose wheel that is mostly used to steer the aircraft on the ground. Gunadnya pushed a rectangular white button on the cockpit panel to trigger the clanking hydraulic sequence that would see the landing gear being lowered in a ballet of moving mechanical parts.

And that's when it happened.

'I waited for the three indicator lights on the cockpit panel to tell me that the landing gear has successfully been lowered. Three green lights in a triangle,' Gunadnya says. 'The nose wheel had lowered fine. Check. The right landing wheel had landed fine. Check. But the cockpit indicator for the left wheel glowed orange, not green.'

Gunadnya's first thought was that it could be an electrical malfunction. It was rare, but not unheard of, for cockpit indicators on old aircraft to return incorrect information owing to something as simple as a short circuit in the wiring. Perhaps the left wheel had indeed lowered successfully, but failed to trigger an electrical signal to the cockpit. Gunadnya quickly leaned towards the cockpit's side windshield, twisting backward while trying to check visually if the plane's left wheel was down. From where he sat, he couldn't tell, so he ordered Flight Engineer Shailendra to go to the flight cabin and look through one of the tiny porthole-style windows. Shailendra returned with the confirmation Flt Lt Gunadnya had been dreading. The entire wheel was indeed stuck in the engine pod in the wing, and had failed to descend.

Sqn Ldr Aditee suggested recycling the undercarriage—raising the wheels and attempting to lower them again, in the hope that all three would descend. It was aviation's equivalent of that old solution to a stubborn computer problem: reboot. In the event of a hydraulic fault, it was possible that the jammed wheel would come unstuck and descend without further trouble. With the wheels whining back up into place, the crew held its breath as Gunadnya waited 15 seconds before pushing the landing gear button again. The hydraulics whined again as the wheels descended into landing position. When Gunadnya turned around in his seat for a confirmation from the flight engineer, he only saw a 'thumbs down' gesture. It hadn't worked.

'We tried everything to lower the undercarriage. Nothing worked,' says Sqn Ldr Aditee. 'And finally we decided to leave the undercarriage in the same position. This was when we all finally came to know that we had to face this situation.'

'This was now a real in-flight emergency,' says Gunadnya. 'We immediately informed ground control that we had an undercarriage emergency and that we were looking at our options.'

Gunadnya and Aditee took a deep breath, looking to each other for a moment. Both knew that an emergency of this kind had never happened in the Air Force. Both also knew that with fuel and time running out, they needed to battle it the only way pilots really can—solve one problem at a time, then move to the next.

'Let's fix this,' Sqn Ldr Aditee said, still smiling. 'Let's take this one step at a time and fix this.'

Eight years earlier, Aditee had dreamed of flying fighters, but had to settle for transport aircraft when she graduated from the Air Force Academy near Hyderabad, since rules stopped women from flying fighters. When journalists chatted with her about whether women would ever be allowed to fly fighters in the IAF, she had confidently declared, 'Only a matter of time.' She was dead-on. Five years later, the very same military flying academy that had no choice but to place Aditee in an An-32 cockpit would graduate its first batch of women fighter pilots after the government changed rules, opening fighter cockpits to women pilots for the first time.

She wasn't flying fighters like she had hoped to, but eight years into service as a Short Service Commission officer, Sqn Ldr Aditee was still living the dream, plunged, alongside the much larger number of male pilots, into the daily rigours of transport aviation, a stream she would quickly learn brought with it its own set of tough challenges. That morning, as she smiled to herself in the cockpit of the K-3060, wondering how she and the pilot were going to land a 20-tonne airplane without all its wheels down, she thought about that old familiar yearning for the thrill of aviation. And like the men on board with her, Sqn Ldr Aditee was fully aware that this could be her final landing.

Except, they couldn't land.

'Looking back, I know that death was, in fact, a sure possibility,' says Sqn Ldr Aditee. 'But there was a lot of communication going on among all of us. All the crew members, the ATC, our CO too had come on RT. We were discussing everything. This kept the atmosphere in the cockpit very positive.'

On the ground at Sulur, the base commander, Group Capt. R.C. Mohile, had rushed to the ATC tower to be in touch with the pilots of K-3060. He knew that it was a highly experienced crew in the aircraft now circling the base. As a proficient An-32 pilot himself, it would have seemed logical for him, an older and more experienced pilot, to guide the aircraft's crew out of their increasingly critical situation. But years of established protocol stipulate that beyond exception, the captain of the aircraft in distress has the final word on the course of action. Group Capt. Mohile could talk to his crew, but he knew that the ultimate decision on what to do lay with young Flt Lt Gunadnya. As it happened, it was an enormously difficult decision.

'How are the hydraulics doing?' Mohile's voice came in through the radio. Gunadnya quickly updated the senior officer on the aircraft's mechanical status.

The An-32's Russian flight manual offered an emergency solution to the landing gear problem. It recommended that if any of the wheels failed to deploy, standard operating procedure would be to retract all wheels and land the aircraft on its belly. Before the actual touchdown, the crew needed to turn off the aircraft's engines 50 feet above the ground. This, according to the manual, was to minimize the possibility of an impact fire.

'So the manual told me to descend slowly, pull back my landing gear, switch off my engines at 50 feet, and kiss the ground,' Gunadnya says. 'Ground control reminded me that that's what the Russians had recommended in such a situation.'

Except, Gunadnya remembers thinking, if he switched off the engines at 50 feet, it was certain to him that the crew would all be dead that day.

'Switching off both the engines simultaneously in the air at 50 feet seemed like a very bad idea,' he says. 'The aircraft would go out of control without engine power. If you switch off the engines, the An-32 becomes one big asymmetric mass of metal and it would become very difficult to control it. And everybody in it would die. It would be practically crashing the aircraft. This was my overwhelming sense.'

A thirty-year-old pilot was basically about to toss a carefully crafted aviation safety manual into the dustbin. With voices from ground control reminding him of the manual's prescription for the mid-air predicament, Gunadnya finally spoke out. Politely, but firmly, he told his crew as well as ground control that he disagreed with what the Russians had prescribed.

Now, An-32s are trusty Cold War-era birds with strong bones and a reassuring ruggedness. That was one of the reasons why the safety manual actually recommended a last-ditch belly landing. The aircraft were tough. But it was the final 50 feet of descent with engines off that had led Gunadnya to dismiss the idea entirely. In those 50 feet, headwinds beating against an unpowered aircraft could easily throw it off course, sending it careening dangerously into the scrub.

The alternative that Gunadnya was proposing was even scarier.

'Everybody was trying to sort out the issue, but the aircraft was in my hands,' says Gunadnya. 'So I announced that we would land the aircraft on one wheel, and hold it up on that wheel for as long as possible.'

If the aircraft were a tricycle, Gunadnya had basically recommended hitting the runway while doing a 'wheelie' on just the right wheel.

A few seconds of silence followed from ground control as the words sunk in. To many in the ATC tower at Sulur, the young pilot had just ditched an admittedly dangerous but prescribed recovery mechanism in favour of what looked like certain suicide. Gunadnya turned to his crew, searching for the tiniest semblance of affirmation. Almost imperceptibly, the others nodded back.

Let's do this.

Gunadnya's 'wheelie' plan was an enormous risk. It would require a magical level of cooperation from several hundred mechanical parts of the aircraft landing precariously on a single wheel. And there would be no practice run. The crew of the K-3060 would have one single chance to get it right. The reason for that was another sharp irony in the air that day. The pilots had enough fuel to last at least 20 more minutes—plenty of time in the normal course to practise-land a few times, if necessary. But for the sort of landing that Gunadnya had decided on, they wouldn't have a second chance. Therefore, they needed to empty their tanks to minimize the possibility of ending up in a big fireball on the ground. The An-32 doesn't have a mechanism to jettison or eject fuel in emergencies, so the crew needed to keep flying until their tanks were nearly dry before they attempted their one shot at a single wheel landing.

'There was consensus in the cockpit that we preferred not to die in a fuel fire,' Gunadnya says. 'Anything but a big fireball.'

The K-3060 needed to stay in the air for another 18 minutes. Gunadnya put the airplane in a long circle around the base, burning off fuel and waiting till the tanks were nearly dry. From the cockpit window, the 'X' of Sulur's two intersecting tarmacs glimmered under the hot afternoon sun. From the crowd that had gathered near the ground control tower, it was clear that the entire base had heard about the K-3060's unprecedented predicament.

From the ground, the base commander and a few other officers used binoculars to look up at the An-32 as it circled the base.

'So he advised us what we should do and we told him what we planned to do,' says Sqn Ldr Gunadnya. 'We did not get a yes or no, but he's in a position where he can't give a yes or a no. My ground controllers were also awesome. They reassured us about everything. When we were flying, we could see that the whole Air Force station had gathered on the tarmac. All the safety services were lined up for any eventuality. It was in a controlled procedural manner, the way it is in the Air Force.'

'It was a very disturbing sight,' says an officer from the Soaring Storks squadron who was at the ATC tower that afternoon. 'Hearing about it from Gunadnya and team was one thing. Seeing the aircraft fly by with just one landing wheel deployed was a shock to everyone who was watching. Many prayers went up in that moment.'

Over those 18 minutes, Gunadnya wondered quietly to himself if he was making a mistake. He had overruled the rule book and chosen to ignore gentle words of advice from ground control, including those of the boss of the Sulur base. This was his call. His crew completely supported his decision. But he knew the decision was his alone.

'Doubt can be a paralysing thing when you're in a cockpit. It can disorient you,' says Gunadnya. 'But there was something inside me that was just totally sure that the path I had chosen was the way to go.'

Sqn Ldr Aditee glanced over at the pilot.

'In my mind, I was very clear that we were not going to die,' she says. 'For me, that day, death was never a possibility. It may be hard to believe and I do not have any explanation for this feeling. But I was very confident. And it was not that I was trying to show calmness.'

'My co-pilot and I shared a very high comfort level. She trusted me completely,' Gunadnya says. 'A crew's lives were in my hands. I couldn't for a moment show fear or hesitation. If I did, it would spread to Aditee and the others. When the emergency first surfaced, Aditee had smiled and said, "Okay, this is new! Let's fix it!" We were a great team.'

Gunadnya worked to control any semblance of agitation that could have surfaced in his manner. Barely a few years into his service, he had been thrust into a difficult leadership position that involved keeping everyone's emotions in control in the face of likely injury and possible death.

'The fear of death cannot persist. You can't sustain fear of death for an hour,' says Gunadnya. 'That old cliché is true. I saw my whole life pass before my eyes when we reached the point of no return. The others were stressed when I looked at them, because they didn't have anything in their control. We're from the same squadron, but in the aircraft, what can you do except trust your captain? We're all personally connected. My crew reassured me and I reassured them.'

Aditee's voice cut through the pilot's thoughts.

'This is going to work,' she said, sensing Gunadnya's quiet tension. 'Time for final approach.'

The aircraft was now on its final minutes of fuel. The margin for error at this point was zero. The K-3060 made a loose final turn from the south-west and headed back towards the base, this time in slow descent from 800 feet. As the aircraft flew lower and slower, turbulence buffeted its frame unequally, since it was slicing through the air with one landing wheel down and another up.

'We turned for the final and then gave a call for the final approach,' says Gunadnya. 'When we descend from 1000 to 800 feet, and then to 500 feet, we give a call. We usually say, "Three greens, landing clearance requested." We only had two greens and one amber. The ground control noted our request for the record. They had never received a request to land in such circumstances before.'

In 40 seconds, the Sulur runway loomed into view. Descending to 100 feet, the crew lowered the aircraft's flaps to slow it down to about 250 kmph, the slowest it could fly for a landing. Gunadnya tightened his fists around the flight control stick and turned to look toward his co-pilot. Aditee gave him one final nod, then took the aircraft down to 50 feet.

'My speed was 250 kmph. What I didn't realize at the time was that I was flying with mixed controls. And now, if one wheel is not there, that means that amount of drag is not there. So, to maintain the same flight profile, you have to apply rudders. There's no autopilot capable of helping you land in such a situation,' Gunadnya says.

There was another reason for Gunadnya's decision not to switch off the engines at 50 feet, as prescribed by the Russian flight manual.

'We needed to reduce the aerodynamic forces on the aircraft at slow speed and increase drag as much as possible,' Gunadnya says. 'We needed the engines on because we needed to "fine" the propellers to increase the aircraft's drag. Therefore, there was no question of switching off the engines as prescribed by the Russian manual. I needed the advantage of drag. Switching off the engines would kill my drag, and the momentum of the aircraft would have thrown it out of control. Drag was our friend. And we needed every ounce of it.'

On any other day, Gunadnya would have had the luxury of touching down further along the runway, even near the middle, since the An-32 didn't need a lot of length to slow down after landing. But on this day, if there was one overwhelming certainty, it was that once K-3060 touched down, it was only a matter of moments before it would veer off the runway. If Gunadnya landed the aircraft too far down the runway, he ran the risk of the aircraft running off the track and ploughing into air base buildings or even the parked aircraft to the left. If the aircraft was going to veer off to the left, it needed to be into the open area filled with red earth near the beginning of the runway.

Twenty feet above the ground, Gunadnya and Aditee could see the full runway before them, the flashing lights of crash vehicles far in the distance. A final prayer went up before the crew took the K-3060 to the tarmac.

'Landing was the most difficult part,' remembers Sqn Ldr Aditee. 'We had spent a lot of time in air. And we had rehearsed the entire procedure. Now there was no turning back.'

From ground control, a video camera filmed the aircraft's approach. Officers present remember the cold silence that descended on the tower and the crowd below as the K-3060 descended the final few feet to the runway on its right wheel. The crew's singular aim was to bring the aircraft down on the tarmac's centre line with the aircraft's nose up and to control it against all the physical forces that would begin to act on it.

Seconds later, the wheel touched the tarmac. A collective gasp went up from the crowd as the aircraft shivered unsteadily for a moment.

'We could not believe what we were seeing. For a few seconds, the aircraft was on one wheel. It was moving at 250 kmph on one single wheel,' says an Air Force Sergeant who had been deployed at the far end of the tarmac to film the landing.

In the cockpit, Gunadnya was leaning forward, his face a grimace of concentration as he fought to hold the aircraft steady while it screamed down the runway, dangerously balanced on one wheel.

'As we roared forward on one wheel after touchdown, there was immediate danger,' Gunadnya says. 'The cantilever beam of the An-32's wings could have turned viciously 180 degrees. It could have broken or toppled over. Anything could have happened in those very tense moments. I was using every bit of my strength to steady the aircraft as much as possible.'

The aircraft screamed down the runway. But as drag began to kick in and shake the aircraft dangerously off its path, Gunadnya shouted to his co-pilot to 'withdraw' the propellers.

'Two seconds after touchdown, I felt myself losing control of the aircraft,' says Gunadnya. 'I could not hold it because as the air speed forces reduced, the ability to control it was gone. Like trying to steer a stationary car. Aditee had withdrawn the propellers. There was nothing more we could do now but hope for the best.'

Control of the aircraft was out of Gunadnya's hands now. For 5 full seconds, the K-3060 shuddered precariously down the runway on its right wheel as the crew battled to keep it steady and slow it down. And then it began to tilt dramatically, as predicted, to the left.

'Kharche was flying very well. He was doing a great job. I was following him on controls,' says Sqn Ldr Aditee.

Gunadnya remembers, 'I had briefed the whole crew that the moment we crash, the flight engineer will jump out of the emergency exit first, followed by the navigators, Aditee and, finally, me. The emergency exit on the An-32 is a small square opening, large enough for barely one person.'

As the aircraft veered with a roar off its course and off the runway into open shrubbery, Gunadnya sent up one final prayer. If the remnants of fuel did spark a fire now, the crew would have just seconds to get out of the aircraft—if the fire didn't cut off their exit entirely, that is. The crash tender trucks would still take at least 2 minutes to safely approach the aircraft.

His hands still clutching the aircraft controls even if there was nothing more he could do, Gunadnya watched in horror as the K-3060 careened to its left, its wing tip hitting the runway in a flash of sparks, turning the aircraft now violently to the left and into the muddy scrubland off the tarmac. The aircraft thudded on for 50 more feet before it finally came to a stop.

Gunadnya immediately received word on his radio that three crash tenders were speeding towards the aircraft. Both Flight Engineer Shailendra and a voice from ground control confirmed that, miraculously, there were no visible fires. Switching off all the aircraft's systems, Gunadnya gave the order for his crew to exit the aircraft from the emergency slot.

'Nobody said a word for the first few minutes,' Gunadnya says. 'We exited the aircraft in silence. Then we walked a few feet away and turned back to look at the plane. It was sitting on its belly, its left wing touching the ground.'

Almost immediately, Gunadnya's mobile phone rang. It was his wife, Shruti. She had chosen to remain in Mumbai at their family home while her husband was deployed with the Soaring Storks in Sulur. Gunadnya heard an anxious voice as he took the call.

'Where are you? I've been calling you for so long. Your sortie should have been over by this time. What happened?' Shruti asked. Her husband chuckled gently into his phone, assuring her that all was well. It would be days before Shruti would find out what he had managed to do that afternoon.

After a medical team quickly attested that the crew had sustained no injuries, a pair of jeeps picked up Gunadnya and his team and transported them to a briefing room with the base commander and other senior squadron staff. Every last person at the Sulur base was sure they had witnessed history that afternoon. But they also knew how close the crew of the K-3060 had come to a fiery death. The Air Force would need to dive deep into every second of what happened in the air. It would check every bolt, every wire on board the K-3060. And it would investigate Gunadnya's own leadership in the cockpit to judge whether his decision to disobey a flight manual was worthy of praise or punishment. But for the rest of the day, after the crew had logged their flight and changed out of their olive-green overalls, ruthless process would have to make way for some welcome revelry.

A large cake was wheeled into the mess that night as the base celebrated the 'rebirth' of the crew of K-3060.

'We were elated. There's a tradition that if you survive an aircraft accident without injury, you celebrate a birthday. A rebirth, really. We cut the cake and there was a huge party that night. There was joy, but also a sense of disbelief that we were safe. It was unreal for both us and those who had received us,' Gunadnya says.

Gunadnya's deadly tribulation in the air may have ended miraculously without damage to life or limb—or even much damage to the aircraft itself—but when the sun rose the following day, he was about to receive a heavy reality check in the form of orders for a Court of Inquiry. This was due process, but on the line would be the young pilot's fledgling career. Gunadnya had been in service just over five years.

An Air Marshal-rank officer, the chief of the Southern Air Command headquartered in neighbouring Kerala, landed in Sulur early on the morning of 7 February, kicking into motion an accident investigation the likes of which the IAF had never handled before. For a whole month, Gunadnya would be summoned to provide detailed testimony and a defence of his actions in the cockpit. He would be cross-examined and interrogated so the Court of Inquiry could draw up a complete, detailed picture of what had happened on board the K-3060. Ten days into the inquiry, Gunadnya was permitted to fly again.

'They basically had to judge whether I was right or wrong. They needed to fix some learnings for the future. And I was at the centre of what happened. Thankfully, the Court of Inquiry gave me a tentative green light to fly after ten days,' says Gunadnya. 'This did not mean that I was in the clear, by any means, but simply that as a pilot, I was allowed to continue my daily flying duties.'

The Court of Inquiry interviewed the other members of Gunadnya's crew, officers who manned the Air Traffic Control at Sulur and senior officers from the Soaring Storks squadron. When the inquiry was complete, Gunadnya waited for their judgement.

'I was expecting trouble. I had deviated from established procedure, but hadn't violated any norms. It was complicated. I ignored the Russian procedure and did what I thought was best. I explained my decision to the best of my abilities. But I was aware that deviating from tried-and-tested procedures is no joke. And no matter, a very serious view can always be taken. I could be in a lot of trouble,' Gunadnya says.

Three days after the Court of Inquiry ended, Gunadnya was headed out of the flight operations room towards the flight line for another sortie, when he received a call on his mobile phone. It was Group Capt. Mohile, the base commander.

'Good news and bad news, Kharche,' Mohile said. 'Which one do you want first?'

'Bad,' Gunadnya said. 'Always bad first, Sir.'

'Okay, the Court of Inquiry has concluded that the decision you took in the cockpit wasn't the correct procedure,' Mohile said.

Gunadnya stopped in his tracks, thunderstruck. Were they really going to bring the axe down on him? After all that?

'And there's good news after that, Sir?' he asked.

'The Court has recommended no action against you. You're in the clear. Congratulations. Have a safe flight,' Mohile said. The Court of Inquiry, as it happened, would make highly nuanced conclusions about the K-3060 incident.

What Gunadnya didn't know was that other wheels had quietly begun to turn, between the Sulur base, the Southern Air Command and the Air Force Headquarters in Delhi. The K-3060 incident had become a hot topic of conversation among aviators and the IAF's senior leadership. By the end of March, despite the Court of Inquiry ruling that the captain of K-3060 hadn't followed established procedure, Flt Lt Gunadnya's name was recommended for a Shaurya Chakra, India's third-highest peacetime gallantry award.

'The Air Force still maintains the diktat that the flight manual is our Bible and that I should have followed it. The rationale was that every other pilot hereafter will have doubts about procedure and instead apply his or her own mind. The Air Force was very mature in making a decision of this kind, and yet awarding me for the decision I took,' Gunadnya says. 'They saw merit in what I had done in my situation. They deemed it worthy of reward.'

Weeks later, Gunadnya would be part of an IAF team visiting Ukraine to inspect the An-32 upgrade programme that had been contracted in 2009. While there, he would get a chance to speak to the Antonov company's test pilots.

'We had a philosophical talk. They agreed with me, mostly,' says Gunadnya. 'They said manuals exist, sure, but when you sit in an aircraft, your decisions are over and above the flight manuals—always. They told me what I did was probably the better way to do it.'

Sqn Ldr Aditee has since retired from the IAF and now lives in Pune. The video of that landing remains on her phone, and she looks at it often.

'Most of the time, aircraft emergencies require immediate action. Like in the case of engine fire or engine failure. The crew has to complete the actions without even thinking. To prepare ourselves for such emergencies, we practise a lot,' she says. 'However, our situation was very different from this. We had to spend a lot of time in air. We had to burn maximum fuel and land with minimum weight. We had never really practised something like this. In this type of scenario, all the crew members had to be in the situation for longer duration. If someone panics, there are chances of the situation going out of control. So it was very important to maintain calmness in the cockpit. Kharche and I kept flying alternately to avoid stress and fatigue. We kept talking with each other and with the ATC. I kept asking our flight engineer to keep calling out engine parameters, to keep carrying out visual checks. The navigator kept watch on fuel and aircraft endurance. This was just to keep everyone occupied. This was a very different and unique situation we were put in. But in the end, teamwork is the most important thing.'

Meanwhile, there was the question of damage to the aircraft itself. The K-3060 had taken an obvious beating as it had landed on one wheel and veered off the runway into muddy no-man's land. But the aircraft had sustained surprisingly little real damage. Test pilots from Kanpur flew down to Sulur to conduct an elaborate check of the aircraft's stress points and mechanical systems. With some minor tinkering at Sulur, the aircraft was flown to Kanpur for more detailed structural testing. Months later, it returned to service.

'It's an amazing aircraft. I actually flew that same aircraft later when I was deployed for a UN Peacekeeping mission in Africa,' Gunadnya says.

Six months later, on India's Independence Day, Flt Lt Gunadnya, who had by then been promoted to the rank of Squadron Leader, arrived in Delhi to receive his Shaurya Chakra gallantry decoration from the President of India. The citation would recount his actions on board the K-3060:

Flt Lt G.R. Kharche carried out a safe emergency landing with only starboard main wheel and nose wheel. The aircraft landing was controlled in the most courageous way and the aircraft suffered minimal damage. After landing, the action to evacuate the personnel on board were carried out most efficiently under his supervision. Flt Lt G.R. Kharche displayed qualities of exceptional courage and extreme professionalism during handling of such a grave emergency inspite of limited experience. His actions not only saved the lives of personnel onboard but also recovered the aircraft with minimal damage. For the display of exemplary courage and composure in handling an extremely rare emergency, Flight Lieutenant Gunadnya Ramesh Kharche has been awarded Shaurya Chakra.

'I can never forget that there were two things that happened that day. We brought the K–3060 down safely,' says Gunadnya. And the other?

'And the K–3060 had also brought us down alive.'

Acknowledgements

In the nearly two years since we wrote the first *India's Most Fearless*, it has become plainly clear that this is no longer just a book series. When we hear from readers across the world every day, we're constantly reminded that these stories have a life of their own beyond the written page.

For this second book in the series, we wish to thank, above all, the chiefs of our three armed forces—General Bipin Rawat, Air Chief Marshal Birender Singh Dhanoa and Admiral Sunil Lanba. They saw how readers received the first in the series and were even more generous to our inquisitive pursuits with the second. Our deepest thanks especially to Major General A.K. Narula and his team at the Additional Directorate General of Public Information (ADGPI), Group Captain Anupam Banerjee and Captain Dalip Kumar Sharma.

Our gratitude to Swati Chopra, our editor at Penguin Random House, who, like us, has been drawn irreversibly into the lives of the heroes you read about in this book.

Our families—Torul, Tavleen, Aryaman, Agastya and Mira—and our parents who literally provide the world necessary to write such a book.

Thank you to our dearest friend Sandeep Unnithan, a powerhouse of information and talent that we are so proud is now a part of India's Most Fearless.

To officers and soldiers, Special Forces men and doctors beyond counting who helped in ways small and big, who cannot be named for reasons operational. You know who you are and what you've done for this book.

Our thanks to our readers who've received *India's Most Fearless* with an unending store of love and praise—even now, we forward your messages of love to the heroes and their families. We can't wait to do the same with this book.

It is to that latter—the heroes and their families—that we give thanks above all. Nothing continues to astonish us more than the graciousness, generosity and modesty of men and women who've accomplished things beyond common understanding. As we tell the heroes or those they leave behind, the best way we can say thanks is to never stop writing.

INDIA'S MOST FEARLESS 3

New Military Stories *of* Unimaginable Courage and Sacrifice

SHIV AROOR | RAHUL SINGH

EBURY
PRESS

An imprint of Penguin Random House

EBURY PRESS
USA | Canada | UK | Ireland | Australia
New Zealand | India | South Africa | China | Singapore

Ebury Press is part of the Penguin Random House group of companies
whose addresses can be found at global.penguinrandomhouse.com

Published by Penguin Random House India Pvt. Ltd
4th Floor, Capital Tower 1, MG Road,
Gurugram 122 002, Haryana, India

Penguin
Random House
India

First published in Ebury Press by Penguin Random House India 2022

Copyright © Shiv Aroor and Rahul Singh 2022

Illustrations by Sandeep Unnithan

10 9 8 7 6 5 4

ISBN 9780143451112

Typeset in Bembo Std by Manipal Technologies Limited, Manipal
Printed at Replika Press Pvt. Ltd, India

www.penguin.co.in

MIX
Paper | Supporting
responsible forestry
FSC™ C016779

EBURY PRESS
INDIA'S MOST FEARLESS 2

Shiv Aroor is an editor and anchor with India Today television, with experience of over a decade covering the Indian military. He has reported from conflict zones that include Kashmir, India's North-east, Sri Lanka and Libya. For the latter, he won two awards for war reporting. As a political reporter on TV, he was also recently awarded for his coverage of the 2018 state elections in his home state, Karnataka. Aroor also runs the popular award-winning military news and analysis site, Livefist, on which he frequently tells the stories of India's military heroes.

Rahul Singh has covered defence and military affairs at the *Hindustan Times* for over a decade, in a career spanning twenty years. Apart from extensive and deep reporting from the world of the Indian military, including several newsbreaks that have set the national news agenda over the years, Singh has reported from conflict zones including Kashmir, the North-east and war-torn Congo.

PRAISE FOR *INDIA'S MOST FEARLESS 1*

'If our nation is to be stronger, the stories of these heroes must spread far, wide and never be forgotten.'—General Bipin Rawat, Chief of the Army Staff

'India's new generation will find it impossible to forget these riveting military tales.'—Air Chief Marshal B.S. Dhanoa, Chief of the Air Staff

'Inspirational accounts of extraordinary courage, fearlessness and heroism of our valiant soldiers under extreme adversity—a must-read.'—Admiral Sunil Lanba, Chief of the Naval Staff

EBURY PRESS
INDIA'S MOST FEARLESS 3

Shiv Aroor is senior executive editor and anchor at India Today TV, and has covered the Indian military for nearly two decades. He has reported from conflict zones that include Kashmir, India's North-east, Sri Lanka and Libya. For the latter, he won two awards for war reporting. Shiv also founded the popular, award-winning military news and analysis site Livefist.

Rahul Singh is senior associate editor at *Hindustan Times* and has covered defence and military affairs for over two decades. Apart from extensive and deep reporting from the world of Indian military, including several newsbreaks that have set the national news agenda over the years, Rahul has reported from conflict zones including Kashmir, India's North-east and the Democratic Republic of Congo. The first story on the ongoing India–China border conflict appeared under his byline in 2020.

*To the heroes who fought and died in the Galwan Valley,
their families and those who lived to tell the story*

Contents

Introduction   ix

1. 'I Had Never Seen Such Fierce Fighting'   1
   The Galwan Clash of June 2020

2. 'I'm Not Leaving This Cockpit'   41
   Group Captain Varun Singh

3. 'Get as Close as Possible before Dark'   63
   Subedar Sanjiv Kumar

4. 'Where the Hell Is His Leg?'   89
   Squadron Leader Ishan Mishra

5. 'I Have Never Touched Anything That Cold'   117
   Major Vibhuti Shankar Dhoundiyal

6. 'They Were Coming to Behead'   141
   Sepoy Karmdeo Oraon

7. 'Whatever We Do, It Has to Be Now'   171
   Lance Naik Sandeep Singh

8. 'The Seas Will Break Your Ship'   197
   Captain Sachin Reuben Sequeira

9. 'This Time Holi Will Be with Blood'   231
   Major Konjengbam Bijendra Singh

10. 'You Have Five Minutes on the Seabed'   267
    CPO Veer Singh and Commander Ashok Kumar

Acknowledgements   293

Introduction

At the time this book goes to print, it has been more than two years since the world's largest modern military standoff began in the icy hellscape of eastern Ladakh. As the two armies remain massed in war-like numbers against each other across frosted plains, frozen lakes and mountains that disappear into the clouds, the standoff between India and China is far from over. As the carnage and close-quarter suspense that defined the first few months of the conflict turn into a slower, calmer but insidiously permanent confrontation, much of what happened has lain hidden in the storm of propaganda, claims and counterclaims.

The incident that sent the unprecedented Ladakh standoff exploding into the minds of millions across India took place on 15 June 2020, on a desolate riverbend in Ladakh's Galwan River valley. Months of coverage uncovered the reality of a terrifying clash between Indian and Chinese soldiers. But the complete story of what happened that night has never been fully told.

And that's why we believe the book you are about to read is so important, not just to the legacy of the India's Most Fearless series, but in terms of giving our countrymen first-hand accounts of the biggest operations and incidents involving the Indian military. This third edition of the series begins with a first-hand account of the Galwan incident by officers and men of the 16th Battalion of the Bihar Regiment, who fought the Chinese that night and survived. Hair-raising as it was, the account proves the many shades of grey that define combat, a far cry from the black-and-white certainties they seem to us from all this distance away.

The Indian Army permitted us generous access to Colonel Ravi Kant— the officer who succeeded 16 Bihar commanding officer Colonel B. Santosh Babu, who was killed in action in the clash and awarded the hallowed Maha Vir Chakra for uncommon courage and leadership—as well as other soldiers from the battalion, allowing us to present the only first-hand account of what happened that night.

The account not only answers many questions but sets the record straight on a central bone of contention, as you will discover. The ghosts of Galwan will continue to haunt a valley that sees a tentative truce amid a larger standoff that is far from over. And the questions remain over whether an expanding hostile force at India's doorstep is a new status quo.

The lead-up to the publication of *India's Most Fearless 3* also saw the Indian military hit by one of its worst peacetime tragedies. On 8 December 2021, a helicopter crash in the Nilgiri Hills of Tamil Nadu killed the country's first chief of defence staff, General Bipin Rawat, an officer who, as chief of the army staff, was an impassioned supporter of this series, having also launched the first book. We were in constant touch with General Rawat during the course of writing this book, and he was looking forward to its release. The last conversation we had with General Rawat about *India's Most Fearless 3* was on 4 December 2021—four days before the Mi-17V5 crash—when he complained that we were taking too much time to bring out this book. He will be deeply missed.

The sole survivor of that horrific helicopter accident was a decorated young fighter pilot, Group Captain Varun Singh, whose face would become iconic across media in the days that followed. His story captivated the nation as he fought to stay alive for a week. Tragically, he didn't make it. But if a man lives on through his actions, then Group Captain Varun is as vibrant as ever in the first-ever account you will read here of his handling of a heart-stopping midair incident that won him a Shaurya Chakra.

Those are the two accounts that kick off what has been a heart-wrenching third volume of this series. You will also, for the first time, read a first-hand account of Operation Randori Behak, a fearsome Special Forces encounter in the mountains of Kashmir's Keran sector that stunned the country, which had just gone into a national COVID-19 lockdown.

You will also read, in the words of his wife, Nitika, the haunting tale of young Major Vibhuti Shankar Dhoundiyal, who was killed in action fighting terrorists in the aftermath of the February 2019 Pulwama terror attack. The tragedy persuaded Nitika to drop her corporate job and join the Indian Army. In her account is embodied the voice and grit of the many proud, grieving military families who fade completely out of view each year in the wake of gallantry awards and official recognition.

From the forests of India's North-east to the depths of the Arabian Sea, from the Line of Control to the unresponsive cockpit of a Sukhoi fighter flying at less than 50 feet, the as-yet-untold stories you're about to read here feature some of India's most fearless military personnel in their finest hours.

When *India's Most Fearless* began three books ago in 2017, we didn't think it would become a living, breathing flame that our readers would never allow to go out. If these stories inspire, thanks must go solely to the men and women we write about. When we began to write about them five years ago, it was because we believed these stories were too compelling not to share. It is, then, into the hands of the families of those we write about, their brave units and to you, the reader, once again, that we commit these ten true stories.

Delhi   Shiv Aroor
June 2022   Rahul Singh

1

'I Had Never Seen Such Fierce Fighting'

The Galwan Clash of June 2020

Even above the loud, steady roar of the Galwan River, he heard the thundering footfalls. The sound of over a thousand men reverberating through the darkness, amplified by the tunnel effect of a narrow valley flanked by steep rising mountains on both sides. Peering into the black void beyond Patrol Point 14, lit only a few metres forward by hand-held torches, the reality of those sounds dawned on Havildar Dharamvir Kumar Singh of the Indian Army's 16 Bihar infantry battalion. He clenched his eyes briefly shut to soak in every vibration. When he opened them again, he knew that the huge horde of men advancing towards his position was not marching.

They weren't even jogging.

They were sprinting.

'There were less than 400 of us,' says Havildar Dharamvir. 'We would soon discover that the number of Chinese Army soldiers running towards us was maybe three times that. We had been fighting smaller numbers of Chinese for two hours before that. But this was their main force. The all-out assault that the Chinese side was launching against us.'

An all-out assault.

Unarmed, as stipulated by decades-old protocol between the two armies, Havildar Dharamvir quickly glanced around at the soldiers with him. Even in the darkness he could tell their expressions. A curious mix of determination and fearlessness, but tinged with an edge of foreboding.

As the soldiers steeled themselves, rallied by their commanding officer and a group of younger officers, Havildar Dharamvir knew what lay ahead would need every ounce of strength the smaller force could muster. But it also made one particular man in the team even more crucial.

A non-combatant with a white suitcase.

Wading through the group of soldiers with him, Havildar Dharamvir emerged on the banks of the gushing Galwan, right where he had last seen the man he was looking for now.

With a big, unmistakable red 'plus' sign painted on to his parka, Naik Deepak Singh wasn't standing. On his knees, his suitcase open with bandages and bottles of tincture, he was crouched over what appeared to be a small group of injured men, all groaning in the darkness. Three were Indian soldiers being administered first aid.

The six other soldiers receiving emergency ministrations from the young Indian Army medic weren't Indian soldiers. They were Chinese Army personnel. Two People's Liberation Army (PLA) officers and four jawans.

'They are badly injured. They need to rest,' Naik Deepak said before Havildar Dharamvir could ask. An hour earlier, the injured Chinese soldiers had been left behind by their retreating force. Naik Deepak, the young nursing assistant, had been summoned to Patrol Point 14 by his commanding officer two hours earlier. Not he, not Havildar Dharamvir and not his commanding officer knew then how crucial his crouched figure would be in the events of that night.

'Is that your blood?' Havildar Dharamvir bent down over Naik Deepak, inspecting a gash just above the nursing assistant's right eyebrow.

'It's nothing. A piece of rock hit me. It's superficial. *Main theek hoon* [I am fine],' said Naik Deepak as he finished bandaging one of the Chinese soldiers, a young man whose face was covered with streams of blood from a head injury.

A short distance behind, at a point where the north-flowing Galwan River abruptly bent westward, Colonel Bikkumalla Santosh Babu, commanding officer of the battalion, had been alerted to the sounds of the Chinese advance. As he began to summon reinforcements and rally his much smaller force to face the arrival of the much larger Chinese advance, one thing was certain to him. No matter what transpired next in that desolate, ravine-like valley at 13,000 feet in Ladakh's Himalayan heights, history had already been made with blood and bone that day.

As word of the lethal Galwan Valley incident shocked the world at 12.21 p.m. the following day, most would see it as a spontaneous flare-up that had ended a healthy forty-five-year run of zero fatal casualties on the India–China frontier. But waiting in the darkness on the banks of the Galwan River the previous night, Naik Deepak and Havildar Dharamvir knew that nothing, including that advancing horde of Chinese soldiers, was unplanned.

* * *

On the morning of 4 May, Naik Deepak was on COVID-19 inspection rounds at the unit barracks in Leh, the capital of the new union territory of Ladakh, when he heard the booming voice of a senior soldier make the announcement.

'Orders received to proceed to KM-120,' Subedar S.R. Sahu said loudly as he poked his head into the dormitory-like hall lined with beds and lockers for the soldiers of 16 Bihar. As a subedar, he was the second senior-most junior commissioned officer (JCO) of the unit. Apart from being a combatant, his seniority also saw him charged with the upkeep and well-being of younger soldiers and men.

'*Aa gaye hain?* [So, they've come?]' Naik Deepak asked, bemused, while he conducted an examination of a soldier with a sore throat. Subedar Sahu smiled, continuing past the dormitory to the other barracks to alert the other soldiers that they would be rolling out soon. The usual bustle of the barracks settled into a quiet rhythm of movements typical of an Army unit that has just been ordered to proceed to an operational area. Boxes and equipment were loaded, bags packed, vehicles fuelled, phone calls made.

Far away in a village near Rewa, Madhya Pradesh, that afternoon, Rekha Singh, a middle school mathematics teacher, was taking a scheduled class. Her phone had been set to stay silent in class, except for calls from one person. And the tone that rang out was the distinct thrum of an incoming video call. Excusing herself, she left the classroom, accepting the call as she did so. A mess of pixelated jerks quickly took the vague shape of the man she had married less than six months earlier.

'I won't have a phone signal for a few days,' Naik Deepak said. '*Aage jaana hai* [We have to proceed to a forward area].'

Rekha tried to read her husband's expression through the glitchy, blocky video, but could only barely make out that bemused half-smile she had come to know well. She also knew not to prolong conversations, given how delicate the signal always was when he made a video call. She quickly muttered the usual hurried plea to take care, stay warm and to call her as soon as he could next. Staring into the phone after the call ended, she repeated her words as a text message and sent it. And with a quiet sigh, she returned to her waiting class.

In Leh, her husband had begun to restock his white first-aid suitcase with fresh rolls of cotton wool, combat bandages, painkillers, suturing kits and several bottles of emergency medication.

KM–120 is an Indian Army post on the western bank of the Shyok River on a highway that connects the Leh area with the more forbidding northern Ladakh. As the crow flies, KM–120 is roughly 85 km from Leh. By road, winding through craggy mountains and cold deserts, the distance is significantly more, about 220 km.

'We knew why we were being sent,' says Subedar Sahu. 'Ahead of Patrol Point 14, Chinese troops had advanced, and they had to be stopped. After getting that order, we moved with men and equipment and reached KM–120 at 1 a.m. on 5 May.'

Later that day at KM–120, the mood was calm and focused, but there was an eerie disquiet in the air. Word had spread that a violent brawl had broken out between Indian and Chinese troops on the north bank of Ladakh's Pangong Tso, the world's highest saltwater lake nestled in the mountains at more than 14,000 feet, and about 130 km south of the Galwan Valley. It was on this precise piece of land that Indian and Chinese troops had previously brawled in August 2017—a clash caught on video, clips of which quickly went viral across the media and Internet. The men at KM–120 heard that the new clash in that same area, involving the Indian Army's 17 Kumaon battalion and a Chinese patrol, was much more violent, with grievous injuries sustained on both sides.

The 5 May clash at Pangong Tso would later be regarded as the trigger point for what rapidly became a wider and far more escalatory standoff. As it happened, the most immediate fallout of the Pangong brawl was about to register in the dim valley that Naik Deepak, Subedar Sahu, Havildar Dharamvir and seventy-two other men of 16 Bihar were about to rumble into.

'It was a bleak, freezing morning on 6 May as seventy-five soldiers moved in vehicles from KM-120. We crossed the Shyok River and advanced eastward into the winding, narrow Galwan Valley to a patrol base just short of Patrol Point 14,' says Subedar Sahu.

With his white suitcase and boxes of medical supplies, Naik Deepak and his senior medic quickly set up a medical inspection (MI) room at the patrolling base—standard operating procedure for combat medics arriving in an operational area. In the brown tented structure with internal heating about 20 metres from the Galwan River, it was equipped with stretchers, gurneys and oxygen supply equipment. Bunking down for the night at the patrolling base, Naik Deepak couldn't have known that his services would be needed the very next day.

At noon on 7 May, shockwaves from the Pangong brawl reached the Galwan Valley when a small group of soldiers headed by 16 Bihar's second-in-command, Lieutenant Colonel Ravi Kant, ventured across from Patrol Point 14 to confront a Chinese patrol that had clearly strayed past its usual limits.

'*Ek chhoti jhadap thi. Baaton baaton mein hathapai ho gayi* [There was a small brawl. But it soon turned into fisticuffs],' says Subedar Sahu. 'There were injuries sustained on both sides. Lieutenant Colonel Ravi Kant was also injured.'

The men were brought to the MI room, some with serious injuries, but none in any danger. One of the injured men, barely conscious but lucid, remembers Naik Deepak administering first aid to the wounds on his arms and neck.

'He was talking to me casually, as if I was sitting up and totally okay,' the soldier recalls. 'He was talking about the weather and asked whether I would be calling my family that day. Mundane stuff to maybe divert my mind from the injuries I had sustained. As I was wheeled away in the stretcher to be taken back to KM-120 for treatment, I remember seeing Naik Deepak. He was smiling and said, "*Ek-do din mein waapis aana* [Come back well in a day or two].'"

The brawl on 7 May might have been brief, but it was only a precursor to what lay in store.

'There was anger among the ranks that night,' remembers Havildar Dharamvir. 'Our second-in-command had been injured. He spoke to all the men to ensure that they remained focused and calm. But all of us, each and every man, knew that the situation was deteriorating thanks to the hostile action from the Chinese side.'

After the 7 May skirmish, Lieutenant Colonel Ravi Kant dashed to Durbuk to report details up the chain of command to the mountain brigade headquartered there. The details were serious enough for the brigade commander to arrive at the Galwan Valley the very next day.

On 9 May, the notion that the Chinese Army, already deployed in unusual numbers close to the Line of Actual Control (LAC) between both sides, had settled into a pattern of aggressive provocation was clear. As the brigade commander arrived in the Galwan Valley for a first-hand assessment of the situation, Indian and Chinese troops violently clashed in an entirely different part of their common frontier—north Sikkim. And as with the brawls at Pangong Tso and Galwan Valley thus far, the clash at Naku La in Sikkim also involved large numbers of troops and injuries on both sides.

A delicate calm descended on the Galwan Valley at Patrol Point 14. Troops and patrol parties on both sides held their positions for a month. But it was clear that the situation had transformed completely into something that could escalate very rapidly.

'Our patrols would go up to Patrol Point 14 and the Chinese troops would also come up to that point from their side. The friction between us was growing steadily,' says Subedar Sahu.

After a series of meetings between officers from both sides across the friction points in the Galwan Valley, Pangong sector—and eventually also in the Gogra Post–Hot Springs area of eastern Ladakh—it soon became abundantly clear that the situation needed a high-level intervention from the senior military brass.

So, on 6 June, the Indian Army's Leh-based Corps Commander Lieutenant General Harinder Singh and his Chinese counterpart, Major General Lin Liu, met for the first time at the desolate border personnel meeting (BPM) hut in the Chushul sector south of the Pangong Lake.

The BPM hut, used regularly by Indian and Chinese officers to hold protocol border meetings, is situated at a site drenched in history. Sitting at the western mouth of the Spanggur Gap, it is close to the famous Rezang La, a narrow mountain pass through which in the 1962 Sino–Indian war, a huge horde comprising thousands of Chinese troops attacked and, after losing huge numbers of their own, ultimately overwhelmed a much smaller 'last stand' by a company of the Indian Army's 13 Kumaon battalion. A memorial to the Indian men who stood and fought against impossible odds, while still inflicting substantial damage to the marauding force, sits on a low ridge less than 5 km away, and is regarded as hallowed ground in the annals of Indian Army war valour. It is also a reminder of the brute strength the Chinese were willing to unleash in their quest for military objectives.

Far removed from the violent clashes that had necessitated the 6 June meeting between the corps commander-ranked officers, the first round of military talks between India and China to cool border tensions turned out to be a very cordial affair— positive enough for the Indian government to issue a statement saying:

> Both sides agreed to peacefully resolve the situation in the border areas in accordance with various bilateral agreements and keeping in view the agreement between the leaders that peace and tranquillity in the India–China border regions is essential for the overall development of bilateral relations. Both sides also noted that this year marked the 70th anniversary of the establishment of diplomatic relations between the two countries and agreed that an early resolution would contribute to the further development of the relationship. Accordingly, the two sides will continue the military and diplomatic engagements to resolve the situation and to ensure peace and tranquillity in the border areas. (Source: Indian MEA Statement on 7 June 2020.)

'The corps commanders had talked, but things kept escalating 7 June onwards,' says Subedar Sahu. 'At Patrol Point 14, the Chinese soldiers came right up to a makeshift helipad on our side of the LAC and even set up tents there. One of our teams confronted them, and after some heated discussions, we were able to send them back. But they were not ready to retreat beyond Patrol Point 14, which is what we consider as the LAC.'

Subedar Sahu was summoned back to KM-120 to oversee combat logistics ahead of a situation that was clearly in freefall.

Over the next week, from 7-14 June, four separate meetings between local commanders in the Galwan Valley would take place, but the Chinese troops and officers refused to budge from Patrol Point 14. In fact, they were reinforcing their position with more tents and equipment on that triangular piece of land circumscribed by the bend in the Galwan River.

'We were deployed in our area and the Chinese side simply refused to listen to any requests not to escalate the situation,' says Havildar Dharamvir. 'Our commanding officer went up to Patrol Point 14 and spoke to his PLA counterpart, telling him that the Chinese soldiers should move back as they were on our (Indian) territory. But they were adamant that we should retreat as the area was theirs. We told them we will not go back, and the Chinese said they will not return either.'

In his bunk at the patrolling base on the night of 9 June, Naik Deepak phoned Rekha.

'He was in an unusually talkative mood, asking me about all kinds of things. He didn't usually do that,' says Rekha. 'I thought maybe he was making up for not calling for a whole week since the time he arrived in the Galwan Valley. I wasn't hard on him. I knew it was a tense situation there. But I did hope he would tell me more. I wanted to share his anxiety, but he did not seem anxious. He seemed upbeat and cheerful. I told him that I was going to my parents' place, and that the next time he called, it would need to be on my sister's phone since my signal behaved erratically there. Before he disconnected, he said yes, please go to your parents' home, I will come and fetch you in a few weeks when I come home on leave.'

Dawn on 15 June was like on any other day that week. But there was one addition to the Chinese position at Patrol Point 14 that was seen by the Indian side as a clear attempt to provoke and escalate the situation. The Chinese Army had set up an observation post (OP) at the river bend.

Colonel Santosh Babu, the commanding officer had had enough. Summoning his top men to an operations room at the patrolling base, he made it clear that the Chinese could under no circumstances be permitted to maintain an OP. It was bad enough that they had squatted with tents and equipment at

Patrol Point 14. But a position from where they could keep an eye on Indian Army movements and provide warning alerts to Chinese positions on their side was quite simply enemy action.

'On 15 June, I was at KM-120 with all the troops and equipment, and was coordinating logistics support. I got a message at around 3 p.m. from Colonel Babu to send seventy-five troops to Patrol Point 14. I had eighty-one troops with me. I arranged seven vehicles and sent seventy-five men to Patrol Point 14 at around 3.30 p.m.,' says Subedar Sahu.

When the seventy-five reinforcement troops arrived at the patrolling base about a kilometre from Patrol Point 14, Colonel Santosh Babu and a clutch of young officers from the battalion were present, including Captain Soiba Maningba. The base was bristling with an operational frisson typical of an infantry unit on the threshold of doing what it was trained for.

'At 3 p.m., we received the orders we were waiting for,' says Havildar Dharamvir. 'Colonel Babu said that a group would be proceeding towards the Chinese position, and that their tents and OP were to be dismantled and their soldiers to be pushed back from the area. We were fully pumped up and ready. We decided that we would go as a battalion . . . *josh aur jazbe ke saath* [in high spirits and morale] . . . and evict the Chinese soldiers from there. Our CO saab was leading from the front, and we said, "Sir, we are all with you on this mission."'

Naik Deepak and a team of combat medics were dispatched from the patrolling base to the MI room closer to Patrol Point 14 that had been set up the previous month.

'Before we rolled out, I saw Deepak leaving with his suitcase,' says one of the soldiers. 'He was loading additional boxes of medical supplies into a vehicle. It struck me as odd at the time that he was carrying so much. One of the men even asked him, "*Itna kyon leke ja rahe ho* [Why are you taking such a large amount of supplies]?" I guess he knew more than we did.'

Colonel Babu and his team of officers and men proceeded towards Patrol Point 14. Emerging from their vehicles, he and the others wasted no time, walking straight up to the site where the Chinese had set up their tents and OP. Spotting the approaching Indian team, a group of Chinese soldiers, led by their commanding officer, Chen Hongjun, began approaching from their side.

'Our CO had fully motivated us to evict the Chinese,' says Havildar Dharamvir, one of the soldiers in the group now face to face with the Chinese.

Colonel Babu addressed his Chinese counterpart, his voice sober, reassuring, yet firm.

'You and your men go back and we will also go back,' he told the Chinese officer.

The reaction from the group of Chinese soldiers was instantaneous and couldn't possibly have been more belligerent—they physically pushed the Indian soldiers. Worse, they dared to push Colonel Babu.

'*Unka behaviour bilkul theek nahi tha. Woh dhakka—mukki karne lage* [Their behaviour was not right, they started to push and jostle us],' says Havildar Dharamvir. He too was pushing back amid an escalating scuffle. 'Even after the pushing started, Colonel Babu was speaking in a calm manner, trying his best to defuse the situation.'

'It doesn't work like this,' Colonel Babu said, his voice momentarily raised. 'I will go back, and you will also go back.'

But every man knew in their hearts that it was too late. No soldier will tolerate his commanding officer being physically assaulted under any circumstances.

'It is like watching your parents being assaulted. Would you stand back and watch? Of course not. Nobody would,' says a soldier who was present.

When the Indian Army men pounced on the Chinese with their bare hands, things rapidly escalated. Chinese soldiers began picking up rocks and flinging them with full force at the Indian troops. Some Chinese soldiers separated from the group and ran up rocky outcrops on the mountain slopes in order to land rocks more accurately from a point of advantage. The 'no firearms' protocol prevailed, but the crossfire of brick-sized rocks was no less injurious. One large brawling group split into smaller groups of Indian and Chinese men attacking each other with their fists and stones.

After several minutes of stone pelting and fearsome hand-to-hand combat, the Indian men prevailed.

'*Bilkul* face to face *tha*,' says Havildar Dharamvir. 'But we finally managed to push them back. *Humne unko peeche dhakel diya* [We pushed them back]. They ran away.'

But in the chaos of the clash, something unthinkable had happened. Six Chinese Army personnel, including their commanding officer, Chen Hongjun, and his interpreter, Chen Xiangrong, were so badly injured that they hadn't been able to retreat with their teams. They remained with the Indian side.

'There were two or three Chinese officers and four–five Chinese jawans *jinko hum ne baitha kar ke rakha* [who were with us]. We told them, "Please do not attack like this, you must return to your area,"' says Havildar Dharamvir.

Colonel Babu knew he had an extremely delicate situation on his hands. The injuries to the Chinese men in his care meant they could only be stretchered back to their side. And the way things were at that moment, a temporary truce seemed practically impossible.

'Get the medics, get Deepak immediately from the battalion aid post,' Colonel Babu ordered, after trying to speak to the injured Chinese men. 'Ask Deepak to meet us at Patrol Point 14. These men need urgent first aid.'

'Colonel Babu was clear that the injured Chinese men needed to be treated quickly,' says Colonel (then lieutenant colonel) Ravi Kant, the unit's second-in-command.

Grim, but determined, the Indian commanding officer and his men walked back along the Galwan River to Patrol Point 14 carrying the injured Chinese men with them. Exhausted from the fighting, but energized by the action, the men proceeded to smash the Chinese observation post and tent structures. Colonel Babu rallied the team. Deep down he knew that the evening was far from over.

As if to confirm his fears, a flurry of stray rocks came hurtling down from the slopes 50 feet above. One rock struck an Indian soldier, Naik Arun, in the chest, throwing him off his feet and injuring him badly.

'I was right there when he was hit,' says Havildar Dharamvir. 'We did not have any stretcher to carry him. Somehow, I lifted him and carried him to the rear to a relatively safer place. Then I gave him some water to drink. Arun told me he was finding it hard to breathe because of the chest injury. Seeing his condition worsening, another jawan and I decided to carry him to the MI room for treatment. I was ordered to drop Arun at the MI room and return immediately, since the fighting was clearly still on. The Chinese CO and his men were all there on the ground. The situation was very volatile.'

Light was fading fast as Havildar Dharamvir and another jawan carried Naik Arun back to the MI room. Near the *nallah*, they ran into Naik Deepak, who was hauling his suitcase and hurrying towards Patrol Point 14.

'Deepak said that the CO had called him to the scuffle site as Indian and Chinese soldiers were injured. I told him I was coming from the spot and requested Deepak to attend to Naik Arun. Deepak quickly examined him and felt it was best to take Arun to the MI room, which was only 100 metres away. A medic was there along with an ambulance and oxygen support if required. Deepak said it was crucial for him to reach the scene of the fighting as ordered,' says Havildar Dharamvir. 'The CO's order was paramount for Deepak, and he wanted to reach the scene of the skirmish as quickly as possible,' adds Subedar Sahu.

'Naik Deepak moved ahead to reach where the fighting was on, and I somehow managed to carry Arun to the MI room. Then I rushed back to the scene of action where stone-pelting had intensified. The first thing I saw was Deepak treating the injured Indian and Chinese soldiers. He was giving them first aid to save their lives. Two of the Chinese men were more seriously injured than the others. He treated them first,' Havildar Dharamvir recalls.

Subedar Sahu was at KM-120 and in constant touch with the Indian soldiers at the centre of action to cater to their needs at that delicate time.

'I was told the first thing Deepak did when he arrived at Patrol Point 14 was to administer first aid to the Chinese commanding officer Chen Hongjung. He also gave first aid to two or three other Chinese PLA officials who were seriously injured. After that he treated our own Indian soldiers,' says Subedar Sahu.

'The Chinese interpreter was almost unconscious from a bad head injury and was being treated and roused by Deepak,' says Colonel Ravi Kant. 'He was barely speaking, that was the main concern. Colonel Babu's priority was to first get him sorted, since the interpreter was the one who could still translate and maybe turn the situation around quickly, or at least help ease the escalating situation.'

The Chinese soldiers pelting stones could clearly see Naik Deepak bandaging and helping the injured Chinese men at Patrol Point 14. To make it even clearer, a young Indian major got on his megaphone to communicate to the Chinese soldiers on the heights as well as in their rear positions that Indian medics were treating the Chinese commanding officer and five other personnel.

'The Chinese had seen in broad daylight that Deepak was bandaging and helping their men,' says Colonel Ravi Kant. 'Every one of their soldiers saw it and recognized that we were helping their men. The trouble was that the interpreter wasn't coming to his senses. He was the fulcrum of how we understand each other. He was crucial in such a volatile situation, and he wasn't gaining consciousness. He was the only man who could make the other side understand that we wanted to calm things down, not escalate. Deepak was trying his best. It was a race against time. Everyone saw that.'

As Naik Deepak gave the injured Chinese personnel some oxygen, a rock came like a frisbee out of the mountainside, splintered on the ground next to him, a piece of it striking him on the forehead, knocking him back. The Indian major with the megaphone once again warned, this time angrily, that the Chinese were targeting a medic administering first aid to injured Chinese personnel.

'Naik Deepak refused to stop, despite being injured,' says Colonel Ravi Kant. 'He was functioning beyond his call as a nursing assistant, treating the enemy, disregarding his own wound. That speaks of his professionalism.'

Naik Deepak understood his responsibility. It was an outside chance, but the Chinese interpreter was literally the only possible key to bringing the escalating situation to a halt. The men of 16 Bihar were well acquainted with the interpreter, having met him several times before during talks between the local battalion commanders.

'I had shared a cigarette with the interpreter on many occasions in the past. He was a pleasant, smiling guy,' says

Colonel Ravi Kant. 'He had told us he was poor, with only parents there at home. He would constantly say "let's have a solution to this dispute so I can also go back home". He always had this desire to go back home.'

And there he lay, bleeding and unresponsive, as the violence continued.

A short while later, abruptly, the stone pelting stopped, with no more projectiles landing for the next ten minutes. A sharp breeze found its way into the narrow valley, whipping itself up into a freezing, steady wind, the sound barely adding to the surging Galwan.

The sun rapidly set thereafter, putting the already extremely tense situation under a pall of total darkness. The temperature plummeted to freezing. In the distance beyond Patrol Point 14, Chinese men could be heard belting out loud threats and still hurling stones at the the Indian soldiers. Colonel Babu used a megaphone to warn the Chinese to stop, to mutually calm things down so both sides could talk peacefully the next day.

'Colonel Babu was being a professional and responsible local commander, trying to negotiate and control the situation at his end rather than involve the higher levels of the Army,' says Colonel Ravi Kant. 'If things could work out, there was nothing like it. It was also the reason why Naik Deepak was summoned—it was a gesture of goodwill by a professional soldier.'

But Colonel Babu knew that the situation had gone way past the point of no return for that night. Staring into the dark void beyond Patrol Point 14, it wasn't immediately clear what would come next. But come it would. Of this, the Indian commanding officer was certain.

It was at this point that Colonel Babu called for reinforcements from KM-120. Troops from 3 Punjab infantry battalion and two artillery units—the 81 Field Regiment and 3 Medium Regiment—would arrive in vehicles shortly thereafter, swelling the Indian force to just over 400 men.

The call for reinforcements was prescient because only minutes later, using darkness as a cover, the Chinese side launched its real assault of the night.

'*Unki main force aayi* [Their main force arrived],' says Havildar Dharamvir, who was part of the front line of troops preparing to face the Chinese. '*Adhik sankhya mein PLA ke log aaye* [PLA men arrived in large numbers].'

That 'main force' was a horde of at least 1200 men—three times the size of the Indian force waiting at Patrol Point 14.

'There was a lot of noise because of the strong water current in the Galwan River. No one could hear one another, but we could still hear the advance. There were that many men,' says Havildar Dharamvir.

'The mountains around the Galwan River are all black, adding to the total darkness that engulfed the battlefield at about 7.30 p.m.,' says Subedar Sahu. 'It was so dark that no one could recognize one another. It was in these circumstances that the huge Chinese reinforcement came from the right side of the mountain and launched an attack on us.'

But the Chinese horde that arrived was armed with more than just rocks, clubs wrapped in barbed wire and their fists. This was an entirely different PLA unit that had been waiting to attack in a sudden wave and not the regular soldiers normally deployed for patrolling the area. These were of an entirely different kind.

The Chinese horde was very well equipped, in padded riot gear and helmets, wielding batons fitted with LED incapacitators—devices that emit sudden bursts of flashing light designed to stun a person in the darkness and temporarily debilitate them. Gear and devices like these had been seen recently in the police crackdown on protesters in Hong Kong.

'They also carried carbon fibre shields which also had that bright flashing light,' says Havildar Dharamvir. 'They would flash that in the dark, and the beam would blind you. That was what they did before the final charge.'

'They charged without any regard for their own advance elements,' says Colonel Ravi Kant. 'It was like a stampede, and I would not be surprised if some of the Chinese men out front were trampled and injured. That was the intensity of the charge in the darkness. Our men were in smaller numbers, but we were ready.'

It is here that the most brutal phase of the Galwan clash began. In groups, 400 Indian Army men collided with 1200 Chinese troops in the dark winding stretch beyond Patrol Point 14 on the Chinese side. An Indian advance party had ventured even deeper into the Chinese side to confront them.

'Our advance elements, *woh aage hi reh gaye* [were left in the front],' says Subedar Sahu. 'The Chinese horde came in the middle cutting off our advance party that continued to fight where it was. *Raat mein bahut zabardast fight hui* [Intense confrontation took place at night]. There were lots of casualties. Some had their heads smashed, some had chest injuries. *Unka bhi wohi haal aur humara bhi wohi haal* [Both sides suffered the same fate].'

Metres from the swelling Galwan River, Naik Deepak continued to feverishly administer first aid. In the minute-long lulls in the savage fighting, Havildar Dharamvir remembers seeing Naik Deepak pick up his suitcase and walk into the scene of the action, following the sounds of injured personnel in order to treat them.

'There were many new injuries on the Chinese side in the fresh fighting that had broken out, and Deepak was busy flitting from one injured man to another, trying to give as much first aid as possible,' says Havildar Dharamvir.

'Imagine this happening in the middle of deafening noise and totally chaotic up-close combat. I don't know how he was managing. Some of us tried to give him cover, but there was too much fighting to do. The Chinese numbers were too large.'

By 8 p.m., when Havildar Dharamvir scanned the scene looking for Deepak, he was nowhere to be seen. The fighting had spread into several groups across a large area, including up the slopes of the mountain and the very edges of the now freezing Galwan River.

'We lost touch with him since he was in one corner busy with his first aid duties,' says Colonel Ravi Kant. 'This corner was right where the fighting was taking place. We don't quite know what happened, but I believe he continued to treat injured Indian and Chinese personnel for maybe another hour or so. There was utter chaos in the darkness. It is very likely that Deepak got encircled by the Chinese while he was doing his work. He did not stop and continued doing it.'

Naik Deepak wasn't seen again for the rest of that night.

'A battlefield nursing assistant (BFNA) buddy was supposed to be with Deepak,' says Colonel Ravi Kant. 'In the army we never leave anyone alone and there's always a buddy system in all situations. Deepak was trained at a proper nursing college. Along with him there was to be his BFNA buddy, trained at the battalion level to assist him and also for any battlefield exigencies. But the current situation was so chaotic, and Deepak was working alone.'

And the terrible cost of that night was about to unfold.

'At around 8.30 p.m. at KM-120, I got a message from the scene of action that our commanding officer was not traceable,' says Subedar Sahu. 'I tried making radio contact with Colonel Babu, but there was no response. Then I called the unit's adjutant but couldn't get through.'

Over the next half hour, a group of soldiers injured in the night fighting was ferried to KM-120 in three vehicles. Some of them had fallen into the Galwan River during the brawl and were rushed to a special 'heating room' for treatment.

'We had four or five soldiers who massaged the feet of the injured troops to stabilize them,' says Subedar Sahu, who was there when the injured men arrived. 'The temperature was well below zero. The water in the Galwan River was so cold that you couldn't be in it for even two seconds. The men were all wet and unconscious. They recovered after half an hour in the heating room. If we had not readied the heating room at KM-120, we would have lost those men.'

It is easy to forget the conditions in which this full-bodied brawl was taking place. While most casualties were caused by direct injury in the fight, weather and altitude were undoubtedly taking their toll. The Galwan Valley is the last place for such devastating energy expenditure.

'You're at over 15,000 feet in sub-zero temperatures. The air is thin, the O2 levels are as it is down by 60 per cent, so you have only 40 per cent in your body, hence with grievous injuries, it becomes ever more difficult to survive. Forget running or charging, even a brisk 200-metre walk puts you out of breath. You can imagine the toll such a brawl was taking on the human bodies on both sides,' says Colonel Ravi Kant.

Back at Patrol Point 14, there was no trace of 16 Bihar's commanding officer. His personal BSNL phone wasn't even ringing. All attempts to contact his group by radio hit a wall of silence. Meanwhile, all through the search for Colonel Babu, the fighting was getting even more fearsome.

'More Chinese reinforcements had come to the area, including their special forces,' says Havildar Dharamvir. 'The Chinese then launched a fresh attack on us. Very intense fighting went on till around 9.30 p.m. This was the fiercest fighting I have ever seen.'

One of the men involved in the 'fiercest' combat was Naib Subedar Nuduram Soren. As he fought, Havildar Dharamvir remembers seeing Soren being injured with a baton blow to his face.

'Some of the other soldiers pulled him to the side, away from the fighting and asked him to return to base,' says Havildar Dharamvir. 'But he was determined to fight. He said, "Whatever has to happen, will happen today and now. *Main peeche nahi jaunga. Mere ko maara hai PLA ne. Mein bhi unko marunga tabhi vaapas aaonga* [I will not step back. The PLA soldiers attacked me. I will return only after I have hit them too]." On the battlefield, you cannot ask for anything more inspiring than that. We advanced and fought even harder under his leadership.'

Naib Subedar Soren plunged back into the fight. Soldiers who fought near him remember his determined screams as he fought. One of them recounts, 'There was no stopping him. Naib Subedar Soren fought extremely bravely against the Chinese despite suffering serious injuries. He kept motivating us in a loud voice. He said the PLA had to be pushed back at any cost. Soren Saab motivated us so much that we thrashed the Chinese soldiers with even greater zeal.'

But the number of Chinese was so great, the soldiers soon lost sight of Naib Subedar Soren as well.

By 10.30 p.m., things fell quiet. There was a sudden shift from the savage chaos of war cries to only the sounds of the Galwan Valley. The last of the brawls broke up, with the Chinese retreating to their side. The sound of the roaring Galwan punctuated now by the sounds of the injured Indian and Chinese men. Many of them had fought to the death.

'After the Chinese were finally pushed back, we sat near the nallah. It was freezing. We were all wet. We sat there all night like that. Our officers asked us to carry out a headcount of the men, how many of us had come for the mission and how many of us were there now. Our main mission now was to find Colonel Babu. Our commanding officer was still nowhere to be seen,' recalls Havildar Dharamvir.

With flashlights, search parties fanned out from the site of the carnage, stepping between the dead and injured, hoping to find Colonel Babu, who had been missing for over two hours. Injured men on the ground were quickly carried to the MI room where a team of eight Army doctors and surgeons had arrived for emergency support.

Shortly after midnight on 16 June, the commanding officer's body was found. He had sustained a grievous head injury, most likely from a rock flung from the heights. Attempts were made to revive him.

Colonel Babu was hurriedly carried to a vehicle and dashed back to KM-120.

'At around 12.45 a.m., Colonel Babu was brought to us in a vehicle,' says Subedar Sahu. 'Doctors tried their best to revive him. *Kissi tarah unko saans aa jaaye* [If only we could make him breathe in some way]. But there was no luck. The doctors then tried to make him sit. The moment they did that, blood started oozing out of Colonel Babu's nose. It dashed our final hopes of reviving him.'

At 1.30 a.m., Colonel Babu was put back in the vehicle and sent to the Army's field hospital in Durbuk, 120 km to the south.

Back at Patrol Point 14, the search for the injured and dead continued in the freezing darkness.

'It was difficult to do a headcount as many of our soldiers were injured, and injured Chinese soldiers were also there. It was difficult to identify who was who in the darkness. We spent the entire night trying to identify our soldiers and those who belonged to the PLA. For the ones we could identify, we carried out the necessary procedure and took them to the MI room for treatment,' says Havildar Dharamvir.

At around 4 a.m., more injured Indian men were transported to KM-120. These included Subedar Anil, who had suffered serious chest injuries. An Indian Army Dhruv helicopter was called to pick up a group of the grievously injured at 6 a.m. from KM-120 and was flown to the better-equipped military hospital in Leh, which was also at a lower altitude.

Colonel Ravi Kant, then the unit's second-in-command, was in Leh on military work just before the incident. When the fighting began, he travelled all night and reached KM-120 at 4 a.m., heading straight to Patrol Point 14. There was no time to acclimatize to the steeply higher altitude. The commanding officer of the unit had been killed in action in an unthinkable sequence of events. It was upon Colonel Ravi Kant to rally his men and keep things together.

'When Lieutenant Colonel Ravi Kant reached, he assembled the battalion, briefed the men and said there was nothing to worry about, whatever had happened could not be changed, and that the battalion should be ready for any task assigned to it,' says Havildar Dharamvir. 'He told the men that their *hausla aur junoon* [courage and passion] should be higher than before. "*Main aa gaya hoon* [I am now here]," he said. He motivated us and said there was no need to be worried. He praised the way we fought against the PLA and drove them away. The battalion was fully motivated and ready to advance further, till the Chinese capital if necessary! We were grieving but that was how galvanized we were. We were prepared to do anything.'

After speaking to the men and reassuring them, Colonel Ravi Kant, now the commanding officer of the 16 Bihar battalion, ordered a team to scale one of the highest points near Patrol Point 14 and swiftly establish an OP. The events of the previous night had made it clear that border protocols and treaties between the two countries were now severely compromised, if not totally destroyed. Foreseeing more uncertainty ahead, he wanted eyes on everything the Chinese did that day. The hours gone by had extracted a terrible cost.

In vehicle after vehicle, the true extent of the human loss was brought home to KM-120, where the tragic duty of dressing wounds of the dead was conducted before the remains were boxed and dispatched down the highway to the 303 Field Hospital in Durbuk.

'At 6.15 a.m., I got a message that a havildar from our unit, Sunil Kumar, has been killed in action, and to "please come and identify him". I immediately rushed to the MI room close to Patrol Point 14,' says Subedar Sahu. 'Havildar Sunil had been in the water, face down all night. It was difficult to identify him. With some difficulty, after verifying all his details and proper identification, he was also sent to Durbuk in a vehicle.'

The next man to be identified was Havildar Palani, an artillery soldier from the 81 Medium Regiment. His body was discovered near the banks of the Galwan. Eyewitnesses had seen him valiantly fight a group of Chinese soldiers to the death, managing to subdue at least three before he fell.

As the sun rose, a search party climbed up one flank of the mountainside, hoping to find more injured men who could be saved. These had been the vantage points from where Chinese soldiers had unleashed attacks with rocks. It was behind a rocky outcrop several feet up on the flanks that the search party found Naib Subedar Nuduram Soren, the soldier whose war cries were heard loudest the previous night in the Galwan Valley. A few feet up the mountainside from his body was a rocky ledge where the search party spotted a rudimentary catapult system. The Chinese had been using a catapult to launch rocks at the Indian Army.

Over the next hour, eight more casualties, including the body of Naib Subedar Soren, arrived at KM-120.

'Soren had suffered severe injuries during the fighting, and blood was oozing out of his head. He fought valiantly and sacrificed his life for the country,' says Subedar Sahu. 'His fearlessness can never be overstated. It was superhuman. Whatever I say about him is inadequate.'

By 9 a.m., several soldiers were still unaccounted for. The brutality of the final phase of fighting had sent the brawling groups far apart, some of them deep into the Chinese side where the Indian Army's advance party had already been cut off earlier. It was almost certain at this time that Indian soldiers were either injured or forcibly held back by the Chinese on their side.

The search continued without pause, with a party led by Havildar Dharamvir venturing deeper into the Chinese side to look for casualties. Against the dark sand on the riverbank he spotted debris of used cotton wool and pieces of broken tincture bottles. He knew it before he saw it.

'Deepak was lying on the ground, his suitcase sprawled open next to him. I knew he was dead, but one never accepts it at first. He had suffered serious injuries and there was blood all over his head. We did all we could, hoping that he could be saved. We carried him to the MI room. The doctor there declared him dead on arrival,' says Havildar Dharamvir.

Just around the time that Naik Deepak's body was found, in Rewa, Madhya Pradesh, his wife had been asked to turn on her television set to catch the first newsbreak on the Galwan incident. At that point, the only confirmed information was that three men from 16 Bihar, including the commanding officer, had been killed in the deadly clash.

'I remember watching the big headlines flashing,' says Rekha. 'First it said three men were killed. A little while later, some of the names began flashing, including Colonel Santosh Babu, Havildar Palani and Sepoy Gurtej Singh (from 3 Punjab battalion). Then the headlines said a total of twenty men had lost their lives. My head was spinning. I kept telling myself that Deepak is a medic, and it's impossible that he would be at the front line in a scenario like this. *Ho hee nahi sakta* [It is not possible].'

When Rekha did not receive any phone call through the day, she called her father-in-law. He hadn't received word either, and Deepak's phone wasn't reachable. The search for the injured and dead in the Galwan Valley would take all of 16 June.

'That night, frantic and anxious, I called a cousin of mine who is also in the Army. *Andar se tension tha* [I was tense],' recalls Rekha. 'I was very emotional and nervous after receiving no news all day. My cousin received my call but was unable to help. I didn't know what to do except wait.'

Rekha didn't have to wait long. Thirty minutes after she spoke to her cousin, Naik Deepak's father finally received a phone call from Leh at 11.30 p.m.

'Deepak's father was unable to digest what he was hearing on the phone,' says Rekha. 'He handed the phone to Deepak's cousin. It was he who then called me to break the news. I thought I was hallucinating for some reason. I was very upset and angry. I disconnected the call. But he called me again. This time I listened. He told me Deepak's unit had called. I still couldn't believe it. I asked him for the number he had received the call from. I dialled, but the call wouldn't go through. You can imagine, I did not sleep that night, praying that there was some mistake. My hopes were raised by a family friend in the Army who called to say he had seen a body box of Deepak but not his body itself, and nothing was confirmed. It was only when the names were released by the Army the next day that my world came crashing down.'

'When things got so brutal, Deepak should have actually moved back but he continued to render first aid to the Indian as well as Chinese soldiers,' says Colonel Ravi Kant. 'He was encircled by the enemy. Even with all the movement of soldiers and chaotic fighting, Deepak was doing what he was trained for. Our boys had fallen, and he continued to render treatment to the maximum extent that he could. He was simply saving lives and honouring his profession. Even though he had helped their men, the Chinese either could not or refused to acknowledge the humanitarian service he was rendering. This boy was giving first aid to their soldiers just a short while ago. But he was still attacked. They could not have mistaken him for a combatant. He was clearly a medic. He did nothing but administer first aid on the battlefield.'

'I have a feeling the Chinese soldiers captured Deepak, used him to treat their injured personnel, and then killed him,' says Subedar Sahu. 'They had witnessed him treating their injured earlier. When Deepak's body was brought to KM-120 at around 1 p.m. on 16 June, the blood from his wounds was fresh. He appeared to have been attacked and killed with some sharp object. In the case of several other injured soldiers who were brought to KM-120, their blood had clotted. But in Deepak's case, the blood was fresh.

'*Deepak ko night mein capture karke rakha hoga* [He must have been held captive at night]. He was from the Army Medical Corps. The PLA must have captured him for their benefit. After getting their men treated, they must have killed him, left him there and gone back. Our boys brought him to KM-120, which is around 7 km behind Patrol Point 14. It takes about an hour and a half to reach KM-120 from Patrol Point 14. Deepak's blood was totally fresh.'

The battlefield was also littered with dead Chinese soldiers. Through the night, the injured Chinese soldiers had been pulled out of the area where the fighting took place to the PLA positions in the depth beyond Patrol Point 14.

'Since the time we had assembled in the area in the morning, we had spotted dead bodies of several Chinese soldiers lying around. Our orders were not to touch them, as the Chinese were expected to retrieve them later,' says Havildar Dharamvir.

Sure enough, late in the afternoon on 16 June, Chinese soldiers arrived with stretchers to take back their casualties. High above them, the Indian OP kept a hawk's eye on the Chinese ambulances taking the men back to a rear position. Through the day, a series of Chinese helicopters were spotted, arriving to collect the dead and wounded.

Silence pervaded KM-120. The loss of men was still sinking in.

'All the twenty fatal casualties that came were handled by me,' says Subedar Sahu. 'I got the doctors to see them and then sent the bodies to Leh and Durbuk. They were the most numbing few hours of my life. Through the night I had been pleading with my commander, saying my battalion was on the front line fighting and I had been assigned logistics support duties, organizing vehicles, rations and other tasks. When the bodies of our men were brought to me, I prayed for their peace, but wished I had been fighting alongside them. It was a painful story. For the next three days, it was almost as if I had forgotten about food, overwhelmed by the situation, the bodies, the blood.'

After all the counting and tragic paperwork, ten men were still unaccounted for. Over a series of emergency, major general-level meetings with the Chinese side, it was confirmed that the Indian Army's advance party had been cut off during the night assault and held back after the clash ended. The ten soldiers, including two majors, two captains and six men, were sent back across Patrol Point 14 at 5.30 p.m. on 18 June, formally bringing to a close the Galwan exchange.

The men of 16 Bihar remained at KM-120 and the Galwan Valley till 19 June, after which they were rotated out of the location and replaced by a new battalion. The men of 16 Bihar had just lost their commanding officer and nineteen men. While every man remained focused and galvanized, Army protocol dictated that they be given some distance, both physically and emotionally, from a location that had completely dismantled border peace and tranquillity mechanisms between the two countries. They had fought back a hostile, premeditated attack by a larger force and had paid a stinging human price.

'It was painful. But our morale was very high,' says Subedar Sahu. 'We were prepared to do anything. We would have sacrificed our lives if necessary. That is how every man in our unit felt. Our families, our wives and children are at home, but equally our families are the faujis who we are posted with. At the age of seventeen we join the Army. We spend our lives together. They are our real friends and every bit our family.'

Honouring the lasting legacy of her husband, Rekha joined the Chennai-based Officers Training Academy in May 2022 and will be commissioned into the Indian Army as a lieutenant in 2023, joining a growing list of proud army wives who have chosen to follow in the footsteps of their husbands after they were killed in the line of duty. She wants to go to the Galwan Valley at least once to see where her husband honourably gave his life treating not just his own but also the enemy.

'Deepak used to tell me he was proud that I was educated and a teacher. He did not get a chance to finish school. Because of his family circumstances, he needed to begin working early, and so he joined the Army,' says Rekha. 'He would say you have the gift of education—"*padho aur mera bhi naam ho, Deepak ki wife hai* [study and make me proud of a wife like you]". He felt he didn't have a chance, so he always encouraged me in the little time that we spent together to learn and climb higher. His English was weak, so he would always say "please only talk to me in English". He really wanted to speak fluently in the language.'

From Ladakh, 16 Bihar was sent far away to Chinthal in Hyderabad, awaiting their next deployment. Eight months later, on 11 February 2021—the raising day of the 16 Bihar Regiment, families of the men killed in action were invited to a memorial event in the city.

'Deepak's wife was also there,' says Subedar Sahu, who was tasked with coordinating logistics for the families. 'I tried to have a word with her. He was my jawan, I was his senior JCO. I was not really able to talk to her at that time. Nor was she able to speak very much. But slowly we shared our stories. Deepak's family is now the family of the unit. They are our responsibility.'

'Deepak and I met for the first time on a video call in March 2018,' Rekha told the senior soldier. 'We had an arranged marriage in December the following year, but I think it turned to love within a week. When the *rishta* came, Deepak couldn't get leave to come down, so his brother Havildar Prakash (from the Indian Army's Armoured Corps) came to see me and to introduce Deepak in some remote manner. But the very next day Deepak and I got on a video call. We met in person only in May 2019. He had strange names for me right from the start. His favourite was "Recruit". I still don't know why, and he refused to tell me.'

'I spoke with Deepak's elder brother, Havildar Prakash,' says Subedar Sahu. 'He said Rekha was like his younger sister, and if she wanted to marry again, move forward in her career or anything she chose, he would fully support her. I am in touch with Deepak's family and tell them often that 16 Bihar is always there for them.'

Havildar Prakash continues to serve in the Indian Army, proud of the stories he hears about his younger brother.

'We were very poor growing up, so both Deepak and I knew we had to take on responsibilities early to take care of the house and family,' says Havildar Prakash. 'Near our childhood home in Farenda village, there is an army range. For years we used to see army cars and cavalcades passing by. We both used to get very excited. We used to feel pride swelling in our hearts as well. Deepak, who was a biology student, enlisted with the Army Medical Corps. He plunged into it with great passion. When he came home for *chhutti*, he was basically the village doctor.'

Deepak's last visit home was in February 2020, barely four months before the Galwan Valley skirmish. The brutal fighting in the remote valley stretched the India–China bilateral relationship to a breaking point, and the trust deficit it triggered still casts a shadow over the ongoing talks to resolve the lingering LAC row that began in May 2020.

'He was to return on 14 February,' says Rekha. 'He called me from the railway station the previous day. I said, *"Jaldi aa jana, kal Valentine's Day hai* [Come soon, tomorrow is Valentine's Day]." There was a pause at the other end. He said, *"Rekha, 14 February ko Pulwama mein attack hua tha pichle saal, aur tum Valentine's Day ke baare mein baat kar rahe ho* [There was an attack in Pulwama on 14 February last year, and you are talking about Valentine's Day]." I felt a little guilty. I quickly said we wouldn't celebrate, but it would be our first Valentine's Day, and a token of our togetherness. I know he felt a little bad for reprimanding me. But I was okay with it. I know he felt strongly about it.'

When he reached home, Rekha would present him with a mobile power bank as a Valentine's Day present.

'Deepak was perplexed. I said, *Desh hai but apna relationship bhi toh hai* [We have to think about the nation, but we also have to think about our relationship]. I told him, sometimes let's talk about us and not only about the army and Pakistan and whatnot,' says Rekha. 'But I had never really felt bad. His devotion to duty was extreme—he would run to the phone every time it rang, wondering if he was being summoned back to his unit urgently. But he was very devoted to me and loved me. He never let me feel otherwise.'

'My wife tells me, if Deepak and she were watching television and if there was news of soldiers being killed in action anywhere, Deepak would simmer with anger and go silent,' says Havildar Prakash. 'And then, when the news channels flashed images of the last rites of a soldier, he would whisper to himself, "*Dekho kitna bada kaam karke gaye hai* [Look at the tremendous work he has done before sacrificing his life]."'

'Once he was talking to an Army friend, and I overheard him describe the cold and the terrain in Durbuk,' says Rekha. 'When I asked him weeks later on the phone, he was surprised. He said, "Stop googling the weather. There's a reason I don't tell you all this." He never shared details of the tough areas he was posted in so that I don't get stressed. What is that, if not love?'

On Republic Day 2021, Naik Deepak and three other men—Naib Subedar Nuduram Soren, Havildar K. Palani and Sepoy Gurtej Singh—were honoured posthumously with the Vir Chakra, India's third-highest wartime gallantry medal. Artillery soldier Havildar Tejinder Singh, who fought fiercely and survived with injuries, also received a Vir Chakra. The 16 Bihar's commanding officer, Colonel Santosh Babu, would be bestowed with a Maha Vir Chakra (MVC), the country's second-highest wartime gallantry medal.

The choice of medals—wartime decorations, as opposed to the peacetime Shaurya Chakra and Kirti Chakra medals—was also a message from the Indian Army and government that the situation in Ladakh was being formally regarded as a live conflict between the two countries.

A further sixteen officers and men from the 16 Bihar battalion would receive Sena Medal (Gallantry) decorations. These include Captain Arjun Deshpande, Subedar Amarendra Singh Parmar, Havildar Birsa Tirkey, Sepoy Nitin Khadke, Sepoy Dheerendra Yadav and Sepoy Bijay Kumar Gond. Posthumous medals were awarded to Havildar Sunil Kumar, Sepoy Ganesh Ram, Sepoy Kundan Kumar, Sepoy Kundan Ojha, Sepoy Chandrakanta Pradhan, Sepoy Rajesh Orang, Sepoy Chandan Kumar, Sepoy Ganesh Hansda and Sepoy Aman Kumar.

Last in the list of decorations to the unit for gallant action in Galwan was a lone Mention-in-Despatches for meritorious service in an operational area. The recipient was a young officer from Manipur, Captain Soiba Maningba Rangnamei. He would inadvertently be thrust into the public eye when China leaked a propaganda video of the Galwan incident on Chinese social media. The Chinese military's apparent attempt to paint the Indian side as aggressors inadvertently put a rare spotlight on the individual gallantry of Indian personnel. One particular sequence showing a defiant Captain Maningba challenging and physically stopping a group of Chinese men went viral. Viral enough that Manipur's chief minister Nongthombam Biren Singh, who hadn't noticed the honour to the young man from his state the previous month, hurriedly organized a meeting to honour the officer.

Authentic details of a military incident rarely emerge very soon after it happens. Which is why much of the detail you've just read has been written about for the first time here.

For instance, how many lives did Naik Deepak actually save that night in the Galwan Valley? His Vir Chakra citation would reveal, 'He was pivotal in rendering treatment and saving lives of more than thirty Indian soldiers, which reflects the epitome of his professional acumen.'

'What was left unsaid was his professionalism to the injured enemy. It is why, when I wrote him up for this decoration, I mentioned that his act of valour surpassed both duty and humanity,' says Colonel Ravi Kant.

But while the Indian Army would release the full list of its twenty fatal casualties on 17 June itself, the cost imposed on the other side would go on to become an object of concerted Chinese propaganda.

Eight months after the Galwan clash, in February 2021, China's official military mouthpiece formally honoured four PLA personnel posthumously—a suggestion that the Chinese Army had lost four men that night in Galwan. Two of the Chinese faces were immediately familiar to men of 16 Bihar.

The first was Chen Hongjun, commanding officer of the PLA unit involved in the first brawl.

'The second face we recognized was Chen Xiangrong, the interpreter. The man who Deepak had treated and tried to revive. Clearly, he had succumbed to his injuries. We saw him as a reasonable, friendly man who could have maybe helped communicate and calm things down. But given what came that night, I doubt it would have been possible even for him. May he find peace,' says Colonel Ravi Kant.

While the Chinese government's propaganda organs made an unseemly attempt to taunt the Indian side on the death toll it had suffered in the clash, the Indian Army maintained a studied, professional silence, unwilling to be pulled into an ugly war of claims over the bodies of the dead. Internally, however, operational reports and logs of course recorded the truth.

The Indian OP on a peak above Patrol Point 14 had been able to peep 7 km into the Chinese side and, right through 16 and 17 June, had witnessed the vehicular and air evacuation of at least thirty-six men on stretchers, though it was impossible to tell for certain how many of these were non-fatal casualties. A conservative estimate suggested about half were highly likely to be fatal, given the savagery of the clash.

While China would never comment or verify officially, a confirmation of sorts would emerge in the officers' lounge of the BPM hut at the Chushul-Moldo meeting point at one of the many rounds of talks that would be held in the weeks and months following the Galwan clash. During a chat away from the main talks, a brigadier-level officer of the PLA, who doubled as an interpreter, is said to have confirmed to an Indian colonel that the number of Chinese fatal casualties was seventeen, with over half succumbing to their injuries in the week that followed the clash.

'We have a number for how many Indian lives Deepak saved, but we don't have a number for how many Chinese men he saved that night,' says Colonel Ravi Kant. 'All I can say is that many of the injured Chinese men who survived that night definitely have Naik Deepak to thank. They were practically abandoned by their forces, while this boy was tending to their wounds. We are trained to take life to protect the country. But what can be higher than saving lives?'

At the time this book was written, the Indian Army and the PLA had held fifteen rounds of military talks to reduce border tensions in the Ladakh sector. A complete resolution is still not on the horizon, even though the ongoing negotiations have led to partial disengagement of rival soldiers from some friction areas on the LAC.

Naik Deepak's niece Palak misses the video calls with her uncle and frequently asks for him.

Her father smiles and replies, *'Woh charon taraf hai, dikhte nahi hai but aas paas hee hai hamare* [We can't see him, but he is constantly near us].'

2

'I'm Not Leaving This Cockpit'

Group Captain Varun Singh

Coonoor, Tamil Nadu
8 December 2021

The sudden roar of helicopter blades forced Joe to jerk his phone camera upward. On a walk along a verdant stretch of railway line near Kattery Park, his phone managed to capture less than two seconds of the helicopter itself, a grey silhouette against a foggy noon sky of even more grey, one half of the image filled by a burst of hillside foliage.

As the helicopter flew out of view beyond the trees, Joe jerked his phone back towards his five fellow travellers walking along the railway track, his friend Naseer, their family and friends craned their necks to follow the sound.

At precisely the eleventh second, the sudden roar of a deafening explosion from across the hill made one of the women snap her head back. Naseer, visibly shaken by the sound, turned to Joe and asked, 'What happened? Did it fall? Has it crashed?'

Still filming in the direction of the sound, Joe said, 'Yes,' unaware of what he had just managed to capture on his mobile phone. Three seconds later, the video stopped.

Hearing the sirens of emergency vehicles echo up the hillside, the six tourists hurried in the direction of the explosion. They couldn't get very near, but it was clear that fire services and the police were on their way.

Less than an hour later, news of the Indian Air Force Mi-17V5 helicopter crash broke across the news media, the story quickly overwhelmed the airwaves, social media and WhatsApp groups like no other military air accident had in living memory. On news channels, anchors and breathless reporters, fighting to digest the unbelievable flash updates, informed the country that India's first chief of defence staff (CDS), General Bipin Rawat, his wife, Madhulika Raje Singh Rawat, and twelve others were in that helicopter.

The Russian-origin helicopter had taken off from the Air Force's Sulur base at 11.48 a.m. and was to land at the helipad at the Wellington Golf Course after a twenty-seven-minute flight. However, the twin-engine helicopter—equipped with latest technology and considered to be extremely reliable—lost contact with air traffic controllers at Sulur twenty minutes after take-off and just seven minutes before it was to land.

It went into radio silence at exactly 12.08 p.m., with Joe making the video a few seconds earlier.

Overwhelmed by the breaking news updates and after word had spread in Coonoor about who was in the helicopter, Joe contacted the police and sent them the video he had filmed on his mobile phone. He wasn't aware at the time that those twenty seconds of footage would become one of the most-watched videos of the year and central to the Air Force's most crushing crash investigations in nearly sixty years.

Hours of public confusion and contradictory media reports spread, with speculation over the death toll and cause of the crash. At 6.03 p.m., nearly six hours after the accident, the Air Force finally announced on Twitter what had been most feared:

With deep regret, it has now been ascertained that Gen. Bipin Rawat, Mrs Madhulika Rawat and 11 other persons on board have died in the unfortunate accident.

There was one final line in the Air Force's statement:

> Gp Capt Varun Singh SC, Directing Staff at DSSC [Defence
> Services Staff College, Wellington] with injuries is currently
> under treatment at Military Hospital, Wellington.

Given that thirteen other persons on board the helicopter
had perished in the crash, there was little doubt about the
seriousness of Group Captain Varun's injuries. And while no
specific information emanated from the military, other than
to his wife Gitanjali and father Colonel K.P. Singh (retd), the
story of this sole survivor quickly and understandably captured
the nation's imagination. Amid the paralysing implications of
the accident, here was one military man who had managed to
evade death.

As Group Captain Varun was transported from the Nilgiri
Hills down to Sulur in the plains near Coimbatore, a running
commentary on his condition dominated the headlines. The
enormous interest and solidarity with the officer, still in a coma,
compelled the media to satisfy audiences with more stories
about Group Captain Varun. The 'SC' after his name stood for
Shaurya Chakra, India's third-highest gallantry decoration for
exceptional acts of valour in peacetime. It was a medal he was
awarded on Independence Day 2021, less than four months
before the crash.

It was clear that this wasn't just any soldier fighting for
his life.

The day after the helicopter tragedy, just around the time
that Group Captain Varun was being wheeled into an Antonov
An-32 transport aircraft at Sulur to fly him to Bengaluru for
top-level intensive care treatment, a copy of a letter he wrote
to the principal of his school in Chandimandir emerged in the
media, rapidly going viral and fanning interest in the officer

to a fever pitch. He had written the letter a month after he had been bestowed with the Shaurya Chakra and barely three months before the crash. He wrote:

> I write to you filled with a sense of pride and humility. On August 15 this year, I have been awarded Shaurya Chakra by the President of India in recognition of an act of gallantry on October 12, 2020. I credit this prestigious award to all those I have been associated with over the years in school, NDA and thereafter the Air Force, as I firmly believe that my actions that day were a result of the grooming and mentoring by my teachers, instructors and peers over the years.

But it was the twelfth paragraph of the letter that became a viral headline story, making Group Captain Varun a hero on social media, even as he fought for his life in the ICU of the Air Force Command Hospital in Bengaluru:

> It is ok to be mediocre. Not everyone will excel at school and not everyone will be able to score in the 90s. If you do, it's an amazing achievement and must be applauded. However, if you don't, do not think that you are meant to be mediocre. You may be mediocre in school but it is by no means a measure of things to come in life. Find your calling, it could be art, music, graphic design, literature etc. Whatever you work towards, be dedicated, do your best. Never go to bed thinking I could have put in more effort.

When Varun, then a wing commander, was decorated with the Shaurya Chakra in August 2021, no one really noticed. He was one among the usual annual list of military personnel recognized for acts of valour and performance in the line of duty. As he lay suspended between life and death, the 297-word citation provided a bare glimpse of what the thirty-nine-year-old pilot had done to earn such a towering honour.

But until now, the full true story had not been told.

* * *

Hindon Air Force Station, Uttar Pradesh

12 October 2020

The vast apron of Asia's largest air force base, on Delhi's eastern suburbs, had been traditionally festooned with different types of combat aircraft just four days earlier for the annual Air Force Day event. The 'static display' which permitted attendees to do up-close walking tours of the fighter jets was a big draw, a thrill on par perhaps with watching those same jets roar overhead in the separate flypast.

But four days after the event, the apron was mostly empty. The Air Force Mirage 2000, Sukhoi Su-30, Jaguar, MiG-29 and other weapon platforms hosted at the Hindon station for spectators had gone back to their home bases. Only one 'visitor' jet remained at the base.

Tail number LA-5006, an indigenous Tejas fighter.

The fighter was among the three Tejas jets that had flown in for the Air Force Day display from their home base at the other end of the country, Sulur in Coimbatore, Tamil Nadu.

Their flight display over Hindon successful, the other two Tejas jets had since been flown back home to Sulur by two pilots, flight commander of the country's first Tejas squadron, Group Captain Syamantak Roy and commanding officer of the country's second Tejas squadron, Group Captain Manish

Tolani. They made refuelling stops in Ozar, Maharashtra, and Bengaluru before touching down in Sulur. It was the third Tejas whose homeward flight was stalled.

At 10.30 a.m., sitting in the single-seat cockpit of the Tejas with the canopy open was fighter pilot Wing Commander Varun Singh, then deputy to Group Captain Syamantak. The early winter fog had mostly lifted, revealing promising patches of blue sky wreathed in wisps of sullen cloud. Standing on the apron next to the aircraft monitoring things was Wing Commander Abhimanyu Chhetri.

Wing Commander Varun had been ferry pilot for this third Tejas, but he hadn't flown in the Air Force Day display four days earlier. At the flypast on 8 October, he was the display supervisor on the ground, tasked with the critical duty of monitoring the flights and providing real-time inputs on bird hazards.* Later that day, after the first two Tejas fighters returned to Sulur, Wing Commander Varun had encountered problems with the third aircraft while he conducted system checks for the intended return flight.

'Varun had taken off shortly after us, but he aborted the flight and landed back at Hindon and called me while still sitting in the cockpit,' says Group Captain Syamantak, who took the call at Sulur. 'While he was at 28,000 feet, there were fluctuations in the pressurization system leading to problems in the flight control performance. He told me there was an issue with the air conditioning system, which maintains temperatures of systems on board, in addition to the cockpit

---

* The Hindon Air Force station is situated close to the enormous Ghazipur garbage landfill, a mountain-sized repository of refuse from the National Capital Region. The landfill attracts large-sized scavenging birds that have posed a threat for years to flight operations at Hindon.

itself. The AC system in Tejas is important because everything is computerized.'

Over three days, the stubborn first-time glitch appeared to have been fixed. But since it was a long north–south ferry flight that had to be flown, Wing Commander Varun had conferred with his Group Captain Syamantak in Sulur and concurred that it was best to conduct a test flight near Hindon to make sure everything was okay with the Tejas before the flight home.

'We were both clear that it was critical to prove things were okay rather than encounter an emergency en route,' says Group Captain Syamantak.

With things seemingly repaired, at 10.50 a.m., Wing Commander Varun taxied the Tejas towards the tarmac, instructed by air traffic control (ATC) to hold. It was a busy morning with aircraft from Hindon—the C-130J Super Hercules and C-17 Globemaster III—departing on logistics flights.

Cleared for take-off at 11 a.m., Wing Commander Varun pushed forward on his throttle, bringing the Tejas's single General Electric F404 turbofan engine to maximum power, roaring into the air, gently banking right to take the aircraft northward into the sector designated for this test flight. Wing Commander Varun was intimately familiar with how agile the Tejas was in the air, but this was a flight to ensure all was well, so there was no need to push the aircraft to its limits.

Nearly 2000 km away in Sulur, Group Captain Syamantak monitored the test flight in real-time, receiving a live feed of what was happening with the aircraft and the pilot.

The Tejas continued to climb steadily, breaking through the cloud deck. Wing Commander Varun reported that everything seemed okay till then. It was when the fighter climbed to 23,000 feet that things began to unravel.

'At 23,000 feet, Varun got a cockpit warning of system failure,' says Group Captain Syamantak. 'He immediately descended to 18,000 feet. The warning alarm stopped. This was a test flight, so obviously he needed to check if it recurred. So, he put the jet once again into a climb.'

Almost instantaneously, another cockpit warning blared out, this time alerting Wing Commander Varun to a 'Level 1' failure in the aircraft's digital flight control computer (DFCC).

'The Tejas flight control computer is very robust, with four separate channels, so there's plenty of redundancy and fallback in case one channel fails,' says Group Captain Syamantak. 'If three channels fail, however, then you're in trouble. Serious trouble.'

While descending from 19,000 feet to 16,000 feet, that's precisely what happened.

'DFCC Level 3 fail!' Wing Commander Varun called out on his radio talkback, though ground control at Hindon and Group Captain Syamantak in Sulur already had that information, thanks to live telemetry.

The Tejas flight manual is crystal clear on courses of action to be followed by pilots for each level of failure in the DFCC:

- Level 1 Failure
  Autopilot trips out, handling becomes tricky, pilot needs to move with additional care
- Level 2 Failure
  A marked degradation in the aircraft's handling
- Level 3 Failure
  If the aircraft is out of control, the pilot must eject

'If three channels fail, the aircraft is as good as gone theoretically, and as per protocol, the pilot is fully within reason to punch out of the plane if it is out of control,' says Group Captain

Syamantak. 'But Varun reported that he wanted to try a few things to regain control of the aircraft. He pushed a switch to reset the aircraft's fly-by-wire* (FBW) system, producing an instantaneous response from the aircraft.'

After wobbling uncomfortably for a second, the Tejas abruptly pitched its nose down violently,† a jolt of a movement that sent the blood rushing to Wing Commander Varun's head and making him experience up to -4.5G, with an average of -3.5G for several seconds.

'Varun had been taken to the human endurance limit of negative G (i.e. -3G). Only display pilots train for negative G beyond that. I can only imagine how it must have felt, especially since it was unplanned and totally sudden,' says Group Captain Syamantak.

'This was a full in-flight emergency now, and Varun was practically thrown out of his seat and against his straps,' says Group Captain Syamantak. 'He was very shaken up but alert and conscious, and he knew he was even more within his rights now to bail out of the aircraft at this point.'

The dive had brought the Tejas rapidly down to 14,000 feet.

'He pulled his stick back to take the Tejas out of the bunt, allowing it a few seconds to settle,' says Group Captain Syamantak.

When Wing Commander Varun had gathered himself after the jolt, he spoke to Wing Commander Chhetri who was at the Hindon ATC.

---

* Fly-by-wire (FBW) is a system that replaces the conventional manual flight controls of an aircraft with electronic signals. The flight control computer processes these signals and carries out manual tweaks to aircraft control surfaces to produce the desired movement or response.
† The movement identified as 'bunt', where the aircraft experiences a rapid, uncontrolled, right forward pitching motion.

'The aircraft has gone out of control, I need vectors for a clear area without habitation ASAP,' came the voice from the cockpit. Ground controllers immediately sent the pilot directions, guiding the Tejas out over a desolate stretch of wilderness north-east of Delhi.

'Ejection planned?' Wing Commander Abhimanyu asked. A clear answer would allow the IAF to set in motion the recovery of the pilot once he had parachuted down.

But Wing Commander Varun replied in the negative, pointing out that he was still flying near the heavily populated Delhi suburb of Ghaziabad. He reported that he wanted to stay with the aircraft and try to nurse it back to normalcy, and only needed uninhabited ground as a precaution. After a few minutes of silence that had led to excruciating tension at both Hindon ATC and Sulur, he finally spoke.

'Aircraft is behaving,' said Wing Commander Varun. 'Will attempt return and landing. Please position me on finals for ILS [instrument landing system] approach.'

'Varun was a softspoken guy, but I could tell from his tone that his adrenaline was at its peak,' says Group Captain Syamantak. 'He had just experienced something very violent in the cockpit. It's something that can make your head spin very badly. He had recovered. He had even been asked if he wanted to punch out, because all criteria for an ejection had been met. Nobody would have blamed him for ditching the jet at that point. But he held her steady and continued to descend, determined to bring the fighter back. And that's when things spun out of control again.'

At 10,000 feet, the Tejas DFCC experienced its second Level 3 failure. This time when Wing Commander Varun reset the flight control system, the fighter aircraft violently rolled on to its back with the nose pitching down—an upside-down bunt that similarly pushed the pilot into a realm of negative G forces for four seconds. Once again, Wing Commander Varun fought to control the jet, rolling it back the right way and pulling back on the stick to get it into level flight. Once again, he initiated controlled descent.

At 8000 feet, the aircraft experienced another Level 3 failure—the third so far on the flight. Again, the plane snapped forward, throwing the pilot against his seat restraints.

'The Tejas flight reference card practically mandates an ejection below 10,000 feet if there is serious loss of control,' says Group Captain Syamantak. 'If there was a time to eject from the aircraft, it was after this third Level 3 failure. I was amazed, but I think I understood what Varun was feeling at that point.'

Wing Commander Varun and Group Captain Syamantak had been at the National Defence Academy (NDA) together at the turn of the millennium. Commissioned as fighter pilots and both going on to fly Jaguar strike jets, the two had been posted together at the Gorakhpur air force station in 2005.

'Varun was always super self-motivated,' says Group Captain Syamantak. 'As a young pilot, he quickly became a qualified flying instructor [QFI] and a test pilot. When the Tejas squadron in Sulur was hunting for a good test pilot, nobody fit the bill better than Varun. Unlike other aircraft, the Tejas is an open book. It's our own jet, designed and built by our own teams. Like all of us Tejas jocks, Varun had huge pride in this indigenous fighter, which had never been in an accident or crash. We had taken great pride in ensuring safety. Varun had therefore come up in that "zero loss" environment. This kind of ownership sentiment we feel for this aircraft is tremendous. I guess Varun was also determined not to be the one crashing a Tejas for the first time. He didn't want to create that kind of history.'

The Tejas was now at 100 feet, with its landing gear down ready for 'short finals' back to Hindon. The jet continued to descend gently, Wing Commander Varun holding it steady, the runway in sight. But 50 feet from the ground, for the fourth time, a Level 3 failure rang out.

If the Tejas pitched down now as it had during the previous failures, it would smash into the ground like a missile in seconds before any possibility of recovery. The decision now was to either eject, sending the jet careening uncontrolled towards the Hindon tarmac and hangars. Or to sign away the final chance of ejection and do everything humanly possible (apart from praying) to gently bring the aircraft down to land. The plane could have rolled or pitched violently as it had earlier in the flight. If that had happened, things would have ended very quickly.

'We were holding our breaths for that moment,' says Group Captain Syamantak. 'The aircraft could have done some other manoeuvre and endangered him. That was the moment of truth for all of us. We could lose a plane and pilot. Or we could lose a plane. For most of those monitoring those were the two main possibilities. But Varun was the pilot. Nobody knows better than the pilot in the cockpit.'

With crash tenders and emergency equipment ready near the tarmac, Hindon ATC watched as the Tejas landed safely on the main runway.

It had been fifteen minutes since the Tejas had encountered its first failure.

The first thing Wing Commander Varun did after taxiing back to the Hindon apron and powering the aircraft down was call his boss in Sulur.

'*Bach gaye* [I have escaped], sir,' said a still-shaken-up Wing Commander Varun. 'This aircraft has really taken me for a ride today.'

The pilot and his troubled jet were back safe, but those fifteen minutes in the air were serious enough for the Air

Force's Southern Air Command to suspend Tejas flights until a fleetwide investigation was conducted. The Level 3 failure had been trained for in simulators but had never happened in a real aircraft. The Air Force leadership was clear it didn't want to take any chances.

Over the following days, a court of inquiry identified the reasons for the malfunction, quickly ordering a bug fix across the two Tejas squadrons in Sulur. Test flights since have satisfactorily shown that the Tejas does not bunt or roll—or move dramatically in any manner—in the event of the DFCC Level 3 failure and reset of the flight control system.

There was a big silver lining to a mid-air incident that could easily have ended in either tragedy or the loss of a prized aircraft.

'The Tejas flight control system is among the best in the world,' says Group Captain Syamantak. 'That's why the aircraft was recoverable after a few seconds. With any other aircraft, there would have been total loss of control. The fleet underwent modifications. This was the time the early Tejas jets were ready to be upgraded with advanced software. When we stopped flying following the incident, we ensured that the upgrade involved the bug fix. That was a silver lining.'

The day after the incident, on 13 October, Wing Commander Varun was flown back to Sulur on one of the service logistics flights that criss-crossed the country. He had to leave the Tejas behind at Hindon until it could be fully fixed. The aircraft stayed for another month, with engineers from manufacturer Hindustan Aeronautics Ltd (HAL) arriving to get a closer look and ensure it was fully fit for its flight back.

'Varun returned to Sulur clearly sobered by what had happened, but he was still in good spirits,' says Group Captain Syamantak. 'That day was spent watching his cockpit and HUD (head-up display) videos of the entire incident. We already knew that his handling of the emergency was nothing short of heroic. But it was after watching the cockpit videos that the base leadership was clear that this guy deserved a gallantry decoration not less than a Shaurya Chakra.'

It was one thing monitoring data and performance of the aircraft as it went through those tribulations earlier in the month. It was entirely another watching the footage of the pilot as he bore it.

'In the second instance, when the Tejas went on its back, it was harrowing to watch,' says Group Captain Syamantak. 'He and I then left for Bengaluru to attend the court of inquiry that had been ordered. Varun had to go through all his medicals again because of the negative G he had experienced more than once. He finally returned to Sulur on 26 October, the day I took over as commanding officer of the first Tejas squadron.'

Before the end of 2020, the first Tejas jets began to be upgraded with new software that included the crucial bug fix. When it was time to test the first 'fixed' Tejas, the new commanding officer of the squadron knew who he should ask.

'Who better than him to do the test flight?' asks Group Captain Syamantak. 'Varun was a fine test pilot, and the flight was smooth and error-free. The aircraft involved in the 12 October flight had also returned to us and was ready for its own upgrade. Tail Number 5006 remains one of the best aircraft in our fleet. We love it even more because it kept our guy safe and landed back safely.'

Far from unnerving pilots and safety crews, the incident and aftermath hugely boosted confidence of Tejas pilots in Sulur. The pervasive sense was that even in such a critical emergency, it was possible to recover the jet. But the confidence was always tempered—the squadron went into deep study mode to 'wargame' what else could go wrong with systems in flight.

What happened may have been an aberration, but it triggered a deep pre-emptive safety sweep in Sulur, making the Tejas even more familiar to pilots than before.

Wing Commander Varun flew the LA-5006 again at Sulur several times with no incident.

'The incident didn't change him—he was more confident in his abilities,' says Group Captain Syamantak. 'He came out stronger. It was a contrast to how he was at that time, shaken, flustered, pumping adrenaline, but super focused.'

In April 2021, Wing Commander Varun received official word that he was being posted to the prestigious DSSC in Wellington near Coonoor in Tamil Nadu's Nilgiri Hills, alongside a promotion to the rank of group captain (equivalent to colonel in the Army).

'He knew it was an honour, but he had mixed feelings because he would not get to fly for some time,' says Group Captain Syamantak. 'But duty calls. We were expecting him to come back to the field and to the Tejas. In June 2022 he would most likely have taken over a Tejas squadron.'

On 15 August 2021, as recommended up the chain of command in the IAF, Wing Commander Varun Singh was decorated with a Shaurya Chakra. Pandemic restrictions following the second COVID-19 wave had forced the Rashtrapati Bhavan to drop plans of a traditional investiture ceremony, keeping the ritual for the following year. But the award citation itself was public:

> Faced with a potential hazard to his own life, he displayed extraordinary courage and skill to safely land the fighter aircraft. The pilot went beyond the call of duty and landed the aircraft taking calculated risks. This allowed an accurate analysis of the fault on the indigenously designed fighter and further institution of preventive measures against

recurrence. Due to his high order of professionalism, composure and quick decision making, even at the peril to his life, he not only averted the loss of an LCA [light combat aircraft], but also safeguarded civilian property and population on ground. For this act of exceptional gallantry, Wg Cdr Varun Singh is conferred with the Shaurya Chakra.

It was less than four months later that a newly promoted Group Captain Varun Singh volunteered to go down to Sulur from Wellington and escort General Rawat and his entourage in a helicopter to the Staff College for a scheduled speech by the country's top military officer.

'On that fateful day, I knew he had volunteered to be the liaison officer for the visit by the CDS because it gave him the opportunity to come down and meet us in Sulur,' says Group Captain Syamantak. 'On 8 December, Varun arrived at 8 a.m. and came straight to meet me in the squadron. He met with all the other guys from the two Tejas squadrons. His people, his buddies. Everyone knew he would be among them again as a commanding officer in six months or so.'

Group Captain Varun had always spoken of being lucky.

'Throughout his life, even with the Tejas incident, he felt he was lucky that everything worked out and he didn't eject,' says Group Captain Syamantak. 'His course mates agree he has been one lucky guy. When news initially broke that fateful afternoon that there were three survivors in the helicopter crash, everyone in the squadron immediately knew Varun had to be one of them. When he was brought to Sulur the next day, I saw a flash of him under a blanket. We prayed so hard for him.'

The Sulur base was already grieving when a comatose Group Captain Varun arrived for the emergency flight to Bengaluru. Four personnel from the base—Mi-17 helicopter pilots Wing Commander Prithvi Singh Chauhan, Squadron Leader Kuldeep Singh and airmen Junior Warrant Officer Rana Pratap Das and Junior Warrant Officer Arakkal Pradeep—had died in the crash.

The CDS's defence assistant Brigadier L.S. Lidder, Lieutenant Colonel Harjinder Singh, Havildar Satpal Rai, Naik Gursewak Singh, Naik Jitendra Kumar, Lance Naik Vivek Kumar and Lance Naik B. Sai Teja were also killed in the horrific crash, which a tri-service inquiry later established as a controlled flight into terrain (CFIT) accident. The probe found that the crash took place after the Mi-17V5 entered clouds in bad weather, leading to spatial disorientation of the pilot, which resulted in the CFIT. It refers to the accidental collision of an airworthy aircraft, under the flight crew's control, with terrain.

'After the crash, when Varun was hanging on to life, it was excruciating for us,' says a Tejas pilot from the second squadron. 'It was truly a collective prayer by us all for his recovery and for his family, including two young children. Many of us who saw him before he was flown from Sulur still have sleepless nights.'

Doctors at the Air Force Command Hospital in Bengaluru combined forces with burn and skin graft specialists from across the country to save Group Captain Varun. In media interviews, his father Colonel K.P. Singh (retd) would say that his son was a fighter and would fight this too.

On the morning of 15 December, seven days after surviving the helicopter crash that killed the other thirteen souls on board, Group Captain Varun succumbed to his injuries.

'Luck ran out for Varun, is how he would have probably described it,' says a maintenance engineer with the first Tejas squadron. 'He always wanted to shine bright, maybe be famous, be known for good.'

As Group Captain Varun had lingered between life and death for an excruciating week, his well-being went beyond just a media story. It felt personal to huge numbers of people, who perhaps felt invested in his recovery. A sliver of hope in a tragedy that could never be fully fathomed.

'A part of Varun will be smiling,' says Group Captain Syamantak. 'He would have loved to know that he is now famous and has inspired an entire nation.'

Five and a half months after Group Captain Varun's unexpected death, his mother, Uma Singh, and wife, Geetanjali, received the fighter pilot's Shaurya Chakra from President Ram Nath Kovind at a defence investiture ceremony held at the Rashtrapati Bhavan on 31 May 2022.

Mixed emotions gripped Group Capain Varun's father, Col K.P. Singh (retd), and his younger brother, Lieutenant Commander Tanuj Singh, as they sat in the imposing Durbar Hall and watched Uma and Geetanjali receive the honour from the President.

'It was a very proud moment for the family, but we missed Varun. We were told by some friends that the cameras at these ceremonies tend to focus on people who applaud the awardees with most enthusiasm. I would have climbed up my chair and clapped wildly had Varun been there to receive his medal,' says Colonel Singh.

Varun didn't really expect very much of himself when he was in school but through hard work and perseverance became one of the Air Force's finest experimental test pilots, a top-notch qualified flying instructor and was shortlisted in 2019 for the country's first crewed spaceflight, Gaganyaan.

Colonel Singh recalls that his son was among the twelve fighter pilots considered for Gaganyaan, but he could not make the cut due to a medical reason—Russian doctors ruled him out for the space mission because of issues related to jawbone density.

Varun found humour in rejection.

'Your destiny is decided by the density of your jaw.' That was Varun's WhatsApp status for the longest time.

3

'Get as Close as Possible before Dark'

Subedar Sanjiv Kumar

OP Randori Behak

The whirl of the Dhruv's rotors filled the helicopter's cabin as it climbed away from the Army helipad nestled between the hills of Trehgam in north Kashmir's Kupwara sector. Rising to only a few hundred feet, the helicopter banked southward towards terrain that rose abruptly into the forested highlands of Jumgund in Jammu and Kashmir's Keran sector.

The destination was a very short 5 km away, a distance that would be covered in a few easy minutes by the chopper. But it was the decision to use the helicopter that was the difficult part. Faced with the prospect of wasting a whole day hiking across high ridges to get to the same place, Subedar Sanjiv Kumar, grim-faced and staring out of one of the Dhruv's side windows, had finally agreed to use the chopper to transport him and his men.

In the cabin with him were five other commandos, including his buddy soldier Paratrooper Amit Kumar, another senior commando Havildar Devendra Singh and two young scouts, paratroopers Bal Krishan and Chhatrapal Singh. There was company in the air too. Flying in pursuit 400 metres behind was a second Dhruv helicopter, carrying six commandos of its own.

The two Special Forces squads of a dozen men from the 4th Battalion of the Parachute Regiment (4 Para SF) knew that this was hardly their style. Helicopters were always a noisy affair, usually unthinkable when it came to the stealthy, unannounced manner in which the men of 4 Para liked conducting their operations.

Or at least starting them.

'Anyone sitting in a valley can see, if not hear, an approaching helicopter. They can tell that helicopter *se kuchh bande aaye huwe hain* [some men have landed],' says one of the men from the second helicopter. 'Therefore, we had to account for a loss of surprise even before we were properly inducted.'

For the Special Forces unit even daunting distances were usually covered in off-road military vehicles that moved by night. But the place the commandos were headed to that morning of 4 April 2020, couldn't be reached by road. And while every man in those two helicopters had mentally factored in loss of valuable stealth, they equally agreed that the swiftness of their arrival would hopefully offset losing that element of surprise.

The two Dhruvs were flying over a country, a world really, that had come to a standstill. Ten days earlier, on 24 March 2020, Prime Minister Narendra Modi had addressed an anxious nation at the beginning of the COVID-19 pandemic. He had declared that India would be entering an unprecedented national lockdown to contain the viral spread. A quiet place like the hinterland of north Kashmir had become even quieter during the lockdown. Mirroring the tiniest detail into which Special Forces planning goes, the decision to use a helicopter also had to account for how lockdown restrictions could mean people with less to do on the ground. This could mean making them perhaps more attentive to a chopper flying nearby, a piece of information they could pass on to four men hiding somewhere on a snow-clad mountaintop. There is no eventuality, crack or weak link that Special Forces planners don't wargame in their heads before setting out for a mission.

But none of this mattered now. If Subedar Sanjiv had green-lit the helicopter ride, that was enough. Nobody needed to have any further apprehensions. Seated across from him in similar winter combat fatigues, paratroopers Amit, Bal Krishan and Chhatrapal, all three in their twenties and among the youngest in the unit, leant in close to be able to hear one another over the rotor noise. They had cut their combat teeth under the direct supervision of Subedar Sanjiv, the sullen-faced commando who wasn't just their senior. Navigating their first years into young adulthood away from their homes and families by voluntarily committing themselves to a difficult life in the Special Forces, it was in the shadow of Subedar Sanjiv that they grew into serious soldiers.

Fifteen minutes into the flight, as the two helicopters cut a lazy turn across a forested ridgeline, a pure white mountaintop came into view. Their destination had arrived. One of the pilots, speaking into the cabin address system, alerted the squad, then gently pitched the helicopter forward and descended towards the mountaintop. All six commandos used the valuable seconds of descent to cast their gaze across the terrain. It is likely that the same thought passed through each of their minds:

A pure white situation.

Miles and miles of snow.

No treeline, no cover.

Having made the difficult decision to arrive in a helicopter, it was near imperative for the commandos that they be able to exit the noisy rotorcraft in a concealed area. But the brief aerial reconnaissance of the mountaintop showed that this would have to be the second major compromise of the day—landing in an open area with plenty of visibility in every direction.

The site was about 12 km east of a stretch of the Line of Control (LoC) that Indian Air Force Mirage 2000 fighters had crossed* to bomb a Jaish-e-Mohammed terror facility near Pakistan's Balakot town just over a year earlier, in February 2019.

Around noon, the two helicopters descended to the general locations identified as Rangdoori and Teen Behak, all high-altitude sites in the Jumgund area. In coordination with Subedar Sanjiv before the flight, the pilots had brought the commando squads to the precise patch of mountaintop where an Indian Air Force Heron surveillance drone had spotted a foot trail in the snow.

The thirty-four-second video clip of the faint footprints filmed by the Heron and beamed to the Army's 68 Mountain Brigade in Trehgam on the morning of 4 April was the trigger for the urgent helicopter mission to get the dozen 4 Para commandos from the brigade helipad to the site as quickly as possible. The video clip was the first breakthrough in days.

The four terrorists had been spotted for the first time in the Rangdoori area of Jumgund three days earlier on 1 April. A patrol party from the Army's 8th Battalion of the Jat Regiment had spotted the terrorist infiltrators at a distance. The brief exchange of fire wasn't enough to eliminate the four men but forced them to abandon their heavy rucksacks and flee into a wooded area on that mountain. This was the first contact.

'The crucial thing was the terrorists had abandoned their bags, so they had no sustenance, apart from maybe some medicines or supplements in their personal pouches,' says Major Abhishek, a young 4 Para SF officer whose unit was already on standby alert by that time to join the hunt.

At dawn the next day, 2 April, troops from the Army's 41 and 57 Rashtriya Rifles counterinsurgency units arrived on site to tighten the pursuit of the terrorists. That afternoon,

---

* See *India's Most Fearless 2*.

at 4.30 p.m., the troops once again spotted the four terrorists from a distance. This time the exchange of fire was longer and more intense. But this time too, slipping off one by one from a mountain ledge, the four infiltrators managed to escape. This tense hunt would see two more exchanges of fire on the morning of 3 April. But for the rest of the day, nothing. As the troop units closed in on what was hoped to be the endgame to a frustrating forty-eight-hour hunt, the terrorists had simply evaporated. There was no sign of them in any direction.

Assisting the troops from the air, drones and helicopters had been deployed on frequent photo reconnaissance dashes to track the terrorists as they furtively moved from one hideout to another. At the 4 Para's team headquarters elsewhere in Kupwara, Subedar Sanjiv and Major Abhishek pored over the latest operational inputs as they streamed in on the morning of 3 April.

'*Jaana toh padhega*,' said Subedar Sanjiv. '*Jaldi jaana padhega, nahi toh woh wapas nikal jayenge* [We will have to get there quickly, if not, they will disappear across the LoC].'

'It had been two days since the boys from 8 Jat had recovered the terrorists' rucksacks, so Subedar Sanjiv had assessed that if we don't eliminate them quickly, they may make an attempt to fall back across the LoC,' says Major Abhishek. 'Temperatures were falling to minus, and there was a lot of snow in that general area. Surviving in those conditions is a Herculean task.'

Subedar Sanjiv knew the area well. He had served in the Jumgund area and operated with the local infantry battalions deployed there. And so, a full twenty-four hours before the 4 Para was actually summoned to help, the senior commando had already got busy with planning the mission down to its last detail.

'He was already studying the maps and recce material that was coming in, choosing the squads and selecting weapons and equipment for the mission,' says Major Abhishek. 'He did not like waiting—he knew by now it was only a matter of hours before the unit would be called in, and he didn't want to waste a minute. He had already stepped up for it.'

When Subedar Sanjiv entered 'planning mode', an unmistakable frisson would pervade the team headquarters. The sudden, hushed bustle of activity was a clear announcement to the men that some of them were about to be handpicked to roll out on their next mission.

Paratrooper Amit Kumar, Subedar Sanjiv's young buddy commando, hovered near the boss. He knew he would be part of whatever was coming next. Havildar Devendra Singh, also sullen-faced and serious, wanted in too, having operated alongside Subedar Sanjiv in dozens of operations. Paratroopers Chhatrapal Singh and Bal Krishan had already begun preparing their personal weapons and equipment. There was no way they weren't going to be on an operation with Subedar Sanjiv.

Just around the time the last firefight was erupting between the terrorist infiltrators and Rashtriya Rifles troops on 3 April, Subedar Sanjiv had finished choosing his dozen men. When the unit was finally summoned, it would take them bare minutes to roll out. This was pre-emptive planning that went to the heart of special operations preparedness.

For the rest of the day, the dozen men went through every bit of input and photographic data that had been received, using it to craft their own battle picture—another compulsion in special operations, since the very need for commandos on the ground meant that a different approach was deemed necessary.

Subedar Sanjiv ordered the squads to turn in early that night. He had made it plainly clear that the summons would arrive the following day. Flying at 12,000 feet, the Heron flying over Jumgund would prove him right.

'On the morning of 4 April, along with the drone video, orders arrived at the team headquarters to deploy as soon as possible to the brigade headquarters in Trehgam for a search-and-destroy operation,' says Major Abhishek. 'Subedar Sanjiv and his eleven men, all trained in mountain warfare, were quickly flown to Trehgam by 10.30 a.m. in two Army Dhruv helicopters.'

At Trehgam, the commando team was quickly escorted into the brigade operations room, where every bit of intelligence collected thus far was on a set of wide panel screens adorning three walls. It was here that the brigade commander first briefed Subedar Sanjiv's team, and then presented them with the choice of proceeding to the Jumgund mountaintop on foot or by helicopter.

'*Chopper se induct honge, sir. Hume jald se jald wahan pahunchne ki zaroorat hai,*' Subedar Sanjiv said. '*Agar wait karenge, toh woh bach niklenge* [The chopper is better since we need to reach at the quickest, or else they might escape].'

It was in the same two Dhruv helicopters that the two squads were now hovering over Jumgund at a drop-off site that couldn't have been more unconcealed for a Special Forces induction.

'Paratrooper Chhatrapal was the first commando to jump out of the first helicopter in low hover mode,' says a commando in the same helicopter. 'When Chhatrapal hit the ground, he sank chest-deep into the snow. He simply went inside—that's how soft and deep the snow cover was. Subedar Sanjiv immediately told the pilot to take the helicopter a little lower to the ground. The helicopter descended and shifted locations by about 10 to 15 metres. The rest of us then jumped out. The snow was very deep. The other squad in the second helicopter jumped out a short distance behind us.'

Induction complete, the two Dhruvs, vulnerable in the open, quickly climbed and peeled away from the drop-off site. On the mountaintop, it would take the dozen commandos thirty minutes to get fully out of the snow and gather their gear.

'We were waist-deep in the snow. We were not walking. We were wading,' says a member of the first squad.

About 20 metres in front were the two team scouts, Paratroopers Chhatrapal and Bal Krishan. Behind them were Subedar Sanjiv and his buddy, Paratrooper Amit. In the rear were Havildar Devendra and his buddy soldier. Just 30 metres behind the first squad were the six commandos of the second squad.

By 2 p.m. they reached the precise point where the Heron drone had spotted the foot trail in the snow. The jumbled mess of shoe impressions, in single file, had survived the light, snowless winds that morning. At regular intervals, it was clear the infiltrators had tried to cover their tracks by kicking at the snow, but the trail was clear enough to follow in a rough direction. The twelve commandos trudged on.

'It was very silent and still, with hardly any breeze,' says a commando from the second squad. 'We were following Subedar Sanjiv's lead while awaiting information from Chhatrapal and Bal Krishan, who were the eyes and ears of the pack. Around 3 p.m., the two scouts spotted something in the distance and signalled to the rest of the squad.'

About 300 metres ahead, Paratroopers Chhatrapal and Bal Krishan saw something move under a small clump of trees. Signalled by the scouts, the squads froze in their tracks.

'*Koi baitha hua hai,*' Chhatrapal signalled back to Subedar Sanjiv. 'Some kind of silhouettes with black hair. It must be them.'

There was excellent visibility in the snow. Now completely still and only signalling to each other or speaking in hushed whispers, Subedar Sanjiv initiated a 'listening drill', a field practice to squeeze every bit of intelligence out of a given situation based solely on sound. But the distance was too much to hear anything other than the dull whistle of the breeze.

'It's definitely them,' Paratrooper Bal Krishan signalled back. 'I can see movement. It looks like there is less snow under those trees and they've taken refuge there.'

Subedar Sanjiv knew the answer already, but he asked anyway, since the two scouts were in the best possible position to judge.

'Can we engage from here, or do we need to get closer?'

'Too far from here. We can't open fire from here. Low chance of hitting them and high chance of their escaping,' said Chhatrapal.

So that was settled. The commandos would need to get closer to the clump of trees to even think of firing the first shot.

'As we readied our assault weapons, Subedar Sanjiv began to plan,' says a commando from the second squad. 'Closing in directly on the location of the trees was ruled out because the sound we would make while wading and trudging through the snow would be simply too loud. If they heard it, they could have been alerted and slipped away down the slope. We needed to play it very carefully.'

With the scouts still out front, Subedar Sanjiv summoned the other nine men towards him so they could hear as he laid out his plan.

'He ordered the second squad to stay in the rear position and try to find some kind of upward slope from which we could get a height advantage to get better eyes on the infiltrators,' says the commando from the second squad. 'Immediately, six of us broke away from the larger team and slowly trudged off in search of a vantage point.'

Leading the advance squad, Subedar Sanjiv and the five other commandos knew that if they were to approach the trees without making a sound, they couldn't trudge through the snow any longer—they would need to crawl.

'We began crawling by centimetres, taking maximum care not to make any noise at all,' says a commando from the advance squad. 'We closed in a little more and stopped at a point 120 metres from the trees, where it was assessed that if we advanced any further, surprise would be lost. It was not feasible to close in any further than this also because there was simply no cover. Six of us were literally out in the open, with nothing to hide behind.'

It was 4.20 p.m. The six commandos were now separated from each other by only a few metres. Subedar Sanjiv looked to the two young scouts.

Chhatrapal and Bal Krishan, both with their TAR-21 rifles now ready, nodded an affirmative. Subedar Sanjiv had his own assault rifle ready. Paratrooper Amit and Havildar Devendra readied their multi-grenade launchers.

'I positioned my Pika machine gun,' says the sixth commando in the squad. 'We awaited word from Subedar Sanjiv to open fire. Generally, we in Special Forces engage at very close distances. In our ambushes, the engagement distance is never more than 25 to 30 metres. But this was 120 metres. But we didn't have a choice.'

Suddenly smashing the silence on that mountaintop at 4.30 p.m., the squad opened a heavy and non-stop volume of fire at the figures they had spotted under the trees ahead of them. Ammunition that included 40 mm grenades and 5.56 mm bullets tore into the target site, whipping up the snow. The firing continued for six minutes without pause.

And then Subedar Sanjiv ordered the squad to hold fire, quickly checking with the scouts for a visual report.

'Because of the distance and lack of an optimal firing position, we were not able to immediately assess the results of the fire,' says the commando with the Pika. 'Still in firing position, we couldn't tell if any of the infiltrators had been hit or had escaped down the mountain. We waited for ten minutes and conducted another listening drill. But there was no sound. Slowly, we started crawling towards the trees again.'

When the six men reached the spot near the trees, there were no bodies. Where the infiltrators had been sitting, they had left behind three syringes and a few pairs of gloves. It was clear that the infiltrators had stopped to rest and had been carrying syringes with chemical cocktails for energy or emergency sustenance. A few feet away from the syringes, the squad spotted something else.

A fresh, dark trail of blood in the snow.

It was now just past 5 p.m.

'In April, last light is about 7.30–8 p.m.,' says the sixth commando. 'Subedar Sanjiv gathered us under that tree and spoke to us in no uncertain terms.'

'*Hume jitna ho sake, unke kareeb jaane ki zaroorat hai,*' said Subedar Sanjiv. '*Andhera hone se pehle* [We need to get as close as possible to them before it gets dark].'

Since one of the terrorists was clearly injured in the firing assault, it was assessed that the infiltration party wouldn't be able to make much progress.

'Blood loss would have compounded three days without any proper food since they had lost their rucksacks. Also, they would likely have been in shock after that sudden and heavy burst of fire,' says the sixth commando. 'It looked like the endgame was in sight, but nothing could be taken for granted. Subedar Sanjiv made it clear that we had less than three hours to close in and finish the mission. We assessed that they couldn't have been more than 200–300 metres down the mountainside.'

The squad reassembled in the snow and reloaded their weapons. Over his radio, Subedar Sanjiv communicated with the rear squad, informing them to advance towards another point on the edge of the mountain and provide cover. Once in position, Subedar Sanjiv and his men began making their way down the mountainside, slowly tracking, following the intermittent spatters of blood in the snow. It was a slow, painstaking descent.

'By 7.30 p.m., light was failing fast,' says the sixth commando. 'We had now reached a steep sixty-five-degree slope.'

About 400 metres downhill from this position was a cordon that had been laid in advance by the Army's infantry and Rashtriya Rifles units that had been hunting the terrorists since 1 April. Subedar Sanjiv used a map with grid references to get a clear picture of where they were and where the 8 Jat infantry positions were down the mountainside. Subedar Sanjiv communicated over radio with the infantry positions further down the mountain, receiving confirmation that they were ready to 'receive' any terrorists attempting to make an escape down the mountain.

It was now reasonably clear that the terrorists were somewhere in the 400-metre stretch down the mountainside that passed through the thick Zurhama forest.

Though 400 metres doesn't sound like much, on a steep, icy slope filled with crags and rocks it is a daunting hunting ground. Even though the squad was taking no chances, it was evident that the terrorists had been trapped—they couldn't climb back because the squad was coming down the mountain after them. And they couldn't move further down the mountain as the infantry would be waiting for them. So, they had no choice but to hide.

The commandos in the squad were now descending the slope keeping a distance between one another. This was to ensure minimum damage in the event of an ambush. The blood trail had gone towards a *pahadi* nallah, an icy, 10-metre-wide snowmelt that streamed down the desolate mountain amid a burst of conifers.

By 8 p.m. it was totally dark. Subedar Sanjiv decided that active pursuit in the darkness was a bad idea—it was better to lay an ambush in the night. The plan was that in the morning the squad would roll down the mountain and quickly eliminate the terrorists.

'It would have been a textbook, easy kind of operation. The terrorists were properly trapped, wherever they were. There was no escape,' says the sixth commando in Subedar Sanjiv's squad.

Under a half moon and the kind of starlight that only lights up such high-altitude areas, Subedar Sanjiv and his squad began crawling out from their positions to begin laying the night ambush along the nallah.

'Try not to engage tonight. We will get them at first light,' said Subedar Sanjiv to the five men. '*Dhyan se*, Chhatrapal, Bal Krishan. *Unko pata hai ki hum yahan hain* [Be careful, they know we are here].'

Dhyan se (Carefully).

The two young scouts had always cherished that warning from their senior commando. It was of course a superfluous instruction—commandos didn't need to be told to be careful during operations. That was their nature, ingrained in their training from the start. But when Subedar Sanjiv said the words, it wasn't just as a senior soldier, but as someone who had watched the two young scouts grow.

Forty-three years old, Subedar Sanjiv had grown up in the neighbouring mountains of Himachal Pradesh. His two scouts, one from the mountains and another from the desert, were half his age. Both were now setting out to survey the nallah for the ambush plan.

But none of their training, none of the pre-emptive planning, none of the minute detailing, nor the word of warning from their senior commando could have prevented what happened next.

As the two scouts began to crawl along a shoulder of the snowmelt, they reached a cornice of snow, an icy ledge over the nallah. It felt solid, except that water from the nallah had eroded away beneath it, making it literally just a precarious overhang of brittle ice. Since the terrain was so steep, there was very little space for manoeuvre and the two scout commandos were crawling along the cornice very close together. Then, with no warning, no creaking of the ice, no heaving of the terrain, it happened.

The cornice abruptly snapped and collapsed, sending Paratroopers Chhatrapal and Bal Krishan 30 feet down into the rocky nallah in the darkness. Twenty metres behind, Subedar Sanjiv and his buddy Paratrooper Amit heard the ice break and the two scouts fall with an uncomfortable series of thuds.

'Subedar Sanjiv wanted to shout out their names to see if they were all right, but he knew that was a dangerous thing to do,' says the sixth commando in the squad who was crawling right behind Havildar Devendra. 'I could tell he was in a huge dilemma. But we needed to act fast and carefully.'

Subedar Sanjiv and Paratrooper Amit used a rope initially to attempt a rapid retrieval of the two scouts who had fallen into the nallah. There was no response or movement from below. So, they both started to crawl down towards the nallah carefully. In the darkness, it was impossible to look down and tell what the condition of the two paratroopers was. As Subedar Sanjiv collected himself as he always did, thinking of his next steps, the silence of the dark mountainside was shattered once again. This time by gunfire.

The unthinkable had happened, and it was impossible to tell in the darkness. Paratrooper Chhatrapal and Bal Krishan had fallen into the nallah and landed, badly injured, just metres from where the terrorists were hiding.

'They fell just five metres from where the infiltrators were crouched,' says the sixth commando. 'Despite the proximity of the incoming fire, they couldn't immediately make out where precisely the terrorists were hiding. Both our guys were already badly injured by the fall. But we listened carefully and could make out that the firing wasn't one-sided.'

Subedar Sanjiv and the other three commandos could clearly hear two different types of rifles being fired. The unmistakable AK-47 clatter from the infiltrators. And the pop-pop echo from the scouts' Tavor rifles. This was a veneer of good news, because it meant the two scouts weren't debilitated during the fall and were still fighting. But the mere sound of rifle fire wasn't enough to inform the rest of the squad just how fearsome the fighting in the nallah down below actually was.

'It was the closest quarter firefight that can be imagined,' says the sixth commando. 'It was all taking place with a separation of 5 to 7 metres. It was impossible to tell who killed who at precisely what point. But we knew that there was no escape from that kind of close combat.'

As the crossfire began to slow down, Subedar Sanjiv and Paratrooper Amit continued their descent. When the firing stopped, Subedar Sanjiv halted, closing his eyes, listening carefully for what the wind would bring to him up that mountainside.

'When the firing stopped, he knew it was possible, even likely, that both Chhatrapal and Bal Krishan had been killed in action,' says the sixth commando. 'But this was a live operation. There's no time to digest such things until the mission is complete. He attempted to establish radio contact with the two scouts but got no response. Even in the darkness, from the sound of his whispers, I could tell this was a very crushing moment for Subedar Sanjiv. He was like a father figure to the young paratroopers.'

In what was likely a combat stress reaction, Subedar Sanjiv refused to give up and hold position, clear in his mind that there was still a chance to get down into the nallah and pull the two scouts out. Navigating between the rocks of the nallah, he followed the lingering gun smoke to the location where the close combat had just ended. In the darkness, he spotted an injured Paratrooper Bal Krishan lying crumpled in the snow. A few metres further down, he spotted Paratrooper Chhatrapal, weapon still in hand, lying face down in the nallah.

This was a very risky exercise since it was still unclear where precisely the terrorists were hiding. If any of them had been killed in the crossfire, there were no visible bodies yet.

Crawling carefully towards the two stricken scouts, Subedar Sanjiv carefully pulled them out of the nallah and up the mountainside. Both the paratroopers were breathing, but weren't making a sound, and were clearly grievously injured. Just as Subedar Sanjiv was pulling them beyond the crossfire zone, firing suddenly burst out again. Under a hail of fire that was landing around him, Subedar Sanjiv sent out a radio message to both his buddy Paratrooper Amit as well as the second squad further up the mountain, that he was bringing the two injured paratroopers up for emergency medical help.

'Subedar Sanjiv's main priority was to pull Chhatrapal and Bal Krishan back and reassess the situation,' says the sixth commando from the squad. 'He passed on a radio message to the other squad that he was pulling the paratroopers back. Just when he had pulled the two scouts to the side, the fire assault began again.'

Positioned about 20 metres higher up the slope, Paratrooper Amit opened 'suppressing' fire in the direction of the terrorists to prevent them from firing or moving. After twenty seconds he stopped, and radioed Subedar Sanjiv for an update on his location.

There was no response.

He tried the squad's second radio, which had been with Paratrooper Chhatrapal. But there was no response from that one either.

At a slightly higher position, Havildar Devendra, the squad's second-in-command (2IC), along with his buddy, the sixth commando of the squad, held position, listening intently through the darkness and scanning the mountainside with infrared night sights on their helmets.

Creeping closer down the mountainside, they spent the night closing in and pinned down the terrorists with periodic fire. Meanwhile further down, intermittent bursts of crossfire between Paratrooper Amit and the terrorists continued at close quarters.

'Havildar Devendra couldn't make out for sure who was firing from where, so he had to be careful. Even with the night sights, it was difficult to conduct effective fire because of the close proximity of the engagement. The radios remained silent. We thought it's even possible that there's no radio response from Chhatrapal and Subedar Sanjiv because they wanted to maintain radio silence as the terrorists were close by.'

In the hail of bullets coming up the mountain, several struck Paratrooper Amit square in the chest and stomach. Bleeding heavily, the young commando trudged back up the mountain, collapsing before Havildar Devendra. He was quickly given first aid and morphine, but Paratrooper Amit had already lost too much blood. As the first rays of the sun filtered through the pines, lying in Havildar Devendra's arms, the young paratrooper took his last breath.

At 4.30 a.m. on 5 April, more firing erupted from down the nallah. The second commando squad which was positioned higher up took a decision to venture down the mountainside as a single AK-47 continued to fire intermittently from behind some rocks in the snowmelt. Morning light made the direction of fire clearer. Scouts from the second squad sent a pair of grenades careening into the hiding place, silencing the sole terrorist who was still firing.

After a night filled with gunfire, the Jumgund mountainside was totally silent once again.

Very little can conclusively be said of what happened from the time Subedar Sanjiv descended into the nallah and the end of the firefight at dawn. But when the second squad descended into the nallah for a final clearing operation, a hair-raising sight lay in wait.

Hair-raising even for commandos.

The squad first passed the body of Paratrooper Amit. The young soldier had received thirteen gunshot wounds but still managed to climb up the mountainside before succumbing to his injuries.

A short distance further down, they found the body of Havildar Devendra. He had spent the night returning fire at the terrorists. Not far from him was the sixth commando. He had sustained two gunshot wounds but was alive. His rifle lay a few metres behind him. It is most likely that it flung from his grip when he was hit. He was immediately taken up the mountainside for medical aid.

The two Dhruv helicopters had been summoned back to the mountaintop for an emergency evacuation. In communication with the second squad, one of the helicopters hovered briefly over the nallah to see if it could capture what happened. From the air, the pilots got the first glimpse of what had happened. They quickly relayed the information to the second squad which was descending to that point in the nallah.

It was when the squad reached the location of the actual crossfire that the truth of the fighting became clear. The bodies of Paratrooper Chhatrapal, Paratrooper Bal Krishan and Subedar Sanjiv were found entangled with the bodies of the terrorists. Injured in the fall and with gunshot wounds, the two young paratroopers had charged at the terrorists and killed them in fearsome hand-to-hand combat down the mountainside. Subedar Sanjiv, too grievously wounded in the crossfire, had pounced on the third terrorist and beaten him to death before life ebbed from him.

'Even for commandos like us, it was a numbing sight,' says a soldier from the second squad. 'These three guys had accepted that their life was over. They had decided that they would make it count for the mission. We still don't know how the two young scouts were able to fight after being so badly injured. We will never clearly know how they managed it.'

Only hours earlier, Subedar Sanjiv's squad had imagined the operation to be a textbook ambush that would see the four terrorists eliminated quickly the following morning.

'We just thought we would roll down and eliminate four to five guys and go back home,' says the sole surviving commando of Subedar Sanjiv's squad. 'But what happened on the ground was different. The paratroopers fell in front of the terrorists. I don't think anybody could have avoided or done better in the circumstances. It just came down to that one mishap. It was the easiest of operations. It was an unplanned contingency that changed the game against us. And the price we had to pay.'

The bodies of the five commandos, including their team leader, were carefully carried up the mountainside where the two waiting Dhruvs flew them first to the Badibai Post nearby, and then straight to the 92 Base Hospital in Srinagar—standard operating procedure even if soldiers are deemed dead at the site of an operation.

Even for a battle-hardened Special Forces unit like the 4 Para, the loss of five men in a single operation was many levels beyond a shock.

'The second squad simply could not comprehend the scale of the loss. They were devastated on that mountainside. The carnage in front of them of their own men. *Unki aatma hil gayi thi* [They were rattled to the core],' says Major Abhishek, who was tracking the operation from the 4 Para team headquarters in Kupwara and was the first to receive the news about the devastating loss.

'I got the news at 7 a.m., and was in total disbelief,' says the young officer. 'My first instinct was to reject the news, telling myself that it was impossible. This could not have happened. We were in a state of shock, it is difficult to explain how we felt. We were in denial for hours. Despite our drills and operational philosophy, this was impossible to digest. I immediately got out my quick reaction team (QRT) and started moving towards Srinagar on the morning of 5 April. It was later, after I saw the five coffins, that it finally hit me.'

'There hadn't been casualties like this since Operation Pawan in 1987,' he recalls. 'We never had such casualties despite being in Kashmir for so long. We had lost Lance Naik Sandeep Singh in 2018.* Otherwise, 4 Para had not been touched. There had been injuries, but we had eliminated so many terrorists. Not just the battalion, the entire Northern Command was in shock. Even now there's disbelief that we lost five men that day just because of a mishap.'

---

* See Chapter 7, on Lance Naik Sandeep Singh, 4 Para Special Forces.

The 4 Para, like the Army's other Special Forces units, is trained fully to prepare for combat losses—for the likelihood of death is an adjunct to operations. Training and probation in units are as raw and realistic as it is possible to be when preparing for near-death situations.

'But no amount of training could prepare us for what happened that day,' says the sixth commando, the only man from the squad who survived. 'We men in the unit live together, eat together, breathe together. Out of fifty men, you suddenly face the fact that five are no more, it's a devastating blow. We are like brothers in a unit. It was not easy to come out of that state.'

But the unit, under the watchful eye of their commanding officer, who needed to rally every ounce of strength and composure to keep his men focused, had to move on. Once the operational debriefs were complete and all information about the mission recorded, the men went back to doing the only thing they do when they're not in an operation—drawing lessons from the previous operation and training for the next one.

'The best thing was to bury ourselves back in training, remind ourselves that we are God's chosen soldiers, built for the toughest of jobs,' says the sixth commando. 'We kept ourselves busy. Telling ourselves that this might happen again, but we cannot falter in our commitment and duty. We have to be on standby for the next infiltration which can happen at any time, so we don't have the luxury to remain in shock. I cannot explain the atmosphere at that time. One could only feel it.'

The phrase 'brothers in arms' is no exaggeration in the military. And it's even more true when it comes to small, tightly knit commando units. The thirst for retribution can be overwhelming. It always comes down to the maturity and focus of the commanding officer to hold his men together and channel their anger.

'The men were full of sorrow, but also full of quiet rage,' says Major Abhishek. 'Some of them were so filled with emotion, they would go to the CO and say, *Saab aap bas aadesh do, dobara jayenge, sar kaatke layenge* [Sir, okay a mission and we will go and bring back their heads]. They wanted revenge. Nobody is a Rambo. There is a human somewhere inside each commando.'

After a numbingly silent wreath-laying ceremony at the Badamibagh cantonment in Srinagar, the remains of the five commandos were flown to their respective hometowns to their families.

On 6 April, twenty-two-year-old Paratrooper Chhatrapal Singh's casket arrived at his home in Chhavasari village in Rajasthan's Jhunjhunu district, where huge crowds from adjoining villages lined the rooftops and streets to bid the commando farewell. Images of his devastated, weeping parents being consoled by two commandos from the 4 Para would numb the nation. Images of medical staff in pandemic PPE kits delivering the body only reminded those watching that the infiltration of Pakistan-sponsored terrorists had been carried out during a global health crisis, and that the young commandos had to be sent in to battle at a time when the world had far more pressing priorities.

At around the same time, Paratrooper Bal Krishan's body arrived at his family's home in a remote part of Himachal Pradesh's Kullu district. A large *shamiana* had been erected outside the house to accommodate the mourners who had gathered there to receive the body and pay their last respects. But it was the image of a tall commando from the 4 Para breaking down while embracing Bal Krishan's grieving mother that would go viral. Special Forces men always hold their composure, trained to control their emotions. The image of the weeping commando was perhaps the most heartbreaking sign of the unit's sense of loss.

On Republic Day 2021, Subedar Sanjiv Kumar was decorated with a posthumous Kirti Chakra, the country's second-highest peacetime gallantry medal, equivalent to the wartime Maha Vir Chakra.

His citation reads, 'Subedar Late Sanjiv Kumar displayed outstanding leadership qualities, raw courage and utmost gallantry in eliminating one hardcore terrorist, injuring two terrorists, evacuating his injured scout and made the supreme sacrifice in the highest traditions of the Indian Army. His selfless action led to the subsequent elimination of five hardcore terrorists. For exhibiting conspicuous courage beyond the call of duty, Subedar Sanjiv Kumar is awarded Kirti Chakra (Posthumous).'

Subedar Sanjiv's wife Sujata Devi received her husband's medal at an investiture ceremony at Delhi's Rashtrapati Bhavan in November 2021. She lives with their teenage son, Kanishka, at their home in Bilaspur, Himachal Pradesh.

The other commandos would be decorated with Sena Medals for fearless gallantry in the operation.

The post-mortem of Subedar Sanjiv's buddy commando Paratrooper Amit Kumar Anthwal showed he had taken fifteen bullets in the operation. He was bid farewell by his parents in Pauri-Garhwal, Uttarakhand. The young commando was engaged to be married six months later in October 2020.

Thirty-nine-year-old Havildar Devendra Singh Rana's body arrived at his village Tinsoli in Uttarakhand's Rudraprayag on 6 April. His mortal remains were consigned to the flames by his parents, wife, Vineeta Devi, daughter, Aanchal, and son, Ayush. His ashes were scattered in the Mandakini River.

Thirty-eight days after the operation, the sixth commando from Subedar Sanjiv's unit was cleared to leave hospital in

Srinagar. A 7.62 mm rifle round had been removed from his shoulder, with the flesh and bone carefully reconstructed to give him back his mobility.

The first call he made was to his commanding officer at the 4 Para team headquarters in Kupwara.

'*Beta, tum kuchh hafte ke liye ghar chale jao* [Son, go home on leave for a few weeks],' said the commanding officer. '*Jab tum tayyar ho, tab wapas aa jao* [Come back when you are ready to join duty].'

The commando didn't take a moment to reply.

'Sir, *main aaj hi waapas aa raha hoon* [Sir, I am getting back to work today itself].'

* * *

Postscript

A commando from the 4 Para, Paratrooper Sonam Tshering Tamang, was decorated with a Shaurya Chakra in November 2021. The award citation noted:

> Paratrooper Sonam Tshering Tamang displayed raw courage and utmost gallantry in eliminating one hardcore terrorist, injuring one other terrorist and evacuating Subedar Sanjiv Kumar to safety. He maintained his composure and fought with nerves of steel under the most adverse circumstances wherein all other squad members of his detachment were killed in action.

4

'Where the Hell Is His Leg?'

Squadron Leader Ishan Mishra

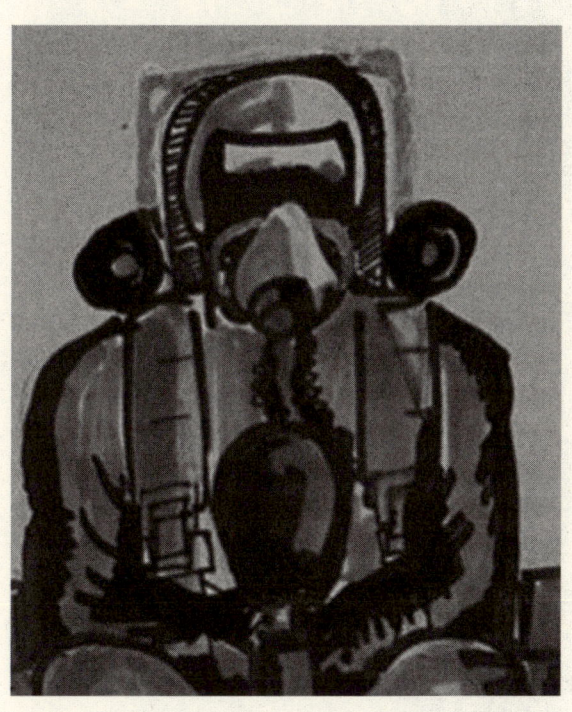

'We're going down! Initiating ejection now!'

The Sukhoi Su-30 fighter jet was barely 23 metres from the ground, flopped on its right side and screaming downward under a moonless night sky near Tezpur, Assam, when Squadron Leader Ishan Mishra made the final call to the pilot in the second seat behind him.

The twin Saturn AL-31FP turbofans—monster engines that powered the big, heavy Su-30 and permitted the jet unusual levels of aerobatic agility for its size—couldn't have failed at a worse time. A night-training flight. A dark sky. And nearly nothing outside for the pilots to look at to orient themselves in a doomed, falling jet fighter. Not even the familiar glint of the Brahmaputra a short distance south.

And at the speed the aircraft was falling, it would smash into the ground in less than three seconds.

His left hand on the glass canopy of the fighter, Squadron Leader Ishan used his right hand to yank the ejection handle. As an explosive charge blew away the canopy and blasted the rear pilot out of the aircraft, Squadron Leader Ishan felt for a microsecond a strange mix of regret and calm.

Regret, that he wouldn't be able to speak to his wife and son one last time and prepare them for what was about to happen.

And calm, in the certainty that there was no way he was going to survive a lopsided ejection from less than 20 metres. In his final moments in the cockpit, he pushed the aircraft's left rudder pedal as hard as he could, in an effort to keep the falling jet nose-up for as long as possible.

The last thing Squadron Leader Ishan felt as the rockets under his seat flung him with bone-crushing force out of the cockpit, was a quick sharp flash of pain across his left wrist and left leg.

Then everything quickly went black.

In a separate corner of the country, 1700 kilometres away, Squadron Leader Shalika Sharma, headset on, sat hunched over a radar console at the ground control station in Punjab's front-line Halwara air base. That night of 8 August 2019, she was on the late shift guiding pilots very much like her husband in similar Su–30 MKI jets in the dark airspace that hugged the international border with Pakistan.

Only six months earlier in February 2019, the Halwara air base, situated just outside the sprawling city of Ludhiana, was on wartime alert following the Indian Air Force's air strikes in Balakot, Pakistan.* While the actual bombing was carried out by Mirage 2000 jets home-based at Gwalior, Su–30 fighters from Halwara had been part of operational support 'packages' conducting flights close to the border. As a forward base, Halwara is constantly in a state of operational alert. But after Balakot, it would be literally war-ready. The night flights being controlled by Squadron Leader Shalika that August night were part of the uninterrupted readiness training for combat pilots at the base, keeping them available to scramble at very short notice.

The night flights, a regular part of base operations, had increased in tempo and frequency, keeping the runway as active at dusk as it was in sunlight. Squadron Leader Shalika wouldn't know until hours later that while she oversaw the

---

* See *India's Most Fearless 2*.

first set of Su-30 fighters on her watch landing safely back at the Halwara base that night, her husband had hit the swampy banks of the Brahmaputra with a barely open parachute, 5 km from the Tezpur air base.

Aware that Squadron Leader Ishan was on a night mission that evening—pilots always inform their spouses or families—the young radar controller also knew that a part of her would only exhale freely once she saw that familiar WhatsApp ping after her shift had ended, and she could leave the 'no smartphones' zone that included the ground control area. A ping that would show up on her lock screen displaying the word 'landed' and that most comforting of emojis in the circumstances—a yellow 'thumbs up'. She had experienced this for nine years, but it never really got easier.

The two had fallen in love as cadets at the Air Force Academy, entering service together in 2010. As young flying officers, their paths would diverge towards the two ends of fighter operations—the cockpit and the ground control station. Not allowing the vagaries of postings in different locations to come in their way, Shalika would be married to Ishan in May 2013 in Pathankot, a short distance from her hometown, Gurdaspur. In the decade of their partnership, she couldn't remember a single instance when she hadn't received a message before and after a fighter flight.

But that never meant it interfered with work. There was more than enough on her plate, especially that night, with heavily armed fighters in the air that were depending on her to track and control.

Above all, there was Ayaan. Not far from the airfield at her family quarters on the Halwara base premises, the toddler son of the two young Air Force officers was being babysat and tucked into bed by Shalika's parents.

It was a particularly busy shift. Thirty minutes later, Squadron Leader Shalika refreshed her radar console, preparing to monitor a third formation of Su-30 jets for low-level night sorties over a stretch of airspace between the rivers Sutlej and Yamuna, unaware that in that other corner of the country, her husband had opened his eyes.

His head swimming and barely conscious, all that Squadron Leader Ishan saw was that same dark sky. He wanted to move to see if he could spot the other pilot, but not a muscle would twitch. To his right, he could feel searing heat from the flaming wreckage of his Su-30, which had impacted the ground right outside what appeared to be a village. And over the crackle of the fire and the hiss of aviation kerosene fuelling the mangled inferno, he heard human voices.

'*Tar bhori khon kat baal*? [Where the hell is his leg?],' one voice asked in Assamese.

'Yes, there's no left leg,' said another.

No left leg.

Woozy from the frightfully low ejection, Squadron Leader Ishan felt himself flinch, his mind leaping to come to terms, forced into a rictus of realization.

Then he heard a third voice from about 30 feet away.

'*Eitu ki? Eikhon niki tar bhori*!? [What's this? Is this his leg!?]'

Squadron Leader Ishan blinked away a mix of blood and sweat that had formed rivulets down his face. He knew just enough of the language to understand what had just been said. And right before things faded to black again, he wondered if, after nine years as a front-line fighter pilot, he ever wanted to wake up again.

* * *

Thirty years old and with 1600 hours of fighter flying in his logbook, Squadron Leader Ishan had arrived in Tezpur early in May 2019. As a freshly minted qualified flying instructor (QFI), he had just spent the previous twenty months training three batches of pilot cadets at the Fighter Training Wing in Hakimpet near Hyderabad. With the end of his tenure and in the tumultuous wake of the Balakot air strikes, Squadron Leader Ishan, like many other combat pilots, received postings back to front-line air bases in the northern, western and eastern sectors.

After nearly two years away from them, Squadron Leader Ishan had hoped to reunite with Shalika and Ayaan. He had barely seen his infant son who was born just months before he received orders to proceed to Hakimpet to train cadets. If video calls had bridged the distance in some small measure, the new parents also knew they couldn't really complain— the Air Force had been more than empathetic, as it usually is with serving personnel married to each other. Helped by the fact that the two were part of the same domain of operations as pilot and controller, the newly-weds had been deployed together at the Bareilly air base in 2013 and had been posted together at the Halwara base from 2016 to 2017. Four years out of six since their marriage spent in the same station was not bad. Until postings allowed them to be in the same place again, Squadron Leader Ishan would fill in the blanks with visits to Halwara to see his wife and baby son.

But once in Tezpur, there was a typically gruelling work schedule at hand. Even though Squadron Leader Ishan had flown Su-30 jets for over five years, with postings to squadron locations that included Pune, Bareilly and Halwara, he still had to 'revalidate' himself on the air superiority fighter, since he had just spent nearly two years training cadets in HJT-16 Kiran intermediate jet trainers. This wasn't a comment on the pilot's abilities by any measure—conversion and revalidation training was a standard fixture to ensure pilots are fully integrated with cockpits and systems that are either new to them or ones they may have lost touch with.

'That night of 8 August was my last revalidation flight,' says Squadron Leader Ishan. 'I had already finished twenty-four flights successfully. This was to be my final flight with a supervisor, a senior pilot from the squadron. After that, I would be back to being full ops on the Su–30 and ready for any mission.'

On an oppressively muggy north Assam evening, Wing Commander Pritam Santra, the squadron's flight commander—second-in-command in the unit—was the supervisor who climbed into the back cockpit of the Su–30. Squadron Leader Ishan wasn't exactly a 'student' here. He had just spent 700 hours in his previous posting certifying his own student pilots to cross the difficult bridge between slow propeller trainers and faster jets. But in the Air Force, training and validation is an unending exercise. Changes in software, avionics, procedures, weapons and moves made by neighbouring nations, place pilots in a constant cycle of keeping pace with technology and doctrine.

Minutes before climbing into the front cockpit, Squadron Leader Ishan called his wife on her mobile. When she didn't answer, he called his mother-in-law quickly, informing her that he was leaving for a night-flying sortie. Shalika was getting ready for her night radar controller shift. As she left home, her mother informed her that Ishan was off on a night mission.

'Yes, he sent me a text as always,' she smiled, before driving out from the quarters towards Halwara's ground control station. She made a mental note about the approximate time the mission was starting, though she knew that Su–30 missions were sometimes unexpectedly lengthy. Just around the time that she arrived at her station for a solo shift on the radar console, her husband and his supervisor were getting airborne from Tezpur.

'We took off and headed east over Chabua in upper Assam,' says Squadron Leader Ishan. 'There we carried out a couple of low-level passes over the airfield. Everything was fine. The aircraft was handling perfectly.'

The pilots then turned their aircraft around, powered up into a climb and headed back west towards Tezpur, where they proceeded to conduct more low passes over the runway.

'This was to get used to the runway at night, the pattern of the facilities and adjoining terrain—crucial elements in case of emergencies,' says Squadron Leader Ishan.

At 8.08 p.m., Wing Commander Pritam instructed the pilot in front to conduct one final overshoot, a manoeuvre in which the Su-30 was to be brought nearly all the way down to the runway as part of a normal landing approach and aborted at the last second, just before touchdown to instead climb away.

'In that last overshoot when I came four metres above the runway, just as I opened both the throttles to accelerate and get airborne again, I instantly felt the weight of the jet simply refusing to pull up properly,' says Squadron Leader Ishan. 'Right on cue, my right engine oil pressure failure indicator blinked on. This made total sense, because this is a very powerful aircraft, and it was refusing to pull up in the nice way that it usually does. We were dangerously low and turning to the right, all the while our speed dropping and altitude rising only very slowly to about 200 metres. But this climb was nothing to feel safe about, since we had no power coming from the engines. I immediately handed control back to my supervisor, Wing Commander Santra.'

Both pilots quickly guessed that if the aircraft wasn't able to climb immediately, it would take no more than eighteen seconds for their Su-30 to hit the ground. Squadron Leader Ishan heard his supervisor quickly send out a calm distress call to ground control at Tezpur, a final notification before the two men would do nothing else but try to save the jet and themselves.

'After handing control back to my supervisor in the rear cockpit, I pushed a button that engaged what is called 'combat mode', which should have given us an additional 1.5 per cent of thrust in the residual engine,' says Squadron Leader Ishan. 'The problem was we were already at very low speed. And in such a regime, opening power doesn't really pay off unless you point the nose down. And I don't need to tell you that pointing the nose down at this altitude would have simply brought things to an end much faster.'

The doomed Su-30 had entered a cruel, paradoxical situation called the 'region of reversed command', a situation where an aircraft losing power and at low speed needs to point its nose down in order to be able to accelerate enough to gain the power required to climb. At high altitudes, such a 'stall' isn't a problem since pilots have enough depth to dive their jet back into realms of control. At 23 metres, the only thing to dive into is the ground.

The fighter jet's landing gear was still down from the attempted overshoot manoeuvre, the wheels adding drag and slowing the aircraft down even further. Squadron Leader Ishan pushed a button to retract the landing gear, hoping the better aerodynamics would allow some vestiges of performance from the plane to power out of this dangerously low emergency. It was at this point that the right engine died completely, with the left engine winding down and simply unable to keep the aircraft level.

'Apart from this being a night sortie, we were also flying in overcast weather, and in the absence of the moon, the only outside feature available to us was light from a small village that happened to be in the flight path of our aircraft,' says Squadron Leader Ishan. 'Our repeated control inputs would push the aircraft to climb only slightly before dropping back down. It was a turbulent oscillation very close to the ground.'

'Ishan, engage reheat!' came the urgent call from Wing Commander Pritam in the rear cockpit, an instruction to activate the engine's afterburners, a system in many fighter aircraft engines that explosively burns additional fuel near the exhaust, adding a sudden punch of thrust to the aircraft.

'The moment I pushed my throttle up to engage reheat, the aircraft started to yaw violently from left to right—this was because we were on residual power from just one engine, while the other had given up,' says Squadron Leader Ishan Mishra. 'From the moment Wing Commander Pritam had called "engage reheat", it would be four seconds until the realization hit me.'

Amid the head rush of decision pressure, and seconds into another stomach-churning lurch downward, the younger pilot suddenly had a moment of total clarity.

'About twelve seconds into the emergency, I was totally sure that if I didn't initiate the punch out now, whatever happened to the aircraft would happen, but there was no way we both would survive,' says Squadron Leader Ishan. 'So instinctively I pulled the ejection handle with my right hand, while my left hand rested on the canopy. I felt my left wrist break in a sudden flash of pain. There was no time to attend to it. The canopy had blown off, and the wind had roared into the cockpit for a split second before Wing Commander Pritam exploded out of the rear cockpit. I used the last ounce of my power to keep the aircraft's nose pointed as far up as possible, while trying to veer it away from the village in our flight path. My eyes momentarily caught the altitude reading just as I was blasted from my cockpit at 8.10 p.m. We were at twenty-three metres and perhaps a second and a half from impact with the ground. I was certain I wouldn't survive. I just said a prayer that my supervisor makes it, and that the aircraft doesn't harm anyone in the village ahead. I felt a deep flash of regret that I could not say goodbye properly to my wife and baby boy. But the rest I had made my peace with.'

The last thing Squadron Leader Ishan saw as he bailed out was the Su-30 ploughing through a grove of bamboo trees and smashing into the ground in under two seconds, about 30 metres from the first house of the village.

'After I punched out, the Su-30 veered slightly to the left and I soared towards the right,' says Squadron Leader Ishan. 'During ejection the amount of force felt by the body is almost 20G, so for that precise moment, everything blacks out. When I hit the ground in the dark, I felt the impact. It was pitch dark and suddenly that blast happened, a big flash in front of my eyes. I landed just a few metres from the aircraft fireball. There was a lot of oil spill. It was flaming hot. Later I learnt I had sustained some burn injuries on my back too, possibly from the burning fuel from the aircraft that came close to me. It was a miracle I didn't land directly in the wreck itself.'

Squadron Leader Ishan wasn't the only one who saw the blast. Alerted to the distress call, two other Su-30 fighters flying in the area had turned towards the location just in time to see the bright orange fireball that momentarily lit up the north banks of the big river.

'When you are flying in places like Tezpur and Chabua, it's away from the big cities and completely dark,' says Squadron Leader Ishan. 'So when a blast takes place anywhere in the visual span of an aircraft, it shows very clearly. That's how the two Su-30s in the air spotted us immediately.'

Wing Commander Pritam, who was lying on his back a short distance away, was in his senses and happened to have his mobile phone with him. With difficulty, he called the Tezpur tower and informed the flight safety officer that he was alive, but that he wasn't able to immediately confirm if the younger pilot had survived the ejection.

Twenty minutes after the crash, news filtered to local TV and then social media that a fighter had crashed near Tezpur, with no information about the condition of the pilots. The reports didn't identify the aircraft type or the names of the pilots, but anyone vaguely familiar with the Indian Air Force's deployments could guess what was involved. Even before emergency response teams had reached the site of the crash, the news had broken by 8.30 p.m.

Squadron Leader Ishan's parents were watching the news at their home in Bhubaneswar when the breaking news flashed across their screens that a fighter jet had crashed in Assam. They froze, immediately picking up their phones to see if they could get more information.

In Halwara, alone in the ground control station, Squadron Leader Shalika couldn't receive calls or messages. But other colleagues had heard about the crash. Two of them rushed to the ground control station.

'Two juniors of mine showed up, and I was not expecting to see anyone on a solo shift, so I was surprised,' says Squadron Leader Shalika. 'They asked me to go home early since I had an exam to study for. I was immediately a little anxious, so I made arrangements for my relieving officer and left for home. Once outside the control station, I checked my phone as I always did, noting that there was no message from Ishan that he had landed. By now I knew something was wrong. But I wasn't panicking. Just as I reached home, the landline phone rang.'

It was Ishan's commanding officer from Tezpur, Group Captain Bopanna.

'Jai Hind, sir, *kaise ho* [how are you]?'

'Shalika, kaise ho?'

'I'm fine, sir. Anything?'

'Shalika, Ishan *ka* ejection *hua hai kuchh der pehle* [Ishan had to eject a short while ago].'

'Okay, sir,' said Squadron Leader Shalika. '*Par woh theek hai na*, sir [But he is okay, isn't he]?'

'*Haan*, we are getting him to hospital.'

She thanked her husband's CO and hung up, quickly collecting her thoughts on the next step to be taken. Most urgently, she needed to tell her parents, Ishan's parents and then travel on the first available flight to see her husband.

As she began to break the news to her mother, a group of senior personnel arrived at her quarters, including the station's chief administrative officer and the commanding officer of the local Su-30 squadron. Calm and composed thus far, Squadron Leader Shalika lost her cool.

'If you have no information about Ishan, why have you come here?' she pleaded with the visitors. 'My parents are here, and I don't want to stress them out unless I have to. Either tell me what is going on or make me speak to Ishan.'

As she waited for more information, she called her husband's uncle, who lived with her in-laws in Bhubaneswar.

'He had spoken separately with some air force people, but was keeping the information to himself,' says Squadron Leader Shalika. 'He told me Ishan is hurt but safe. "*Ghabrana mat* [Don't panic]." He switched off Ishan's parents' phones. They were made aware that Ishan had ejected from an aircraft but would not get to know any details about their only child until the following day.'

At the site of the crash, a group of locals from the village, who had just about escaped the flaming Su-30, were now gathered around the crumpled body of Squadron Leader Ishan, his flight suit caked with blood and slush. Among the villagers was a man who worked at the mess in the Tezpur air base.

Someone from the base had called him and asked him to rush to the crash site to secure the two fallen pilots.

'The last thing I heard was that voice saying they had found my leg some thirty feet away from me,' says Squadron Leader Ishan. 'After that I didn't have much consciousness. When I woke up, I was at the hospital late that night, having been brought in an Innova from the crash site. The only thing I remember was one of my squadron mates standing next to my bed with his phone in front of my face. I could vaguely make out Shalika at the other end of the video call. Luckily my face wasn't damaged.'

'Ishan, don't say anything, just give a thumbs up,' she said softly.

'Shalika, *main theek hoon*. I'm alive. Don't worry. *But tu aaja yahan par* [Please come here].'

Of course, Squadron Leader Ishan was far from fine. His spine was miraculously intact—unusual, given that compression injuries to the spine are among the commonest afflictions from an ejection—but the young pilot had suffered other terrifying injuries. He had to be given forty-two units of blood to make up for the amount he had lost at the crash site. A huge number of personnel from the Tezpur station were already there to donate blood for the injured pilot.

And the reason he couldn't give Shalika a 'thumbs up' on the video call was because all four of his limbs were shattered, broken to pieces.

And his left leg had been completely ripped off below the knee. The severed limb had been retrieved by a separate team seven hours later from the boggy swamp and placed on ice at the hospital.

'My leg got separated from my body during the ejection,' says Squadron Leader Ishan. 'It was a sleek cut like someone had cut it off with a saw. When I landed, my left leg was not connected. That is why my knee sustained injuries. I was not conscious when they brought the leg to the hospital. Had the severed leg remained inside the aircraft, it would never have been found.'

Shortly after Squadron Leader Ishan, his supervisor Wing Commander Pritam was also brought to the hospital. The senior officer had suffered injuries too, but mercifully of a less serious nature. Both men would be sedated for pain and remain unconscious through the following day as doctors dressed their wounds and stabilized the vitals of two bodies in shock. Senior medics from the Command Hospital in Kolkata and Air Force Hospital in Assam's Jorhat also arrived early on the morning of 9 August to monitor Squadron Leader Ishan and decide on the next course of action.

The doctors were unanimous—Squadron Leader Ishan needed to be flown as quickly as possible to the Command Hospital in Kolkata. So, twenty-four hours after the crash, the sedated pilot was wheeled into the cabin of an An-32, which would fly him and an icebox containing his left leg, to the eastern metropolis, better equipped to handle the injuries he had sustained.

By the time Squadron Leader Ishan was rushed into the intensive care unit (ICU) at the Eastern Army Command Hospital in Kolkata, the court of inquiry (CoI) investigating the accident had deduced that the pilot had sustained his life-altering injuries because his parachute hadn't had enough time to slow his descent fully before he hit the ground.

'When I arrived in Kolkata late at night on 9 August, two officers were there to receive me,' says Squadron Leader Shalika. 'Throughout the flight, I was preparing myself for the worst. When I saw the two officers there, I wondered if they would take me to the hospital or a wreath-laying ceremony.'

Heavily medicated, Squadron Leader Ishan vaguely remembers seeing his wife next to his bed in the ICU.

'He was barely conscious and asked me to go to the mess and take rest and come back in the morning,' she says. 'Imagine asking me to go and take rest while in that state! I was convinced that Ishan wasn't fully aware of what had happened to his leg. I was so relieved to see him alive and talking to me that the news of his lost limb didn't quite sink in. His brain was okay, his blood was flowing, he could communicate. I tried to shut out the lost limb from my mind, focus on the positive and not break down. I know how much Ishan loved his life as a pilot. I knew his heart would break, but for that moment, I was too busy thanking our stars that he wasn't dead.'

She was wrong about one thing, though. Her husband was fully aware he had lost a limb.

'When they were dressing my stump in the ICU, they would put a curtain that hid my legs from me,' says Squadron Leader Ishan. 'Everybody thought I had no idea about what had happened. But I knew from day one because *pain toh mujhe ho raha tha* [I was feeling the pain]. Initially they were giving me local morphine injections and not general anaesthesia, because the latter ran the risk of sending me into a coma.'

For five days, doctors at the Command Hospital investigated the possibility of reattaching the pilot's severed left leg in the hope that the blood vessels would work again. But the damage to the stump was too severe. And therefore, on 14 August, six days after the crash, it was decided that Squadron Leader Ishan's left leg would be amputated further, to excise the smashed portion.

That morning, the commandant of the hospital, Major General S.R. Ghosh, and a senior Army orthopaedic surgeon, Lieutenant General R.S. Parmar, came to visit the pilot. Both wanted to mentally prepare Squadron Leader Ishan for the surgery, break the news to him that he wouldn't get his leg back.

'The irony is everyone was talking to me carefully, like I didn't know what had happened,' says Squadron Leader Ishan. 'On the other hand, there I was stopping myself from discussing my injuries with my wife or squadron mates in the ICU because I thought they didn't know I'd lost a limb and didn't want to upset them. They were being strong for me. I was being strong for them. I guess that's what love and solidarity is all about.'

The advice to the injured pilot from the senior doctors was clear. Reconstructing his shattered knee would take time, and there was little or no guarantee that it would sustain. A prosthetic knee also had low chances of success given the complexity of his injury. There was only one option left.

Squadron Leader Ishan and Squadron Leader Shalika listened intently. Neither wanted to look the other in the eye in the room that day as the two doctors offered their final verdict.

'Absolutely, sir,' said Squadron Leader Ishan, careful not to allow a pause to betray any hesitation whatsoever. 'You do what you think is best. I don't have any problem.'

The amputation was scheduled for 14 August. As he waited in his hospital bed, Squadron Leader Ishan received an uninterrupted stream of visitors, mostly course mates and fellow personnel from his current and past squadrons. One course mate from Hyderabad had arrived at a time when the injured pilot had been prohibited fluids of any kind, since he was about to be operated on.

'I was simply dying of thirst but had been barred from touching any fluids,' says Squadron Leader Ishan. 'I was being given a lot of saline, so I was maddeningly thirsty. My mouth was dry, it was a horrible feeling. My course mate Sumit was there in my room. I said, "*Bhai mujhe bahut pyaas lag rahi hai*

[I am very thirsty]." He said, "*Doctor ne bola hai main nahi de sakta, teri zindagi ka sawal hai* [The doctor has said that I can't give you liquids, it's a question of your life]." I pleaded with him, saying, "*Tujhe hamare NDA days ka vasta paani pila de iss bhai ko* [For the sake of our NDA days please give your brother some water]." I was literally begging him with folded bandaged hands. He gave in and picked up a bottle and was pouring some into my mouth when the doctor walked in and saw what was going on. The doctor was furious. He said, "Don't you want your friend to remain alive during the operation? Do you want him to die?" Then he was thrown out of my room, poor guy.'

A twelve-hour surgery followed. And around the same time, the pilot's severed left leg was taken out of its icebox and sent into the hospital's incinerator. The leg that had pushed on the aircraft's rudder pedal in an effort to pull it away from the path of the village, was turned to ashes.

If the loss of his left leg wasn't traumatic enough, Squadron Leader Ishan came terrifyingly close to losing his right leg too. The latter limb had contracted a condition called acute compartment syndrome (ACS) in which elevated pressure within the leg compartment was stifling blood supply to the tissue in the limb, thereby endangering it as a whole. Amputating the right leg was discussed as a possible compulsion if things didn't improve, especially since the condition was seen to be affecting the pilot's kidneys as well. Mercifully, the doctors were able to operate on the leg, relieve the pressure and save the limb. It was still shattered in places, but the verdict was it didn't need to be amputated.

The CoI was now a week into its investigation, but nobody was clear yet as to why the accident had taken place. Whether a technical fault or engine failure, it remained unclear for a while. That a young fighter pilot's front-line combat career had ended so abruptly imbued the probe with urgency and a painful veneer.

'People were not ready to conclude why I had lost my leg,' says Squadron Leader Ishan. 'As my anaesthesia wore off, and I began to get my faculties back, the horror of realization came in waves. After my initial rehabilitation, I received my posting back to Halwara. Before proceeding, I travelled back to Tezpur in December 2019.'

Missing a leg now and in a wheelchair, the pilot was adamant that he wanted to visit the crash site.

'I asked a duty medical officer from the Tezpur base hospital to accompany me there,' says Squadron Leader Ishan. 'As we stood there, looking at where my aircraft had hit the ground and where they had found me, the medical officer got a little emotional.'

'You don't know how good it is to see you well,' said the medical officer. 'When we were cutting through your flying overalls and doing your MRI [magnetic resonance imaging] for an initial look at whether there was internal damage to your organs, you were screaming in pain. I was holding you down, trying to calm you and ask what I could give you to make you feel better.'

In convulsions of agony, Squadron Leader Ishan had screamed back, 'Sir, *mujhe* cold coffee *pila do*, I am feeling very thirsty.'

Cold coffee.

Standing at that crash site, the two men allowed themselves a laugh.

'I remember just thinking, how long will it be before I get to fly again,' says Squadron Leader Ishan. 'I was refusing to accept the reality of what had happened. I found that I was telling myself to simply be patient. And that I would get my wings again. The mind tries to protect you from the trauma you have faced. But I also knew that the most difficult part of the journey wasn't losing the leg in that ejection. It was the fight that lay ahead.'

Squadron Leader Shalika had been up for a posting to Tezpur in the weeks ahead. But after the crash, her husband would be going back to Halwara to her instead.

'Life had changed so suddenly for Ishan,' she says. 'Everyone thought he would be depressed. Who wouldn't be in such a situation? It was deeply traumatizing for us all. But I wanted to tell Ishan that I was there to be strong for us. "You can vent your emotions for a while. You don't have to put up a brave face for our sake." But he truly was extremely strong. I was worried sometimes about how matter-of-fact he was about how his life had changed. Yes, it is true that life must go on. But for a moment imagine losing a leg suddenly, doing what you love.'

When the couple returned to Halwara at the end of 2019, the first thing doctors at the base asked Squadron Leader Ishan was whether he needed more sick leave. The pilot was certain he didn't want any.

'I said no, please allow me to join back on ground duty,' I told the doctors. 'I'll do office work till the time I get medically fit. I knew in my mind that if I stayed at home, I wouldn't be able to sustain things. *Ghar mein baitha rahunga toh mein baitha hee rahunga because kaam hai jaan hai* [If I sit at home, I will always sit at home. If there is work, there is life].'

Like the morphine that had been administered to him after the accident, the self-preservatory impulse of acceptance melted in the weeks that followed. This would often manifest in loud nightmares and waking up agonized in the middle of the night.

'I would spend my days shifting between the wheelchair and bed,' says Squadron Leader Ishan. 'I was at a fighter base where I used to fly with all these guys. So, I was watching them every day, and the reality of my situation just sank in more and more.'

For months, the two young Squadron Leaders hadn't wept in each other's presence, careful to hold themselves together for the sake of the other. But back now in their private dwelling in Halwara, it was finally possible to give vent to the pain and sadness that had built up. Both knew it was important for their well-being too.

'There were times when I would cry out loud if I wasn't able to do something owing to my condition,' says Squadron Leader Ishan. '*Jab raat ko mera beta so jata hai mujhe bahut kharab lagta hai—uss samay mein apni wife ke saath phoot phoot ke rota hoon* [After my son would go to bed, strong emotions would overcome me. My wife and I would break down and cry our hearts out]. Ayaan wouldn't get up because he was small and slept deeply. But then my wife would also start sobbing. After crying I would feel a little better. *Jo hua maybe kuch aur reason tha* [There must be a reason for what happened].'

With his mother and father in Kolkata, little Ayaan had been taken by his grandparents to their native Gurdaspur. In December, before reporting to Halwara, the child's parents came to fetch him.

'As we drove to Gurdaspur, the only thought in my mind was how was I going to face my little boy,' says Squadron Leader Ishan. 'I was in a wheelchair. As expected, he seemed puzzled when he saw me, wondering, "*Arre Daddy, yeh kya gaddi gaddi* [Daddy, what is this vehicle you are on]?" I couldn't pick him up, as my hands were still heavily bandaged. Ayaan then lifted my shawl and noticed I didn't have a left leg. Children are so innocent. He simply asked me where the leg was. That was the most difficult moment for me. I had to hold back my tears until he ran off to play.'

From cockpit to wheelchair, the abruptness of his new situation was frequently overwhelming. But like his training for emergencies in the air, the pilot knew that the only thing he could do was to solve the problem in front of him. Taking on more than he could manage was an invitation to despondence, depression and frustration. He decided he needed to summon all of his pilot-like qualities to solve what lay in front him first. And only then move to the next problem.

And what was that first step?

'Going to the washroom,' he says. 'It was an enormous challenge for me. I hope nobody ever has to go through it. It is the most traumatic thing to get used to and accept. That you need help with the washroom. So, my first aim was to gain full independence in that department.'

The slow rhythm of a violently altered life was taking root when in early 2020, Squadron Leader Ishan was asked to proceed to Pune, a city he had served in and where he had flown Su-30 jets for the first time. But he wasn't being sent to the famous Lohegaon air base in the city, but an equally famous institution in the armed forces—the Artificial Limb Centre. Established in 1944, the centre provided wide-ranging rehabilitative care to injured armed forces personnel, including prosthetic limbs. And it had come a long way since the days of wooden legs strapped to amputated stumps.

'When I reached Pune, I was starting from scratch,' says Squadron Leader Ishan. 'I had foot drop, characterized by difficulty in lifting the front part of the foot, essential for any mobility. Initially I was given a rudimentary pneumatic leg.'

In April 2020, eight months after the crash, Squadron Leader Ishan stood up from his wheelchair for the first time. It was a moment of triumph, and a milestone the pilot won't forget.

'I was standing up on my old leg and new leg,' he says. 'A leg I had almost lost. I just stood there for a while, allowing my legs to take my full weight. Everything weighed a ton. I wobbled a bit, but I maintained my balance. It was a small moment, but once again, I was overwhelmed. Shalika was in Pune with me that day. She saw me, she smiled and clapped. I know she was overwhelmed too. We celebrated that day.'

A steady course of physiotherapy, counselling and exercise followed. Giving the pilot a pneumatic leg was one thing. Making him meaningfully mobile and independent was still a way off. Hopeful but impatient, Squadron Leader Ishan turned to social media for help in speeding things along on his road to recovery.

'Shalika and I came across a woman named Arpita Roy on Instagram,' he says. 'She was a double amputee yoga practitioner and fitness trainer. We put a great deal of faith in her. I signed up for a three-month course with Arpita. By December 2020, sixteen months after my accident, I was able to put my crutches aside. I started cycling. I was permitted to drive a car within the premises of the centre. But there was something else that I wanted to do, that I had been hoping against hope I would be able to.'

At the end of 2020, Shalika brought Ayaan to Pune to visit his father.

'The three of us went to the basketball court. My son played for half an hour with me. Initially Ayaan was a little concerned and didn't want me to strain myself. But then the child in him kicked in, and he enjoyed the game. The delight in his face at being able to play with me again was unforgettable. I wanted to cry, but I was too happy to do that. Shalika was also very emotional.'

That night, after Ayaan slept, husband and wife stood in the small balcony of their quarters, silently letting the glow of the day wash over them. Shalika took Ishan's hand.

'*Jo ho gaya ho gaya*,' she said. '*Tu zinda hai, sab theek ho jayega* [Whatever happened has happened. You are alive and everything will be fine].'

In early 2021, just as the COVID-19 pandemic hit, Squadron Leader Ishan returned to Halwara. The pilot's caregivers and physiotherapists wanted him to return to his place of work and get used to living in his real-world setting using his prosthetic limb. He would return every few weeks to have his prosthetic limb calibrated since the sockets would get loose with use. Each time he would return to Pune accompanied by Squadron Leader Shalika. In mid-2021, when he was summoned back to Pune for a check-up, he decided he wanted to travel alone.

'Everyone said, don't travel alone, take Shalika with you,' he says. 'But I said, "*Bhai ek baar akele jaane do* so at least I know my limitations. When there is always someone at hand, I take it for granted. If I want water, someone brings it. I need to be fully alone to know just how independent I am now. There is no other way." And so, I flew alone in that scheduled An-32 flight to Pune. It took a while to get used to getting everything myself. But I cannot tell you how exhilarating and empowering it was. The water I got up and got for myself actually tasted sweeter.'

Nearly two years since he received his prosthetic limb, Squadron Leader Ishan has begun to nurse a dream he thought would never be possible again—climbing back into an aircraft cockpit.

'I won't ever be able to fly a fighter aircraft again, I had made my peace with that,' the pilot says. 'But the Artificial Limb Centre was doing everything it could to get me fit enough to perhaps fly transport aircraft, like the An-32s that have been flying me everywhere since the accident. It would be a huge thing if that happens. I don't carry any delusions, but

I do live in hope. I cannot sit around and feel sorry for myself or angry about what happened. Ultimately, I am a national servant. It was my choice to be a fighter pilot, and I was aware of the risks. If I am paying the price for wanting to be a *fauji*, then so be it. What happened, happened in the line of duty. But I am not finished with my duties.'

Squadron Leader Ishan's hopes have remained especially high ever since a course mate sent him an article about Captain Christy Wise, a US Air Force pilot who returned to flying the C-130J Hercules transport aircraft a year after she lost her right leg in a boat accident. Captain Wise spent eight months in similar rehabilitation before she was approved to return to flying duties.

'I am doing everything I can to get back to flying,' says Squadron Leader Ishan. 'After I receive an upgraded prosthetic limb with microprocessors, I believe my case will be even stronger. I miss the cockpit. That's where I belong. I will work as hard as I can to get back into one.'

Before his more gruelling physical exercises and physiotherapy, the pilot now begins his day with multiple surya namaskars and an hour of specialized yoga. In the early days of his rehabilitation, he had taken to meditation to help organize his thoughts and channel his emotions in order to sleep better. But more recently, he finds he is much more confident and able to get rest without too much trouble.

'The Air Force has protected me like a family member,' says Squadron Leader Ishan. 'This is the real face of a war-waging organization. It isn't just about combat, but the ability to take care of your people when they are in trouble. I could not have asked for more. I'm fighting it out. I always feel the end is in sight. But I know the Air Force is looking out for me, no matter what.'

Back in Halwara in 2021, Squadron Leader Ishan met with an old friend, perhaps the current generation's most famous fighter pilot—Wing Commander Abhinandan Varthaman. The Vir Chakra-winning aviator had been shot down over Pakistan-occupied Kashmir (PoK) on 27 February 2019, after shooting down a Pakistan Air Force fighter in an air battle the morning after the Balakot air strikes.

'Once Ayaan is slightly older, people will say many things to him—"see your daddy doesn't have a leg", like my child is still told, "*Tere baap ko Pakistani maar rahe hai* [Your father was hit by the Pakistanis],"' Wing Commander Abhinandan said to Squadron Leader Ishan. '*Usko ek mazak bana* [Turn it into a joke] so that he understands that whatever happened has happened and it happened for the good.'

When Ayaan now asks his father about the missing leg, he is informed that a tiger took it away. Probed further, a full story is narrated for the child, his eyes shining in morbid fascination, tinged with the kind of open-mouthed pride only a four-year-old can show.

'He has now been told by people at the base that his father has a "Terminator" leg,' says Squadron Leader Ishan. 'I've heard him boasting about it to his friends, saying, "*Mera* dad has a Terminator leg and he can finish enemies with it." He will soon know what happened. He's a bright kid. But you know how kids live in their own world.'

As he awaits an advanced new prosthetic limb and the vistas it could reopen for him, Squadron Leader Ishan hates offering advice. But the trauma of the last two years has given him a priceless worldview he often shares with his close friends in service.

'You'll always meet people who aren't fully satisfied because *mera rank nahi aaya, mera course nahi aaya, mujhe commendation nahi mila, mujhe Shaurya Chakra nahi mila*

[I haven't been promoted or received any commendation or won a Shaurya Chakra]. I always tell them, be happy to be alive and work hard. Everything else will take care of itself.'

One of the pieces of advice Squadron Leader Ishan hears most often from civilian friends is to quit the Indian Air Force and move to the private sector.

'*Mujhe pata hai mujhe fauj nahi chhodni hai* [I don't want to quit the forces],' he says. 'I can leave, and I will probably get a disability pension as well. But I don't want people to pity me. *Main kissi ke samne haath nahi phelana chahta.* The kind of training we've undergone in NDA, we cannot plead in front of anyone or ask for anything.'

Does he miss his leg?

'I don't miss it. It went away for a reason. I don't know what that reason is. And I don't want to know. I can't see anything but what's in front of me.'

'I Have Never Touched Anything That Cold'

Major Vibhuti Shankar Dhoundiyal

'*Kabhi kabhi bhagwan se kuchh maang liya karo, Vibhu* [Vibhu, ask God for something once in a while].'

The WhatsApp video call had glitched out in a mess of pixels but leaning over the railing of her balcony in Dehradun, Nitika Kaul Dhoundiyal could sense her husband was smiling at the other end. She waited for the connection to strengthen, mentally reminding herself that even if it didn't, she had managed to squeeze a few seconds of a video conversation with him before he left his base. His voice came across in patches, cut by the dying mobile data signal. It didn't matter. She had learnt to piece together the snatches of speech.

'*Yaar, zindagi main bas do cheez mangi hai* [I have only asked for two things in life],' said Major Vibhuti Shankar Dhoundiyal (Vibhu), a company commander with the 55 Rashtriya Rifles in Pulwama, Jammu and Kashmir.

'*Kya* [What]?'

Nitika knew the answer, but she never tired of hearing it.

'*Ek,* I wanted to get into the Army. And two, I wanted you in my life. Both prayers have been answered. What more can I ask for?' he said.

Nitika said a quick, quiet prayer of her own.

'Be protected and be safe always,' she said with a touch of exasperation.

It was anything but an unreasonable plea. The two were talking on the evening of 15 February 2019, barely twenty-four hours after a suicide bomber had targeted a Central

Reserve Police Force (CRPF) convoy on a stretch of the Jammu–Srinagar highway that runs through south Kashmir's Pulwama district, where Major Vibhuti operated as an officer on counterinsurgency duties. Forty CRPF men had perished in the vehicle-borne attack planned and executed by the Pakistan-sponsored terrorist group Jaish-e-Mohammed. Not surprisingly, units like the 55 Rashtriya Rifles that were based in the area were on full operational alert to carry out post-attack search operations. The terrorist, a local from the district, couldn't possibly have acted alone, and it was on groups of soldiers like the one under Major Vibhuti's command to move quickly before terrorist accomplices melted back into the hinterland, or even back across the Line of Control (LoC).

Knowing she would be sick with worry after news of the suicide bombing broke on television and social media, Major Vibhu had quickly called his wife to tell her he was okay and still at his base. The scale of the loss was still sinking in.

'*Yaar,* it's a part of life,' Major Vibhu said. '*Jo hona hai woh hona hee hai* [What has to happen will happen], so no need to mess up your sleep over that.'

'Why do you always do this? Stop it.'

Nitika hated it when he spoke this way, but she also knew it was probably her husband's way of dealing with the fragility of everything around him. It had been just over a year since he had been deployed to the Kashmir Valley. Nitika had hoped her anxiety would settle but quickly learnt there was no question of that. Ever since Major Vibhuti arrived in J&K in January 2018, he had requested his commanding officer not to saddle him with desk duties. As an officer from the Army's Corps of Electronics and Mechanical Engineers (EME), his training was tuned towards the upkeep of weapons, vehicles and other hardware.

'When he came to the battalion, he made it clear from the start that he wanted operations,' says Colonel Pradeep Duggal, then commanding officer of the 55 Rashtriya Rifles. 'As a young officer from the EME, I wanted him as the unit adjutant for some time. But he said, "*Nahi sir, kuchh bhi kara do par desk pe mat bitha do mujhe* [Give me any assignment except on the desk]."'

Colonel Pradeep knew that Major Vibhuti was to be married three months later in April. He suggested to the new young officer to settle into the unit for the time being and perhaps join some operations after his wedding. But Major Vibhuti had pleaded that he didn't want to spend a moment behind a desk.

'Unknown to each other at the time, Vibhu and I took the same flight from Delhi to Srinagar on 22 January 2018, and reported to our new Rashtriya Rifles unit together,' says Major Saurabh Patni. 'We had a "dining-in" at the unit that night. Vibhu being from EME, it was unusual how keen he was on operations. It was unlike his work arm, but it was the first thing I noticed about him. "*Jo bhi ho jaye, operations mein ghusna hai* [Whatever happens, I want to be in operations]," he would insist.'

The two young majors became roommates through February as part of the mandatory pre-induction training.

Colonel Pradeep had been apprehensive about pushing Major Vibhu directly into operations, but by the middle of that year, it was clear he had a nose and heart for what was, even during a lull, a supremely difficult and dangerous job in the most restive sector of a troubled state. In October 2018, his talent in the field could no longer be ignored and he was made commander of one of the companies within the 55 Rashtriya Rifles. Nitika had known Vibhu for years and was fully aware of his eagerness to lead troops in the field, but as his newly wedded wife of six months, she received the 'good news' with mixed feelings.

'There was nothing he wanted more,' says Nitika. 'It was the beginning and end of what drove him forward. And yes, he fit me into that somewhere.'

Courtship, Wedding and Back in the Unit as Company Commander

'From the word go, with a company under his command, this boy was out to prove that we had taken the right decision to put him in operations,' recalls Colonel Pradeep. 'In November 2018, there was an operation away from his company area, but he insisted on being sent. "Sir, I can't be sitting here when my guys are out there even if it's a different location," he told me. So, I sent him. Vibhu managed to sniff out and engage a terrorist in the very first house that he searched. I immediately noticed that he had a nose for where they were hiding.'

The following month, amid an unusual flurry of confusing intelligence inputs, Major Vibhu had pieced together an operation that ended with the interception of an over-ground worker (OGW) who was transporting 200 kg of explosives in the Pulwama area.

By the end of 2018, after just three months as company commander, Major Vibhu had a reputation that stretched across the unit.

'*Jahan Vibhuti Saab honge, wahan contact ho gaya hoga* [If Vibhuti Sir was involved in an operation, it meant we have zeroed in on something],' says a soldier who served in a neighbouring company of the 55 Rashtriya Rifles. '*Agar woh operations pe nikal gaye, iska matlab tha ki kuchh hai input mein. Aur yeh baat kabhi bhi galat sabit nahi hui* [If he was out on an operation, it meant that the intelligence input was correct. His instinct has never been proven wrong].'

The reputation was based on an uncanny consistency.

'You cannot get a more challenging environment than south Kashmir,' notes Major Saurabh. 'In the unit there would be long discussions to make sense of inputs, whether they are good or bad. Vibhu used to stay quiet. It was when he began planning and embarked on an operation that it became a reliable indication that the input was solid. I don't know how to explain it—he never went wrong.'

In January 2019, a piece of intelligence had come in suggesting that two armed terrorists were hiding in a village in Pulwama that had 120 houses.

'Six officers, including Vibhu and I, began a sweep of the houses in that village at about 1 a.m.,' says Major Saurabh. 'Four other officers and I had completed the search of around fifteen houses each. Vibhu was stuck on his first house. He simply wasn't moving from there. That turned out to be the house where the two terrorists were hiding. I can't explain how he sensed it. I can only give such examples.'

Days later, another piece of intelligence trickled in. It was the kind of input that was deemed solid—it was specific, mentioned precise coordinates, timings and more. While Major Saurabh's company was dispatched on the operation, Major Vibhu was ordered to remain at his base and provide backup if necessary.

'So, the rest of us were involved, but not him,' says Major Saurabh. 'We rolled out in an already celebratory mood since the input was that good. We were certain it would click. While we were at the location, a second intelligence input dropped, and that muddied the waters a little for the deployed men. Vibhu was dispatched to check out the second input, which was about a kilometre from our location. Moments after Vibhu and his men arrived there, we heard firing. It turned out there was nothing in our input, but in Vibhu's they found terrorists and engaged them. I can imagine this happening once or twice. How does one explain this happening every single time?'

Major Vibhu's reputation had filtered back to his home as well, triggering a familiar mix of pride and anxiety.

'I remember once I told him, *Yaar tera phir se jaana zaroori hai andar* [Do you really need to go inside]?' remembers Nitika. 'I would keep saying *dekh ke ja* [be careful].'

Usually, he calmed his young wife's anxiety with a few loving words. But when the two were on a call a few days before the Pulwama suicide bombing, Major Vibhu found himself triggered enough to respond.

'Nikki, *ek baat bata. Main andar nahi jaunga toh kaun jayega? Jo mere peeche hain, jo mere bande hain, jo meri team hai, agar main andar nahi gaya toh unka honsla kaun badhayega? Marna-warna kya hai, marna sabko hai. Sharmindagi se marne se achha toh yeh hai, aise hee maut mil jaye, na ki team wale bole kya bhagoda banda tha, apni team ko marwane chhod ke khud chala gaya* [Tell me Nikki, if I don't go in first, who will? If I shy away who will boost the morale of my men, my team? Everyone has to die someday. I would rather die than have my men say that I left them in the lurch and ran away].'

Nitika had winced. Major Vibhu had spoken gently as he always did, but his wife could sense there was a strain in his tone. She listened in silence. Holding the phone that night she knew she would need to dig even deeper into her reserves of empathy.

'I didn't need to tread gently,' she says. 'But I knew I needed to be more aware of the constant pressure on officers and soldiers in the field. They don't get much of a break from stressful situations.'

In February 2019, the two had been married for a little over nine months but Nitika had known Major Vibhu for over seven years. The two had met through a common friend shortly after Vibhuti was commissioned into the Army in December 2011 after graduating from the same business school in Punjab that Nitika had just enrolled herself in.

'He started off really funny—on our third meeting strictly as friends, he said he really wanted to tell me something,' says Nitika. 'I was expecting something silly. But he said, "You know we are going to be married one day." I laughed it off, feigning shock and outrage, but there was something in his manner that said he wasn't really joking.'

As a young lieutenant fresh into the Army, Vibhu had remained in touch with his friend. Through text messages over the next few months, he made it clear how he felt, even if Nitika still baulked at the thought.

'After all our chats, he would end by saying, "Baby, I love you,"' says Nitika. 'I would not reply, not knowing how to reply, and he would say, "It's fine, you don't need to say anything, I'm just portraying my love for you, yaar."'

Nitika graduated and soon began her career with a major consulting firm. As she navigated a whole new corporate world in Delhi, the one thing she found herself counting on was a daily phone call from Vibhu.

'If I missed his call, he would text me. This carried on for a few years,' she says. 'One day he didn't call, and then he didn't text. This was while he was on a weekend trip to Goa with his buddies. And then there was silence from his side for three days. When he finally called, I found myself leaping out of my chair in excitement and relief. He had dropped his phone into the sea at Anjuna Beach. He was very apologetic. It was weird. It was clear I missed him, but it was still a one-way thing.'

In the summer of 2017, Nitika found herself looking up dermatologists for a skin condition on her face. At a clinic in Delhi, while undergoing a procedure, her skin was singed, making her condition visibly worse. Nitika was gripped by panic, and her phone rang, Vibhu's name flashing across the screen. She picked up and burst out crying on the phone.

'Between sobs, I told him I had destroyed my skin and nobody would ever look at me again,' says Nitika. 'I was very distraught, and it was the first time he had heard me cry.'

The next day, while at the dermatologist to fix the damage to her skin, she received another shock. Captain Vibhu had dashed to Delhi from his military station to see his friend, arriving at the clinic that afternoon.

'I covered my face and screamed for him not to be allowed in, but he was already inside the treatment room,' she says. 'I broke down again, begging him to leave. But he sat down on the bed and prised my hands off my face. I was a mess, and he was just sitting there with a bemused look on his face.'

'What are you so worried about?' he asked, laughing and taking her hand. 'I've been asking you to marry me for years now. You're beautiful to me no matter what. And you're the person I love and want to be with.'

Nitika cried even more.

'He asked me why are you doing this? For whom are you putting yourself through all this? He could tell I was doing the treatment as a paranoid twenty-something, and he saw right through it.'

'I'm here for you. Stop all this,' Vibhu had said.

'If ever there was a moment when things changed between us, it was this,' says Nitika.

Things moved quickly thereafter. Less than a year later, in April 2018, with three months of operations with the 55 Rashtriya Rifles under his belt, Major Vibhuti returned to Delhi to marry Nitika.

On a quick honeymoon to Goa, strolling on Anjuna Beach, the newly-weds promised each other that their real honeymoon would be in the Maldives a year after on their first anniversary. The two then plunged themselves back into their work, separated by distance, but speaking many times a day. Major Vibhu immersed himself into his operations, and Nitika was back into the slipstream of her ongoing rise in the corporate world.

In December 2018, Major Vibhu came home for only the second time since the wedding. It was a last-minute plan and Nitika didn't have time to postpone the work she had scheduled for the day, a meeting with an important client.

At the meeting, as Nitika briefed the people present, a member from the client's team asked her how her husband was, aware that he was in the Army and deployed in J&K. A little embarrassed, Nitika mentioned that he was fine and had actually just arrived in town for a break. The client was shocked.

'What are you doing here?' he had said. 'Work doesn't stop. Stop losing precious time with your husband. He knows he can't always be here because he's on duty protecting the country. So please don't put anything above family time. And please thank him for his service.'

Nitika had rushed home in the metro, leaping into Major Vibhu's arms on the platform where he said he would wait. After a few blissful days in Delhi's glorious winter, the officer once again returned to duties in south Kashmir.

February began ominously and with a tragedy. On the first day of the month, as Major Vibhuti and his company conducted a morning search operation in a Pulwama village, a Mirage 2000 fighter aircraft crashed in a ball of flame during take-off at Bengaluru's HAL airport. Two young test pilots, Squadron Leader Samir Abrol and Squadron Leader Siddhartha Negi, had died in the accident. Negi had been Major Vibhuti's classmate in school in Dehradun and a close friend.

On 16 February, two days after the Pulwama suicide attack, while Major Vibhuti and his company were preparing for a gruelling week of cordon-and-search operations, news arrived that another childhood friend and classmate from Dehradun, Major Chitresh Bisht, had been killed while attempting to defuse an improvised explosive device (IED) in the Rajouri sector along the LoC.

'I was very worried about his mental state,' says Nitika, who had travelled to Dehradun for the weekend to see her mother-in-law. 'This was a lot of personal and professional tragedy to deal with in a month. We spoke many times that day. What could I say apart from the usual comforting words? He was very stoic about it.'

On the night of 16 February, Major Vibhuti called again. Speaking first to Nitika and then his mother. But before disconnecting, he asked for Nitika to be put back on the line.

'Nikki, promise me no matter what happens to me, you will not stop living your life,' he said, his voice still gentle and unstrained. 'You are not going to spoil your life. Promise me that you will move forward, find love again, marry and have children.'

'Are you stupid? Don't talk such rubbish,' Nitika shouted back. She could feel her hairs standing on end. 'You please go to sleep. I hope there's no going out right now?'

'No, winding down for the day, will call you tomorrow,' he said.

'I love you, baby.'

Late next night, on 17 February, the familiar thrum of an incoming video call woke up Nitika. She had slept early to catch the morning train back to Delhi the following day.

'I'm missing you a lot,' Major Vibhu said.

Nitika could make out even in the grainy visual that her husband was in his field fatigues.

'We are going out for an operation soon. I love you so much.'

Nitika rubbed her eyes sleepily, telling her husband to be safe and take all precautions like she had said so many times before. Major Vibhuti paused, staring into the phone, his face soft but serious.

'Nikki, you take care, and make sure you take care of the family also.'

Nitika felt that familiar feeling of her skin crawling, holding her breath for a second, staring into the phone with no words to say. Major Vibhuti smiled, waved and hung up.

'For the first time, he forgot to say I'll call you back,' says Nitika. 'He never forgot to say that.'

Returning to bed, Nitika could barely sleep. Fifteen minutes later, she turned to her bedside table, picked up her phone and dialled her husband.

'Nikki, *kya hua* [what happened]?' he asked softly. He realized what he had forgotten to say.

'I love you, Vibhu.'

Major Vibhuti paused, smiling.

'Nikki, you know what I do before going for an operation? I take my phone out, and I look at a picture of you and then of my mother.'

Nitika felt tears well up in her eyes.

'Don't talk like this, my love.'

'If someday something happens to me and if I haven't seen your face, then I won't be at peace.'

Nitika was crying now but said nothing.

'Go to sleep, baby. I'll call you in the morning.'

Earlier that day, the 55 Rashtriya Rifles had received fresh intelligence on the whereabouts of three terrorists who were believed to be hiding in Pinglana, a neighbourhood near Kakapora, the home village of Adil Ahmed Dar, the

Jaish-e-Mohammed suicide bomber who had carried out the devastating attack three days earlier.

'Our commanding officer personally briefed us that day,' says Major Saurabh. 'There was palpable excitement in the ranks. After three days of night searches and interrogations we finally had a chance to get the guys who had facilitated such terrible damage to the country.'

Before departing from the battalion headquarters with his team, Major Vibhuti went to seek his commanding officer's permission to move.

'Just go all out, take care of yourself, don't be in a hurry,' Colonel Pradeep told his company commander.

These brief meetings were common, but they always weighed heavily on the unit boss.

'Whenever someone moves out on an operation, you know it could be their last. I remember Vibhu's last words before he left that night: "Sir, *bas main nikal raha hoon, aur kuchh hai toh batao* [Sir, I am leaving now, is there anything else you want to tell me]." There was no drama. We do this every day.'

Major Vibhu and Major Saurabh rolled out with their respective teams at 10.30 p.m. and headed straight towards Pinglana.

Totally dark in the hinterland, the landscape was a luminescent white that night with many inches of freshly fallen snow as the vehicles rumbled towards the village. Pinglana consisted of two rectangular clusters of about sixty houses divided by a road running through the middle.

Arriving on site, Major Vibhu and his team threw a cordon on one side of the road, preparing to begin their side of the search, as Major Saurabh and his team did the same on the other side of the road.

'Vibhu and I had a quick word before setting up the cordon—he simply said, search carefully,' recalls Major Saurabh.

'We were pumped up and ready to get this done. And as always, Vibhu's nose for terrorists didn't fail him that night.'

Of the thirty-odd houses that comprised his cordon-and-search area, Major Vibhuti carefully approached the second house in the row.

'Our teams were in touch via radio, so we were updating each other in real time,' says Major Saurabh. 'I received an update that Vibhu and his team had moved the family out of the second house and taken them a safe distance away. And just as we received word that Vibhu and his men were starting their search in that house, I heard the firing begin.'

After forty seconds of heavy assault-rifle fire from inside the house, Vibhu's team sent a radio confirmation to Major Saurabh and other units on standby in the area that contact had been established with the terrorists.

'Now everyone was in the picture as to where the action was,' says Major Saurabh. 'Once again, Vibhu had homed in perfectly. It was around 11.30 p.m., and the temperature was minus 8 degrees Celsius. It had begun to snow heavily. Minutes later, another radio update came that one terrorist had been killed while two terrorists had been injured. But both were still firing.'

Four minutes later, a search party led by 55 Rashtriya Rifles second-in-command, Lieutenant Colonel Rahul Gupta, closed in on the firefight, acquiring first visual contact of the encounter. They noticed that both Major Vibhuti as well as three soldiers from his team were flat on their stomachs, clearly injured, but still firing back. Major Vibhuti had crawled forward, gesturing to the other men to find cover.

'He was leading from the front,' says a soldier who was part of the six-member team inside the compound of the house. 'He was telling us to move to the left and out of the line of fire. He refused to turn back.'

The sound of rifle fire punctuated a steady howling wind, bringing with it even more snow.

Colonel Pradeep was also in radio contact nearby, tracking each move.

'As Vibhu was clearing the house, the terrorists shifted to a cowshed in the same compound,' says Colonel Pradeep. 'After Vibhu cleared the house with his buddy, they approached the cowshed. Suddenly there was a volley of fire from inside the cowshed and Vibhu took a few bullets. Badly injured, he still held his ground, pushed his buddy away to safety, and he returned fire. He did not let the terrorist break contact and escape. He maintained his composure, continuously passing on information to the other team members. Once hit, people usually give up. But badly injured, Vibhu still remained strong and eliminated the terrorist. Then he changed position with his buddy. There was another terrorist behind the first one, who he also managed to injure with a few shots. But by then he had lost too much blood.'

During a brief lull in the firing, as both sides paused to reload, Lieutenant Colonel Rahul dashed into the house and pulled out the six men who had gone in to engage the terrorists. Major Saurabh's team had now also arrived on site and was providing cover fire to facilitate the evacuation. Four, including Major Vibhuti, had multiple gunshot wounds.

'A helicopter evacuation was impossible in that weather,' says Major Saurabh. 'So, I was coordinating vehicles to dash them immediately to the 92 Base Hospital in Srinagar. Three of the men were motionless. But Vibhu was still breathing.'

In Dehradun, at half past midnight, Nitika stirred awake.

'When I opened my eyes, I could not see anything. You know how you can still see things when it's dark. But I couldn't. It was pitch black. I felt for my phone and switched on its light. I couldn't see that either. I thought I may be dreaming, but then I heard and smelt something that made my blood run cold. I could hear Vibhu's voice whispering, "Nikki, I love you." And I smelt cigarette smoke. Vibhu and I used to smoke sometimes. Nobody else in the house did. My head was spinning, and I was very scared. I was sure I was dreaming. I just whispered *I love you* twice and then pulled the sheet over my head. I mumbled the Gayatri Mantra to myself. When I opened my eyes again, I could see pale light streaming in from the window. I got up and had a glass of water. There was no smell in the room. I tried my best to sleep the rest of the night.'

In Pinglana, Major Saurabh and another soldier carried Major Vibhuti to a waiting stretcher.

His eyes were almost closed, and he was breathing very slowly. There was no movement. All we could do was hold his hand. The team's nursing assistant came and started administering first aid. Vibhu and injured soldiers Havildar Sheoram and Sepoy Ajay Kumar were in one vehicle. The fourth man, Sepoy Hari Kumar, had already succumbed to his gunshot wounds. In a furious snow blizzard, the vehicles rumbled north in the darkness towards Srinagar's 92 Base Hospital.

'There was no time to rest, the operation was still on,' says Major Saurabh, who was now in front of the house and directly engaging fire with the terrorists. 'At 1.15 a.m., I received a radio input that Vibhu and his three men had been pronounced dead. Standing there in the snow, crouched behind a parapet wall, I felt the wind being knocked out of me. My friend. My men. I had commanded the same company just a few months earlier before Vibhu had taken it over.'

The house now surrounded by troops, Major Saurabh allowed himself a few minutes to digest what had just happened.

'It was in the middle of an operation. We could not stop. It was the first time I had faced such a situation. While the firing paused, I gave a call to one of our unit officers who was on leave at the time—Lieutenant Colonel Vipul Narain. He was a mentor to Vibhu and me. I just needed some strength.'

Firing resumed an hour later and continued through the night and much of the next morning. Major Saurabh, who attempted to advance towards the house at dawn, received gunshot wounds in both legs. Lying in the snow, in great pain, he crawled to safety before passing out. By morning, the two remaining terrorists were finally cornered and killed, one near the cowshed and the other two houses away.

In Dehradun, Nitika was up early after a night of fitful sleep. Bleary-eyed, she began packing her suitcase for the journey back to Delhi. She tapped on her phone to see if she had missed any calls or messages. It was unlikely, her phone was always kept on loud volume. There were no messages or calls from her husband.

While she sipped from a cup of tea before leaving for the railway station, Vibhu's mother said, 'I had a funny dream last night, of Vibhu massaging my feet.'

Vibhu's sister, who was also in the room, said she had dreamt of Vibhu as well.

Nitika smiled, finished her tea and then got into the car so the family could drop her to the station.

'That morning I didn't feel right,' says Nitika. 'Vibhu always texted me the morning after an operation saying, "Baby I'm back, I love you." A friend of mine was on the same train. He could see how agitated I was getting and suggested I call someone from the unit to reassure myself. There could be many reasons why Vibhu had not called. I wasn't panicking, but my nerves were shot.'

At 7.30 a.m., while she was on the train, Nitika's phone rang. It was from an officer at the 55 Rashtriya Rifles.

'He has received injuries, we are trying to revive him, ma'am,' said the voice at the other end.

Nitika felt a flood of anger rise from within.

'Please just tell me the truth,' she said. 'I can already sense what has happened. Just give it to me straight.'

The officer at the other end paused before breaking down.

'Please don't worry about me, just give me the information I am asking for,' said Nitika. 'When will I see his body? Give me a proper time.'

'Vibhu Sir's body is in Srinagar. It will be flown to Dehradun via Palam (Delhi) by this evening.'

'Okay, thank you,' Nitika said before hanging up.

The train had left Dehradun and was picking up speed.

'I didn't know what to do,' remembers Nitika. 'I asked a ticket collector where the next station was. Thirty minutes later, as the train slowed down, I got off. The first thing I did was call Vibhu's sister. She was closest to him. She broke down and wept uncontrollably. I remember staying very calm on that empty railway platform.'

Next, Nitika called her mother in Delhi.

'My mother's painful screams tore my heart apart,' she says. 'I was petrified of one thing. How was I going to break the news to Vibhu's mother?'

On her way back to Dehradun in a taxi, she remained in a daze.

'I was wearing my *chura* and *mangalsutra* [wedding bangles and necklace] at that time. Suddenly, I just couldn't bear the weight of those bangles and other adornments of marriage. I found myself removing my mangalsutra, *payal* [anklets] and my *dejihor* [earrings]. I just couldn't bear anything on me. *Mera dil beth jata hai yeh yaad karke* [My heart sinks when I remember that time].'

Arriving back at Vibhu's family home, Nitika stood in front of the house just as a few close relatives and media personnel began to arrive.

'I didn't have the heart to break the news to Vibhu's mother,' she says. 'It was only after she was told the news that I went into the house. We ran to each other, hugged and fell down to the ground wailing.'

Major Vibhuti Shankar Dhoundiyal's flag-draped casket arrived at the family home at 8 p.m. that day.

The next morning, the casket was opened for the family to have one final look at Vibhu.

'I have never touched anything as cold as my husband's face,' says Nitika. 'When they put the lid back on the casket, my family tells me I was knocking on it. I guess I was hoping for some miracle. Maybe he would wake up. I recall how my grandmother had once lost her pulse but had been revived. I kept knocking on the wood of that casket. I wanted to see him again, but I didn't have the heart to open it again. I wanted him to rest.'

The 55 Rashtriya Rifles were grieving too.

'I lost four brave men that night,' says Colonel Pradeep. 'Vibhuti's spirit stays with us. He always took care of his men. Even in Pinglana, two of the men in his team survived because he led from the front and ensured they remained out of the line of fire. I remember how Vibhuti would sometimes be harsh on troops during training. Back at the base, while eating together, he would tell the men, "My anger can save our lives, that's why I lose my cool, don't take it any other way." This was always appreciated by the men. They would say, "*Vibhuti Saab hain toh dhyan rakhenge hamara* [If Vibhuti sir is there, he will take care of us]."'

Major Saurabh Patni, who received bullet injuries in both legs, remained unconscious for two days. It would be seven months before he walked again. Metal rods were inserted into both his legs, with one of the rods only removed in late 2021. The officer's WhatsApp DP hasn't changed since he regained consciousness after the encounter—a happy picture of him with Vibhu in their barracks during pre-induction training.

'I don't think I will ever change that picture,' says Major Saurabh. 'One of the things I keep going back to is something that happened during that training period before we were thrown into ops. There was a 5 km run we had to do weekly with a bulletproof jacket, *patka* and full load of ammunition. Most of us preferred to run as a group, because it's easier to blend in and keep up. But the area also had a night duty checkpost. The person manning that post had to run a day later alone, which nobody wanted to do. We decided to draw lots. But before we could do that, Vibhu volunteered and ran alone. It was never visible on his face that he was sacrificing anything.'

Back in Delhi, Nitika returned to work two weeks later.

'For a time, I had the profoundest hallucinations in which Vibhu would manifest in front of me almost every day,' she says. 'One of the most common visions I saw was Vibhu and me shackled in different rooms, struggling to break free and run to each other. We were scratching the walls, desperate to be together.'

Her grief would also lead Nitika to immerse herself in her phone, obsessively texting her husband on WhatsApp.

'I would send him texts like, "Where have you reached, baby, please call back." Nothing would be delivered. I tried to call him, but the phone was switched off. Then when the trance ended, it would hit me that he is no longer with me, and I would break down. My parents were very worried about me. I didn't feel like talking to anyone. Like so many people, I drowned myself in my work. I would demand even more work from my clients. I buried myself in my career.'

Nitika continued to be haunted by visions of Vibhu in shackles. The vision wore her down mentally. On a particularly low day, as she stood on her balcony, thinking about whether life was worth living any longer, she saw her phone screen blink.

'When I tapped open the phone, it went straight to a voice note that Vibhu had sent me many months earlier,' she says. 'It was him singing a song. I can't tell if I'm imagining all of this, but I know it happened.'

When April 2019 arrived, the month of their first anniversary, Nitika took seriously ill.

'I stopped eating. I was like a skeleton. I was admitted to a hospital for a few days. I was administered shots to help me sleep. In that state, I remember having a vision of Vibhu and me sitting on what looked like a white cloud. He is holding my hand and saying, "Yaar, what are you doing to yourself." And I say, "*Yaar mazaa nahi aa raha hai, mujhe bhi saath le chalo* [Nothing feels good. Take me with you]." And then he refuses.'

Such conversations would continue, says Nitika.

'In one particular dream or vision, Vibhu appeared to read my mind. He said you must be wondering how much pain I felt *jab goli lagi* [when I was shot]. He said he felt no pain and it was very quick. The first bullet itself ended everything. I know this wasn't true, but he needed to reassure me, and I have since accepted that he's telling me the truth.'

The visions would sometimes have very specific conversations, says Nitika.

'In one of these dreams, Vibhu told me I would soon be speaking with a woman who had lost her husband. And that I needed to tell her that he was safe and sound with Vibhu. Some days later, towards the end of April, I got a call from a naval officer who asked if I could speak with Karuna Chauhan, a newly-wed lady who had lost her husband, Lieutenant Commander Dharmendra Singh Chauhan, in a fire accident on board the aircraft carrier INS *Vikramaditya*. I realized she was the one Vibhu was telling me about.'

During another bout of ill health and hospitalization, Nitika dreamt of Vibhu sitting in the chair next to her bed.

'He said, "Nikki, when you wake up there will be a nurse who will be holding your hands. She will be reviving you. Just tell her I said thank you and kiss her hands. Say your husband is eternally grateful to her for saving your life." When I woke up, I did exactly that. And when I was alone in the room again, I whispered to Vibhu that he is free. I told him that he doesn't need to worry about me any more. I was sure I wanted to live.'

In August 2019, the Indian Army announced that Major Vibhuti Shankar Dhoundiyal would be decorated with a posthumous Shaurya Chakra. The medal citation noted his 'unparalleled courage, conspicuous bravery and exceptional leadership qualities'.

'When I was discharged from hospital, I saw Vibhu again only once in my visions,' says Nitika. 'But I also had total clarity on what I wanted to do next.'

Taking the fourteenth Services Selection Board (SSB) examination, Nitika joined the Officers Training Academy (OTA) in Chennai.

'During physical endurance training one day, I was exhausted and about to give up on a long run. I didn't even look up, and I knew he was standing there with a big grin on his face. I heard him whisper, "*Arre bhaag, moti, bhaag, fail ho jaogi* [Run, or else you will fail]!"'

In May 2021, Lieutenant Nitika Kaul Dhoundiyal was commissioned into the Indian Army, her officer stars pipped on her shoulders by the then northern commander of the Indian Army, Lieutenant General Yogesh Kumar Joshi. In November, in full uniform and accompanied by Vibhu's mother, Nitika received her husband's Shaurya Chakra from the President at a Rashtrapati Bhavan investiture ceremony.

'I don't get visions any more,' says Nitika. 'Vibhu never wanted death. He loved our life together. We had so many plans. He wanted a certain house. We talked about children. We were thinking about life.'

With her unit now, Lieutenant Nitika hopes Vibhu is aware of the tribute she's trying to pay him.

'And I hope he isn't laughing,' she says.

6

'They Were Coming to Behead'

Sepoy Karmdeo Oraon

Line of Control, Naugam Sector
4.25 p.m.
29 December 2018

The only sound that could be heard from inside the light machine gun nest was a slow, cold breeze through the pines that lined the track leading down the mountainside to a bend 100 metres away. The only smell was the familiar tingle of frozen conifer-scented air mixed with gun grease.

Sepoy Karmdeo Oraon strained at the peephole, holding his breath so he could look as steadily and clearly through the little square of vision as possible. His hands rested instinctively on the weapon that gave the mini bunker its identity. A battered black machine gun with a distinct orangish brown buttstock. One hand curled around the pistol grip of the weapon. The other on the handguard.

He shifted his gaze up over the trees, scanning a brief horizon that held territory just across the Line of Control (LoC). And he knew even before they landed that weapons had been fired on the other side, the frozen calm broken by a series of whistling and hissing sounds that ended in earth-shaking thuds around the bunker. A fearsome, uninterrupted mortar and rocket-propelled grenade attack had begun on India's Karalkot Post, a cluster of five bunkers—including Sepoy Karmdeo's light machine gun (LMG) nest—spread across a 50-metre frontage at LoC, Naugam.

Even though it pushed him instantly into ready-to-retaliate mode, the incoming attack didn't startle Sepoy Karmdeo. The battalion was already on high alert following an exchange that had taken place in the same area thirteen days earlier. Ceasefire violations by the Pakistan Army at the LoC were more than routine at the time. The year 2018 had seen a staggering 1600 ceasefire violations by the Pakistan Army. That December alone had already witnessed 175, an average of nearly six each day. The soldier from the Charlie Company of the 8 Bihar infantry battalion adjusted his protective headgear and bulletproof vest and turned his head for a moment to count the number of hand grenades he had stacked in a wooden ammo box.

The firing was coming from hundreds of metres away across the LoC, too far to humanly fling a hand grenade. But there was a good reason why the grenades—a French word meaning pomegranate—were available in Sepoy Karmdeo's bunker. The post was perched atop a hill that straddled a route notorious for infiltration by Pakistan Army-aided terrorists. And only ninety seconds into the high-tempo bombardment from the other side, Sepoy Karmdeo saw them.

Five shadowy figures, 100 metres away.

In single file and in a crouched manner, they emerged from around the mountain bend. Their Kalashnikov rifles in firing posture, the five men edged forward under protection from a fresh blaze of cover fire that was now raining down on the Indian machine gun nest from two hill ridges emanating outward.

Apart from the mortars and rockets still smashing into the ground around the bunker, rifle fire tore into the bunker's walls, filling the air with a deafening clatter. Squinting through his peephole but holding fire, Sepoy Karmdeo noticed flashes of light from the combat fatigues of two of the men creeping forward.

It was only after they had crept closer that the Indian soldier realized what that was. Tucked into the belts of two of the approaching figures, glinting in the early evening sun, were 12-inch-long hunting daggers.

They were coming not just to kill and overrun the Karalkot Post.

They were coming to behead.

* * *

Thirteen days earlier
Line of Control, Naugam Sector
16 December 2018

The wireless intercepts couldn't have been clearer. Two Pakistan Army personnel had been seriously wounded on the other side of the LoC. The chaotic static and crackle of the intercepts also indicated that the injured men were officers—a brigadier who commanded Pakistan's 75 Infantry Brigade and was visiting forward positions to meet troops, and a colonel who commanded Pakistan's 37 Punjab infantry battalion. In a blood-drenched year that had seen multiple soldier casualties on both sides, the wounding of two senior Pakistan Army officers in an Indian retaliation was deemed an escalation—an invitation for the cycle of violence to climb to the next level.

Not that the location wasn't a raw nerve anyway.

The Karalkot Post was located 2 km ahead of the 8 Bihar battalion headquarters in the Kaiyan bowl, an area that had been grabbed back from the Pakistan Army during the 1971 war. For a half century since, the Pakistan Army had been attempting to either take back the Kaiyan bowl, or at least dominate it in such a manner as to make it simply too dangerous for Indian posts to remain. The men of 8 Bihar therefore not only had to defend the frontage in that area, but constantly retaliate to ceasefire violations. The Army unit that had been deployed in the area before 8 Bihar arrived eleven months earlier, had suffered several casualties and bore the scars of a particularly ill-tempered sector on an aggressively lit-up LoC.

Corroborating the wireless intercepts, informers in Pakistan-occupied Kashmir confirmed to the 8 Bihar battalion headquarters that a Pakistani brigadier and colonel, who had been hit on 16 December, were battling for their lives in a military hospital not far from the location. The confirmation triggered an instantaneous level of higher alert, courtesy 8 Bihar's commanding officer.

'They will try to even the score,' said Colonel Debashis Nath in a signal to his front-line posts, including Karalkot. 'Don't let your guard down against counter-retaliatory action by the Pakistan Army and terrorist infiltrators. Not even for a second. Be alert at all times for a strike on your posts.'

'Leave of all soldiers was cancelled to beef up the strength at each post. Multiple inputs were received from higher headquarters too that Pakistan was planning to retaliate and strike Indian Army posts on the LoC. Our unit was deployed ahead of the LoC fence, and it had been on red alert. Drills to counter any contingency were rehearsed and troops were fully prepared,' says an officer from the battalion.

This LoC fence, built in early 2000, is a meandering 550 km fence mounted with modern electronic surveillance gadgets along stretches of the 740 km LoC on the Indian side. The reason the fence isn't continuous is because of the forbidding high-altitude terrain in several sectors. Built with two rows of fencing and concertina wire, it is designed to obstruct infiltrators. The Indian Army mans posts on either side of the fence. The Karalkot Post, for instance, is located 900 metres ahead of the fence, making it the first line of

India's defence against the Pakistan Army in this pocket of the Naugam sector.

The alert was more detailed than just a general call to a higher state of readiness. It predicted the possibility of Pakistani retaliation in the form of a Border Action Team (BAT) operation, referring to an attack launched from across the LoC by mixed squads of Pakistan Army Special Service Group (SSG) commandos and terrorist infiltrators. In BAT actions, the commandos and specially trained terrorists would work together, performing a military-style covert assault that was notorious for the intended endgame—the decapitation of any Indian soldiers they managed to kill—and going back across the LoC with their heads. These attacks were designed not just to exact revenge, but to instil a deep sense of fear and unease in the minds of the units deployed at the LoC. The most infamous beheading, one that made national headlines, took place six years earlier in January 2013, when an Indian Army patrol party was ambushed, and one of the soldiers decapitated and mutilated.

Line of Control, Naugam Sector
4.30 p.m.
29 December 2018

The five men were slowly creeping closer, their movement covered by a hail of fire directed at Sepoy Karmdeo's bunker and the four other bunkers a few metres behind his, up the mountainside.

But it was Sepoy Karmdeo's bunker that was nearest to the advancing terrorists. They would make the first close-range 'contact' with the Indian soldier and attempt to overrun the post, trying to kill as many soldiers as possible. The other Pakistani attack squads—it was now clear there were at least three more in addition to the terrorist squad—would focus on providing an incessant volley of cover fire from the mountain flanks and from across the LoC.

About 2 km away at his battalion headquarters, Colonel Debashis heard the explosions and the sound of machine gun fire. None of these were unusual sounds in the sector under his watch, but as he prepared to take the daily 'all OK' report from his second-in-command (2IC) over the phone, the forty-year-old commanding officer could tell this wasn't just another ceasefire violation.

'The moment my 2IC wished me "good evening" and started giving me the all-OK report, I could hear unusually intense firing in the area. But it wasn't immediately clear which of my posts was bearing the brunt of the multiple explosions and the raging, high-intensity firing producing powerful echoes in the mountain side. However, I could make out that this was happening nearby, in my backyard,' Colonel Debashis says.

Backyard, indeed.

Twelve minutes into the firing, Colonel Debashis received an urgent signal from one of his company commanders manning a post not far from Karalkot. The message also contained a crucial input, mentioning that this appeared to be a BAT action, likely vengeance for the 16 December injuries inflicted on the two senior Pakistan Army officers.

The Pakistani brigadier was back to work a week after his injuries, but the Pakistani colonel, initially presumed dead, was still fighting for his life in a military hospital. But the Pakistani revenge mission was now in full flow.

'We tried to speak to the Karalkot Post but the telephone lines were cut off by the raiding teams. It took us a while to connect with the post commander over the radio set to find out what was happening,' the CO recalls.

Naik Deepak Singh at an Army base

Naik Deepak Singh with his wife, Rekha Singh

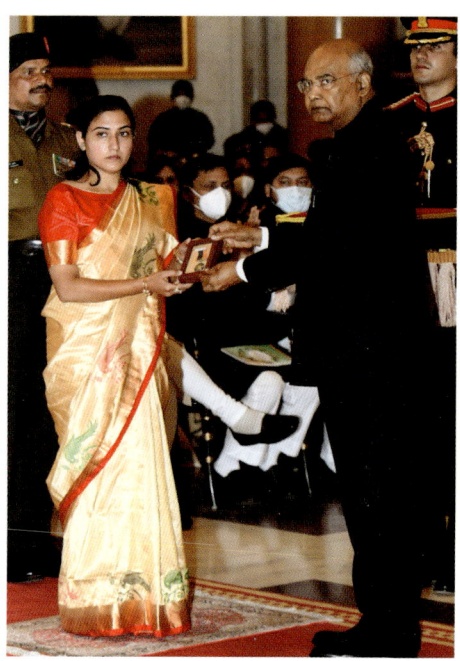

Rekha receiving Naik Deepak Singh's Vir Chakra from President Ram Nath Kovind at an investiture ceremony at Rashtrapati Bhavan in New Delhi

Colonel B. Santosh Babu's wife, Santoshi, and mother, Manjula, receiving his Maha Vir Chakra from the President of India

Naib Subedar Nuduram Soren's wife, Laxmi Soren, receiving his Vir Chakra from President Ram Nath Kovind

Havildar K. Palani's wife, Vanathi Devi, receiving his Vir Chakra from the President of India

Sepoy Gurtej Singh's parents, Prakash Kaur and Virsa Singh, receiving his Vir Chakra from the President of India

President Ram Nath Kovind presenting the Vir Chakra to Havildar Tejinder Singh

Still from a video released by the Chinese; Captain Soiba Maningba Rangnamei of 16 Bihar challenges a Chinese officer during the Galwan confrontation

Group Captain Varun Singh's mother, Uma Singh, and wife, Geetanjali, receiving his Shaurya Chakra from the President of India

Group Captain Varun Singh

Subedar Sanjiv Kumar (*centre*); (*clockwise from top left*): Paratrooper Bal Krishan, Paratrooper Amit Kumar, Havildar Devendra Singh and Paratrooper Chhatrapal Singh

Paratrooper Sonam Tshering Tamang was decorated with a Shaurya Chakra for his role in Operation Randori Behak

Subedar Sanjiv Kumar

Squadron Leader Ishan Mishra with his son, Ayaan

Squadron Leader Ishan Mishra

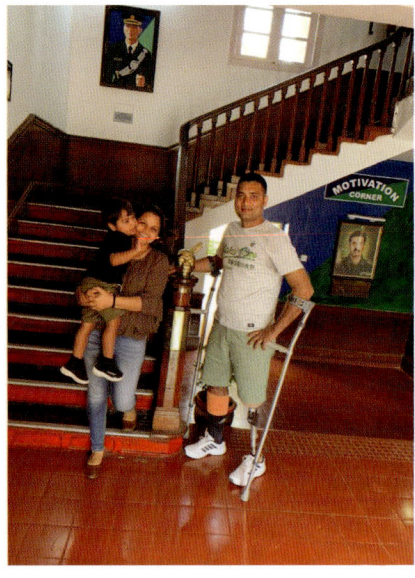

Squadron Leader Ishan Mishra with his
wife, Squadron Leader Shalika Sharma,
and son, Ayaan

Major Vibhuti Shankar Dhoundiyal
with his wife, Nitika Kaul
Dhoundiyal

Major Vibhuti Shankar Dhoundiyal at
a forward Army base

Major Vibhuti Shankar Dhoundiyal's mother,
Saroj, and wife, Nitika, receiving his Shaurya
Chakra from the President of India

Nitika keeps this picture of Major Vibhuti and her in
the back of her phone

Lance Naik Karmdeo Oraon receiving his Shaurya Chakra from President Ram Nath Kovind

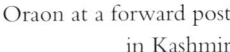

Oraon at a forward post in Kashmir

Lance Naik Sandeep Singh's son, Abhinav, with officers of 4 Para at his father's funeral

Lance Naik Sandeep Singh

Defence Minister Rajnath Singh shares a warm moment with Lance Naik Sandeep Singh's son, Abhinav

Lance Naik Sandeep Singh

Lance Naik Sandeep Singh's wife, Gurpreet, receiving his Shaurya Chakra from the President of India

Captain (now Commodore) Sachin Sequeira receiving his Shaurya
Chakra from President Ram Nath Kovind

Commander Bipin Panikar with his family

Commander Bipin Panikar poses in front of his
Sea King helicopter

Major Konjengbam Bijendra Singh with his wife, Jenipher, and daughter, Jasmine

Major Konjengbam Bijendra Singh receiving his Shaurya Chakra from President Ram Nath Kovind

Chief Petty Officer Veer Singh and Commander Ashok Kumar
in the Arabian Sea

Commander Ashok Kumar

Chief Petty Officer Veer Singh with his family

At Karalkot, things were getting worse as the ferocity of the attack intensified each minute. The post commander, Havildar Jitendra Singh, was somehow able to establish radio contact with his commanding officer and communicate the nature and gravity of the furious attack that had completely pinned down the men holding out against it.

'*Saab, yahan halat theek nahin hain. Dushman ne humko gher rakha hai, par hum jawabi karyawahi kar rahe hain* [The enemy has surrounded us, but we are taking retaliatory action],' Havildar Jitendra told Colonel Debashis over the radio.

'Karalkot Post was under a very heavy barrage of rocket and mortar fire from a Pakistan Army post,' says Sepoy Karmdeo. 'We were also subjected to automatic fire from the three other squads that had crossed the LoC and taken positions on nearby heights. It was difficult to pinpoint the directions from where the fire was coming.'

The five infiltrators were now 100 metres away and closing in on the bunkers of Karalkot Post. As bullets from the Pakistani support squads ricocheted off his bunker, Sepoy Karmdeo waited, watching the advancing figures closely.

They were 70 metres away.

As they closed in, the infiltrators became more visible. They were gesturing to each other, using tactical sign language typical of a special operations unit, a clear sign of some level of military-style training. From what Sepoy Karmdeo could make out, it seemed the infiltrators were perhaps under the impression that the furious fire assault launched minutes ago from all directions had in all likelihood paralyzed the men holding the post and maybe even killed a few of them.

Now, 60 metres.

Soon, 50 metres.

The five infiltrators stopped for a moment. After a quick gesture to each other, they opened up their assault rifles, firing directly at Sepoy Karmdeo's machine gun nest, the bullets smashing into the sandbags piled up in front of the bunker as a protective barrier. But the Indian soldier did not return fire. Not yet.

Taking deep breaths and cautiously watching their every step, Sepoy Karmdeo's position was now peppered by the five blazing Kalashnikovs right ahead, adding to the enormous volume of fire already pouring in from the other Pakistani squads. But he had decided to wait.

In the midst of the deafening clatter, another realization dawned on the solitary Indian soldier with his LMG and box of grenades. While the other bunkers at Karalkot Post were also under fire, it was unlikely that they knew about the infiltrators now just 30 metres from Sepoy Karmdeo's position. The sequence of events had moved so quickly that there was no time for an emergency communication. And the deafening chaos surrounding the bunker now made it impossible to communicate the situation clearly. And anyway, the infiltrators had just stepped even closer, the barrels of their Kalashnikovs flashing from just 20 metres away.

'For that moment, I needed to be oblivious to what was going on behind me at the post, or of any counter action being taken by our side to hit back,' says Sepoy Karmdeo. 'It was a moment for panic, but we know what panic can do in such situations.'

Closing in, 15 metres.

In a sudden move, two of the five infiltrators peeled away from the group and took positions on a mountain flank off the left overlooking Sepoy Karmdeo's position. It was clearly a move intended to split his focus and engage him from another direction. The three infiltrators firing from the trail ahead knew what they had to do.

'I wanted them to get closer. I could see the three terrorists firing at my post from barely 15 metres. And then, in the blink of an eye, they darted towards my post while the two men on the heights started firing rockets from a rocket-propelled grenade launcher and also turned a Pika belt-fed machine gun at my position,' says Sepoy Karmdeo.

Now, 10 metres.

With total focus on the three infiltrators now just a few steps away from his bunker, Sepoy Karmdeo's hand tightened around the pistol grip of his machine gun. He remembers skipping into a rare moment of total clarity, not uncommon with front-line troops in defensive close engagement situations. A difficult-to-describe lucidity where survival instinct and emotional attachments give way to a sudden, startlingly clear immediate picture.

'For me that moment of total clarity was that I wasn't going to allow these terrorists to storm the post on my watch at any cost,' says Sepoy Karmdeo. 'I was sitting on that desolate hill for a purpose. They would have to do it over my dead body. *Aur mujhe marne ki koi jaldi nahi thi* [And I was in no hurry to die].'

As the Indian soldier had not fired a single bullet yet, the three advancing infiltrators appeared certain that they were on the path of least resistance and would soon be celebrating inside an overrun Indian bunker. If the unremitting hail of fire hadn't torn the occupant's body to shreds already, they would have the chance to separate his head from his torso and return with the setting sun across the LoC. They began their final stealthy steps towards the bunker.

It was at this point, with the infiltrators less than 10 metres from his position, that Sepoy Karmdeo decided it was time to respond.

'I had been itching to do it the moment I spotted them but kept waiting as I knew I may only have one chance,' says Sepoy Karmdeo. 'I opened fire with my LMG, hitting two of the infiltrators at close range even as they continued to get cover fire from the two infiltrators who had separated from them minutes earlier to take positions looking down on my position. The two I shot dropped to the ground instantly.'

It was at this point that Sepoy Karmdeo was able to get his clearest view of the infiltrators. Dressed in black battle fatigues, they were wearing Kevlar body armour and helmet–mounted optical devices.

'*Jaise filmon mein dikhate hain* [Just like in the movies],' he recalls. 'They were equipped like special operations soldiers, not like the usual terrorists you see. Our input was totally correct—this appeared to be a BAT action by Pakistan.'

And as the two infiltrators lay motionless on the mountainside a few metres from the bunker, Sepoy Karmdeo once again saw the combat daggers tucked into their belts, left visibly unholstered.

But this was far from over. Only two men had apparently been dropped. There was a third infiltrator on the path ahead, two more on the mountain flank to the left, and three more squads of infiltrators in different positions who were still firing at the Karalkot Post.

'I was encountering something like this for the first time in my army career, although I have been involved in gunfights before,' says Sepoy Karmdeo. 'After I shot the two terrorists, the third one retreated and started running zigzag towards the mountain bend where all five of them had come from. I think I got him too but couldn't be sure at the time. Worse, when I scanned the path to ensure that the two infiltrators I had shot were dead, I noticed they were still moving. *Woh haath-paer hila rahe the* [They were moving their limbs].'

The Kevlar body armour the infiltrators were wearing had done its job. Not only were the two men not dead, they were not even badly injured. In a flash, they got to their feet and tumbled behind a boulder just 15 metres from the Indian machine gun position. With the rock now covering them, they resumed their attack, firing quick Kalashnikov bursts at Sepoy Karmdeo's bunker. The terrorists were following a simple but effective pattern in the close-quarter fight to pin down the Indian sepoy, by taking turns to take shots at him and providing cover fire to each other.

This was now a significantly more perilous situation. If the advancing infiltrators had earlier assumed the huge volume of fire had killed or incapacitated the Indian bunker's occupants, Sepoy Karmdeo had now signalled with his retaliatory fire that he was alive, pushing the terrorists into a stealthier attack mode.

That was how quickly things had changed. And the training and equipment given to the terrorists had made all the difference. In a matter of seconds, from being sprawled on the mountainside and practically assumed neutralized, the advantage was now fully back with them. The implications of this new dynamic hit Sepoy Karmdeo instantly. With a rock for cover and some breathing space, the infiltrators could summon more Pakistani squad members from the mountain flanks for a final assault on the Karalkot bunkers and overrun them. There was no option left, and the Indian soldier, his breath now coming in wintry wisps in the dying light, knew what needed to be done.

As the Kalashnikov rounds thudded into the sandbags that wreathed his bunker, Sepoy Karmdeo crept out of his post, armed with only an INSAS assault rifle. Positioning himself to the left side of his bunker and now almost fully exposed, he returned fire at the two infiltrators, emptying two full magazines of ammunition. It had been a daring move, but as the terrorists realized where the fresh fire was coming from, and now training their weapons directly at him and sending bullets whizzing past his head, Sepoy Karmdeo, on his elbows, dragged himself back into the bunker.

The LMG and assault rifle were now practically useless against the hiding terrorists. Sepoy Karmdeo looked over at the wooden ammo box containing hand grenades.

Meanwhile, an Indian Army post just north of Karalkot had been activated. The trail that disappeared behind the mountainside ahead of Karalkot led to a series of LoC-hugging posts held by the 8 Bihar battalion's Delta Company. Receiving urgent orders from his commanding officer at the Kaiyan Bowl, the Giani Post commander, twenty-four-year-old Captain Talisunep Longkumer, was placed on standby to rush much-needed reinforcements as quickly as possible to Karalkot.

But there was a problem, and it wasn't a small one as it would later emerge. The raiding Pakistani squads had placed ambush teams on all three mountain routes leading to the Karalkot Post, including the one from Captain Talisunep's Giani Post. The young officer had his orders, and he needed to figure out as quickly as possible a way to rush his troops to help Karalkot fight off the Pakistani assault.

Back at Karalkot Post, Sepoy Karmdeo was ready to use his grenades. But just as he picked one up from the box, a Kalashnikov round fired by one of the infiltrators found its way into the bunker and hit the Indian soldier squarely in the forehead, knocking him violently to the ground.

'I was down for a few seconds, and then staring at the bunker ceiling,' says Sepoy Karmdeo. 'I could move my eyes. Could still hear the firing. I moved my fingers. I was still alive.'

The 7.62 mm rifle round had impacted the Indian soldier on the steel plate of his bulletproof headgear. An inch below, it would have hit him straight in the face, killing him instantly. It was a providential escape, and Sepoy Karmdeo remembers thinking that the patka had done precisely what it was designed to do—give attackers a few less square inches to hit. Square inches that made all the difference in such fearsome close combat.

But as bullets continued to rain down on the bunker, Sepoy Karmdeo's patience turned into absolute fury.

'There was no blood. I felt no pain. I was alive. The firing was so intense from their side that I didn't even realize that a bullet had hit me on the forehead. I only felt a big *jhatka* [jolt] and fell. I decided to go after them with all my might. This firefight had gone on for longer than it should have, and I had to end it no matter what,' recalls Sepoy Karmdeo.

Certain now that there was nearly no chance of eliminating the terrorists from inside his bunker, the sepoy did something the infiltrators probably didn't expect.

The box of grenades in his hands, Sepoy Karmdeo crept out of his bunker into the fury of the incoming automatic fire once again. Taking position a few metres to the right of his bunker, he quickly flung all ten grenades in the direction of the two terrorists hiding behind the rock. Then he dived back into his bunker, hearing the grenades detonate in bright fiery thuds.

'I was not able to take them on with my LMG. The effort with the rifle also failed. The grenades were the only option. It worked. The only thing on my mind was that there was grave danger to the lives of my comrades if these terrorists slipped past me. I would not have let that happen at any cost,' Sepoy Karmdeo says.

The grenades had exploded at the intended mark, but the proceedings of that evening had made it clear that no assumptions could be made. Whatever the condition of the two terrorists behind the rock, Sepoy Karmdeo would only be able to check personally by going there. And at that point, it was out of the question—his post was still taking intermittent fire.

He looked out to the left for the two terrorists who were covering the manoeuvres of the 'dead' pair from the height, but there was no trace of them. Firing from that direction had stopped. The two appeared to have unexpectedly melted away into the forest behind them, the first solid indication that Sepoy Karmdeo's grenades had done their work. With the attack squad dead, the cover fire served no purpose. It was possible that they would attack from another direction, but for the moment, they had retreated into the darkness.

'Those two on the mountain flank were most likely Pakistani SSG commandos, who were not supposed to make contact but only provide cover fire to the three terrorists from a position of relative safety,' says Sepoy Karmdeo.

It had been twenty minutes since the first bullets flew, with the grenades bringing a blanket of silence back on the post by 4.50 p.m.

Back at the battalion headquarters operations room, Colonel Debashis's unwavering focus was on obtaining latest inputs on the overwhelming attack on the Karalkot Post. He was receiving a steady supply of information from three neighbouring locations in the area that had an unimpeded view of the Karalkot Post. It was about 4.45 p.m. when the firefight involving Oraon was in its closing stages.

'The picture was hazy, but I had managed to glean from the other posts that at least three men were trying to storm the post from where Karmdeo was standing guard and five men had positioned themselves between Karalkot and Giani to block the closest reinforcements,' says the commanding officer.

In minutes, it would be completely dark at Karalkot Post. The echoes of firing had cascaded away, allowing back the sound of the breeze through the pines, the air now tinged with gun smoke. There was no question of Sepoy Karmdeo lowering his guard at this time. He remained in position, fully aware from his training that the threat of terrorist squads mounting a second wave of attack was more than just possible.

It was last light at 5.30 p.m. when the unit's commanding officer ordered Captain Talisunep to depart from the Giani Post along with a seven-man quick reaction team (QRT) armed with multi-grenade launchers (MGLs), LMGs and Kalashnikovs, and rush to Karalkot in the cover afforded by darkness.

The unremitting severity of the Pakistani attack on Karalkot Post wasn't entirely over. The BAT action by infiltration squads may have temporarily been hit, but the post was now the target of heavy shelling and firing from Pakistan Army posts across the LoC. This time, the picture was being beamed live to the battalion headquarters. Specialized artillery observation equipment deployed by a neighbouring battalion at a height overlooking Karalkot was transmitting video feeds to the headquarter operations room. The real-time feeds were the first solid inputs illustrating the severity of the attack.

'Troop commander Havildar Jitendra Singh's description of the post having been surrounded by the enemy was corroborated by the video feeds being transmitted by the thermal imaging intensification observation equipment (TIIOE) deployed at the neighbouring battalion's Key Post,' says Colonel Debashis. 'The equipment can see up to 4 km at night. From my headquarters, finally I knew *kahan kya harkat ho rahi hai* [what movement was taking place where]. I could make out how our boys were moving within the post and what the enemy was up to outside. The squads attacking the post consisted of fifteen to sixteen men, including the ones who engaged Karmdeo.'

The images transmitted confirmed what Sepoy Karmdeo had personally seen as he shot the infiltrators. The raiding squads had all the trappings of a military team. The target chosen, Karalkot, was a remote and vulnerable post. The sure-footedness of the intruders showed that the Pakistani team had clearly carried out thorough reconnaissance of the area before the attack. Pakistani squads had positioned themselves to encircle and cut off the post and only one team attempted to storm it by making direct contact even as the rest launched coordinated assaults from stand-off distances. Military-style tactics.

Even the route taken by the infiltration squad that made direct contact with Sepoy Karmdeo was a surprise choice, intended to catch sentries off guard. Terrorist infiltrators were least likely to take that particular route to mount an assault as it was frequently used for administrative tasks such as transporting rations and other stores to forward posts scattered further up in the mountains. The route was also frequented by specialized teams dispatched to construct and repair bunkers, rotation of soldiers and evacuation of casualties. It was the last route that terrorist infiltrators would be expected to use.

And that's precisely what they did.

'Every method employed by the attackers was a clear indication that this was a well-planned operation,' Colonel Debashis says. '*Aisa nahin tha ki woh aise hi utth kar aa gaye wahan* [It was not as if they were there in the area by happenstance]. They must have carried out close reconnaissance days before the attack because the routes within a dense jungle are hard to tell from a distance. The terrain is such that they must have come closer undetected at some stage.'

The Pakistani teams positioned to intercept reinforcements attempting to reach Karalkot from nearby posts was another tell-tale element of military planning. On his way to Karalkot from Giani Post in the darkness, Captain Talisunep was expecting stiff resistance. Naugam was the young officer's second posting after he was commissioned into the unit in June 2016. That night was his fourth major operation along the LoC.

En route to Karalkot Post, the young officer spoke to his men. By now they had heard about the heavy attack and were raring to reach the location and join the fight. In a short radio message from Karalkot, troop commander Havildar Jitendra relayed that not an inch of his post had been spared by the devastating volleys of incoming fire from mortars, rocket-propelled grenade launchers, LMGs, Pika machine guns and other automatic weapons. The bunkers were almost completely destroyed. Captain Talisunep's QRT was speeding towards Karalkot in the nick of time.

'The boys in my QRT were highly trained and among the best in the company,' he says. 'We had been out on operations earlier too. We conduct drills to respond to different scenarios almost on a daily basis. Every man knew his job, what weapon he would carry and how he would move with the squad.'

As Captain Talisunep and his men manoeuvred down the snow-covered slope as fast as they could to answer Havildar Jitendra's desperate call for backup, the latter crept relentlessly between the scattered bunkers at Karalkot Post, personally reassuring each of the dozen men under his command, including Sepoy Karmdeo, that reinforcements were on their way, motivating them to give their all to defend the post.

Apart from Sepoy Karmdeo and Havildar Jitendra, the men holding Karalkot Post included three other soldiers from 8 Bihar, four men from the Border Security Force (BSF troops are co-located with the Indian Army in forward areas like Naugam) and three soldiers from an air defence (AD) unit. Each of the twelve men had a critical role to play that desolately cold night after the infiltration by the BAT team had been rebuffed.

The two AD soldiers used their ageing but highly effective Soviet origin Zu-23-2B guns to target the Pakistan Army posts that were providing cover fire to the infiltration squad. The other troops at Karalkot fired their medium machine guns (MMGs), LMGs, under-barrel grenade launchers and Kalashnikovs in retaliation.

'*Uss din karo ya maro wali baat thi* [It was a do-or-die situation that day],' says Havildar Jitendra. '*Aamne-saamne ki ladai thi* [We were in an eyeball-to-eyeball confrontation with the attackers]. The scale of the attack was overwhelming, and I had never seen anything like it in my twenty-one-year army service. I was able to locate their positions with inputs from the higher headquarters that were getting live feeds. We were concentrating our fire on those areas.'

The aftermath was turning out to be much less of a one-way affair. Karalkot was now getting critical support from other Indian Army posts in the area which were not only targeting the Pakistan Army posts with heavy weapons, but also firing at the likely locations of the retreating Pakistani squads that were estimated to be at distances of 100 to 200 metres from the nearly overrun post. But the men of Karalkot were still pinned down in their damaged bunkers, with shells still crashing into the ground around them.

The fearsome and sudden attack had become a full-blown crossfire across the LoC, with tracer rounds from both sides peppering the darkness with points of red light. How long would this go on for? What was the endgame now? Havildar Jitendra and the men under his command were prepared to fight to the last man but knew that only the timely arrival of reinforcements could save the post. Each shell that impacted was damaging the bunkers little by little. A handful of well-aimed shells would kill them all instantly.

Hurrying on foot, Captain Talisunep's team was now approaching the Pakistani squad that had been positioned between Karalkot and Giani posts to fight off any reinforcements. Pakistani shells hit the ground not far from them, sending showers of loose earth at the QRT team as it trudged down the mountainside in the darkness towards Karalkot.

One of the artillery observation posts had relayed imagery of the terrorist squad lying in wait for Captain Talisunep's QRT along with coordinates of their precise location. Now within minutes of that location, the officer issued final instructions to his men not to engage the terrorists until they were within a range of 150 metres, the range at which the 40 mm multi-grenade launcher, the most potent weapon the QRT was armed with that night, could be used in a precision role to strike a target at night in fully hostile conditions.

Since Captain Talisunep's QRT was moving down the mountainside, it was approaching the terrorist ambush from higher ground, a precious advantage in the circumstances.

'We finally stopped, took position, and the moment we opened up the multi-grenade launcher on them, the five men just fled,' says Captain Talisunep. 'They even left some of their weapons behind. It was evident they were not expecting us to be there. In my assessment, their operation was well-planned but not well-executed. Frankly, I was expecting a fight to the finish from them.'

It took the Captain's QRT an hour to reach Karalkot Post located 1200 metres downhill from Giani Post as the men were cautiously led to their destination by the officer whose top priority was to keep them out of harm's way.

'We were moving into the unknown in the midst of heavy shelling and automatic fire that was hampering our advance. I will not be stating the truth if I say there was absolutely no fear in my mind. I was concerned about the safety of my troops and taking all possible precautions on the way to Karalkot where we were needed. *Mere ko kuch nahi hoga syndrome khatarnak saabit ho sakta hai* [The nothing-will-happen-to-me syndrome can prove to be dangerous],' says Captain Talisunep.

By 6.30 p.m., when the reinforcements from Giani Post arrived at Karalkot, the intensity of fire had dwindled. Havildar Jitendra, still energized, was able to direct accurate fire at the two remaining infiltration squads, which were now making rapid manouevres to evade the fire coming from Karalkot. The precision of the fire directed at them was owed to a steady stream of real-time inputs picked up by Indian thermal imagers, relayed straight to gunnery positions from the battalion headquarters.

As company commander of 8 Bihar's Charlie Company that manned Karalkot too, Captain Talisunep immediately took charge. The arrival of the reinforcements led by an officer, their own company commander, immediately boosted the morale of the dozen men at the post. With twenty minutes of aggressive retaliatory fire under the captain's leadership, the two remaining Pakistani squads, most likely composed of the SSG commandos and terrorists, abandoned their mission and quickly retreated across the LoC. The exchange of fire finally ended at around 8.30 p.m., two hours after the backup reached the post, four hours after the first shot was fired that day.

With the greatest of care, Captain Talisunep and Havildar Jitendra crept between the bunkers to assess the damage and check on the dozen men. They still needed to be careful. The Pakistan Army posts would still be watching closely, and it was well within the realm of possibility that a few rocket-propelled grenades could be fired if a human target presented itself.

'I wasn't aware of Karmdeo's kills or how he staved off the attack on his position until later that night,' says Captain Talisunep. 'I learnt of his exploits while I was checking on the men, asking them about their well-being, the condition of their weapons and the ammunition they were left with. I got to know that Karmdeo had killed at least two terrorists and their bodies were lying outside the post.'

The captain told the men to stay vigilant through the night as the Pakistan Army could send squads to retrieve the bodies of the two terrorists and possibly launch fresh attacks. As he communed with his men, the officer knew it was a miracle that each one of them was alive.

'You have done a splendid job and made the unit proud. But there is no room for complacency. Do not forget the enemy can return with a vengeance,' he warned his men.

The maximum state of alert continued the following day. A failed high-energy assault of the kind that had been thwarted the previous evening meant that the Pakistan Army would try again at some point. The men of 8 Bihar also busied themselves decoding the assault.

The unit never did recover evidence to establish that the attacking squads consisted of Pakistan Army SSG commandos and regular troops, but Colonel Debashis and other officers of 8 Bihar, who studied the operation in minute detail, were fully convinced of their involvement and certain that it was a military operation. Every aspect of it, from start to finish, screamed commando assault.

'The way the attacks were planned, coordinated and executed was a dead giveaway of SSG involvement,' says Colonel Debashis. 'The men who made contact with Karmdeo could have been terrorists. The others most certainly included Pakistan Army personnel who cut off the post, provided close support with a variety of automatic weapons and vanished when the mission failed. They didn't come close enough for the obvious reason that if they were caught or killed, their involvement would stand exposed.'

The centrepiece of the assault was clearly an infiltration by the Pakistani raiding squads, but this wasn't an infiltration attempt in the general sense of the phrase associated with LoC operations. Terrorists usually infiltrate across the LoC to join terror cells deeper inside Jammu and Kashmir and replenish force levels of terror groups operating there. The 29 December infiltrators had no intention to remain in J&K.

'Their aim was to capture Karalkot Post, and if possible, return with the heads of a few of the boys,' says one of the soldiers who defended the post that night. 'All the terrorists were wearing bulletproof headgear. Have you heard of terrorists wearing such gear?'

In such a situation, it would always be near impossible to get a fix on the exact composition of the four Pakistani infiltration teams that mounted the assaults on Karalkot. But what there is total clarity about is Sepoy Karmdeo's actions.

'He showed extraordinary courage and composure even when his attackers were barely ten metres away,' says his commanding officer. 'He killed them and halted their advance. This changed everything. Had he frozen for some reason— and believe me it can happen to the best in such heavy-fire scenarios—the Pakistanis would have stormed the post. And remember, at this time, other men at the post were not aware of this group.'

In the morning following the incident, Captain Talisunep's QRT crept close to the rock behind which the two terrorists had taken cover. As expected, the grenades flung by Sepoy Karmdeo hadn't given them a chance to escape. Their bodies, clad in clear, military-supplied fatigues and vests, lay in the muddy snow.

'We found the daggers in their belts, and plastic bags in their rucksacks,' says Captain Talisunep. 'They had come with the intention of chopping off heads and taking them back as trophies. The fight would have been over long before reinforcements reached if it were not for Sepoy Karmdeo. And the mutilation of men has an unspeakable effect. It instils an anger that cannot be described.'

Recoveries made from the two terrorists killed included AK-47 assault rifles, a dozen magazines, grenades, improvised explosive devices, 9mm Chinese pistols, navigation equipment and incendiary devices.

'They came with the intention of burning down the post, killing us and returning with our severed heads,' says Sepoy Karmdeo. 'Why else were they carrying the incendiary devices and daggers? Had I become defensive that evening, the story may have been vastly different, and I may not have been the one telling it.'

The body of the third terrorist was never recovered, but the unit believes that Sepoy Karmdeo's fire in all likelihood killed him. Search parties found copious amounts of blood on a trail down the mountain path. It is possible that he (or his body) was picked up by the retreating Pakistani squads later that evening.

What happened during those two hours at Karalkot Post is now the stuff of Bihar Regiment's legends.

'Havildar Jitendra kept rallying his men to fight when things looked very bleak. The inputs he gave us over the radio after the telephone lines were cut off were critical and helped us with our initial planning. He also coordinated the movement of his company commander who was bringing reinforcements,' says Colonel Debashis.

Forty-one-year-old Havildar Jitendra is eagerly looking forward to upholding 8 Bihar's motto *Himmat hi jeet hai*

[Courage is victory] on foreign soil. The battalion has been selected to deploy as a UN Peacekeeping force in volatile South Sudan in 2022.

Six months after the defence of Karalkot, Sepoy Karmdeo Oraon retired from the Indian Army as a lance naik, returning to his village outside Ranchi in Jharkhand to begin the next chapter of his life as a farmer along with his wife and three children. He was eighteen when he joined the Army in 2002 and had never been in such a close-range firefight in his sixteen years of service. After a life on the front lines, he was ready for a new start in his fields.

On Republic Day 2020, the government announced Shaurya Chakra, India's third-highest peacetime gallantry medal, for the soldier. His company commander Captain Tenisulap Longkumer was awarded a Commendation Card by the Army chief for his indispensable role in the operation.

While the retired soldier has his hands full now with the ginger, tomatoes, beans and chillies that he lovingly cultivates on a 10 acre farm, the defence of Karalkot Post is never very far from his thoughts.

'There were some nerve-wracking moments. *Mujhe yakeen tha ki agar meine unko khatam nahin kiya toh woh mujhe bhi maarenge aur mere saathiyon ko bhi* [I was sure that if I didn't finish them off, they would kill my comrades and me]. It was a matter of my unit's honour. I trusted myself, I trusted my training and felt an instant rush of strength and confidence to take them on. I was shouting the Bihar Regiment's war cry "*Jai Bajrang Bali*" and targeting them,' he says.

He also realizes, with a smile, how closely death passed him by that evening.

'The first few seconds matter most. If you are able to stay on top of things during those crucial moments, chances are you will be triumphant. If not, you may be going home in a coffin.

The right training and discipline make all the difference. I was fully trained to do what I did, and so were the others holding the post,' he says.

Very little about the Indian Army's training regimen ever emerges in the public sphere, other than familiar images of fitness training and obstacle courses. While physical fitness is enormously important, the events of 29 December 2018 demonstrate the potency of mental conditioning and strength imparted to military personnel. A well-trained squad succeeded in fighting off a relentless commando-style wave of attacks without suffering a single casualty may boil down to a few strokes of happenstance, but is a huge measure of training.

The families of soldiers, however, cannot be trained to insulate themselves from tormenting thoughts about the perils and uncertainties their men unhesitatingly embrace as a military way of life. Only a day before the attack on Karalkot Post, Sepoy Karmdeo had managed to place a call to his wife.

'I reminded him that he was nearing his retirement, and that we, his family, were expecting him to return to our village in one piece,' says Priti. Nearly all conversations would end with earnest appeals not to take risks and to think of the family.

'I told her I was fine and there was no reason to get stressed. What else could I have said to reassure her?' says Sepoy Karmdeo. '*Duty toh karni hi hai aur achhe se karni hai* [I have to perform my duty and perform it well]. I told her that my superiors often tell other soldiers to carry out their duties like I do. She said the only thing that mattered to her was that I stayed safe. Neither she nor I had an inkling of what would unfold the next day. That's life. Isn't it?'

A full forty-eight hours would transpire after the Karalkot incident before Priti received a call from her husband. It was New Year's Eve.

'Families have a way of getting an inkling about what is going on but telling Priti about what happened at the post was out of the question as I did not want my family to get worked up and start imagining the worst,' says Sepoy Karmdeo.

What he did not know was that Priti was already aware of the Karalkot incident, courtesy her brother, Naik Somra Oraon, who happened to be in the same infantry unit as her husband but deployed in a different post.

'I got a ten-minute slot to make my call to Priti. There was no mobile network in the area. I was making small talk when she cut me short and began questioning me about the Karalkot operation. My brother-in-law had passed on the information that I wanted to keep from her. "*Jijaji ne aatankwadiyon ko maar giraya operation mein* [Brother-in-law killed terrorists in the operation]," Somra had told her,' says Sepoy Karmdeo.

Priti's question also brought a measure of relief. If she already knew, then he didn't have to endure the pain of seeing her jolted by the information. But Priti was relentless during that phone call, demanding that she be briefed with all details. In loving whispers, Sepoy Karmdeo shared a few nuggets.

'I told her the terrorists came to kill us but ended up dead themselves. I killed them. *Yeh sunn kar woh bahut khush hui* [She was delighted to hear that]. *Maine Priti se kaha ki agar meri duty ke dauran kuch ho jaata toh meri badnami hoti* [I told Priti if something untoward had happened under my watch, my honour would have been tarnished],' he says.

Priti still asks her husband about the Karalkot incident, each conversation bringing out fresh layers in the story. Hearing him tell it is a form of catharsis for her, a symbol somewhere that her anxious, earnest pleas for Sepoy Karmdeo to return to her in one piece had been obeyed by a dutiful husband.

'Whenever I was on sentry duty, I would tell myself there is a reason I am sitting here, far away from my home and family. That reason is my duty and I have to make myself count. When I learnt that there was a possibility of an attack in my area, my readiness was based on the premise that my post was the likely target. I was forever alert. I knew someone had my back and I had theirs,' says Sepoy Karmdeo.

President Ram Nath Kovind awarded him Shaurya Chakra at a ceremony held at Rashtrapati Bhavan in November 2021. The investiture ceremony at which the decoration was conferred on him was delayed because of the COVID-19 pandemic.

In February 2021, with both sides honouring it at least till the time this book went to print, India and Pakistan renewed their 2003 ceasefire on the LOC. How long it will last and how many more Karalkot-type incidents lie ahead is best answered by the retributive cycles that define the infiltration-heavy sectors of the LoC.

'I would get goosebumps watching TV and hearing about the daring exploits of soldiers being awarded gallantry medals at the Rashtrapati Bhavan. It felt surreal to be on that stage, standing in front of the President of India. Hopefully, soldiers from my unit watched me and thought, "If Karmdeo can reach there, so can we,"' he says.

His Shaurya Chakra citation speaks of his 'stupendous gallant actions, acute presence of mind, focused aggression and willingness to make the supreme sacrifice in pursuance of the operational task'.

The soldier may have made a new life in the fields, but his connection with the Bihar Regiment won't ever be severed. Not just because those ties never evaporate, but because Sepoy Karmdeo's two sons, aged eighteen and sixteen, are hoping to enlist and join their father's unit—8 Bihar.

'Nothing would bring me more joy than my children donning the uniform that I once did. I mean it, nothing. It's not an easy way of life, but I can't think of anything more honourable than it. The best part is that Priti agrees,' says the soldier. 'The overwhelming feeling of belonging to an army unit and the *josh* to uphold its undying traditions and ethos is something I hope my children can experience.'

As he waits for that day, Sepoy Karmdeo's life is as far as it can possibly be from the LoC. Gone are the familiar echoes of automatic weapons, the ominous whizzing whistle of shells as they careen in both directions across the frontier. Gone is the ever-clinging feeling that sudden and certain death is always just an inch this way or that. Gone is the terrifying need for decisions to end life or have yours ended instead.

'On that patch I killed, on this patch I grow. Little has changed, really. I was a proud soldier then. I am a proud farmer now.'

7

'Whatever We Do, It Has to Be Now'

Lance Naik Sandeep Singh

'*Saab, bahut chhota baccha hai, bahut din se shakal nahi dekhi hai* [Sir, I have a very small child and I haven't seen him for a long time].'

Lance Naik Sandeep Singh was wearing a white vest, black combat trousers and what he hoped was his most lovable grin. It was September 2018, and at 12,000 feet in the Tangdhar sector of north Kashmir, it was anything but warm. But the thirty-year-old soldier had just finished a two-hour workout, and he had no intention of hiding the results of his toil behind clothing.

He had spent two weeks the previous month with his five-year-old son, Abhinav, and wife, Gurpreet, in his village, Kotla Khurd, a few kilometres outside Amritsar. It was the longest leave he had been granted in eleven years in the Army. It had been a restful, idyllic fortnight, spent mostly playing with his little son and showing him selfies of the breathtaking Kashmir landscape he had been deployed to. Lance Naik Sandeep had been grateful for the long break. But now back with his unit, the 4 Para Special Forces, in a forward area close to the Line of Control (LoC), he missed his family more than ever.

Thankfully, unlike Lance Naik Sandeep's earlier years in the Army, video calls were now a welcome salve for homesickness, even though these were nowhere near the smells and sounds of home. But high up in the Shamshabari mountains, a mobile signal is a dear but stubbornly elusive friend. And that's why the soldier now stood at the foot of the bed of his squad commander, Captain Gurjeet Singh Saini, a young Special Forces officer who had joined the 4 Para a few years before.

'Mobile network is scarce in the higher reaches of Tangdhar,' recalls Captain Gurjeet. 'There are barely a few pockets where you can make a voice call, maybe one or two points where signal strength is just enough for a very shaky video call. In our unit, we would go maybe once a week to make these calls. But that September, Sandeep would come to me three to four times a week, saying, *Saab, chalte hain* [Sir, can we go and try?]. How does one explain this soldier–family bond?'

Men from 4 Para had been forward-deployed from their unit base elsewhere in Jammu and Kashmir's Kupwara district to a 'staging base' in Tangdhar, a small camp situated close to an area of expected activity. And in Tangdhar, which sits on the LoC, there is only one kind of activity the Indian Army deploys its Special Forces to deal with: terrorist infiltration.

The officer leant back in his bunk, watching the soldier for a moment. He knew it was going to be futile trying to convince Lance Naik Sandeep that they should wait a couple of days, since they had made their calls just two days earlier.

Climbing out of his bunk and throwing on a jacket, Captain Gurjeet and Lance Naik Sandeep set out from their tiny Himalayan forward base nestled in a mountain pass. They didn't need a map to find their way around in that terrain. Remembering features and trails in the mountains was a standard requirement for soldiers in the area. After a twenty-minute trek up to a favoured 'signal point', the two men took turns to make calls home. They had to take turns because 'signal point' was literally that—a spot a few metres across. Lance Naik Sandeep, of course, went first.

It was usually Gurpreet who answered, but that morning, to Sandeep's even greater delight, it was little Abhinav. The call lasted just over six minutes. And then, as it usually happened with 'signal point', it was impossible to reconnect.

'*Mera ho gaya, aap karlo, saab* [I have finished. Sir, you go ahead],' Lance Naik Sandeep said, waving to Captain Gurjeet a short distance away.

'*Bacche ki shakal dekh li* [Were you able to see your child]?'

'*Haan, saab, bacche ne hi phone uthaya aaj* [Yes, sir, the child picked up the phone]!' said Lance Naik Sandeep, handing over his mobile phone to Captain Gurjeet.

Displayed on the phone was a screenshot Lance Naik Sandeep had taken from the video call. In the top right was a stamp-sized image of the soldier beaming into the phone, a perfect blue sky behind him. The rest of the screen was filled by a chubby little boy wearing a blue *patka*, his eyes wide and mouth open mid-sentence.

'*Bahut naughty shakal hai bacche ki* [He looks very naughty],' said Captain Gurjeet, returning the phone to Lance Naik Sandeep.

'*Naughty toh hai hi . . . Aapko ghar call nahi karni saab* [He is indeed naughty . . . Don't you want to call home sir]?'

'*Nahi, phir aayenge kissi aur din* [No, we'll come another day].'

The two men trekked down the mountainside to their staging base in silence, Lance Naik Sandeep immersed in the call he had just made. Captain Gurjeet slapped the soldier's back fondly.

'*Fit lag rahe ho, Sandeep, workout theek thaak kar rahe ho* [You look fit, Sandeep, seems you have been taking your workout seriously].'

'*Saab, fauji munda hoon, body banani hai, maintain karni hai, woh impression jo hai para wale bande ka, woh rehna chahiye* [Since I am a soldier, I have to maintain my body and keep up the good impression people have of the para forces].'

Lance Naik Sandeep didn't lose an opportunity to exude the commando look. A few days earlier at Kupwara, Captain Gurjeet had bumped into Lance Naik Sandeep returning from the unit's wet canteen with two large bags. Asked what he was ferrying with so much enthusiasm, the soldier sheepishly opened the bags. They were mostly filled with bags of dry fruits. But nestled underneath was a strip of Streax hair colour for a deep-brown tinge.

'Till those few weeks, my interaction with Sandeep had been very limited,' says Captain Gurjeet. 'Before we were deployed at our staging base, there were multiple inputs about infiltration attempts in the Tangdhar area. However, none of these were concrete enough to warrant deployment. But as a para unit, we rapidly prepared ourselves, selecting the team that would be sent forward once a specific input came through. It was at a final briefing of this team in Kupwara that I was taken aback to see Sandeep.'

Lance Naik Sandeep, who joined the Army in 2007, had been a mechanical transport (MT) driver before joining the 4 Para Special Forces in the same role. He would spend years transporting commandos and weapons between their bases. But with exposure to operations and training within the unit, he was soon convinced that he must volunteer as a scout—a high-risk commando position that forms the eyes and ears of a Special Forces squad in an operational area. If driving military vehicles in a hostile and demanding environment wasn't perilous enough, the role of a scout was literally in the line of fire.

'I was a little confused about how a guy from MT could suddenly become a scout,' says Captain Gurjeet. 'Being a scout requires youth and physical fitness. You are the eyes and ears of your squad. What you do impacts everyone else. You walk in the front and provide the information necessary to complete an operation. It is not a joke. What settled my doubts and concerns was when he told me he had volunteered to be a scout. Now remember the Special Forces are 100 per cent voluntary. Nobody can be forced to join the Special Forces. And within it, if you volunteer to be a scout, it says a lot about you. It is a dangerous thing to volunteer for. He had put in enough service and seen a lot of action from close quarters, so I was confident about Sandeep.'

This thought would remain at the back of Captain Gurjeet's mind for a few days as he contemplated the mission ahead. Much was being placed on his young shoulders, and he knew he had the unit's formidable recent legacy to uphold. It was this very unit, the 4 Para, that had conducted the September 2016 surgical strikes* on terror launchpads in Pakistan–occupied Kashmir, an operation in which 4 Para commandos had managed to eliminate nearly forty terrorists and return safely without a single casualty to their own. In the storm of public attention following the announcement of the operation, the reputation of 4 Para, already hallowed within the forces, reached legendary status. The Special Forces work in secrecy, necessarily away from the public glare. But the stupendousness of the 2016 operation meant that the unit would forever, going forward, have soaring standards to match.

'When the standby team was being assembled, Lance Naik Sandeep could have easily remained silent,' recalls Captain Gurjeet. 'Or if he was asked, he could have simply said, "*Nahi, saab*, I'm from MT, I prefer not to come for this operation." But he was fully ready and willing. That spoke to me about his self–belief and professionalism. It was also his first scouting mission after coming from MT.'

---

* The only first-hand account of the 2016 surgical strikes, by the officer who led the operation, is a chapter in *India's Most Fearless 1*.

With Captain Gurjeet as the standby team's leader, the team waited for its orders, using the time to train and prepare for what would be a much higher altitude operation. Over the next three weeks, the inputs varied widely, making it persistently unclear if an infiltration would or would not happen in the Tangdhar sector.

On 19 September, the team finally received an intelligence input that was detailed enough to trigger the much-awaited forward staging to Tangdhar. Since the team was ready in most respects, it was only a question of loading their weapons and gear and rolling out. The team of nine men (one and a half squads) with Lance Naik Sandeep as scout, his buddy Lance Naik Surendra Singh as second scout and Captain Gurjeet as team leader departed the unit base at 11 p.m. on foot. The movement of Special Forces is masked to the maximum extent in forward areas in order to deceive powerful human intelligence assets that terrorist groups maintain in the area. If terror groups receive word about the arrival of Special Forces in an area, they are able to switch tactics, methods and sometimes even the composition of infiltration parties.

'There were no proper tracks leading from our base to the staging area, so we had to make the trail ourselves in the dark,' says Captain Gurjeet. 'Our lights were off as we made our way forward. We used our night-vision devices to navigate the very steep climb to a high-altitude, boulder-strewn terrain.'

With Lance Naik Sandeep out in front, his priority was to get the team to the staging area before sunrise.

'Sandeep got us there before first light, by 4 a.m.,' says Captain Gurjeet. 'This was his first mission, and he was proving his physical capabilities as a scout.'

At the staging base in that small mountain pass in Tangdhar, the intelligence input abruptly ran cold. It was something the team was prepared for, but it also meant they had no choice but to wait.

On the night of 20 September, Lance Naik Sandeep, his buddy scout and the team leader stepped out of the staging camp for some air after supper. The moon was bright enough to cast shadows of the mountains and of the three figures who now stood on the gentle grassy slope. Captain Gurjeet was looking closely at a tablet device with a map displayed on it. He pulled and zoomed across different parts of the screen in silence. The men knew where they were, but the conflicting intelligence inputs, including the one that had just been retracted, meant they could be setting off in any of the three or four directions up the mountains of Tangdhar.

Captain Gurjeet could tell that Lance Naik Sandeep was restless that night.

'*Jaldi input aajaye to achha hai, saab* [Sir, it will be good if we get the input quickly],' he said. His team knew what the scout was feeling. A familiar, enormously ironic discomposure typical in Special Forces units deployed on an operation. They would remain restless until the mission began properly, along with everything that came with it.

Captain Gurjeet remembered the words of his esteemed senior in 4 Para, who had led the surgical strike mission less than two years earlier. Asked about his mental state as his squads awaited final orders to roll out from Uri into PoK for their assault, Major 'Mike Tango', who went on to be decorated with the Kirti Chakra for the mission, had said, 'For us, there is infinitely more disturbance in calm than in an actual firefight. Once "contact" is made and bullets begin to fly, that's when calm truly returns.'*

---

* See *India's Most Fearless 1*.

Every Special Forces man understood the essence of those lines that captured precisely what the men of Captain Gurjeet's squad felt that night as they turned in.

As always, the men expended their anxious energy training and working out at the staging base. They couldn't go far from the site, lest they be spotted and reported by the eyes and ears of the enemy. So they had to make do with what they had—tables, chairs or each other for a rigorous group workout. There was nothing else they could do but wait for that order with a direction to head out in.

It was in this cloud of hair-trigger anticipation that the next morning, 20 September, Lance Naik Sandeep had woken up his team leader, requesting a trek to 'signal point' to phone his family. This was a welcome distraction from the numbing drudgery of the commando's wait. The rest of the day had passed by even more slowly.

The next morning, Lance Naik Sandeep appeared even more restless than usual. Once again, Captain Gurjeet knew why. The squad in Tangdhar would likely be 'deinducted' from the forward base back to Kupwara if a concrete infiltration input didn't come in the next forty-eight hours. This was a high possibility, given just how many inputs had run cold over the last few weeks.

Before turning in that night, Captain Gurjeet went to check on Lance Naik Sandeep. The soldier was busy neatening up his bunk. It was a familiar sight. He had a compulsive obsession with tidiness, something that gave his hearty, jovial personality a studious, measured side.

'*Kal kuchh milega, saab* [We should get something tomorrow],' Lance Naik Sandeep said to his team leader. '*Woh feeling aa rahi hai, ki kucch tagdi input aayegi* [I get the feeling that we will get a major input].'

Early the next morning, noticing the particularly crisp sunshine, Lance Naik Sandeep stepped outside, stretching and rubbing the sleep from his eyes in the kind of rejuvenating air only found in those mountains. His hair dishevelled, he pulled out his phone and took a picture of himself. At 8.15 a.m., he sent it to Gurpreet.

'*Bahut oonche pahad hain* [These mountains are very high].'

The soldier had a separate mobile phone just to talk to his family. When he got it, he had told Gurpreet, '*Jadu wee time mile, mein tenu phone karanga, phone apne kol hee rakh* [Whenever I get time, I will call you, keep your phone with you always].'

At Kotla Khurd village Gurpreet saw the message immediately. Texts from her husband had a separate notification tone.

She beamed at the picture and the message. He loved showing her the landscape that was now his second home. Gurpreet's phone was filled with such pictures, of the man she had fallen in love with at her cousin's wedding in 2011. She was nineteen, he was twenty-three and four years into Army service. They would marry the following year, and their son Abhinav would arrive the year after, in August 2013.

'*Aap risk kyu le rahe ho, mobile rehne do abhi* [Why are you taking a risk, leave the mobile alone], focus and be safe,' she replied.

'*Abhi kaam pe jayenge thodi der mein* [I will soon be on duty].'

'*Zyada fit lag rahe ho aap* [You are looking nice and fit].'

He couldn't continue the conversation. Something had just come in.

Early in the morning on 22 September, in Awantipur, about 100 km away from the staging base as the crow flies, an Indian Air Force Heron drone took off on a surveillance flight. A pair of Israel-built unmanned aircraft operating from the base south-east of Srinagar was being constantly used over days in the hope that conflicting inputs about infiltration plans from Pakistan-occupied Kashmir in any of the LoC sectors could be nailed down with unshakeable visual intelligence.

The Heron's flightpath that morning took it over the Kashmir valley, west over the Kupwara sector, and then a sharp turn south parallel to the LoC. As it flew along the LoC, the drone's sensors scanned large swathes of mountainous terrain on both sides of the line, the live video stream being watched closely by imagery analysts in an operations centre in Srinagar. At 8.04 a.m., the drone found what it had been sent out to look for. Two tiny dark figures stood out against the rest of the terrain on a desolate ridge in the Shamshabari mountains. The drone operators at Awantipur were quickly signalled to put the Heron in a tight circuit, to keep that frame firmly in the view of the drone's cameras. For a whole minute the camera zoomed slowly, attempting to get a clearer image. But as it did so, and while the analysts watched, the figures moved a short distance. And then they abruptly disappeared in the shadow of a rock feature.

Formatted and encrypted, it was relayed in minutes via secure communication channels to the 4 Para base in Kupwara and onward to Captain Gurjeet and his team at Tangdhar, where it was more than sufficient to trigger an immediate roll-out.

Lance Naik Sandeep scanned the Heron's images. Obsessed with maps and charts, part of his training, he immediately recognized the location from a set of hill features captured within the camera's frame.

The climb this time was even more steep, and it would involve scaling one rock face after another to save time. Out front once again, Lance Naik Sandeep led the way, providing a clear path to the other eight men to follow. After scaling one particular rock face in Tangdhar's Retnar area, the scout felt his silent mobile phone vibrate in his pocket. *An active mobile signal!* Instinctively he first looked in every direction, a commando habit. He fished out the phone. A pending message had delivered. But the real notification was that the phone had picked up a tiny strand of unexpected mobile signal. Quickly he dialled Gurpreet.

She picked up on the second ring.

'Found a lucky mobile signal, thought I would call and tell you I'm leaving for some work, I will call you tomorrow.'

'Okay, *apna khayal rakho* [take care].'

Gurpreet sighed and hung up. She knew what getting a phone call through meant to her husband, so there was no way he would want to end a call if he didn't really have to. Abhinav had been pestering his mother to try calling his father. 'You'll have to wait,' Gurpreet told him after the ten second call, 'your father is busy.'

Mobile phone back in his pocket, TAR-21 rifle back in ready position, Lance Naik Sandeep resumed his scout duties.

'The input we got was that some movement was seen across some passes along the LoC, so we immediately deployed forward and began an initial search. We reached the ridgeline by afternoon because, again, the climb was very steep,' recalls Captain Gurjeet.

At over 12,500 feet, the air perceptibly thinner, the sun was beating down sharply on the ridgeline. To the east, the mountain descended sharply into a small snow-blown valley, and then endless green dappled hills rolled across the LoC. They were now close to a location in the mountains where the dark figures spotted by the drone earlier that day could reasonably have reached. If their destination was the Kashmir Valley, then there weren't too many different paths they could take.

The timing of the squad's arrival in the Retnar area of Tangdhar to intercept two terrorists—either Pakistani nationals, or Pakistan-armed—couldn't have been more ironic. At 3.01 p.m. that day, Pakistan's prime minister, Imran

Khan, had publicly lashed out at the Indian government for cancelling a scheduled meeting between the foreign ministers of both countries on the sidelines of the UN General Assembly in New York City the following week. While Khan bristled with indignation, India's reasons were clear: first, the abduction of three policemen in Jammu and Kashmir's Shopian; and second, in a move that displayed a flagrant disregard for Indian sentiments, Pakistan had officially, and provocatively, issued a postage stamp with the face of Burhan Wani, a Hizbul Mujahideen terrorist killed by security forces in 2016.

It was an incredible repeat of a similar provocation that had been made two years earlier. On 21 September 2016, speaking at the UN General Assembly just three days after the terror attack in Uri, Pakistan's then prime minister, Nawaz Sharif, would take the stage and exalt terrorist Burhan Wani. A little over a week after Sharif's speech, men from 4 Para conducted the surgical strikes.

Two years later at Tangdhar, Captain Gurjeet's squad had no orders to cross the LoC, but the relentless cycle of history between the two countries would become starkly apparent in the days that followed.

Diplomatic channels lit up over the fresh crossfire, and Prime Minister Khan made a grand display of anger and despair that his peace gesture had been declined by India. Unplugged from the news whirl, on a remote mountainside in north Kashmir, the Indian commando team prepared to make contact with Pakistan's newest exports from across the LoC.

Spread out but held together by their scout and team leader, the nine commandos began an initial search, a careful surveillance of the area for tell-tale signs either of the presence of the infiltrators or of their having passed that way. Forty minutes into the initial search, Lance Naik Sandeep raised his hand, signalling the squad to stop in its tracks. Whispering, he motioned Captain Gurjeet to move towards him. All the men, armed with Tavor TAR-21 assault rifles, held their weapons in firing position as their team leader stepped towards the scout.

'Sandeep was pointing at a patch of ground which had what looked like some footprints,' says Captain Gurjeet. 'Because of a light drizzle that day, it was difficult to verify if those really were footprints or just some scattered mud. But Sandeep seemed certain that it was a track and insisted we follow it.'

The squad moved as quietly as possible behind Lance Naik Sandeep as he climbed up the rocky mountain slope. They were now around 100 metres from where the scout had sighted the footprints when Lance Naik Sandeep once again raised his hand to signal another tell-tale find to the commando squad. This time he looked back, nodding and smiling.

'Sandeep had identified the footprints correctly. What he had now found was some food wrappers, aluminium foil and fruit peels,' remembers Captain Gurjeet. 'This was Sandeep's crucial confirmation of the input we had received earlier. It also meant we needed to be even more cautious as we continued our search. The infiltrators were breathing the same air as us on that mountaintop.'

The same air.

The painfully slow sequence of events was over, and things were unfolding fast now as the 4 Para squad continued its climb. Four minutes later, Lance Naik Sandeep stopped the squad for the third time. This time it wasn't for footprints or food wrappers. He had spotted the infiltrators.

'*Banda dikh raha hai* [I can see someone],' Lance Naik Sandeep whispered in an urgent hushed tone from out front.

'*Kitna dur hai* [How far is he]?' Captain Gurjeet whispered back.

'*Bees metre* [Twenty metres].'

The same air.

'Sandeep was out front and could see him, but I did not have a direct view of the infiltrator because in that terrain boulders and nooks provided natural hideouts,' says Captain Gurjeet. 'Sandeep carefully gestured to me to come towards him and also conveyed that we should not yet open fire. I asked Sandeep to wait and cautiously moved ahead till I could see the infiltrator for myself. And I now saw what Sandeep had seen. A sliver of a black jacket was visible behind a boulder. The path to that location was across a very narrow crag with deep gorges on both sides.'

Captain Gurjeet signalled to the rest of the squad to hold their positions. It had to be ensured that the commandos weren't being lured into a trap. They waited, hoping for some movement or activity—anything that gave away a little more about the man wearing the black jacket or those with him.

'*Saab, hum wait nahi kar sakte* [Sir, we cannot wait],' Lance Naik Sandeep whispered to his team leader. '*Jo karna hai, abhi karna hai. Andhera ho jayega toh khel khatam* [We need to act right now, otherwise it will become too dark and we will lose the game].'

Captain Gurjeet knew he was right.

'Darkness would make things easy for the terrorists and difficult for us. I agreed with Sandeep, we needed to wrap up the operation as quickly as possible.'

With whatever planning was possible in such a close engagement, the nimble-footed scout moved forward along the crag, his team leader a few steps behind. The gorges on both sides fell steeply down a jagged cliff with rocky outcrops.

'I was a little nervous about Sandeep, since he did not have real experience in this kind of terrain,' recalls Captain Gurjeet. 'But everything he had done that day was already of a very high calibre, so there was absolutely no reason to doubt his abilities. He was sure-footed and calm. And this helped the team keep its focus steady.'

Half-way across the crag towards the hideout, a burst of fire flashed, hitting the ridge a few feet in front of Lance Naik Sandeep. The 'black jacket' had now stood up and opened fire with his AK-47 at the approaching Indian squad. Lance Naik Sandeep immediately sprang forward, letting loose a long burst of 5.56 mm rounds from his TAR-21, 'dropping' the black jacket instantly. On his elbows, he turned quickly to signal the rest of the squad to stop, and that he had managed to drop the first terrorist.

Just as he turned back to face the hideout, a grenade came flying from behind the boulder, exploding against a jagged edge on the crag. Before Lance Naik Sandeep could flatten himself, shrapnel flew into his chest and neck, injuring him severely. For a few seconds, he lay there motionless.

The grenade thrower—the second terrorist—now opened a blaze of AK-47 fire at the rest of the squad as they stood exposed on the crag, but Captain Gurjeet and the squad's second scout, Lance Naik Surendra Singh, buddy soldier to Lance Naik Sandeep, returned fire, forcing the second terrorist to remain behind the boulder that served as his hideout.

'Sandeep had neutralized the first guy in a very close firefight, just about ten to fifteen metres,' says Captain Gurjeet. 'Everything happened in a matter of a few seconds.'

During the exchange of fire, Lance Naik Surendra noticed the lead scout moving. Lance Naik Sandeep, bleeding profusely, had revived himself from the grenade blast, and was now crawling along the crag towards the hideout. Making the final few metres on his elbows as bullets flew between the hideout and the rest of the squad, Lance Naik Sandeep then leapt to his feet as he turned the corner behind the boulder.

'He was so close, he initially grappled with the terrorist in hand-to-hand combat. Then pulling himself loose, he shot the second terrorist from literally point-blank range, but sustained bullet injuries himself,' says one of the squad members. 'As he fell, his assault rifle came loose, rolled and fell into the gorge. He then turned towards us for a brief second and screamed, *Ek aur banda hai* [There is one more fellow]!'

There was a third terrorist.

'The third guy was in the hideout, and we didn't expect that,' says Captain Gurjeet. 'We could not see who or what was inside. Even the initial surveillance had only seen two terrorists. Sandeep had also said, "*Do hi bande honge, do hi bande ki movement dikh rahi hain* [I have seen only two]." And the hideout looked very small. So, we didn't expect a third guy inside.'

Lance Naik Surendra immediately fell flat on his stomach and began crawling towards his buddy, grievously injured on the other side of the crag near the hideout. When Captain Gurjeet motioned him to stop, he turned to say, '*Saab usko goli lagi hai, usko iss taraf kheench lete hai* [Sir, he has been hit by a bullet, let us pull him to our side].'

Injured but conscious, Lance Naik Sandeep heard his buddy's words and noticed Lance Naik Surendra crawling across the crag towards him.

'*Meri fikar mat karo . . . Bas inko na jaane dena* [Don't bother about me . . . just don't let that man go]!' he screamed.

Lance Naik Sandeep had been hit on his body by a grenade shrapnel and at least one 7.62 mm rifle bullet. Nobody in the squad expected him to do what he did next.

Unarmed, he rose once again to his feet with the last ounce of his strength and charged at the third terrorist in the hideout, but he was met with a burst of fire that hit his helmet, violently throwing him backwards and into the deep gorge where his rifle had fallen.

'It happened right in front of my eyes,' says Captain Gurjeet. 'It took me a moment to register what had happened. It was a moment of shock for everyone. We quickly took our positions. It was still not confirmed what kind of condition Sandeep was in. And there was a third terrorist still in the hideout.'

Keeping four men from the squad on the ridgeline, Captain Gurjeet along with Lance Naik Surendra and two other commandos descended into the perilous gorge to look for Sandeep.

'I hoped and prayed that he may have just fallen a short distance away and could be saved,' says Lance Naik Surendra. 'As we were climbing down the gorge, the third terrorist opened fire on us because we were down in a nallah and he was at a dominating height. This was the most difficult part. Every time we tried to get closer to Sandeep, he would fire at us.'

'I remember thinking as we were climbing down, how rapidly things can change in an operation like this where in a flash the priority shifts from taking a life to saving a life,' says Captain Gurjeet.

In the hail of fire from up at the hideout, one 7.62 mm round tore into Lance Naik Surendra's thigh, a flesh wound that wasn't enough to debilitate the commando. Provided cover fire by the four commandos still on the crag above, Captain Gurjeet and the other three finally managed to reach Lance Naik Sandeep.

'When we saw his body, I said another prayer,' says Lance Naik Surendra. 'I prayed that he would only be injured. *Jo bhi hai, hum fix kara denge* [Whatever the injury, we will take care of it].'

'But Sandeep was gone, we had lost him,' says Captain Gurjeet. 'The fall had been very intense. The gorge was full of boulders, and Sandeep had fallen over sixty metres. His body had taken a grenade hit, many Kalashnikov rounds and a sixty-metre fall. It was only that last thing that killed him though. He fought till the end.'

Injured by the third terrorist, Lance Naik Surendra volunteered to remain with his buddy soldier's body to secure it. It was 6 p.m. and the sun was about to set. There was one terrorist still in a dominating position in his hideout.

As Captain Gurjeet and the other two commandos climbed back up towards the crag, the 4 Para unit base had been updated about the sequence of events. Three more commando squads from the unit—eighteen men, including Captain Gurjeet's team commander—arrived on site shortly thereafter to surround the hideout and ensure the third terrorist wouldn't escape in the darkness.

'With the reinforcements, we established surveillance points to monitor the third terrorist. We needed to be careful not to remain in the same place, or else he would have had a killing zone,' remembers Captain Gurjeet.

Over four squads of the 4 Para Special Forces remained deployed overnight on that desolate mountain. The loss of Lance Naik Sandeep, the first such operational loss the unit had suffered, had galvanized the men, but also made them supremely aware that more men could drop if they didn't take every last precaution in eliminating the third terrorist. There was also the possibility that the terrorist had summoned his own reinforcements via radio, so no chances could be taken. The surveillance points provided a wide, but tight net around the deadly hideout.

The caution would mean the hideout couldn't be cleared the next day either. Scouts from the squads would venture close to the hideout and return. The third terrorist was inside, but not firing at the commando squads. This could either have been a trap or he was injured and buying time to recuperate for a final blaze. Either way, the unit had decided it would take no more fatalities in that operation, no matter how long it took to get the third terrorist.

On the morning of 24 September, the team commander decided that the operation could not be allowed to stretch any further. A scout was tasked with firing a pair of incendiary devices into the hideout.

'The third terrorist was injured but not killed. We didn't know what was going on in his mind. He came out of his hideout, probably knowing that he had no escape, and opened a burst of fire that narrowly missed my team commander and his scout, who were making their way towards the hideout for a final clearing mission. They fired back, taking cover. When the terrorist crept out a little later, one of the men at a surveillance point managed to hit him,' says a member of one of the squads.

For forty-five minutes, the commando squads conducted a meticulous search on the mountain to make sure there was no fourth, fifth or sixth terrorist waiting for an opportunity to strike. A Heron in the air provided a sweep of the area, confirming no further suspicious presence or movement. In the Srinagar operations room, analysts would receive images of the commandos as they wrapped up the mission and began their return.

All twenty-six men climbed down the gorge to accompany Lance Naik Sandeep's body down the mountainside to the Tangdhar staging area. An Army Dhruv helicopter picked up the soldier's body and flew it straight to Srinagar where the 92 Base Hospital would officially declare Lance Naik Sandeep Singh dead. This would pave the way for numb ceremonials.

'It was the first time the unit had suffered an op casualty,' says Captain Gurjeet. 'Our unit had been blessed with never having had such a tragedy. It felt personal for everyone in the unit. Everyone felt like they had lost their own family member.'

At the 4 Para Special Forces unit base in Kupwara, the usual raucous camaraderie had given way to a grieving, but galvanized silence. When the initial operational debrief concluded, there was nothing to do but train endlessly for such missions in a hostile high-altitude terrain.

The team commander, who had narrowly dodged the third terrorist's bullet a few hours earlier, ended the briefing with an exhortation to the men.

'This is what we are here for. These things will happen in the future as well. We should be ready for it. We must train more to ensure we don't lose another man. Our morale should always be high. Sandeep died for us. We must pay him back by training even more fiercely every day. The only way to overcome any feelings of doubt, and to keep Sandeep's memory alive, is to train for such missions and to overcome anything we face.'

After a wreath-laying ceremony in Srinagar's Badami Bagh cantonment, Lance Naik Sandeep's body was flown by helicopter to Kotla Khurd village, where images of a grieving Gurpreet and a stoic Abhinav would be beamed by television cameras across the country.

'A good soldier was lost. Usually when you lose someone like this, there is something to blame. But in this case, I don't know. I still go over the operation minute by minute, wondering if something different could have been done, whether I could have approached it differently, but there are no clear answers,' says Captain Gurjeet.

If the young officer had doubts about his performance, the Army didn't. His role in the operation, despite just two years of combat experience, would be widely commended by his seniors. The Army concluded that his 'tactical acumen, intelligence generation and source handling' had been impeccable. The Army would also observe that despite frictions imposed by terrain, climate, lack of cover and a limited number of troops at the operation site, Captain Gurjeet successfully led his squad to eliminate the first two terrorists. The message from the leadership was clear—the young officer had demonstrated every quality of a fine Special Forces warrior.

While every other soldier in the squad would similarly be commended, the Army made it a point to recognize Lance Naik Surendra's 'nerves of steel', standing guard over the mortal remains of his buddy while braving a hail of bullets from a terrorist holding a position of enormous advantage.

Captain Gurjeet couldn't travel to Kotla Khurd with Lance Naik Sandeep's flag-draped casket. It was important for him to remain with his squad as they dealt with the aftermath of the operation and readied themselves mentally for whatever came their way next.

While Special Forces units are immensely tight-knit and familial, the jolt of losing Lance Naik Sandeep was visible as his body was consigned to flames. Not only was the 4 Para's then commanding officer present, but two of his immediate predecessors also attended Lance Naik Sandeep's last rites and grieved with his family.

One of the former 4 Para commanding officers who attended noted, 'To demonstrate the reach of our nation's capability in ensuring that the last line of defence is never breached or exploited by our neighbour, it is men like him with the capacity of a thousand hearts who operate along and ahead of the LoC to deny the enemy that space to hurt our nation. He was one of the tigers I led, who gave me the largeness to roar from the pits of my cave.'

Two men from Lance Naik Sandeep's earlier squad were permitted to attend his funeral. One of them says, 'If Sandeep had not spotted those footprints and that trail, and correctly identified them, we would have walked on ahead unclear about the presence of terrorists nearby. We would not have been alive today. It was his astuteness as a scout that gave us that crucial information before the encounter. His scouting skills saved us. He took all the risks to keep us juniors safe like an elder brother.'

On Independence Day 2019, Lance Naik Sandeep Singh would be decorated with a posthumous Shaurya Chakra, India's third-highest peacetime gallantry medal.

The 4 Para's current commanding officer, a commando who had operated alongside Lance Naik Sandeep on numerous operations, will never forget the soldier.

'He lived his life for the *paltan*. And left this world for his paltan. He will be unforgettable to us as the *joshila Khalsa* [bold and courageous Sikh], whose constant refrain was, "*Ohh, saab ji, tension di koi lod ni, sabb ho jao gaa* [Sir, there is no need to be tense, everything will be done]." His pet dialogue, I can never forget, was, "*Sahab, kaam koi bhi karaa lo, chhutti time ton bhejj deyo* [Give me any amount of work but send me home for a holiday]." Devoted to the fight and devoted to his family. He lived, he died, and now he lives again through this book,' says the commanding officer of 4 Para.

As he was consigned to flames, an image would go viral across the country. Little Abhinav, his face serious, in a checked, black-and-white shirt, cracking a full salute, his mother by his side.

'He always told me, "*Officer banayenge Abhinav nu* [We will make Abhinav an officer]." I feel even more lonely now, as Abhinav is studying in a hostel. Every evening Sandeep's parents and I speak about him,' says Gurpreet.

Each day, she scrolls through her most prized chat window where her husband beams at her from a never-ending stream of images. The photos over the years capturing the soldier's transformation as a commando.

Gurpreet smiles at that last photograph from the staging base in Tangdhar.

'He really took care of his fitness,' she recalls. 'He used to tell me constantly *ki fitness par dhyan doh and theek raho* [be fit and be healthy]. A month before whenever he was supposed to come home on leave, I would sometimes go on walks to improve my fitness. I couldn't tell if it worked because he was always so happy to see me.'

Gurpreet now hopes to fulfil Lance Naik Sandeep's dream for their son.

When Abhinav made that final salute, he was wearing the same blue patka that he had worn in that last video call to his father.

8

'The Seas Will Break Your Ship'

Captain Sachin Reuben Sequeira

16 March 2021
Heera Oil Field
Arabian Sea

'Return to your bunks immediately.'

Dilip Kumar looked quizzically at the foreman standing guard at a heavy door that led out on to the deck of the P-305—the large accommodation barge* that for both men, along with 259 others, was home for weeks on end out in an oil-rich patch of ocean south-west of the bustling western metropolis.

'Very heavy wind and big waves, it's not safe on the deck.'

That wasn't a big deal, Dilip thought. The men on board weren't unused to the deck being closed, but it was an unusually cool night with a steady breeze that had blown through every opening of the barge, bringing some welcome respite to what was otherwise an existence that toggled between fierce, unrelenting air conditioning and sticky salty humidity. A walk on the deck on such a cool night was the best possible way to wind down after a day of labour.

---

* A large, flat-bottomed vessel with accommodation for personnel working on nearby offshore platforms. The barge is unpowered but held in place by a series of heavy anchors. Personnel are ferried to and from the rigs either by boat or helicopter.

The 261 men on board the P-305 all worked at the giant Oil and Natural Gas Corporation (ONGC) oil rig that towered 200 metres out of the sea. Dilip had spent a typical day as one of the many welders working on the upkeep of an endless network of pipes that made up the heart of the rig. Barred from any deck time, he and his mates walked back down the steps towards their bunks. Half-way down the flight of stairs, one of the men caught a glimpse of the sea outside.

'The waves were easily nine metres high and crashed into the side of the barge,' he recalls. 'And in the distance, I could see the lights on the rig just as it began to rain.'

This wasn't unexpected weather. Barge P-305, along with three other barges and a tug boat had all received a weather alert two days earlier about a deep depression in the Arabian Sea that had gorged on warm coastal air, whipped itself into cyclonic strength and had been christened Tauktae. On the day when Dilip and his fellow workers were making their way back down to their bunks, Tauktae had strengthened into a very severe cyclonic storm.

From the bottom of the stairs, Dilip turned back up towards the foreman, who was watching them from the deck door.

'We're moving around a little bit. Are all the anchors holding steady?' Dilip asked.

The foreman shot a quick glance out through the thick pane of glass on the heavy door that separated him from a deck that was now being lashed with heavy rain falling at a sharp angle, almost horizontally. He stared back down into the stairwell at the men who were waiting for an answer.

'I don't know, but I hear six of the anchors have snapped.'

Barge P-305 had eight anchors.

16 March 2021
Naval Dockyard, Mumbai

Well past midnight, 70 nautical miles (129 km; 1 nm at sea equals 1.85 km on land) away on the mainland, Captain Sachin Reuben Sequeira walked hurriedly through a corridor on the upper deck of INS *Kochi*, the over 7500-tonne Indian Navy destroyer under his command. Six years old, 163 metres long and formidably armed, the warship had returned to her dock the previous day after weeks of sailing across the Arabian Sea, delivering hundreds of tonnes of COVID relief supplies to Persian Gulf countries, and on the return leg had brought back oxygen cylinders[*] to Mangaluru on India's west coast.

The ship was at her berth, but the full crew—250 men—was on board that night. Rough sea swells rolling into the harbour as a result of the gathering storm meant that warships docked there needed to be on scramble alert to sail out if the weather deteriorated beyond a certain limit. Captain Sequeira was rushing to the bridge of his ship to get a full view of what things were looking like.

The officers and men on the bridge were at their stations, already ordered by their commanding officer to be prepared to move at very short notice. The ship may have been docked, but things were far less than comfortable for the crew, because INS *Kochi* was berthed next to—literally rubbing shoulders with—her two sister ships, INS *Kolkata* in the middle and INS *Chennai* alongside the actual dock. This wasn't ideal even in calm weather. And that night, it was a dangerous risk the crew of all three ships had to take, thanks to the well-known congestion in Mumbai harbour.

---

[*] This was during India's second COVID-19 wave, one marked by severe shortages of medical oxygen.

'So, there were three of us, three massive ships closely berthed alongside,' says Captain Sequeira. 'When weather goes bad, the basin takes a lot of swell and the ships tend to move a lot, scratching and rubbing each other. Double berthing is not uncommon because of limited space. But when you do that, your ropes and bollards are not meant to take so much stress. And here we were with three warships on one berth, that's nearly 23,000 tonnes held together by ropes. The crew of INS *Chennai* complained that there was a lot of strain, and her ropes might just tear apart. We were on tenterhooks that night.'

Briefing the officers and sailors on night shift on the bridge and reminding them that the ship was to be ready to move out at very short notice, Captain Sequeira retired to his cabin below a deck that was being lashed more fiercely with each passing minute.

At 8.30 a.m. the following day, the rain now ruthless, an emergency message from the Western Fleet office flashed on the ship's communication system, ordering INS *Kochi* to prepare for sailing out towards a location where an unspecified but dangerous 'situation' was evolving in the Arabian Sea.

'No further inputs were available on what the situation was,' says Captain Sequeira. 'In the next sixty minutes, there were two more calls from the fleet office saying that something was happening and a barge seemed to be in distress and may require some sort of assistance.'

By 10 a.m., the orders became definite, but the crew of INS *Kochi* still didn't have a clear picture of just how bad things were beginning to get at the Heera Oil Field.

'The conditions were quite bad because Tauktae was about 100 nautical miles south-west of Mumbai,' says Captain Sequeira. 'The port had shut down completely. There was no movement on the Mumbai Port Trust side. Ships outside from the previous night were already in distress with some dragging anchors and not under control. In such conditions, harbours generally don't allow you to move in and out to minimize the possibility of damage due to collision.'

But INS *Kochi* had to get out of the harbour, and fast.

The lake-like calmness of the harbour on normal days was shattered by a swell that now rose 3 metres, churned by winds gusting at speeds of 35 knots. The warship needed to get out before things got worse at the dockyard. Even the slightest deterioration in weather could potentially lead to the heavily-armed warships smashing against each other, risking the prohibitively expensive vessels and their crews.

INS *Kochi* had just been assigned a mission at sea. But the first big challenge was to safely get the destroyer out of the harbour. With the spiralling outer arms of Tauktae now heaving coastal waters into a boiling churn, the ship was about to commence it's most dangerous exit from its berth.

'When the port is not operating and winds are that high, it becomes extremely hard to leave the harbour,' says Captain Sequeira. 'Prudence tells you not to try any stunts because the risks involved are huge. But there was an SOS out at sea. And it was grave enough for us to have to take every risk possible to move.'

'Can you do it, Sequeira?' asked a voice from the fleet office on the ship's comms.

'We will do it. We absolutely will,' replied the commanding officer before ordering his men to unmoor and sail out.

As the crew got busy with the symphony of tasks and manoeuvres required to detach their ship from its berth and point it out of the harbour, Captain Sequeira glanced around at the young officers and sailors on the bridge. It was late in the morning, but the weather was a deep, foreboding grey.

'The safety of my men and my ship is my responsibility,' says Captain Sequeira. 'I was the captain, and the buck stopped with me. Those things were weighing on my mind, all the more so because we weren't sure what was out there. You only hear of cyclones and how people try to dodge them. You don't hear of people going *into* cyclones at sea. Even for a 7500 tonne sturdy warship like ours, we knew this was going to be a day requiring our fullest.'

One of the newer ships in the Western Fleet, INS *Kochi* was in excellent condition, with the crew having just 'worked up' the vessel after its first scheduled refit.

Now standing on the ship's deck to cast her off, Captain Sequeira remembers the rapidly changing sight outside.

'Nothing could be seen,' he says. 'Visibility was so bad that I could not see the front part of my ship from the bridge.'

On a good day, the warship could have cast off and manoeuvred itself out of the harbour without help. But on the morning of 17 May, every conceivable impediment to safe passage was ringing an alarm bell.

'Most of us take pride in our manoeuvring skills and usually use the powerful and agile engines to manoeuvre the ship out of harbour,' says Captain Sequeira. 'But this was out of the question that morning as we had no control. We got three tugs to pull us out of our berth. After that, it was entirely up to us.'

The crew of INS *Kochi* gave their vessel a massive burst of power to charge out of Mumbai harbour.

'This was risky because we didn't have enough speed, and the elements could take charge,' says one of the crew members on the bridge that morning. 'Even in normal conditions, if winds reach a speed of 15 knots or more, it can become tricky. The winds start pushing you one way or the other, and you don't have power as yet to pull her out. And here we are talking of a wind speed of 35 knots, which is off the charts.'

With the ship's four Ukrainian-built Zorya-Mashproekt DT-95 gas turbines now roaring at near maximum power, Captain Sequeira estimated that the powerful winds produced by Tauktae were blowing in precisely from the direction in which the ship was headed.

'I had no choice but to push full power and charge into the wind,' says Captain Sequeira. 'I had to make sure that I didn't bang into the jetty, berths and walls that flanked the harbour. The winds were coming right at us.'

As the ship sailed out in the extreme weather, a tourist on the coast captured the destroyer's silhouette disappearing into a firmament of grey. Tauktae's rain was coming down in copious torrents now.

'I took cover behind a radar antenna and ducked below it to look ahead,' says Captain Sequeira. 'I told my navigator to just stand behind me and hold the mic next to his face and just repeat what I was saying. I was looking from below the radar. It was hard to even stand there. We somehow pulled the ship out, turned it, gave it a burst of power and charged out.'

It may have seemed difficult then, but the INS *Kochi* crew had just completed the easiest part of their mission. Once out of the harbour, there was still ship traffic to look out for, especially since the cyclonic winds were likely to have thrown smaller merchant ships off their course and caused them to 'drag' anchor.

'What we essentially did was blind piloting,' says a sailor who was assisting with navigation duty that morning. 'That meant piloting the ship without any visual inputs and relying only on our radars and other electronic sensors to safely navigate the ship out of the harbour because the visibility was very bad.'

'We realized that the wind was so powerful, and the swell was so strong that we simply couldn't sail straight,' says another sailor on radar duty. 'We were actually going about 25–30 degrees off our intended course. We had the ship in a zigzag manoeuvre to maintain this narrow path.'

On a good day, Captain Sequeira would have issued orders and attended to other tasks on board his ship. But with his vessel headed into the heart of Tauktae, he knew there was no question of stepping back.

Manoeuvring around buoys was also critical.

Moored with heavy anchors and chains, large floating buoys help ships navigate safely through a channel, but they were anything but helpful that morning. Brightly coloured to enhance their visibility, the overwhelming weather had even blurred the buoys out. Breaking free from their moorings, the buoys with their chains could severely damage a ship's propellers.

'One nick from a metal object can cause enormous damage because the shafts are moving at about 100 rpm (revolutions per minute). Here we are talking about massive cylinders. If my ship were to get damaged, I would have to return to the dock and then it would take forever. It would also involve huge costs,' says Captain Sequeira. 'The buoys outline the channel, and in normal conditions, there is enough space for a ship to move. But in stormy conditions runaway buoys can pose a great danger. We were relying on our eyes to pick them up because it was quite likely that the radar wouldn't detect them in that bleak and worsening weather.'

Radar operators tend to optimize the parameters of the detection system during stormy passage to pick up bigger vessels to prevent collisions at sea. Echoes coming from fishing trawlers, smaller boats, buoys and the like tend to get lost as the main focus is on clutter reduction. 'The smaller boats and buoys go off the picture as you want to see less clutter. You have to play a very careful game on your radar settings to detect the valid echoes,' he says.

The only way to keep an eye on what was ahead of and around the ship was to deploy more men to lookout positions on the deck.

'In such conditions you come back to your basics, which is eyesight, hearing, monitoring every possible input inside the ship, outside the ship, radio, anybody seeing something—everybody has to be on the ball,' says Captain Sequeira.

Finally, out of Mumbai harbour after a tense ninety minutes of white-knuckled navigation, INS *Kochi* began to encounter a more vicious aspect of Tauktae, with Sea State 7, identified by waves up to 30 feet high.

'Once we were out at sea, it was rougher than anticipated, with the waves picking up and a massive swell coming in,' says a sailor who manned one of the lookout positions. 'You could not stand exposed on the deck. There was no question. The first thing we did as we came out of the harbour was to go and batten down the hatches to ensure there was no flooding of the ship. Our ships are designed to be watertight, but the situation was perilous and we couldn't take any chances.'

Captain Sequeira ordered all men on the deck back inside the ship.

'As soon as we were in open water, I ordered everyone inside,' he says. 'These huge waves were breaking over our fo'c'sle, the forward part of the ship, sweeping across and coming over two decks high and almost hitting the bridge. I was on the bridge and could see the towering walls of water smacking us.'

Late in the afternoon, an emergency update flashed on board INS *Kochi* informing the crew that barge P–305 was now adrift. A frantic call from on board the barge had finally been relayed to the Western Fleet.

The P-305's last two anchors had snapped. The barge was now like a piece of cardboard being tossed around by the monstrous waves that were rising up to 40 feet.

'Our information was that the barge had taken in water and was moving without control,' says Captain Sequeira. 'There were 261 people on board, we were told, but there was no direct communication between us and the barge crew. We checked all parameters and decided to increase speed towards the location of the barge. P-305 was drifting fast and was about seventy nautical miles away.'

The P-305 was now being battered by wall-sized waves. No longer moored to the seabed, the barge had been flung by a particularly large swell towards the oil rig, colliding with a piece of machinery, before spinning away, its hull damaged and now taking even more water. On board, there was total panic.

Dilip and a group of his mates had descended to the lowest deck of the barge to help clear out the water ingress, though this soon proved to be futile. Sea water had begun to gush into the barge through a gaping hole torn in the hull by the brief collision with the rig. And the enormous waves were now sending huge amounts of water down all open vents and hatches into the lower decks of the vessel.

Things rapidly worsened when the men on P-305, now convinced they needed to abandon ship, found that most of the twenty-four life rafts attached to the sides of their barge were damaged in the swell, with many torn into useless shreds by razor-sharp barnacles that encrusted the sides of the vessel.

Over 50 nautical miles out, INS *Kochi* was also getting rocked like never before by one of the strongest cyclones ever to churn the Arabian Sea.

'We realized that we were rolling about 35–40 degrees at times, and as we were tossed around, things inside the warship started falling all over the place,' says Captain Sequeira. 'It was difficult for one to even stand. We normally stand at 3–4 degrees of rolling. People start getting sick between 8–10 degrees. Most can't stand at 35–40 degrees. I remember taking a moment to wonder, can I handle her or not? If you get seas on the side, the waves breaking on the side, she will tend to roll. If you get the waves from the front, from where you are headed, you slam into the waves and then you come out. So, either way, you choose what damage you want to take.'

The commanding officer remembered the words of a mentor from his days as a cadet:

'If you go into the seas, seas will try to break your ship.'

Forty of the 250 men on board were new personnel who hadn't spent much time out at sea in rough weather, and quickly began to experience the effects of a violently rolling and lurching ship. Young men in their twenties on their first tenure in the open ocean. But under the care of the more experienced men, the affected young sailors were back on their feet. They would never be that sick again.

But INS *Kochi* was taking a true beating that afternoon.

'The sea was so rough that it was now pushing us back with a strength of about four knots,' says Captain Sequeira. 'So, let's say if I was using engine power to give me what is normally twelve knots, I was only getting about eight knots. My effective power was getting struck down by almost 30 per cent.'

The crew was confident of their warship. This was a superbly sturdy vessel with solid, indigenously built bones. But this was a sea the crew had never encountered before. Not even the most experienced men on board.

'She was getting slammed badly, and it put a lot of stress on her,' says a senior sailor. 'Our ships are built to take a certain amount of beating. But we had to be mindful that we were subjecting Kochi to a lot of stress. How much could she take? You are exposing your machinery to a lot of beating, which is not normal. All we could do was try our best to see how we could minimize this roll and pitch motion of the ship.'

As barge P-305 continued to drift and take in more water, INS *Kochi* continued to get inputs from the shore about its last known position. Dilip and a small group of men were now huddled in the radio control room on the barge, desperately trying to send out more messages.

'At around 2.30 p.m., we got a call and we could hear a guy from P-305 calling us on Channel 16, the common calling channel for all merchant vessels,' says a sailor on INS *Kochi* manning comms in the operations room. 'He got to know we were coming; he must have been in touch with his people. He was a radio operator. We asked him where the barge was now, and he gave us some coordinates. We realized that the barge had drifted further north and was now north-west of us.'

The mission had been a proverbial race against time from the beginning. The radio call and fresh coordinates brought the first realization on board INS *Kochi* that the mission could fail badly. The barge, with 261 men on board, was being violently tossed around. How would an over 7500 tonne warship be able to safely manoeuvre anywhere close to it? How would it rescue anyone at all? The ship hadn't even reached the drifting, sinking barge, but it already seemed like an extremely difficult and dangerous mission.

'As we turned, the ship got slammed right in the beam, sending us rolling well over forty degrees,' says Captain Sequeira. 'That was very risky. We were finding it hard to maintain course and the seas were tossing the warship around. If I could barely stand on my feet, I could only imagine the impact on my machinery, my radar, my equipment. The basics would go for a toss. And that would have risked the whole ship. So, we were doing the desperate zigzag manoeuvre again to somehow reach the waters where the distressed barge said it was. And this position itself was changing dramatically by the minute.'

Another call from the P-305 crackled through on Channel 16. It was Dilip.

'I remember his voice very clearly,' says the sailor on INS *Kochi* who received the call. 'I could now make out there was enormous desperation in the voice. We got inputs that the barge had taken in water, and it would go down. And in their mind, there was full panic that they were going to die before we reached them. They said everybody was up, on the part of the barge that was out of water, and said, "*Jaldi aao, jaldi aao* [Hurry, come quickly]." I told them to have faith that we would reach them soon and would not spare any effort to get there as quickly as possible.'

The P-305 was now tilted at an angle and sliding slowly into the water. With over 100 men jumping into the sea, the other 161 perched themselves on the end of the barge that was sticking out of the water.

'Now that we knew the barge was sinking, it was plainly life and death,' says Captain Sequeira. 'As we were approaching, we kept telling them over Channel 16 that we were coming, *aa rahe hain*.'

But it was impossible to go any faster. There was a limit to how much power the ship could be cranked up to.

'We had all the four gas turbines running right from the start,' says a sailor from INS *Kochi*'s engine room. 'In fact, we never stopped the engines at all. As you push more power, you get more impact on the ship. We had to trade off between how much speed we could achieve and how much beating we were willing to subject the ship to. The higher the power, the greater the beating. So, there was also an increasing strain on the propulsion system.'

At 4 p.m., five hours after INS *Kochi* had set sail from Mumbai, the naval vessel finally reached the general location of the sinking barge, slowing down when it spotted the P-305 from just under two nautical miles away. The barge was leaning at a steepening angle into the aggressive swell.

'Using binoculars, we saw that the remaining on board the barge were all clustered in the part that was protruding out of the water,' says Captain Sequeira. 'Over Channel 16, they said they had their life jackets on and pleaded they be rescued at the earliest. In my mind, that was the moment of truth.'

One of the lookout sailors tasked with assessing the approach says, 'Their accommodation block was in the aft structure. The barge was hit in the front, so the front part was now below the water. There was one huge crane-like structure that was sticking out.'

'As I looked out at the barge, it was quickly confirmed that the barge was huge, almost as big as my ship,' says Captain Sequeira. 'And I am nearly 7500 tonnes. There is no way I can take my ship anywhere close to the barge and let's not forget, I can't fully control my ship. At one moment I am here and at another moment, I am there. The angry sea is just tossing me around. Those guys were getting panicky, asking us to reach quickly but at that point there was little we could do. Like I said, moment of truth.'

If this had been a routine rescue of a distressed ship in fair weather, the crew of INS *Kochi* would have lowered lifeboats to bring the men on the P-305 to safety. But with the conditions now worsening to Sea State 8 with waves over 45 feet high, anything lowered into the sea would simply be swallowed. No boat would have remained afloat in that swell.

The next possibility was the warship's life rafts, emergency twenty-man rafts reserved for the ship's own crew in the extreme event that they needed to abandon the vessel. These had a better chance of surviving that maniacal sea. But how would the crew get those rafts out to the men marooned on a rapidly sinking P-305?

'Whatever we threw into the sea at that point would be eaten up by the sea. And I was not prepared to send divers into the water either. Nobody could swim in that water,' says Captain Sequeira.

The crew of INS *Kochi* then conducted a sweep for other vessels in the area, quickly latching on to an offshore support vessel (OSV) that operated at the Heera Oil Field to support and service barges. Fitted with bow-thrusters, OSVs move at a slower speed and are more manoeuvrable, unlike warships that rely on brute power to roar through the water in aggressive forward motion.

'We asked the OSV if it could approach the barge and transfer the life rafts,' says Captain Sequeira. 'The guy who responded said it was not possible at all. So, we were in a very difficult situation. We were less than a nautical mile away from the sinking barge and kept telling the men to hold on.'

Just before 5 p.m., INS *Kochi*'s comms room could make out that there was an abrupt spike in the desperation of the voices reaching them from the barge.

'There was total panic now, and the men on the barge clearly felt that it was going under,' says one of the sailors who was responding to the desperate calls. 'It's not easy to calm somebody whose life is on edge. We were there for an hour. We also felt that the situation may only worsen. We were in Sea State 8 and the sea was deteriorating further.'

Armed with fourteen perfectly good life rafts, Captain Sequeira was determined to get them to the men on P-305. But the rafts needed to be released into the water. And given how violently their ship was being buffeted, that in itself was a death wish.

'These life rafts are meant to operate automatically when they hit water, by pressure release,' says Captain Sequeira. 'They're supposed to open on their own. But in this case, I decided to send five of my men to the upper deck to release the rafts. At this point the ship was still being tossed around badly, and at any instant, there was the possibility of these guys getting knocked into the rough seas. One wave comes and they slip and go, that's it. I will never see them again. But we took the chance. There was no other way. We lowered fourteen life rafts and hoped that they would get somewhere close to the barge.'

When the men returned below deck, the commanding officer instantly knew that this had failed too. The men revealed that as the life rafts went down, they were all ruptured, torn and got lost in the sea.

At 5.30 p.m., the comms room on INS *Kochi* received a call from the barge, strangely calm and therefore even more disturbing, informing that the remaining men were abandoning the P-305 and jumping into the sea.

'It was a very gentle voice, without any panic,' says the sailor who responded. 'It was very unsettling. It was the voice of someone who had totally given up. There were many men on that barge who did not know how to swim.'

Minutes later, one of the sailors on lookout on the upper deck of INS *Kochi* shouted through the ship intercom to the bridge that he had spotted groups of men in life jackets from the barge in the water not far from the ship.

'So, the men were now in the water, which changed things,' says Captain Sequeira. 'But we still couldn't go near them as our warship could hit them and knock them dead. There was literally nothing else we could do at this point.'

It was then that they thought of a final last resort.

Scramble nets.

These were cargo nets that could be slung down the side of the ship into the water, giving the men in the water the chance to somehow clamber aboard the destroyer.

'You roll it down, the guy can climb or hold on to it, just average fitness is good enough,' says Captain Sequeira. 'We said let's try the scramble nets. That was the only thing we could think of. But again, the problem was the same. How do we get close enough to them so that they can reach the nets and climb up? The sea was still tossing *Kochi* around. The men in the water were also getting tossed around everywhere.'

Desperate now to close the gap between his big destroyer and the men bobbing desperately in the sea, the commanding officer of INS *Kochi* began a process of manoeuvring his ship with a series of power stops and starts. It would be the most delicate navigation he had ever done.

'We saw them coming in a certain direction, so we weren't really moving in that direction but manoeuvring in a way with massive power surges and stopping and trying to see how we can get anywhere close to them,' says Captain Sequeira.

Four teams of men from INS *Kochi* were now positioned on the decks to handle the four scramble nets, one giant net and three smaller ones.

'Frankly, at that point nobody believed it would work as we had to get the ship close to these small groups of survivors when we ourselves were not able to control how we were moving,' says a sailor on one of the scramble net deck teams. 'But the way we were manoeuvring helped and we finally came closer to the survivors. We had to be very careful as a large ship. One small move and our hull could kill everyone in the water.'

An agonizing hour of painfully delicate manoeuvring ensued as dusk fell over the Arabian Sea. Just before 8 p.m., the deck team manning the big scramble net saw something that it was least expecting.

An exhausted survivor had reached the scramble net and was now slowly clambering up the side of INS *Kochi*.

'It was just one guy, but it was a huge thing for us,' says the sailor who spotted the first survivor. 'We suddenly realized this was working. A few minutes ago, we were not even able to stand on the deck. And now we had got one of the survivors on board. So, we just kept doing what we were doing.'

'Even in normal conditions, climbing the net is difficult,' says Captain Sequeira. 'Even one or two metres of up and down motion can make it very difficult for a person to climb. It sounds simple but it isn't. The rope in the net is quite thick and has wooden beams in it. One of them can hit your head and break it open.'

The men of INS *Kochi* had thus far seen how ruthless the sea could be. In a twist right after the first rescue, they began to witness what they regarded as the other side of the ocean's dual nature.

'The sea abruptly calmed down a bit,' says Captain Sequeira. 'Suddenly it wasn't that bad. The swell had come down to five to six metres from the earlier eight to nine metres. We were now able to manage the rescue slightly better.'

With more survivors from the barge managing to reach the scramble nets and hoist themselves to safety, the rescue settled into a hopeful rhythm.

The sea, which had brutalized the barge, appeared to be willing to make up for it in at least some of the cases.

'There was this one survivor who could not grab the net. He drifted in the current and ended up behind our ship. But a powerful wave threw him right on to the ship's deck. It was a miracle,' says Captain Sequeira.

By 9 p.m., INS *Kochi* had rescued forty-five men from the P-305. In the darkness lit up by the ship's floodlights, the crew watched the barge finally sink beneath the waves. The stray sounds of groans and whimpers from the sea kept the warship's crew on full alert. With 216 men still theoretically out there in the sea, the night would afford no break to the rescue crews. In fact, it would provide an opportunity for individual acts of heroism.

'I saw my anti-submarine warfare officer, Lieutenant Commander R.K. Manu, struggling to balance himself on the warship's deck,' says Captain Sequeira. 'He peered down the starboard side and was gesturing to a group of middle-aged offshore workers to climb up the scramble net. These men were fatigued after being in water for hours, hopeless and gripped by the fear of imminent death. They could not climb the net. They could get to the net, but they just could not climb. That's when Manu secured himself with a rope and climbed down the net to get these survivors to safety. He held the survivors by their collars and pulled them up.'

The sea fell ominously silent, with no further survivors visible or audible. By 11 p.m., INS *Kochi* received a call from another OSV a few kilometres away in the area saying it had spotted more survivors in the water.

By 5 a.m. on 18 May, INS *Kochi* had rescued 115 persons from the sea. Right around this time, before first light, things were stirring at a separate location back on shore.

INS Shikra Naval Air Station
Colaba, Mumbai

A part of the main hangar roof had blown away in Cyclone Tauktae's powerful winds, with rain flooding the insides of the hangar where Sea King Mk.42B helicopters of the Harpoons squadron were parked.

Commander Bipin Panikar was up while it was still dark. He had received orders from the Western Naval Command late the previous night to fly out on a search-and-rescue (SAR) mission at first light to the Heera Oil Field, where INS *Kochi* had managed to rescue a number of survivors from the sunken barge.

Like they had with INS *Kochi* at the Mumbai dockyard, Tauktae's winds were doing everything they could to keep the navy's Sea King helicopters firmly on the ground.

'We were on the clock, but there was a real issue,' says Commander Panikar. 'There's a wind envelope in which you can take off and fly. The Sea King has an automatic blade folding system. Wind speeds have to be less than 45 knots for the blades to spread. If the wind speeds are more than that, not only will the blades not spread, but there could be damage to the helicopter too. And then we can forget about flying.'

That was the first problem.

'Secondly, during very high winds, if you ask a chopper guy to engage rotors, when the rotors start turning at low rpm, there is a huge chance of the blades striking the body of the aircraft,' says Commander Panikar. 'It's a big aircraft and the blades are big too. The first thing was to get the chopper started.'

The pilot's predicament was unprecedented. The Indian Navy's trusty old Sea Kings had proven worthy of launch and flight in very poor weather before. But Tauktae had presented new limits to what the machines could be forced to go up against.

At 5.30 a.m., the squadron duty officer called Commander Panikar to inform him that there was no way a Sea King could even be wheeled out of its hangar—the blades were banging against each other even inside the hangar, he said. Not launching a helicopter mission wasn't an option that morning, given the situation with barge P-305. And so, with the squadron's commanding officer in tow, Commander Panikar walked across INS *Shikra*'s helipad to the Sea King hangar.

'The Sea King is a very robust aircraft but the winds were 45–50 knots, just on the margins of its operating capabilities,' says Commander Panikar. 'We needed to act fast because I should have been airborne by that time. I knew I would not be able to take off from the standard take-off position close to the sea. That was not possible as the chopper would encounter more wind than is considered safe. So, I picked a spot between two hangars for lower wind impact.'

With some difficulty, the big hulking helicopter was wheeled to a strip of asphalt between two large hangars. Climbing into the cockpit with his co-pilot, Commander Panikar carefully started the first of the two engines.

'The wind was still too strong and the main blades started rattling with each other,' says Commander Panikar, 'So we had to put one of our technical guys on top of the aircraft, where he held the blades. He was a brave guy. The blades were earlier colliding with each other. Once you start one engine, it starts all the drives, and they activate the blade fold system through hydraulics. Normally, you wouldn't spread wings or engage rotors in these conditions. The aircraft can take much more once it's airborne. The most important part was the rotor engagement.'

After a measure of struggle, the Sea King was finally running its rotors at full power. What was usually a routine helicopter launch sequence had turned out to be an hour-long battle with Tauktae's winds.

'The entire air station, crew and commanding officer had gathered at the take-off area and it felt as if I was going out for batting in a cricket match,' says Commander Panikar. 'Some guys told me not to go far out in the sea if the conditions were too bad. If it's not looking possible, come back, they said. In aviation, you don't chase the weather. I knew the weather picture, it was moving northwards and I was trying to chase the weather. This was not normal. Ask any aviator, he will try to steer a course which is opposite to such weather conditions.'

As with INS *Kochi* a day earlier, the Sea King's mission that day was literally to fly into trouble.

'I was to be that guy who went there, understood the weather, the situation and tried to rescue whoever he could. I was to come back and brief the other guys on the situation so that subsequent launches could happen,' says Commander Panikar.

The commander had flown Sea King missions during the milder Cyclone Ockhi in 2017, but the moment he got airborne from INS *Shikra* and headed into Tauktae, he knew this was something else altogether.

'The abnormal movement of the control stick made me realize we were up against a daunting task of having to factor in wind effects on the helicopter,' he says. 'The stick was shaking. When the winds are strong and changing direction very rapidly, you have to continuously move the stick to nullify that wind effect. There was a lot of stick movement.'

Banking over the grey Arabian Sea, the Sea King headed past Prong's lighthouse on the southernmost point of Mumbai, four miles from INS *Shikra*. There were no other aircraft in the sky that morning thanks to Tauktae, affording rare permission to Commander Panikar from Mumbai air traffic control (ATC) to fly wherever he wished.

'"Nobody is flying, and we are heading into bad weather"—that's the first thought that crossed my mind and it was a bit unsettling,' says Commander Panikar. 'But we also knew we had to reach the rescue site at the earliest and expand the scope of the SAR before it was too late. When I looked out at the sea and the weather, it looked really terrible. We had to fly ninety to hundred nautical miles to reach the SAR location. Dark patches of low-level clouds at 200–300 feet were visible, and my radar was showing me a lot of red, warning me about the weather I was wilfully flying towards.'

The Sea King's navigator piped in.

'Do you really want to fly into this?' he asked. 'It looks very bad.'

'Let's try and skirt those clouds,' said Commander Panikar. 'And the moment we feel it is not doable at all we will turn back. Let's do some assessment so that we give a "sitrep" to the base at least.'

Cruising at 1000 feet, the Sea King descended to 500 feet. Almost immediately, inside the cockpit, it felt like day had turned to night.

'We were totally and suddenly engulfed in clouds,' says Commander Panikar. 'That hit all of us in the aircraft—my co-pilot, our navigator and my aircrew diver Prahlad. Each of us thought the exact same thing. That this was like night flying.'

At 500 feet, and with the pilots flying the Sea King on instrument readings rather than outside visibility, things took an even more surreal turn.

'There was sudden lightning all around,' says Commander Panikar. 'The lightning lit up the insides of the cockpit. And then it would be dark again, so I switched on the cockpit lights.'

At 500 feet over a surging sea, the pilots were forced to trust their instruments. Things in the cockpit took a hit with pressure instruments that provided crucial airspeed and altitude data becoming unreliable.

'The weather was making them go wonky,' says Commander Panikar. 'One electronic instrument that was still working was the radio altimeter (radalt) which was telling me the exact height from the sea surface. I turned a course forty-five degrees away from my destination to get away from the dark clouds.'

The pilots flew a course that took them in and out of the low, dark cloud deck, the helicopter shuddering with each repeated dive into the clouds. The crew exchanged glances with each jolt, agreeing that this was one sturdy helicopter they were flying.

'Every two or three minutes we got such bad jolts that we actually asked each other if everything was okay,' says Commander Panikar. 'There was crazy wind, rain and this thick cloud deck we were flying through. The pressure instruments were giving us incorrect and unreliable data, so we had to fly cautiously. For instance, my airspeed reading showed sixty knots when my ground speed was only thirty knots. That cyclone was really testing us that day.'

Pilots are trained to trust their instruments when visibility goes to zero or all else fails. The crew of the Sea King were flying without visibility and reliable instruments just 500 feet from the ocean surface. The risk of controlled flight into the sea couldn't possibly have been higher.

It was the helicopter's sole non-pressure instrument that kept things on even keel.

'The radalt came to my rescue,' says Commander Panikar. 'It transmits a radio beam down, it hits the ground or sea surface and comes back. I was relying on it. There are so many instances of controlled flight into terrain in bad weather. When you are flying and visibility is not good, you feel you are at 1000 feet and are comfortable and then suddenly you are in the water. To avoid that, it was very important for me to know at what height I was flying. These things were weighing on my mind.'

Like INS *Kochi* on its voyage towards the Heera Oil Field, the Sea King was being flown in a zigzag flight path to avoid the worst of the bad weather. Since this was a vanguard mission, the pilots were carefully recording their flightpath in order to share it with other Sea King crews who would fly subsequent missions later in the day.

Ninety minutes later, the Sea King arrived at its target position, coordinates that had been shared by the Indian Navy's maritime operations centre.

'We were at 200 feet and could see nothing in the sea,' says Commander Panikar. 'I was thinking, we've come ninety nautical miles and there's nothing here and nobody to rescue.'

After an uncomfortable and dangerous flight over the Arabian Sea, there was disappointment and anxiety in the cockpit as it hovered at 200 feet.

'We wondered if we had come to the wrong position. We checked the coordinates again and it was clear we were at the position given to us. Visibility ahead was not even one nautical mile. I tried calling the ships in the area, including INS *Kochi*, over the radio frequency for the scene of action, but there was no response. It was imperative for the ships to respond to me because I had been tasked with dropping body bags on their decks. And we couldn't stay airborne indefinitely.'

The Sea King's navigator started scanning the area again on radar, desperate to locate something—anything—before it was time for the Sea King, forced by fuel endurance, to swivel around and head back to base.

'We needed to find something very urgently,' says Commander Panikar. 'I was circling around in an environment that was rapidly changing, the winds, the clouds. All of us in that cockpit knew it could lead to disorientation. About eight minutes into the latest loop of low-level flight, my observer spotted something on the radar.'

But it was 30 nautical miles out.

'I was constantly concerned about losing endurance,' says Commander Panikar. 'I asked my observer if he was sure. He said he was. With every passing minute, we were losing the ability to stay airborne longer. I was concerned about having enough fuel to return. How much time will I take to return? Thirty nautical miles meant another fifteen minutes of transit time. The observer was very confident that he had picked up something. I was praying he hadn't picked up some merchant vessels.'

The helicopter flew low, but it was still in a thick envelope of clouds. As the Sea King approached the coordinates identified by the navigator, its radar started picking up a sudden burst of contacts. Commander Panikar prayed once again that it wasn't a pod of whales or dolphins that had been spotted. As much as he loved animals, he wasn't here on a sightseeing sortie. He pitched the Sea King forward so it could descend through the clouds.

And there they were.

'When we broke cloud, we saw a large number of people floating in the sea,' says Commander Panikar. 'We came down to 150 feet for a clearer picture. It was a sea of survivors in orange jackets, and we could also see INS *Kochi* and other ships conducting their rescue operation at some distance.'

It was 9 a.m., and there were a lot of men to be rescued from the sea. The Sea King descended further. And as they got nearer to the sea surface, the true horror of the picture became apparent.

'Many of them didn't seem to be alive,' says Commander Panikar. 'But we didn't have time to process the tragedy. We had to find those who were alive and get them out of the water. We were now keeping a hawk's eye on any movement in the water.'

Hovering gently above the surging sea surface, the crew of the Sea King carefully scanned the cluster of orange below them. The co-pilot quickly spotted one man struggling to keep his head above the water, barely alive.

'Normally you have an automatic hover system which you engage and the helicopter tries to maintain its altitude with respect to the sea which is crucial for such rescue missions,' says Commander Panikar. 'But that was out of the question because of the swell of the sea. And that meant we had to do the hover manually. Hovering over sea is harder than hovering over land because of lack of visual reference. A manual hover is difficult in such situations, besides the swell was tossing the survivors around. There I was trying to position myself over the guy, and in a flash, the swell displaced him. Next, I would find him fifteen to twenty metres behind me. All this kept happening. And all this when wind speeds were touching fifty knots.'

With time rapidly running out, and wind speeds only picking up, Commander Panikar asked master chief aviation (flight diver) II Prahlad to get into the water and bring the survivor up.

'I told Prahlad not to disconnect from his harness under any circumstances,' says Commander Panikar. 'If he disconnected to pick someone up, I could lose him in the choppy sea.'

It would take twenty minutes to winch up the first survivor.

Remaining at that dangerously low altitude, the Sea King began to hunt for more survivors. As it did, INS *Kochi* came into view once again, reminding the pilots about their secondary task. In a careful manoeuvre that had its own share of challenges, the Sea King delivered body bags to INS *Kochi* and the other ships rescuing men from the P-305.

'After a second rescue, we were about to turn back,' says Commander Panikar. 'We were running very close to the required fuel levels for the return flight. While we were on our way back, we spotted a guy folding his hands and shouting for help in the sea. This time we were hovering at thirty feet. That was the height of the waves in the area. It was a surreal sight from the cockpit.'

For this third rescue, diver Prahlad decided he had to disconnect from his harness as the survivor was totally exhausted.

'We were very low on fuel but I had seen this guy, and we knew we couldn't leave him behind,' says Commander Panikar. 'And the other two survivors we had rescued urgently needed medical support. They were in very bad shape, and I felt they could die in the aircraft. I had seen that happening before in an earlier mission.'

Winched up along with Prahlad, the Sea King rose above the cloud and flew at maximum speed back to the INS *Shikra* air base.

'We were dangerously low on fuel when we landed,' says Commander Panikar. 'It was important for me to reach back and brief the other crews on the mission and the shortest route. The briefing room was set up next to the helipad to save time.'

The next Sea King missions began immediately and continued through the day, contributing to a string of additional rescues from the Heera Oil Field site.

Diver Prahlad accompanied other Sea King missions that day, rescuing a total of twenty-eight survivors, including three who were on the verge of drowning and rescued during the first mission.

Commander Panikar flew more sorties himself, rescuing a dozen more marooned persons from another barge. Sea Kings from INS *Shikra* rescued thirty-four survivors in all.

Operating an aircraft to the extreme of its limits is an art pilots can learn only when they fly missions like this . . . In normal day-to-day flying, pilots can't push aircraft to the extreme because if something goes wrong, they will be held accountable.

By noon on 18 May, ten more survivors were rescued by INS *Kochi*. Another sixty-one were rescued by other Indian Navy ships, the Sea King helicopters and OSVs.

That evening, INS *Kochi* called off its search and set course back for Mumbai. On board, along with the 125 rescued men, were the bodies of seven of their mates who had drowned.

'We were in touch with Sea King choppers right through the rescue missions,' says Captain Sequeira. 'The warships and the helicopters worked in tandem to carry out the SAR. We saw people floating in life jackets but found that many of them were dead. We reached Mumbai on the morning of 19 May, delivered the survivors and the bodies of the dead and then returned to the Heera Oil Field area the same day. In the second round, we found only bodies and no survivors.'

While Captain Sequeira was decorated with a Shaurya Chakra, Commander Panikar received a Nao Sena Medal for gallantry. In many ways, both officers came away from Tauktae with similar experiences.

Both faced crushing challenges to get their missions started.

Both were forced to deal with the possible mortality of their own men.

Both were forced to think out of the box while a life-and-death clock ticked ominously.

Both also came away with an uncannily similar lesson from their respective missions: that humans are capable of achieving much more than they believe they can.

Captain Sequeira says, 'Everybody on my ship was worried about what will happen. Not only to the men on the barge but also to us. What would our fate be? How will this end? But we made it. At the end of the day, you have to give it your best shot no matter what the odds are. You really have to try to make some sort of a difference. If you just keep sitting and thinking, "What do I do and how is this going to play out?", you are not likely to achieve anything. When we left the Mumbai harbour on 17 May, even I was thinking of what exactly I was going to do there. There were no answers then. But my philosophy seems to have changed a bit: at least try, do something, it might just work.'

The warship captain's gallantry citation reads, 'The successful and daring execution of the SAR operation under unprecedented stress and extreme rough weather highlights the extraordinary leadership, professionalism and fortitude of the officer. This brave act of leadership and courage was in keeping with the ethos of the Navy and in service of the nation in times of peril and adversity. For his extraordinary bravery, courageous leadership, exemplary grit and determination, in the face of extremely challenging conditions, the officer is recommended for the award of Shaurya Chakra.'

INS *Kochi*'s anti-submarine warfare officer, Lieutenant Commander Manu, who went into the water down the side of his ship to pull exhausted survivors from the sea, was decorated with a Nao Sena Medal for gallantry. His citation reads, 'He displayed extraordinary leadership skills and courage with complete disregard to his personal safety during the SAR operation undertaken by the ship, during Cyclone Tauktae, resulting in the saving of 125 lives. Through personal example, the officer led his team of men during the rescue mission for 36 hrs at a stretch, under extremely adverse weather conditions.'

Sea King pilot Commander Bipin Panikar's gallantry citation reads, 'As the first pilot and Captain of the aircraft, the exceptional courage, bravery and fortitude displayed by him in the face of extremely dangerous, hazardous and unforgiving weather conditions resulted in the successful rescue of three lives of sunken Barge P 305. For utmost exceptional courage, fortitude, skill and display of valour during the SAR operation of survivors of sunken barge P305, Cdr Bipin Panikar is very recommended for the award of Nao Sena Medal (Gallantry).'

And finally, commando diver Prahlad, who saved twenty-eight lives, was also decorated with a Nao Sena Medal for gallantry. His citation reads, 'Prahlad, with absolute disregard to his own safety and life, undertook these unparalleled rescue missions single-handedly in extremely hazardous and hostile sea state. It is his commendable swimming skills and years of experience as a flight diver that saved the lives of 28 personnel from a near-death situation. The exceptional bravery, initiative and nerve displayed by him merits recognition and he is strongly recommended for the award of Nao Sena Medal (Gallantry).'

Dilip the welder from barge P-305 survived and was rescued.

9

'This Time Holi Will Be with Blood'

Major Konjengbam Bijendra Singh

Dibong village
Manipur
March 1999

The eight-year-old boy usually slept soundly, the breeze gently blowing through this verdant village near the Assam border a reassuring lullaby. But that night, a loud midnight knock on the front door of the small mud house had made Konjengbam Bijendra Singh sit up, rubbing his eyes. To his side, his two siblings were still asleep nearly four hours after he had tucked them in. The children had gone to bed earlier than usual that night so they could be up early to celebrate Yaosang, the five-day Manipuri equivalent of Holi, a festival to usher in spring starting on a full-moon day.

And like in Holi, the children would play with colours.

Bijendra heard hushed, anxious voices in the Meitei language as his parents shuffled out of bed. Then he saw the shadow of his father, Konjengbam Binod Kumar, a part-time teacher in a local government school, stumble half-awake to the door. Bijendra stepped out of his bed and squatted furtively near the door of the room, so he could see what was happening.

As soon as his father unbolted the door, three men pushed their way into the house. They had assault rifles slung around their shoulders. One of the men wielded a pistol of some kind.

'Serve us dinner, we're hungry like animals,' said one of the men in a rough tone as he looked around the tiny dwelling.

Bijendra's mother, Konjengbam Tilotama Devi, now shuffled into view, walking up to her husband as he faced the three men. His father turned around briefly to check on the room in which the three children were supposed to be asleep. He saw three sleepy but wide-awake faces peering back at him from the shadows of the bedroom. He gestured to them to return to bed immediately, but Bijendra and his brother and sister stayed where they were, transfixed by the late-night invasion.

After a few seconds of inaudible conversation, Bijendra's mother proceeded towards a corner of the main room that served as the kitchen. Turning on a kerosene stove, she prepared a quick meal of leftover rice and pork curry from the night's dinner, serving it in bowls to the three men who were now sitting on the floor, their weapons leaning against a wall in the dimly lit room.

It wasn't the first time Bijendra and his siblings had had their sleep interrupted this way. Armed insurgents were a common sight during those decades in Manipur, knocking at doors in village homes for a meal, shelter or to hide from security forces. Amid a rising tide of militancy in the North-eastern state, armed groups frequently extorted from innocent families that had nothing to do with the insurgency. To villagers like young Bijendra and his family, the sound of those three men gorging noisily on the rice and pork was still a gentler aspect in an insurgency that was otherwise drenched in blood.

'We should sleep now,' Bijendra said to his siblings, the three of them still watching as their parents sat down with their uninvited guests. 'We have to be up early for Yaosang.'

Back into bed, Bijendra remained awake, still listening. The smell of the rice and pork had wafted into the children's room, triggering a pang of hunger in the boy. But he knew there was no way he was going to see if he could get a bowl of his own. He lay in bed, waiting for the narrow triangle of light to go off, so he could relax and fall back asleep. Then he heard their voices again.

'Thank you for the food,' said one of the men. 'We are likely to be back tomorrow. And the day after that. Cook extra so there's enough. There should be more rice and curry next time.'

Then they left, and Bijendra heard the front door being bolted. His parents didn't say a word as they cleaned up, switched off the light and went back to their room. As he waited for sleep to overcome him, Bijendra was surprised by the spasm of rage he felt. This wasn't the first time he had seen armed men arrive at their house in the darkness and demand food or shelter. But it was the first time he hadn't heard his parents talk about it after the men left. Their silence numbed the boy. Only eight years old, he felt a tide of anger over whether his parents were so worn out by the unpredictability that hovered over the safety of their home with three little children, that they simply had nothing left to say.

It was with that resentment that Konjengbam Bijendra Singh fell into the deepest sleep he ever remembered.

* * *

Assam Rifles base
West Manipur
March 2019

'Are you sure you want to launch the operation on Holi?'

Colonel Rajkumar Bishnoi, commanding officer of the 23 Assam Rifles unit, was talking to his company commander.

Now twenty-eight, Major Konjengbam Bijendra Singh found himself in Manipur on the eve of another Yaosang.

'I have no doubt in my mind, sir,' Major Bijendra said. 'Haopu and his men will be least expecting the Assam Rifles to come after them on Holi. I have handpicked my best men, trained them hard.'

'So, *khoon se Holi khelne ja rahe ho* [So, you will be playing Holi with blood]?' the commanding officer said with a grin.

Holi was six days away, on 21 March.

Major Bijendra smiled. The very same words had been used by a soldier on the team a few hours earlier when plans were being finalized.

Khoon se khelenge Holi.

It wasn't a facetious comment. Major Bijendra, whose job it was to lead his company of men and their charged emotions, knew that the blood of innocent villagers was already being spilled in villages and the hunt was on for one of the principal insurgents responsible for local terror and kidnappings in that part of Manipur. Their target was Haopu Khongsai, the self-styled commander-in-chief of the Liberation Tigers of Tribals (LTT), an armed Kuki group.

The hunting ground was a vast one. Major Bijendra's area of responsibility (AoR) covered the expanse between the Barak and Makru rivers and was infested with various small militant groups owing allegiance to a handful of larger insurgent outfits, including the two primary factions of the National Socialist Council of Nagalim (NSCN), Zeliangrong United Front (ZUF), United Tribal Liberation Army (UTLA) and Haopu's LTT.

Major Bijendra's company had been on Haopu's tail for over a year. And on that March afternoon in 2019, months of often frustrating intelligence collection and analysis had been distilled down into a pattern of Haopu's movements, crucially narrowed down to a jungle stronghold 75 km from the company operating base (COB) at Kamai.

'We are very close, sir,' Major Bijendra said, pointing to an area on the map that pinpointed the jungle stronghold.

'If the intel fits, go for it,' said his commanding officer. 'But I don't need to tell you. They'll have innocents with them for cover. Don't come back with any innocent's *khoon* [blood] on you.'

The intelligence Major Bijendra was armed with was as precise and actionable as it was possible to obtain in the circumstances. It captured details of when and which village Haopu and his militant group were in, where they went next, who led them, what weapons they carried and the uniforms they wore.

'We shouldn't delay, sir,' Major Bijendra said. He knew he was as close as he could get to getting a militant who had been in his crosshairs since October 2018 following a spate of killings and kidnappings of villagers in the Kamai area. This was just months after the young officer, fresh on deputation from the Corps of Army Air Defence, had assumed command of 23 Assam Rifles' Alpha Company. In each encounter, Haopu's intricate network of informers had managed to ensure he escaped before the Assam Rifles arrived. And after each escape, Haopu would send a message by cranking up the violence and kidnappings. Villages would be crippled by fear every time Haopu managed to escape. They knew he would be back soon to announce his presence with blood.

The intelligence Major Bijendra now had even detailed when and at which house the militants showed up unannounced for a meal or shelter, very much like the late-night visits to his family home two decades earlier. Nothing had changed. Things had, in fact, become worse.

His commanding officer's go-ahead itself was a breakthrough. Major Bijendra had accumulated and waded through piles of intelligence data for almost six months before approaching his superiors to get the green light for the operation he was planning. All previous attempts to get approval for hunting the militant from the Assam Rifles leadership had failed. His superiors had ordered Major Bijendra to revisit and fine-tune the intelligence. They told him that it simply wasn't enough to avoid either collateral damage or major casualties to the Assam Rifles.

Weeks before the day his commanding officer finally approved the mission to hunt down Haopu, Major Bijendra had failed to convince his seniors at a security conference chaired at Khongsang, another 23 Assam Rifles company operating base, by Brigadier Ravroop Singh, the commander of Indian Army's 59 Mountain Brigade.

At the conference in early February 2019, Major Bijendra made a comprehensive presentation on the security situation in his area and detailed it with the intelligence he had gathered on the militants and their activities, the fourteen-man squad he had hand-picked and readied for the mission and a broad description of how and where they planned to ambush Haopu and his band of militants. He had displayed a sense of impatience, even exasperation.

In a testament to just how meticulous planning, checks and balances are almost always in place in the Army, Major Bijendra's pitch hit a wall. The brigadier and the commanding officer, while impressed by the major's unsparing approach, single-mindedness and vigour, still declined approval. A disappointed but disciplined Major Bijendra was ordered to keep preparing for the operation, further verify the leads he had gathered, make sure there were no loose ends and wait for authorization.

There was no time to be despondent. Days after the security conference, Major Bijendra, along with a few Manipuri members of his squad, disguised themselves as villagers to carry out reconnaissance of their area of interest to obtain first-hand information about the terrain, routes and possible vantage points for the planned mission. At the time, the soldiers weren't fully aware of the purpose of the reconnaissance missions they had been part of, though they vaguely knew that something big was afoot. Their company commander had dropped perhaps the biggest hint a few months earlier.

'No Holi holiday this time,' he had told them. 'You can all take leave after that.'

As it turned out, the reconnaissance missions gave the Alpha Company precisely the inputs it needed to fill in the blanks in the operation plan. Coupled with large amounts of intelligence that poured in subsequently, the overall accuracy of Major Bijendra's existing assessment stood confirmed.

By early March, he was convinced that there were big risks in keeping the operation on hold for too long. And with some crucial finishing touches that he asked for from his informers in the target area, he had the final operational plan that checked all the boxes in his commanding officer's requirement.

As frustrating as the process was, Major Bijendra knew that intelligence can be a tricky beast at the best of times when tackling militancy. There's no shortage of incidents, both big and small, where flawed intelligence has resulted in things going horribly and tragically wrong.

Colonel Bishnoi was clear. He knew that no operational plan could be 100 per cent foolproof. But he knew he had worked the young officer to the bone, building a mission blueprint that was close to actionable as it was possible under the circumstances. It was one of the sharpest pieces of intelligence he had seen for a high-level militant. He was impressed, but he knew he also needed to be apprehensive. Even the most well-constructed missions had a nasty habit of going south.

'KB, I am fine with your Holi plan,' Colonel Bishnoi said. 'You have been at it for months and I trust your judgement. Go ahead and get those guys.'

The commanding officer then departed from the Alpha Company base to head back to the 23 Assam Rifles headquarters in Tamenglong in the Naga Hills of Manipur.

The boss had left. His approval was in.

'Finally,' Major Bijendra said aloud to himself in the operations room of his base. Looking around at the maps and charts that had finally won him approval to proceed, he went to prepare his men.

There were loud cheers from the barracks as Major Bijendra broke the news to them. Every single soldier at the Kamai company operating base knew the name of Khaikhohao Khongsai, alias Haopu, the ruthless and slippery commander-in-chief of the LTT, and his trigger-happy cadres who had brazenly opened fire at the base at least twice in the past six months.

The Alpha Company's hunt for Haopu began with the daylight abduction of two employees of a private company associated with the construction of a crucial Indian Railways project in the North-east, the 111-km Jiribam–Imphal rail line. The Rs 12,265 crore project, expected to be completed by December 2023, is intended to significantly improve accessibility to Manipur and involves the construction of forty-seven tunnels and 156 bridges, including the world's tallest pier bridge that will rise 141 metres over the Ijei River. India plans to further extend the railway line to Moreh on the Myanmar border to link up with the Trans-Asian Railway Network in line with its Act East Policy to make the North-east a gateway to Southeast Asian countries.

Extortion and abduction by militants like Haopu had slowed down the Jiribam–Imphal rail line project in the preceding years, throwing a shadow over the intended benefits to local communities in remote parts of Manipur and its neighbouring states.

When news of the abductions filtered in on the Assam Rifles intelligence network, Major Bijendra had been particularly enraged. As a student, it would take him around twelve hours to travel from his village in Jiribam district to Jawahar Navodaya Vidyalaya in Manipur's capital Imphal by bus. The railway line would now allow children from his village to cover the distance in just over two hours. Haopu and his militant group were specifically targeting development projects because they knew development meant prosperity, movement of communities and a loss of territory to manipulate. To a young Army officer who had struggled to break barriers of access to wear the uniform, a more sinister, cynical, self-defeating set of actions couldn't be imagined.

Images from his childhood in Dibong would flash frequently in his mind as he got to work on the hunt for Haopu. It was men like these, Major Bijendra thought with a familiar pang of resentment, who pushed their way into people's homes, demanded pork and rice, and then actively destroyed the hopes and dreams of those very families.

Bijendra's father had worked as a government school teacher in the 1990s, but lost his job as the situation deteriorated in the area. Through that decade he drove lorries along the dangerous highways of Manipur to support his young family. There were no jobs, and there were no choices.

The internecine warfare between militant groups had destroyed decades of development in Manipur and its sister states in the North-east.

It saddened, infuriated and galvanized Bijendra. As he and his men broke bread together that night, they knew Haopu was but one of several such men who paralysed the North-east and its people. But to their tiny paramilitary company, he was the face they needed to hunt. That was their bit in the larger fight for peace and tranquillity in one of the most beautiful corners of the country.

'*Iss baar Holi thodi alag hogi* [This time it will be a different Holi],' Major Bijendra said as the men ate together. '*LTT ke saath khelenge. Yaad rakhna, sab ki nazar humare operation par hogi* [We will be playing Holi with the LTT. Remember, all eyes will be on how we conduct our operation].'

At sunrise the next day, 16 March, Major Bijendra ordered his squad to the operations room at the company base. While the soldiers knew that Haopu was their quarry, this would be the first time the soldiers would be given the full picture of what they were up against. The maps on the military easels made it immediately clear to every man in the room that the plan was to hunt Haopu and his men down in their very backyard, on their turf. Major Bijendra didn't need to explain how dangerous this was.

'We depart at first light on 19 March,' said Major Bijendra. 'Carry rations to sustain ourselves for at least a week. No other tasks or duties except this operation now.'

He sensed questions in the silence of the operations room.

'The intel is airtight,' he said. 'I'm vouching for it personally.'

Not that the men needed convincing—they had been waiting for this mission for six months—but Major Bijendra knew they were now committing themselves to a mission that each man knew could be his last.

'I am not exaggerating when I say I can visualize where Haopu is and what his men are doing right now,' he said. 'The militants won't know what hit them. This is the operation you and I have been waiting for.'

Being a Manipuri, it had helped Major Bijendra build a strong rapport with locals in his area of responsibility, as well as cultivate a fairly solid network of informants, who he often turned to for leads in the constantly shifting cat-and-mouse world of anti-militancy operations. On the night of 18 March, he called them for an update on the latest developments in and around Longphailum village near Nungba town where intelligence had picked up significant militant movements in the preceding weeks.

The Alpha Company squad planned to set up its ambush in a jungle not too far from Longphailum seventy-two hours later.

But being a Manipuri had a deeply dangerous side. One that radiated outward from the company operating base to Major Bijendra's home village and the village of every Manipuri soldier in the squad. Both he and his men would often receive phone calls from militant groups, warning them against conducting operations unless they wanted to see their families harmed.

'Doesn't your mother still live in Dibong?'

'Isn't 4698 the number of your brother's motorcycle?'

'Don't you want to see them ever again? Aren't you afraid of the consequences?'

Threat calls were common, but it wasn't easy to dismiss them. Militants went to great lengths to gather information about the families of security forces. If not to actively harm them, then to play mind games with them over the phone. Mind games that could have a telling effect, especially on younger soldiers new in the force and freshly away from their families.

Major Bijendra had trained himself not to be rattled by the calls. Sometimes they would mention things about families that nobody could have possibly known unless they were physically present near them. That was the extent of the network and the lengths to which militants like Haopu often went to intimidate the men who were now hunting him.

Growing up during the militancy, Major Bijendra was largely inured to the atmosphere of threat. But he would never forget how close those men came to his family. Their weapons casually leaning against that mud wall as they ordered his parents around in their own home. The sound of boots crunching the gravel and bullets whistling through the air and hitting their home's front wall on more than one occasion only to be anxiously, quietly prised out and discarded at first light with an iron spike by his sleepless father.

As he got older, the resentment grew, churned by a sense of helplessness over not being able to do anything about it. It was in this brooding phase that a teenaged Bijendra was returning to his boarding school in Imphal after spending the summer vacation with his family in Dibong in 2004.

As the bus rumbled down the Imphal–Jiribam highway, Bijendra noticed a band of armed men on the road ahead, using oil drums to block the way. When the bus stopped, the men ordered every passenger out with their hands up. Over the next few minutes, each passenger was robbed of their money and belongings at gunpoint. From Bijendra, they took the little money his father had given him for the journey and a bag of pickle and home treats his mother had packed for him to enjoy in the hostel dormitory with his friends.

Such looting of passengers was a frequent occurrence in those dark days. But what happened on that highway would fan a Manipuri boy's indignation into an unremitting rage. He would never forget the faces of those militants who grabbed his belongings. At his hostel that night, especially homesick and alone, Bijendra decided he would 'do something'. His friends would tease him, wondering why he had taken a commonplace highway robbery so much to heart. What they didn't know at the time was that the incident had made a quiet, unusually focused teenager out of Bijendra.

Seven years later, Bijendra had grown into a confident young man. His commitment to 'doing something' would remain unwavering, but he was confused like so many young adults about precisely what to do. As he waited for an epiphany, he enrolled for an engineering degree at the North Eastern Regional Institute of Science and Technology (NERIST) in Arunachal Pradesh right after graduating from school. But before classes could start, a chance meeting with a cousin in the elite National Security Guard (NSG) finally made Bijendra's path clear.

He quickly withdrew his college admission deposit from NERIST and used it to pay for a crash course before appearing for the Services Selection Board (SSB) interview. With an anxious family firmly behind him from distant Dibong, Bijendra saw to it that he was admitted into the hallowed campus of the Indian Military Academy (IMA) in Dehradun.

A year later, in December 2012, Lieutenant Konjengbam Bijendra Singh was commissioned into the Corps of Army Air Defence. For his parents, who had arrived from Manipur for the ceremony, the moment couldn't have been prouder. Bijendra would be only the second man from their village to be commissioned into the Army. If there was any anxiety about their son's career choice, they didn't let it show on that day, only hugging him and wishing him a safe, peaceful journey in uniform likely to be far away from them.

And now, seven years later and two ranks senior, Major Bijendra was back in his home state as part of the Assam Rifles. In fact, the Alpha Company base was less than 50 km from his village near the Assam border.

At dinner with the men, Major Bijendra heard his phone vibrate in a low thrum on the Sunmica table. He knew who it was even before he looked.

'You haven't seen your texts, everything okay?'

It was Jenipher, Major Bijendra's girlfriend in Imphal. The young officer had created a shadow account on Instagram to keep track of militant propaganda and pictures sometimes posted by younger cadres. Joining conversations on the app, he and Jenipher had begun talking. She was in her final semester of a BSc zoology course at the GP Women's College in Imphal. After due diligence from both sides, the two had met offline, their Instagram affection blossoming into a real-world relationship. Not that he didn't trust her, but Major Bijendra made it a point never to discuss his work in any detail with his loved ones. If they ever asked, he would simply say it was for their own safety.

'Sorry, Jen, was a little tied up with some planning work. Everything okay?'

'That's fine, but please don't ignore my texts, it's not easy for me.'

'Sorry, love. Also, Jen, I may be a little offline for the next few days, so if messages don't deliver or if I don't answer, don't worry. I'll call you soon.'

It had been two years since their relationship had begun. Jenipher's early insistence on details had given way to acceptance that this was how it needed to be. She understood that her Army boyfriend's reluctance to speak about his work was out of concern, and not based on trust issues of any kind. She and Major Bijendra's family were simply better off not knowing.

'OK, please text me whenever you get a signal.'

Major Bijendra knew he wanted to marry Jenipher. He had decided he would pop the question after the mission was over. It would also give him the opportunity to take some leave and spend time with her.

But for the next five days, the officer knew he would need to be laser-focused on his mission.

Glancing over at the men who were still eating, Major Bijendra knew he didn't need to worry about their abilities.

He and these men had been on multiple missions together in the lush jungles of Manipur, where they had forged deep one-on-one relationships to create a cohesive team that was supremely disciplined, well-trained and fierce.

One of the men held up a leg of chicken.

'Mission *ke baad*, this is them,' he said with a smirk.

After dinner, with less than forty-eight hours left for the squad to move to the ambush site 75 km away, the men conducted a final check of weapons and ammunition, bullet-proof gear, night-vision devices, communication equipment, field rations and first-aid kits.

Before turning in for the night, Major Bijendra dialled one of his informants located not far from the target site.

'No change. All normal,' said a hushed voice, before disconnecting.

* * *

Kamai COB
19 March 2019

Fully kitted out for the mission, the squad was on parade at the designated staging area of the company base at 4 a.m. as ordered, the sky still dark and pouring a steady drizzle.

The men split into two teams with their equipment and the soldiers were about to climb into the two Tata 407 pick-up trucks when Major Bijendra signalled to them to stop for a quick final briefing, more of a ritual before any mission he led. The men knew what their company commander was about to say.

'We play to our strengths, we fight as a team, we make sure there is no collateral damage and we come back victorious. We have done it before, and we will do it again.'

With a hoot of approval, the men bundled into the trucks, their company commander in the front seat of the first. The vehicles would ferry them until just ahead of Nungba, 50 km away. The remainder of the 25 km journey to the ambush point would be a trudge through a jumble of jungle routes.

As the pick-ups rumbled over hilly terrain at moderate speeds, Major Bijendra's mind was occupied with three dominant concerns. First, he needed to make sure the squad didn't squander the element of surprise. Second, the squad needed to make contact with the militants at the earliest. And third, all fourteen men needed to make it back alive and if possible unhurt. It was a tall order, but anything less would have meant second thoughts about the sharpness of the operation plan.

As planned, the 407s dropped the soldiers a few kilometres ahead of Nungba from where, with heavy bags, AK-47 rifles and light machine guns (LMGs) on their backs, they started their journey on foot to the ambush site. Led by two scout soldiers who were part of the reconnaissance missions in the previous weeks, the Alpha Company men trekked through rough and rolling tracks in the hilly jungles, moving as stealthily as they could. It wasn't, by any measure, a casual march. Each man had his weapon unlocked and ready, moving in a single file with extreme caution as militants were known to have observation posts and small hideouts in the woods.

'We were there to surprise them and not the other way round,' says Warrant Officer Shivram, the second-in-command for this mission. 'Hence the caution. The immediate goal was to reach the ambush site in time to pitch a few small tents and set up a small base before it got dark. We weren't sure how long we'd be there.'

Fifty-four years old at the time, Shivram was the oldest man in the squad. With maximum exposure to combat, he was a soldier whose experience the squad depended on. He had operated against the Liberation Tigers of Tamil Eelam (LTTE) in the late 1980s as part of the Indian Peace Keeping Force (IPKF) in Sri Lanka and also operated in Jammu and Kashmir as well as other North-eastern states in his three-decade military career.

'If I could fight and survive the LTTE, then I could definitely hold my own against the LTT under KB Saab's command,' says Shivram.

Slicing through jungle vines and bridging boggy swamps churned by the rain, the squad steadily inched towards the ambush zone, making the final approach as surreptitiously as possible, their postures now crouched. In the failing light, the men arrived at their destination, a small hilltop looking down on a swathe of jungle and a network of narrow dirt tracks. It was here that Major Bijendra ordered the men to establish their temporary base. It was from here that they would lie in wait for Haopu and the other militants.

By 8 p.m., under driving rain that spattered the orange mud of the hill, the tents were up. One of the men had sliced down some banana tree leaves enroute to keep over their heads in the downpour. Deep orange rivulets flowed down the hillside.

Khoon ki Holi.

The squad was again split into two teams. The first, including Major Bijendra and Shivram, took concealed positions along an arc on the hilltop to keep an eye on the main track below, a small stretch of which lay hidden from the base camp's view because of the way the hill jutted out.

Rifleman K.D. Burman manned the LMG position, the powerful weapon with a very useful rate of fire for the vagaries of jungle warfare, where targets often melded into the foliage.

The second team pitched pup tents using jungle bamboo a short distance behind, its crucial task to provide backup and ensure sufficient quantities of ammunition, food and water. The men had arrived equipped to stay for a week. If the mission stretched beyond that, the men of Alpha Company would need to abort and leave empty-handed.

'We were carrying more ammunition and rations than anything else. No metal support for the tents, using bamboo instead. A 4.5 kg tent weighed only 1.5 kg without the support gear. The planning was that detailed,' says one of the soldiers.

If jungle warfare is about fighting an often unseen enemy who has harnessed a wealth of hiding places and camouflage, it is equally about survival and endurance in a setting that is almost always unforgiving. Training at the Indian Army's jungle warfare schools dwelt deeply on living off the land to conserve rations.

And so, on that hill, pairs of soldiers would be dispatched into the woods, variously bringing back wild sweet potatoes, banana stems and berries. A dried fish pickle the men carried as a compulsory ration added a spicy, protein-rich tang to forest-given chow. Five men were capable of finishing a kilogram of rice in a single meal. It needed to be stretched.

'The dried fish achar was among the staple items we carried during all our missions, apart from Maggi noodles, dry fruits, protein bars and some other high-calorie items. The pickle never disappointed, it could make anything taste good. While we were carrying rations, wild treats were common as the food had to be conserved to cater for an extended mission,' says Rifleman Burman.

Snakes are a recommended food item on jungle missions, given the numbers in which they're encountered in the deep, rain-drenched woods. They're considered nutritious and easy to prepare over a simple campfire. Strangely, Major Bijendra and his squad found no snakes on the hill or nearby.

'You need to adapt and improvise to accomplish the mission. Use the terrain to your advantage. Staying out in jungles for long and operating in those conditions had become second nature to us,' points out another soldier from the squad.

Under the dense canopy of the forest, the soldiers settled into their positions, harvesting rainwater using their raincoats and keeping a watchful eye on possible movements in the undergrowth below. All this while they battled an expected battalion of leeches, mosquitoes and numerous other creepy crawlies that make Manipur's forests some of the richest and most splendid in the country.

'The leeches don't matter much when you are on the move. But leech attachments can be very irksome when your mission involves lying in wait for days,' says Shivram. 'We spent a good amount of time flicking them away,' he says.

Up on the arc in his own tent, Major Bijendra bunked down for the night. Four soldiers at separate lookouts would remain awake through the night on watch. As he wound down, he pulled out his phone for the first time since they had got out of their pick-up trucks earlier in the day. A single message from Jenipher flashed on the lock screen.

'Just be safe and come back, that's all.'

While there was a single bar of mobile signal, he thought it best not to reply immediately. Switching the phone off, he tossed it into his rucksack and lay back to sleep. Pictures from the day washed over him. The jungles and terrain of Manipur had instantly reminded him of Vietnam war films like *Platoon*, which he had seen in school.

Treacherous and stunning, he thought, as he drifted into an exhausted sleep.

Another day passed on that hill with no sign of any movement on the rain-spattered dirt tracks below. All through, Major Bijendra remained in contact with his informants on a second mobile, a primitive old push-button phone with no data or Internet. The informants steadily confirmed that Haopu and the other militants were still at Longphailum village and didn't appear to be suspicious about any imminent action. One of the informants provided a crucial tip, the kind the squad was depending on.

'They will cross the patch of jungle you are in anytime now. Don't take your eyes off the track.'

Major Bijendra mostly spoke directly to his informants. But on this mission, he had assigned the important responsibility to two Manipuri soldiers brought along for the specific purpose of staying in touch and updating the rest of the squad.

The concrete tip-off put the squad on full alert, with every man at his station and with his weapon. But nothing happened on 20 March. After hours in combat mode, the men exhaled and returned to their ambush posts. Every man knew that the next day was Holi.

'There was no compulsion that we had to strike on Holi, but symbols are very important to us, and we knew it was important to KB Saab,' says Shivram. 'The only fear was that sometimes you returned empty-handed after putting in one hell of an effort. It had happened before with our squad. *Kuch nahi mila aur hum vaapas base ja kar agle operation ki tyaari shuru karte the* [We found nothing and returned to our base to prepare for the next operation]. And then we were back in the jungles after one or two months for another hunt. Holi or no Holi, we were just hoping that was not the case now.'

Holi arrived with a distinctly ironic lack of colour. A low fog and steady rain painting the otherwise verdant landscape all the many shades of grey. But if the weather wasn't going to be festive, the day would bring some crucial gifts to the men of Alpha Company in the form of the first direct evidence of the presence of Haopu and the other militants in Longphailum village. Early in the afternoon, as the sky cleared a little to bring a semblance of colour back to the forest, Major Bijendra and his men heard the echoing clatter of automatic gunfire from the direction of the village.

Informers conveyed to the squad that the gunfire—once in the afternoon and the second time later that night—was to celebrate the construction of Haopu's new home in the village.

As the gunfire continued into the night, Major Bijendra quickly assembled a five-man team, leading them down the hill and towards Longphailum village. The team stopped 3 km from the village, carefully avoiding any further advance.

'There was a lot of firing and blaring music. Haopu and his men were there for sure,' says one of the men.

The team remained on site near the village for an hour, conducting 'listening' drills and soaking in everything they could. The number of firearms, the stray hoots of laughter, the shouts. The militants didn't appear to be masking their presence in the village. But like all things in guerrilla warfare, this could mean two things. One, it could mean that Haopu and his men were clueless about the Assam Rifles squad lying in wait for them just a few kilometres away. Or they could be celebrating as a decoy to lull the Alpha Company men into complacency, only to either quietly escape or strike unexpectedly.

'Haopu's informer network was too solid to discount that second possibility,' says Shivram. 'So if we were already being careful, we needed to double our guard now.'

Holi had ended, but no blood had been spilt. Back at their hill camp by 11 p.m., Major Bijendra gathered the men of Alpha Company. He was convinced that the mission was about to reach its peak.

'*Tayyar raho, militants se contact hone wala hai,*' he said to them. '*Shayad kal hi. Holi ek din late manayenge* [Stay ready, we may make contact with them not later than tomorrow. We will celebrate Holi a day later].'

'We fully trusted our commander's instincts. If he was saying contact was imminent, it had to be,' Shivram says.

Their commander's sense of certainty infused fresh energy into the squad. There was still an hour of Holi left. And so, on that desolate hilltop, intoxicated by the prospect of the encounter finally in sight, the men decided on a small celebration.

Generous helpings of khichdi, the one-pot meal cooked with fistfuls of rice, dal and water followed, in a welcome break to the monotony thus far of Maggi noodles and forest produce. The dried fish pickle once again worked its magic.

'We had been eating Maggi and the jungle fare as our orders were to conserve food. Khichdi had never tasted as good before. I will say it had the makings of a princely meal for us,' says Rifleman A. Brijit Singh, the then thirty-five-year-old buddy soldier of the company commander. The battle buddy system followed by the Indian Army involves pairing soldiers in combat to watch over each other.

Their bellies full of rice, the men should have been exhausted that night. But not a single man slept. The moment Major Bijendra and his men had been waiting for was now less than fifteen hours away.

Jungle near Longphailum
22 March 2019

It had been pouring non-stop for three days, but the day after Holi brought with it crisp, glinting sunshine. Bereft of mist and rain clouds, the squad on the hilltop had its first sweeping view of the woods that stretched all the way towards Longphailum village. It wasn't just an omen. It was also literally better weather for the militants to move.

At 1 p.m., Major Bijendra used his binoculars to scan the dirt tracks that snaked their way into the foliage. As he panned, he stopped abruptly. In the distance, maybe 3 km away, he saw a giant JCB earthmover rumbling slowly through the undergrowth with what looked like six men in the excavator's big mouth bucket. It was near the area he had scouted the previous night with four other soldiers, after the first round of celebratory firing.

'It's them,' Singh called in a hushed tone audible to the men holding the arc on the hilltop. 'We will still wait for an airtight visual confirmation. Let them come as close as they can. No one will open fire till I order it.'

The possibility of local villagers accompanying any militants in the earthmover couldn't be ruled out. The true dilemmas of jungle warfare were now throbbing. The near impossibility of identifying targets until they were very close and a major risk to the team lying in wait. Not a sound emerged from the hilltop as the men of Alpha Company held their breaths, not a word more was spoken after Major Bijendra's last order not to open fire without his order.

A squad scout sitting on the arc then observed through his binoculars the JCB machine turning to the right and making its way back into the woods. The men they had spotted in its mouth bucket minutes ago, now mysteriously gone. Every man in the squad now either had a pair of eyes or a weapon pointed at the dirt track down from the hill. For twenty minutes, nothing moved, and there was no sign of the men from the JCB.

Then they appeared.

The faces were familiar. Each one of the men, a known LTT militant. The intelligence had been spot on. But where was Haopu, Major Bijendra asked himself as he searched from one face to the next.

By 1.30 p.m., the first five militants were within rifle range of every man in the squad. Except the last man.

Where the hell is Haopu?

The squad knew it couldn't wait longer. With the militants in range, it was imperative that they act quickly. Any closer and the militants would either spot the ambush site and escape or be in a better position to return fire.

'I was ready to pull the trigger on my AK-47 but held back as the company commander's instructions were unambiguous,' says Shivram. 'We were to open fire only on his command. I had waited three days for this moment, a few more minutes were fine by me.'

Their automatic weapons slung around their necks, the militants walked single file down the meandering track into the Alpha Company squad's 'kill zone'. Their advance neatly spaced, each man walking around 40 metres behind the one in front.

It was an obvious tactic militants deployed to keep casualties low in the event of an ambush by security forces or even by rival militant outfits. The tactic also provided militants further back in the line valuable time to slip away or take cover to fire back at the ambush team. Additionally, it keeps the ambush team in suspense about the exact number of men making the advance.

The distance between the first militant, carrying a Kalashnikov with a tripod, and the sixth man, who was still perplexingly not in view of the ambush party, appeared to be more than 200 metres.

'After I spotted the first militant on the track, the second man came into my view after twenty-five-odd seconds and then the others in the same time gap,' says Rifleman Brijit. 'We waited as I wanted to make sure that all of them were within the range of our weapons.'

The first militant crossed the blind spot on the track just below Major Bijendra's position and reappeared after a few seconds. The firing positions were barely 10 metres above the track. The first man was about to disappear behind the hill, but the sixth militant had still not appeared.

'Saab, fire kholein nahi toh yeh nikal jayega. [Let's open fire before the first man disappears],' the Major's buddy whispered directly into his ear, articulating the immediate concern of the squad.

Holding his breath, Major Bijendra, with his AK-47 aimed at the first man, waited two more seconds. Then he gave a hand signal to Shivram to ask the militants to surrender, a standard protocol during anti-militancy operations. It was a contentious combat rule, but needed to be followed, since the militants, for all their destructive ways, were Indian citizens with rights.

The call to surrender was the most dangerous element of an ambush. It meant that any squad lying in wait needed to voluntarily sacrifice its element of surprise in the hope that it had done everything necessary to pin down the militant into a 'no escape' situation. In encounters such as this one with most-wanted militants, the possibility of the targets dropping their weapons and raising their hands was next to nil.

As expected, once Shivram shouted to the militants to give themselves up, the militants responded with immediate bursts from their firearms. The silence of the forest quickly erupted in a fierce assault-rifle and light-machine-gun crossfire.

Major Bijendra fired shots at the first militant as the rest of the ambush squad engaged the remaining militants in a firefight. The first man tumbled down the hill after being hit but the shot wasn't fatal. He was definitely still alive.

Major Bijendra leapt out of his ambush position and scrambled down the hill, four soldiers from the squad following him in pursuit of the injured militant. The blind spot on the track offered the soldiers temporary cover as they emptied their Kalashnikov magazines in the most likely direction of the militant.

Quickly regrouping in the gun smoke, Major Bijendra made a quick assessment with his men, deciding to quit pursuing the injured militant any further, and instead train their sights on the other militants. Including the sixth, who by now would have stopped in his tracks and taken cover from a distance.

Haopu?

But climbing back up the hill and returning to the safety of his ambush position was no longer an option. Doing that would make him and the four men with him vulnerable to a well-aimed fire from the hiding militants. They would need to remain on the dirt track now, exposed and on the same level as the men who were likely to hunt them back.

'We will now confront them on the track. Let's hit them from an unexpected direction,' the officer told the four soldiers as they slammed fresh magazines into their assault rifles.

Like with much else in jungle warfare, decisions were frequently the result of a total lack of options.

'We trusted KB Saab's judgement. We knew he had a plan,' says Rifleman Brijit. 'And there wasn't a moment to waste.'

Flat on the ground, the men could hear bursts of fire whizzing overhead and towards the hilltop behind them. This was a stroke of luck. The militants were aiming up, not aware that five army men were on the dirt track and edging closer to them.

Major Bijendra and the team emerged from behind the blind spot on to the track in a dash-down-crawl-fire military move typical of close-combat warfare. The move involved quick sprints, throwing oneself down to the ground, crawling on one's elbows and firing a few shots. It allowed the men to close the gap between themselves and the militants, while still maintaining a firing posture.

'We didn't have a good view from the top. That was no longer the case after we approached them from below and directly took them on from barely a few metres. KB Saab had trained us for such situations,' says Rifleman Brijit.

The militants, especially No. 2 and No. 3, were now facing the full fury of the Alpha Company's assault weapons, both from the hilltop LMGs and the AK-47s now just metres away from them.

'From above, I could see KB Saab get very close to the militants, dangerously close,' says Shivram who focused his fire from up on the hilltop. 'I could tell that he wasn't planning on retreating. He simply did not stop crawling forward. It was a little unnerving to those of us who could see. But he was leading from the front and doing what we had come here for.'

As the rest of the Alpha Company squad provided cover fire, Major Bijendra made a final hair-raising dash towards the two militants within his reach, killing both with a spray of close-quarter rifle fire. Falling back to his elbows, Major Bijendra screamed for the squad to blaze their weapons into the forest, determined not to let No. 4, No. 5 and, crucially, No. 6 escape.

'Even from that distance I could read KB Saab's mind,' says Shivram. 'We had come this far. We had succeeded. But where was Haopu?'

Over the next four minutes, return fire from the militants tapered off, with only the Alpha Company squad still firing into the forest. After two more minutes of intermittent bursts, Major Bijendra asked his men to hold fire, the echoing clatter taking many seconds to be engulfed once again by the silence of the woods. He needed to use the lull—it if was a lull—to get his men back to the safety of the hilltop. Still careful not to present themselves as targets, the five men crawled back towards the hill, quickly scaling the slope in a crouched jog, single file, spaced 20 metres apart.

Up on the hilltop, Major Bijendra grabbed his binoculars once again to scan the woods and dirt track below. There was nothing. Not even signs that there had just been a gunfight. The only giveaways were a streak of blood on the dirt track where militant No. 1 had been hit, and then the bodies of No. 2 and No. 3. That night, a pair of scouts went down the hill to look for No. 1, but he was nowhere to be found. It was concluded that he managed to get some distance, but in all likelihood died of his wounds. Militants No.4 and No. 5 managed to flee the encounter site.

No. 6 didn't even come into view.

As his scouts retrieved the bodies of No. 2 and No. 3 from the track below, Major Bijendra called his informant at Longphailum village.

'We got two of them, maybe three.'

'Yes, the news has just filtered in.'

'Was the sixth man Haopu?'

'Yes, it was Haopu. He has not returned.'

The militant squad's tail, the man this mission aimed to eliminate, had made a clean getaway. Shivram and the other soldiers knew this was a delicate moment, but they knew they needed to stand with their commander as he processed the events of the day. No man said a word. It was Major Bijendra himself who broke the silence.

'Had I known for sure that Haopu was the last man, I would have waited for him to enter the kill zone, and allowed the first three or four guys to pass,' said Major Bijendra to the men.

'*Unke ghar mein ghus ke maara hai saab* [We have entered their territory and killed them],' Rifleman Brijit said. 'It was never going to be easy. We still scored at least two kills. Haopu now knows he cannot rest in his own backyard. We have stolen his sleep forever. His time will soon come.'

Now that the ambush site was known to the militants, the Alpha Company squad needed to move quickly. Packing up their tents and equipment in the darkness, the fifteen men began the 25 km trek back to the point where they would be collected the next morning by their vehicles.

On the long march back, Major Bijendra dug his phone out of his rucksack. Switching it on, the lock screen displayed a stack of messages from Jenipher. The last one simply read, 'Missing you, please tell me all is okay.'

This time he texted back: 'All okay. Back at base shortly. Will call.'

Then Major Bijendra dialled his commanding officer, who by this time had received the operation report.

'We didn't get Haopu, sir. He got away.'

'Never mind, very good show, KB. You got them in their own playground. Next time.'

'I'm sorry, sir. He was there and we didn't get him.'

'No time for sorry. Stay focused.'

But the escape of Haopu would continue to haunt Major Bijendra. There was nothing personal in the hunt for the individual. But his escape was a reminder of how difficult it was to eliminate those holding Manipur hostage to their terrifying ways.

By late morning on 23 March, the Alpha Company men were back at their operating base in Kamai. As they returned to their barracks to rest, Major Bijendra dialled his mother.

'Is everything okay?'

'I'm okay, just called to check on you.'

'Yes, please come and see us soon.'

'I will, very soon.'

And then, after a day of debriefings and post-mission stock-taking, just before he turned in for the night, Major Bijendra texted Jenipher again.

'I'm coming to see you.'

* * *

Ten months after the mission, in January 2020, it was announced that Major Konjengbam Bijendra Singh would be decorated with a Shaurya Chakra, India's third highest peacetime gallantry medal. The citation would credit Major Bijendra's 'conspicuous bravery and stout leadership in going beyond the call of duty in the face of grave and imminent danger to his men'.

A month after the Shaurya Chakra announcement, in February 2020 Bijendra and Jenipher were married in Imphal; their daughter Jasmine arrived in October.

Nearly a year later, while Major Bijendra was in Imphal with his family on leave, he received a call from Warrant Officer Shivram.

'*Haopu khatam ho gaya, KB Saab* [KB Sir, Haopu has met his end].'

The elusive LTT commander-in-chief, who had completely dropped off the map for over two and a half years since the Alpha Company encounter, had finally met his end. Haopu and two other LTT militants were eliminated by security forces in Maojang village in Manipur's Sadar Hills area. The slippery militant had managed to move large distances within the state, finding a safe haven in the Saikul area, a testament to his resources and ability to survive.

'We didn't get him, but we put a huge target on his back,' says Major Bijendra. 'My men are right in what they say— Haopu must have spent his final years constantly on edge, constantly running, never being able to rest in one place. Hopefully he felt something like what the people of Manipur felt thanks to the actions of outfits like his.'

Three months after the killing of Haopu, with the COVID-19 pandemic slowing down in the country, a long-postponed defence investiture ceremony was called at the Rashtrapati Bhawan in Delhi. Those whose names had been announced for decorations could finally receive their medals from the President of India.

As Major Bijendra, now thirty years old, walked down the maroon carpet to receive his Shaurya Chakra, twenty-six-year-old Jenipher took in the crowd's thunderous applause.

'It was an indescribable moment,' she says. 'But my heart equally went out to the brave women whose husbands were posthumously honoured for their heroism. The way these women held themselves together when the citations of their husbands were being read out gave me goosebumps. We in the army are a close-knit family. Like our pride, we share our grief too.'

Major Bijendra's Shaurya Chakra citation, read out at the Rashtrapati Bhavan ceremony, was televised live and watched by his mother, Konjengbam Tilotama Devi, who was babysitting the couple's thirteen-month-old daughter at a Delhi hotel. Her eyes welled up, thinking of what an extraordinary achievement this was for her son, who, as a teenager, struggled to arrange money to travel to Allahabad in 2008 for his SSB interview, the final hurdle before joining the armed forces.

'My heart was full,' says Konjengbam Tilotama Devi. 'Bijendra could rarely afford to buy lunch when he was undergoing coaching at a private academy in New Delhi ahead of the SSB interview. He would tell me how even a paratha served by a street hawker was a luxury thirteen years ago. And here he was on the television screen, looking proudly into the camera as the supreme commander of the armed forces was pinning the medal on him. What could be a greater moment for a mother? I only wish his father was alive to see this.'

Major Bijendra will never forget the hard times the family fell on after his father was forced to drive lorries to make ends meet.

'After I became an officer, I made sure my father stayed with me till he breathed his last a few years ago. I wish he was here to see this medal,' says Major Bijendra. 'I may be wearing the medal, but it belongs to each of the fourteen men who were under my direct command during the operation. We were a cohesive unit and in the driver's seat all along. The men followed the instructions they were given to a tee. We did well but for not getting Haopu. It took time for us to digest that. That file is now closed. We have moved on.'

Several men from the Alpha Company of 23 Assam Rifles also received honours for their roles in the March 2019 mission.

Warrant Officer Shivram was awarded the Director General Assam Rifles' (DGAR) commendation card; Rifleman K.D. Burman's contribution was recognized with a similar commendation by the Eastern Army commander; Rifleman Kazhingmei, the scout, got a commendation card from the division commander; and Singh's buddy, Rifleman Brijit, was awarded the governor's gold medal.

Militant activity in the Kamai area dropped drastically after the March 2019 operation.

'I left Kamai in December 2020. There was no firing incident in the area till then,' says Major Bijendra. 'The militants got the message that we had the means and the will to hunt them no matter where they hide.'

The frequency of militant activity may have reduced over the years thanks to persistent operations, community building and development work in the state. But militancy itself is far from over.

Just nine days before the award ceremony at the Rashtrapati Bhavan, the dangers that soldiers operating in Manipur face came terrifyingly to the fore once again. On 13 November 2021, a group of heavily armed militants ambushed an Assam Rifles convoy in broad daylight near the porous 398 km India–Myanmar border in Manipur's Churachandpur, killing five soldiers, including the commanding officer of 46 Assam Rifles, Colonel Viplav Tripathi, along with his wife and their eight-year-old son.

A crucial piece of India's modern military history is associated with the North-eastern state—India's first publicly acknowledged 'surgical strike' on foreign soil was launched from here in June 2015.

On 9 June 2015, a team from the 21 Para Special Forces infiltrated from Manipur* into Myanmar to hunt down the militants responsible for a brutal attack on Indian soldiers in Manipur's Chandel district just days earlier. The operation had been supervised by Lieutenant General Bipin Rawat, who at the time commanded the Army's 3 Corps, headquartered in Nagaland's Dimapur. The general, who rose to become chief of the army staff and then India's first chief of defence staff, was killed in a helicopter crash along with thirteen others near Coonoor in Tamil Nadu in December 2021.

Major Bijendra was posted out of Manipur at the time this book went to print. But he welcomes the opportunity to go back if the Army requires him to. As a decorated officer, his name and face are now well known across the landscape.

---

* See *India's Most Fearless 1*.

A source of pride, but also a source of danger for him and his loved ones.

'As a child, I saw militants intimidating simple, hard-working Manipuri people,' says Major Bijendra. '*Hero ban kar ghoomte the woh* [They used to roam around like heroes]. There has to be a breaking point where we say enough is enough. For Manipur. For India.'

10

'You Have Five Minutes on the Seabed'

CPO Veer Singh and Commander Ashok Kumar

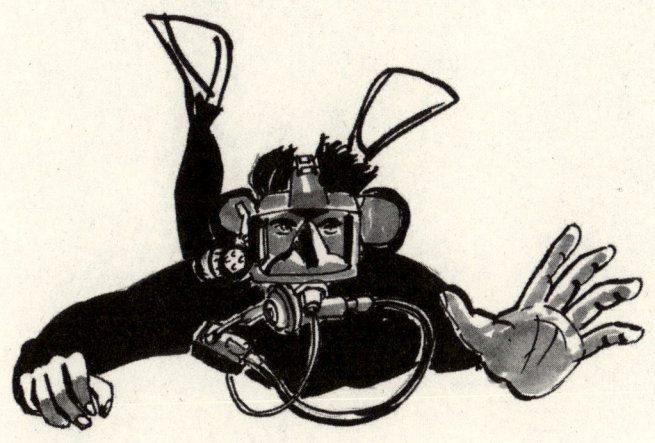

25 March 2015
Arabian Sea
25 nautical miles off the Konkan coast

'It's been twelve minutes. Time's up. Initiate ascent, immediately.'

The dive supervisor's voice had acquired an urgent tone as he knelt by the edge of a Gemini boat, a tiny black speck far out in the Arabian Sea, off the coast of Karnataka.

'It's way too dangerous to remain there any more,' he said.

Master Chief Petty Officer (MCPO) N.S. Dahiya was speaking into a handset that sent his voice through a communication wire, nestled in a dive 'umbilical' that went deep into the sea to the two men it was his job to warn.

Sixty metres below the sea surface, two Indian Navy divers armed with 12-inch combat knives heard the call from the boat above through the cable that fed into their diving harnesses and headsets. The warning came again after ten seconds.

'Abort NOW! Do you read me? Initiate ascent without further delay.'

MCPO Dahiya was now nearly screaming over the whip of sea air. The two divers knew that warnings from the man in the boat were non-negotiable. Their lives were quite literally in his hands as they scoured the inky depths of the Arabian Sea.

But on that day in March 2015, the two men—both among the Navy's finest divers and at the very top of their game—had decided to push the limits.

In their full combat diving suits, Chief Petty Officer (CPO) Veer Singh and the officer leading the mission, Commander Ashok Kumar, were certain they were not prepared just yet to begin their ascent towards the surface.

'This is your second ascent warning, do not ignore this,' the dive supervisor called now, his voice softer but firm.

As the two divers descended further, they needed to be sure what they were looking at before they sent word up through the umbilical to justify not heeding a critical safety warning.

'I see the watch on her wrist,' CPO Veer said.

'I see her hand.'

Less than 2 metres away in the murk, the two divers had finally spotted what they were searching for—the pale, lifeless body of Lieutenant Kiran Shekhawat.

* * *

24 March 2015
INS Hansa Naval Air Station
Goa

The previous evening, twenty-seven-year-old Lieutenant Kiran had climbed into the observer's seat of an Indian Navy Dornier Do-228 maritime patrol aircraft at the INS Hansa Air Station in Goa. The aircraft, part of the Navy's most decorated unit, Indian Naval Air Squadron (INAS) 310, nicknamed the Cobras, was being flown that night by pilots Commander Nikhil Joshi and Sub Lieutenant Abhinav Nagori, the mission was a familiar, routine night mission over the Arabian Sea to track suspicious movements by potentially hostile vessels looking to land in the darkness.

The Dornier took off at 6.30 p.m., with the pilots steering it straight out over the Arabian Sea to begin the night surveillance mission. On board, Lieutenant Kiran got busy with her observation duties, keeping a watch through the plane's night sensors and maritime patrol radar for anything that needed to be red-flagged for either urgent action or follow-up the next day. Assigned a sector of the ocean to surveil, the plane flew a large circular path, tightening with each instance. Missions of this kind were designed to last up to five hours where necessary, giving the surveillance crew 'time on station' to get a proper fix on the ocean surface. But just over three hours after the mission began, the Dornier flew into major technical trouble. The pilots tried to keep the plane flying, relaying their position and situation over the talk-back.

Ground control at the INS Hansa base lost radar contact at 10.08 p.m., around the time the Dornier had crashed into the Arabian Sea.

Minutes later, an alert sounded at Mumbai's naval dockyard, where some of India's top combat divers had just finished dinner and were preparing to retire to their barracks. While details of the accident were still sketchy, the alert quickly triggered emergency protocols that saw eleven men from the Navy's elite Command Clearance Diving Team (CCDT)—including CPO Veer and Commander Ashok—on full readiness to roll out for a search-and-rescue (SAR) mission.

Well before midnight, the divers' team set sail south towards the Karwar coast on board the ocean-going tug INS *Matanga*, a 68-metre vessel that was deployed on a variety of emergency missions off the Konkan coast.

'We knew there was no time to waste,' says CPO Veer. 'Each one of us in that team knew that the essence of our job was to scramble against time and complete the mission. We couldn't wait to get there and begin our work.'

INS *Matanga*, one among a dozen vessels activated for the emergency search mission, was now speeding towards the crash area. Another vessel, INS *Makar*, a hydrographic survey vessel equipped with sonar equipment capable of gazing into the murky oceanic depths, was also on its way.

No one slept on the vessels that night. Not after the teams on the ships received word from the Western Naval Command control centre that the pilot of the Dornier, Commander Nikhil, had miraculously survived the crash. He had reportedly drifted nearly 20 km from where the plane hit the water and had been rescued by a fishing boat. How did the Indian Navy get to know this so quickly? The crew of the fishing boat had dialled a toll-free hotline that had been introduced after the 2008 26/11 Mumbai terror attacks. The pilot was soon picked up by a fast interceptor boat and rushed to naval hospital INHS Patanjali in Karwar, south of Goa.

'The news of the rescue made this an even more time-critical rescue mission,' says an officer who was at the Western Naval Command control centre at the time. 'If the pilot had been rescued from the sea surface, then we had to fully account for the possibility that the other two may have survived too. And may have been drifting.'

Other Dornier aircraft and Sea King helicopters were already flying missions in rotation both by night and the following day, scanning the sea surface as meticulously as possible in the hope of spotting bobbing heads. An alert was broadcast to fishing boats in the area, including the one that had rescued Commander Joshi, requesting them to keep a close eye out for Lieutenant Kiran and Sub Lieutenant Nagori in the hope that they had also, by some miracle, survived.

As the aircraft skimmed over the sea, hunting in tightening flight paths for the two missing officers, the surface ships went all out to locate the doomed Dornier. Criss-crossing each other in wide arcs over the Arabian Sea, a dozen ships worked uninterruptedly to triangulate and pin-point the crash site. Combining inputs from last detected location, the direction in which the aircraft was headed and last radar contact, the vessels quickly narrowed things down. But it wouldn't be until the morning of 25 March that something finally glinted on INS *Makar*'s side-scan sonar from the murky, heaving seabed 55 to 60 metres below.

The commanding officer of INS *Matanga*, Commander Nikhil Srivatsa, dropped anchor around 200 metres away from the coordinates given by INS *Makar* and the dive teams were launched on a pair of Gemini boats, along with the equipment they had carried from Mumbai.

'INS *Makar* had given the coordinates of the main chunk of the wreckage or what it called a large metallic object. But the debris was spread over a wide area. The skipper of INS *Matanga* anchored the ship at a safe distance to avoid any further accidental damage to the wreckage,' says a Navy officer who was involved in the operation.

The two Geminis were above the possible crash site by 10 a.m., and the first pair of divers was launched into the sea an hour later to confirm the exact location of the Dornier wreckage. The Navy follows a buddy system in which divers operate in pairs during such missions.

Before the first two divers slipped into the warm Arabian Sea on a particularly crisp, sunny day, a shot-line was lowered down from one of the Gemini boats to the seabed to calculate its precise depth as INS *Makar* had only presented a broad estimate. The depth turned out to be 60 metres. Every diver on those two boats immediately had the same thought.

60 metres.

What followed was a quick satellite phone call to the Western Naval Command headquarters in Mumbai. The chain of command had to be informed about the depth at which the divers would be required to operate and how the discovery of the wreckage had suddenly raised the stakes.

'Underwater visibility was unbelievably good that day. It was around twenty metres,' says CPO Veer. 'I have been part of many operations where I could barely see more than a few inches ahead of me. The first two divers were instructed not to descend beyond forty metres as from there, they could easily see a further twenty metres below, the depth at which the wreckage possibly lay. The operational constraints were weighing on our minds.'

The first two divers returned to the surface without sighting anything. Commander Ashok then ordered the two boats to a distance 50 metres further south of where the first two divers were launched into the water. From there, the second pair of divers plunged into the sea to look for the wreckage, holding a shot-line fastened to a 100 kg sinker earlier dropped into the sea. The shot-line allows divers to head straight to the intended location at the bottom of the sea without drifting.

Minutes into that second dive and from a depth of 40 metres, the two men sighted the Dornier sitting on the seabed around 20 metres below them and instantly communicated the critical discovery to their supervisor. The second dive team, having done its job of spotting the wreckage, was instructed not to descend any further and return to their Gemini immediately.

For all the technology involved in detecting aircraft wreckage at sea, finally sighting it is something of a magical process. Unlike an air crash on land, finding a crash site at sea is an exponentially more difficult proposition. Everything about an accident is loaded against those looking for where it happened. Consider this: when an aircraft crashes into the sea, it quickly sinks. Any floating parts or patches of tell-tale oil or fuel quickly drift from the point of the crash. Once the aircraft sinks below the surface, it may not always go down straight, its planar surfaces and angles twisting it in different directions, possibly at considerable distance from where it entered the water. The aircraft could sink to unmanageable depths depending on how deep the sea is at that point.* It is a common misconception that the seabed is one endlessly flat plain, when it is in fact not only full of dramatic terrain with mountains, trenches, valleys and crevasses, but also dynamic, in that the seabed constantly shifts with the currents. So theoretically, the place where aircraft debris settles after sinking could shift significantly with the shifting seabed.

'It's a difficult, moving target where nothing is in your favour, literally nothing,' says a sailor who served on INS *Makar* at the time the Dornier was detected. 'A very competent and committed search, combined with some favourable depth parameters allowed us to locate the wreckage quickly.'

With the wreckage finally sighted by human eyes, it was time for the decisive third dive. CPO Veer and Commander Ashok were already in their dive suits and in a Gemini boat with dive supervisor Dahiya when the second team had ascended.

---

* Air France Flight 447, a wide-body Airbus A330 airliner which crashed in June 2009 in the Atlantic Ocean took nearly two years to locate. The aircraft wreckage was found nearly 4000 metres deep. Malaysian Airlines Flight 370, which disappeared over the Indian Ocean in March 2014, has still not been located. In July 2016, an Indian Air Force Antonov An-32 aircraft went missing over the Bay of Bengal with twenty-nine people on board. It has not been located to date either.

The dive team had changed for this third descent. What hadn't changed was the depth at which the wreckage lay. Commander Ashok and CPO Veer were certain about one thing when they planned their descent, though. If someone had to take the risk to pull it off, it would have to be them. They had more diving experience than anyone else in the squad.

'It is not possible to measure risks involved in such operations. But it is reasonable to assume that someone with lesser experience may not be comfortable operating at depths they are not qualified for. Manoeuvring with a body isn't easy either. My foremost priority was to keep the mission incident-free,' says Commander Ashok.

Like so much in the military, 25 March 2015 involved a collision of emotions. On the one hand, things had been upbeat for the rescue teams. The Dornier's pilot had survived. And the aircraft debris had been located. On the other hand, with several hours of airborne search for the other two aviators drawing a blank, every sailor in the rescue team knew deep inside that a tragedy was unfolding.

'You have fifteen minutes,' said Dahiya as CPO Veer and Commander Ashok lowered themselves into the sea. Carefully descending into the murk, the two divers reached a depth of fifty metres in under three minutes. The location data they had been provided was spot on. Ten metres below them, through the murk lit vaguely by filtered sunlight, they saw it.

The twisted, mangled fuselage of the Dornier sat on a sloping seabed dune at 60 metres. But there was a problem. Stringent diving protocol normally does not allow Navy divers to descend to depths below 55 metres. The effectiveness of their unpressured diving sets diminishes significantly beyond 50 metres.

'Ten more metres may not sound like much, but with those constraints, it's all the distance in the world,' says CPO

Veer. 'But the wreckage was right there below us. We could see the whole thing, and we reported this to Dahiya.'

It was at this moment that the dive supervisor issued his first warning, asking the two men to terminate the mission and return to the surface. The instruction wasn't unreasonable. As per well-defined operational standards, the two divers were literally operating out of their depth with potentially dangerous consequences. Keeping the two divers safe and focused was Dahiya's chief responsibility. He had assessed risk and decided that a further 10-metre descent, along with an unpredictable amount of time that would be required at that depth, meant it was a clear no-no. Dahiya also knew he had a pair of very committed divers on his hands, including Commander Ashok who headed the squadron's salvage unit.

'Dahiya, we're proceeding with descent,' Commander Ashok said into the dive mask microphone. The dive supervisor knew he had to acknowledge.

The two divers gave each other a thumbs up as they began their final descent past 50 metres, approaching the wreckage as it rocked gently in the current. At 2.30 p.m., with the sun high, CPO Veer and Commander Ashok reached the broken Dornier. In their earpieces, the two divers received a second warning on their depth.

A risk had been taken, and it needed to pay off as quickly as possible. The two divers were far from oblivious to the double threat they faced—human limitation and the very real possibility of equipment failing since it had been taken beyond its prescribed parameters.

Both divers unsheathed their combat knives, steering clear of the sharp metal edges protruding from the mangled heap of the Dornier. They used their knives to cut the mess of wires and cords that had been ripped from the aircraft's innards to allow passage into the cabin and cockpit area. At every step, they relayed their position and what they were looking at to their dive supervisor. Up on the Gemini boat, Dahiya acknowledged.

It had taken the divers just over three minutes to reach the wreckage. Of the fifteen-minute window they were allowed, they now had under twelve minutes to search the wreckage and begin their ascent. Dahiya had already allowed the divers a depth risk. He knew he couldn't permit a 'bottom time'* risk, or what is also known as 'limiting line' in the Navy's safety manual.

The men, focused on the mission's execution, had not only breached their safe operating envelope but were now clearly flirting with danger in an alien environment by staying there longer—even if it was only for a minute or two. And the risk to their safety was mounting with every passing second.

'We knew that if we returned without retrieving Lieutenant Kiran's body, we would not be able to go back down the same day,' says CPO Veer. 'The safety manual does not permit back-to-back diving missions beyond a depth of forty-two metres. It prescribes a surface interval of six hours between two deep-sea dives.'

In love with the sea from an early age, CPO Veer had joined the Navy in July 1997 so he could spend as much time either in it or close to it. When the opportunity to train at the prestigious Diving School in Kochi presented itself, a young Veer hadn't thought twice.

With Dahiya providing a stopwatch update every thirty seconds, the two men had to rely on brute strength to pull apart the jammed cargo door on the port side of the aircraft in order to finally gain entry into the main cabin.

---

* Bottom time is calculated from the moment a diver descends from the surface till the point the diver initiates his/her ascent.

The two divers had taken crucial inputs on the structure of the Dornier aircraft and its design from two men. The first was Commander Srivatsa, an aviator who had himself served as an observer in the Cobras squadron in a previous role.

'It was a stroke of good luck that the commanding officer of INS *Matanga* had served in a Dornier squadron,' says Commander Ashok. 'With his inputs, I felt I knew the aircraft well. It was most critical to planning the mission, finding entry into the aircraft and locating Lieutenant Kiran's body.'

The second man with useful inputs was Petty Officer R.K. Singh, a sailor from the Navy's aviation branch with years of experience in a Dornier squadron. He was a close friend of CPO Veer's who had also served in the aviation wing before he earned the coveted diver badge.

'In my naval aviation days, I worked as an electrical technician on the Sea Harrier fighter aircraft,' says CPO Veer. 'But I had no clue about the Dornier. Apart from Commander Srivatsa, my friend also shared information on the plane, cockpit layout, observer's workstation, emergency exits, location of the black box, and the like.'

The inputs were critical for the mission.

CPO Veer now carefully slipped into the shattered cabin of the Dornier. It was dark inside, the wreckage closing out any sunlight filtering down to the surrounding depth. Stepping carefully forward through the murk, he switched on his underwater flashlight. The observer's seat in the Dornier is located right outside the main cockpit. He stepped towards it, twisting his body to avoid protrusions of jagged debris inside the aircraft's cabin.

'For some reason, at that point I remembered the words of one of my former commanding officers,' says CPO Veer. 'Whenever he addressed us he would say, "We are here for war and we are here to win." Those words rang in my ears when I was about to enter the wreckage. Every mission, to my mind, is akin to war and the only choice is to emerge victorious.'

The aircraft was totally destroyed, with no clear paths to exit.

That's when he saw Lieutenant Kiran's wrist, with her watch still on. With Dahiya reporting time in an ever-elevated tone, the two divers took their last few steps towards the observer's seat to free her body.

'It was a surreal sight. She looked peaceful,' says CPO Veer. 'Her unfastened seatbelt seemed to suggest that at some stage, she tried to get out of there. It's hard to imagine what the last moments would have been like.'

Both divers knew instantly what they were looking at— the first woman naval officer to be killed on duty.

'Extricating her body from the seat was very painful,' says CPO Veer. 'But there was no time to stop and think. Processing the tragedy would have to wait till later.'

'We had to use our knives again to clear more wires and cords to bring Lieutenant Kiran's body out from the cargo door. We manoeuvred with extreme caution as sharp metal parts could have not only damaged the body but also our diving gear,' he says.

Lieutenant Kiran had been part of the Navy's first ever all-woman marching contingent at the sixty-sixth Republic Day parade two months before the crash. Two years before, she had married fellow naval officer Lieutenant Vivek Singh Chokker. At the time of the crash, he was an instructor at the elite Indian Naval Academy in Ezhimala, Kerala. Operational postings had kept the young couple in different locations since their wedding, though both were finally to be posted in Kochi together later that year.

The two divers had been made aware of all these details as they made their way to the crash site the previous day.

Up on the Gemini boat, MCPO Dahiya hadn't received the customary acknowledgement from the divers of his time warnings for three whole minutes. The two men had been so focused on removing Lieutenant Kiran's body from the aircraft, they had forgotten to acknowledge the last six countdown warnings, driving Dahiya into a frenzy of worry over whether the rescuers would now need to be rescued themselves.

CPO Veer and Commander Ashok had been on the Arabian seabed for fourteen minutes. Counting their descent time of three minutes, the men were well past the fifteen-minute bottom time considered safe for a diving operation at that depth. This was no longer a calculated risk, Dahiya thought. This was life critical. And there was total silence from the depths. Sensing that he needed to be prepared for the worst, Dahiya put the dive teams in the two Geminis on standby, and if needed, the men would plunge down to 60 metres for a rescue.

Had the two divers, so immersed in their mission, slowly lost their faculties and drifted into a paralytic daze? Had the additional pressure slumped their bodies? Had they been incapacitated—alive, but unable to communicate? All possibilities crossed Dahiya's mind as he screamed for a reply through the umbilical.*

Deep-sea diving is fraught with danger.

Nitrogen narcosis during descent and decompression sickness on the way up are deadly conditions that can develop very fast in the underwater environment and have well-

---

* The surface-supplied diving umbilicals—a lifeline for divers— consist of a breathing gas hose, communications cable and a pneumofathometer or pneumo hose to monitor the operating depths. The divers were using umbilical cables measuring 75 metres in length that day, with their supervisor feeding the line to them as they went down.

documented debilitating effects on divers. Variously described as 'raptures of the deep' or the 'Martini effect' in the world of diving, the first condition refers to a dangerous altered state of consciousness triggered by breathing nitrogen under increased pressure, usually kicking in at depths of more than 40 metres. That's why the Navy's safety manual prescribes a limiting line for different depths.*

Nitrogen narcosis can cause disorientation, dizziness and vision trouble, significantly impairing a diver's ability to make quick and considered decisions in testing conditions. CPO Veer and Commander Ashok, with hundreds of diving hours between them, managed to evade nitrogen narcosis and its associated complications during their descent.

On the eighteenth minute, right before Dahiya had planned to set off emergency rescue protocols, a voice came in from 60 metres.

'We have her, Dahiya,' shouted Commander Ashok. 'We are about to initiate ascent. Please inform INS *Matanga* that we are coming up with Lieutenant Shekhawat's body.'

Dahiya and the eight men in two Gemini boats finally exhaled.

The two divers carried Lieutenant Kiran's body to the shot-line a few metres away from the wreckage before they began their ascent. Right before they set off for the surface, Commander Ashok sent up a crucial update.

'No sign of Nagori.'

But as they ascended, well past their time and depth safety margins, the threat of decompression sickness loomed even larger. The buoyancy of Lieutenant Kiran's body

---

* The atmospheric pressure at sea level is 14.7 pounds per square inch, or 1 atmosphere. Pressure mounts as a diver descends deeper into the sea, increasing by 1 atmosphere with every 10 metres.

made CPO Veer's ascent faster than the stipulated rate. Surfacing too quickly can cause decompression sickness due to a sudden drop in pressure, a condition called 'the bends'. The sickness is the consequence of a sudden switch from a high-pressure environment to one with low pressure. It causes the formation of potentially fatal nitrogen bubbles in the tissues and bloodstream of deep-sea divers. The bubbles can block arteries and trigger a stroke or heart attack in extreme cases.

Before the completion of every dive, divers are mandated to make safety stops at prescribed depths to control their ascent rate and expel the extra nitrogen built up in their bodies.

Standard procedure for a dive up to 60 metres prescribes three safety stops during the ascent—at 9, 6 and 3 metres from the surface. The ascent has to be paused for five minutes at 9 metres, another five minutes at 6 metres and fifteen minutes at 3 metres. The number of safety stops and their duration, spelt out in the Navy's dive table, vary with the depths at which the divers are operating.

CPO Veer had decided against making the mandatory safety stops during ascent that day as he was in a rush to take Lieutenant Kiran's body to the surface. INS *Matanga*, alerted by dive supervisor Dahiya, had launched another boat minutes earlier to bring the body back and arrange for it to be flown to a naval hospital in Goa by helicopter.

If a diver chooses to ascend without making safety stops for any reason, as CPO Veer did, he either has to return to the prescribed depths for the stops within nine minutes or complete the drill in a recompression chamber on board the diving ship to rid the body of extra nitrogen.

'I went straight up and delivered the body, and within five minutes I was back in the sea at the 9 metre depth for the safety stops. We didn't have a recompression chamber on the

Gemini. It was on board INS *Matanga* and I needed to get it done quickly,' says CPO Veer.

By 3.30 p.m., the dive teams were back on board INS *Matanga*, still processing the speed with which they had managed to complete the most critical part of their mission. The divers watched in silence from the deck of their vessel as an Indian Navy Sea King helicopter picked up the body of Lieutenant Kiran, rising and then peeling away towards Goa.

It was a quiet night on board INS *Matanga*. Responding to questions from the younger divers on the team, Commander Ashok and CPO Veer shared lessons from the day's dive.

'If I can do it, so can you,' Commander Ashok said. 'Trust your instincts and always remember this is what we train for. This is the life we have chosen. Unexpected twists will be there. But the best part is when the time comes, you will know exactly what to do.'

It had been a day of deep success and deep tragedy. After a briefing on follow-through protocols and missions for the next day, the squad retired to their bunks for the night.

In his bunk, CPO Veer couldn't sleep. He knew he would never forget those images from the shattered cabin of the Dornier. That hand. That wrist. That watch. He had learnt over dinner that night that Lieutenant Kiran traced her roots to Khetri town in Rajasthan's Jhunjhunu district.

'What are the odds of retrieving someone's body from the bottom of the sea and that someone turns out to be from your own town,' says CPO Veer.

As sleep finally overcame the exhausted squad, its final thought was most likely a single one. The Dornier's co-pilot Sub Lieutenant Nagori was still a missing piece in the tragic story, and the truth about what happened to the aircraft would only be known if the dive team was able to locate and retrieve the aircraft's black box.

The squad on his vessel was up before sunrise, Commander Srivatsa remembers, fully charged up and ready for a second day of dives, determined to complete the mission.

The operation to locate the missing Dornier and the two crew members was being monitored at the highest possible level, with then chief of the naval staff, Admiral Robin Dhowan, rushing to Goa from Delhi to monitor the mission firsthand. Aircraft were on standby at INS *Hansa* for Admiral Dhowan, to carry out reconnaissance of the crash site, one of the first things he did after landing in Goa. That the chief was personally on the scene to keep tabs wasn't just an act of solidarity, but a message of urgency that filtered down and energized all ranks involved.

The previous day's dive had been a successful one. But that was no guarantee that further descents would turn out the same way. Each mission down to those depths carried with them the same as well as additional risks. With each passing hour, the Dornier wreckage, sitting on loose diatomaceous earth on the seabed, could potentially drift down to further depths.

At the dawn briefing, Commander Ashok made it fully clear to the squad that optimism had to be balanced with caution for the mission's success.

'You remember what I said last night. When the time comes, you will know what to do. Just focus on what you are doing and have a great dive. The bottom time is non-negotiable,' Commander Ashok told the two men in the build-up to the dive.

By 9 a.m., the first pair of divers was given the thumbs up by MCPO Dahiya to flip backwards into the sea from their Gemini. The dive team was cleared to go down to the wreckage and find and recover Sub Lieutenant Nagori's body as well as the Dornier's black box.

Based on inputs provided by Commander Srivatsa, the divers were also equipped with information on the two possible ways to access the cockpit in which Sub Lieutenant Nagori's body was likely to be trapped. Fortunately for the team, underwater visibility was as good as it was the previous day.

If visibility hadn't been that good, the divers would have had to use what is known as a distance line, a coiled reel fastened to the sinker or shot-line, to navigate through the murky waters and return safely to the starting point. A distance line also serves as a critical dive-site navigation tool when the sea current is high.

'The latest dive team was sent down with a Nikon camera for underwater photography. The photographs and footage would be of enormous help to other divers to understand the conditions in which the operation was being conducted as not all of them would get a chance to explore the wreckage,' says CPO Veer.

But for all the preparation and benefit of first-hand advice, that first dive attempt on day two was unsuccessful.

The two divers could neither locate Sub Lieutenant Nagori's body nor the black box. They entered the wreckage from the Dornier's port side taking the same route as CPO Veer and Commander Ashok had after forcing the cargo door open. What had seemed like a straightforward task when the dive team set out from INS Matanga for the crash site that morning was turning out to be far more complicated than the divers might have imagined.

'Sir, I don't think Sub Lieutenant Nagori's body is there in the wreckage. We did a thorough search. It may have drifted away,' said one of the divers. The limiting line of fifteen minutes at 60 metres left the two divers with no time to scan the wreckage further.

While they had hoped that the divers would make some headway in at least locating Sub Lieutenant Nagori's body, if not retrieving it, CPO Veer and Commander Ashok weren't really surprised. They had themselves not detected any signs of the co-pilot's body in the mangled aircraft the previous day, though their focus on Lieutenant Kiran left them with little time for a thorough look at the cockpit area.

Back on the Gemini, the dive team downloaded the photographs and videos they had recorded during the dive on to a laptop for closer analysis. Commander Ashok and CPO Veer looked closely at every frame.

'There were no leads in the photographs or the video footage. We didn't have enough time to look for Sub Lieutenant Nagori's body the previous day, but I was convinced it was in the wreckage. I don't think the divers made it to the cockpit or whatever was left of it. The bottom time restriction may have played on their minds and diluted their focus. It can happen. But we still had a fighting chance,' says CPO Veer.

There was no question of calling off the mission. It was tantalizingly close to completion, but there was no telling how long this final stretch could wind on for.

Within fifteen minutes of the first two men resurfacing, a second pair of divers was launched into the sea. Before they disappeared beneath the surface of the water, CPO Veer had some advice.

'The body has to be in the cockpit. The fuselage is badly crushed, and the cockpit will not look like how you imagine it to be. Go for the cockpit door on the port side. This is your chance,' he said to them.

But the second dive also failed to locate the co-pilot. This was a disconcerting turn of events, with the mission appearing to hit a wall of questions: where could Sub Lieutenant Nagori's body possibly be? Was it trapped inside the wreckage? Had it drifted away? Did he make it out of the cockpit?

The team was fully focused on Nagori. The black box would have to wait.

A third pair of divers volunteered to go next but after deliberations on board, Commander Ashok decided that he and CPO Veer would make the third descent themselves.

'Let's stick to the limiting line this time,' MCPO Dahiya said, minutes before the two men dived. He made little effort to hide his scowl. It was his duty to make sure that the divers conformed to the safety protocol to the maximum extent possible, but he knew what he was dealing with.

The only silver lining to the diving operations till that point on the second day was that the divers, who had operated at depths where they weren't even supposed to be in the first place, were not involved in any incident that could have diverted the team's focus and delayed the search.

Familiar with the wreckage site, CPO Veer and Commander Ashok descended fast and hit the bottom in exactly three minutes. They wanted to squeeze every minute they could of bottom time to search for Sub Lieutenant Nagori.

After spending a full minute examining the battered fuselage from outside, the two divers noticed the Dornier was not sitting upright on the seabed, something that had escaped their attention earlier and that of the other dive teams. It hadn't been noted in discussions and debriefs since the wreckage had been located either. They were looking at the wreckage afresh now.

The front portion of the fuselage housing the cockpit had slammed right into the seabed and was buried a few feet in the mud. It was hardly surprising therefore that attempts to approach the cockpit from the cabin had proven futile. 'It was buried in the seabed. The co-pilot's body and the black box had to be in there,' thought CPO Veer.

The two men quickly made for the fractured cockpit door that was partly buried in the soft soil, once again using sheer strength to try and remove the big, mangled piece of metal. From the Gemini above, MCPO Dahiya reminded the team not to go silent for any period of time. The two divers had been given a respectful earful from Dahiya the previous day, and fully understood his consternation. They reported with updates every twenty seconds on this dive.

Time moves quickly when you're at 60 metres and busy as hell, using every ounce of your strength to break free a jagged, heavy piece of metal jammed into the seabed with shattering force. It took seven whole minutes for the two divers to break the door free.

'You have five minutes of bottom time left. Do you see him?' shouted Dahiya from above.

CPO Veer and Commander Ashok exchanged a quick glance of ironic relief. They both saw the body at the same time.

Sub Lieutenant Nagori was in the co-pilot's seat, still strapped in with his seatbelt.

'His seat was also partly buried in the seabed. We unfastened the safety belt and gently pulled him from the wreckage. As Commander Ashok held the officer's body, I returned to the cockpit to retrieve the black box,' says CPO Veer.

The black box was entangled in a jumble of wires that CPO Veer had to cut away with his knife. He inserted the device into his safety harness and quickly exited the cockpit. Nearly seventeen minutes had passed since the two men made the dive. The prescribed bottom time had elapsed again, just like it had the previous day. But this time Dahiya was calmer since the team had kept him posted at every step. No paralysing silences from the depths this time.

The two divers completed the recovery of Sub Lieutenant Nagori's body in exactly the same manner as Lieutenant Kiran's.

'The only difference was that the buoyancy of the co-pilot's body was taking me back to the surface even faster this time,' says CPO Veer. 'I clearly recall Commander Ashok holding my legs to slow my ascent. I could have released the body closer to the surface but decided against it as it could have drifted.'

When they broke the surface, a third Gemini boat was on site from INS *Matanga* to carry the body back to the ship from where it would similarly be flown to Goa in a helicopter.

Triumph and tragedy mixing like blood in water, the divers returned to INS *Matanga* early that evening. The sorrow of losing two personnel would always be the core of what had happened. But with two grieving families awaiting news after the shock of the accident, it was upon the shoulders of the divers to deliver the possibility of closure quickly. An endless search would have protracted their agony and bereavement. The divers weren't celebrating the recovery of two corpses from the depths of the ocean. But they knew they had the right to bask in the simple glow of a job well done.

So upbeat was the mood on board that night that the dive squad discussed the possibility of recovering the Dornier wreckage using salvage bags the following day.

'The team's morale was sky high,' says CPO Veer. 'We talked about how we would go about recovering the wreckage and the role each squad member would play. However, we were later informed that the responsibility of salvaging the wreckage had been given to the Oil and Natural Gas

Corporation (ONGC), which has specialized diving crews and sophisticated equipment.'

The Navy honoured CPO Veer and Commander Ashok with the Nao Sena Medal for gallantry for rising above and beyond the call of duty and leading the operations under the most trying circumstances. Their citations for the military honour brought out in detail the tough and many challenges the men encountered and how they overcame them to earn a place in the Navy's diving history.

'He dived repeatedly and even exceeded the stipulated limiting line, which exposed him to great danger. He personally recovered the bodies as well as the black box by innovatively gaining access through the damaged fuselage by sheer strength and brute force. His courage and determination percolated to other members of his team leading to the accomplishment of the task and brought closure to the families of the deceased aircrew,' reads CPO Veer's citation.

CPO Veer, since promoted up the ranks, has gone on to serve with a submarine rescue unit based in Visakhapatnam.

Commander Ashok's citation also acknowledged the courage, selflessness and perseverance he demonstrated while recovering the two bodies and the black box from the seabed. 'His valiant efforts at the age of fifty-plus led to recovery of the bodies of two officers and the flight data recorder (black box) in adverse circumstances,' his citation added.

Commander Ashok retired in 2018 and settled down in his hometown Pathankot, Punjab. The memories of the mission, which was an extreme test of physical and mental strength, are still fresh in his mind.

'It's my firm belief that when you lead from the front, your men will give their 100 per cent. They will not hesitate to even put their lives on the line for the mission. We were honest in the pursuit of our goal and constantly thinking of what the families of the two officers must be enduring. I believe that the sincerity of our approach even aligned the forces of nature in our favour. You get what I am saying?' he asks.

'The sea was as calm as a swimming pool when we were out there,' he concludes.

Acknowledgements

To our readers, our deepest thanks. They've taken what began as a collection of stories in 2017 and transformed it into a series that has a life of its own.

Thanks to the chiefs of the armed forces, General Manoj Pande, Air Chief Marshal Vivek Ram Chaudhari and Admiral R. Hari Kumar, and their predecessors, General Manoj Mukund Naravane, Air Chief Marshal R.K.S. Bhadauria and Admiral Karam Bir Singh, who permitted us unfettered access to units, documents and front-line personnel—the very core of what makes India's Most Fearless perhaps the only authentic series on Indian military operations.

Thanks to the several military personnel who helped in the journey, many of whom cannot be named for operational reasons. But among those who can, our deepest thanks to the Indian Army's additional director general (strategic communication), Major General Mohit Malhotra, and his team of officers, including Colonel Pankaj Gaur and Lieutenant Colonel Vivek Tripathi, as well as the spokespersons of the Indian Air Force and Indian Navy,

Wing Commander Ashish Moghe and Commander Vivek Madhwal, for seeing our vision.

Our friend and comrade Sandeep Unnithan, whose gritty, heartfelt illustrations of the heroes adorn this book. You are one of the most talented, knowledgeable and generous people we know.

To Sanjay Simha, one of India's greatest military and aviation photographers, our respect and gratitude. He gave us the stunning photograph you see on this book's cover.

To Milee Ashwarya, our friend and comrade at Penguin Random House India who, along with her amazing team, has ensured *India's Most Fearless 3* is every bit the book we hoped it would be when we committed it to her hands.

This book would not have been possible without the steady hand of our organizations. Thanks to the India Today Group leadership, Aroon Purie, Kalli Purie and Supriya Prasad; and the HT Media leadership, Shobhana Bhartia and Sukumar Ranganathan.

Our incredible families that have watched us disappear for months—now years—to finish these books, ever patient, ever supportive. Tavleen, Torul, Aryaman, Agastya and Mira: you know where you stand in the endless river of this writing.

Thanks to our parents—Prakash and Usha, Mahabir and Archana—who've been proud ambassadors of our work, like all Indian parents!

Above all, our humblest thanks go to the soldiers and officers we've written about. No words—certainly nothing in a few sentences—can express how deeply grateful we are for their deeds, their fortitude, their sacrifice and their time for us. As we've promised in the earlier books of this series, their stories will never be forgotten.